The Handbook
of Environmental Chemistry

Volume 5 Water Pollution
Part M

O. Hutzinger
Editor-in-Chief

Advisory Board:
T.A. Kassim • D. Barceló • P. Fabian
H. Fiedler • H. Frank • J.P. Giesy • R. Hites
M.A.K. Khalil • D. Mackay • A.H. Neilson
J. Paasivirta • H. Parlar • S.H. Safe
P.J. Wangersky

Environmental Effects of Marine Finfish Aquaculture

Volume Editor: Barry T. Hargrave

With contributions by

M.R. Anderson · S.M. Armstrong · L.M. Auffrey · M.A. Barbeau
K.D. Black · J.B.C. Bugden · L.E. Burridge · J. Chaffey · B. Chang
C.J. Cromey · I.M. Davies · A. Ervik · D. Greenberg
P. Kupka Hansen · B.T. Hargrave · W.G. Harrison · K. Haya
D. Higgs · M. Holmer · P.D. Keizer · P.E. Kepkay · B.A. Law
P. Lawton · C.D. Levings · W.K.W. Li · R. Losier · P. McCurdy
T.G. Milligan · F.H. Page · V.A. Pepper · T. Perry · G.W. Pohle
S.M.C. Robinson · M.T. Schaanning · W. Silvert · J.A. Smith
J.N. Smith · J.W. Sowles · J.E. Stewart · P.M. Strain · D. Stucchi
T.F. Sutherland · M.F. Tlusty · D.J. Wildish · P.A. Yeats

Volume Editor

Dr. Barry T. Hargrave
Ecosystem Research Division
Science Branch
Department of Fisheries and Oceans
Bedford Institute of Oceanography
P.O. Box 1006
B2Y 4A2 Dartmouth, Nova Scotia, Canada
hargraveb@rogers.com

Library of Congress Control Number: 2005924134

ISSN 1433-6863
ISBN-10 3-540-25269-X
ISBN-13 978-3-540-25269-6
DOI 10.1007/b12227
Springer Berlin Heidelberg New York

This work is subject to copyright. All rights are reserved, whether the whole or part of the material is concerned, specifically the rights of translation, reprinting, reuse of illustrations, recitation, broadcasting, reproduction on microfilm or in any other way, and storage in data banks. Duplication of this publication or parts thereof is permitted only under the provisions of the German Copyright Law of September 9, 1965, in its current version, and permission for use must always be obtained from Springer. Violations are liable for prosecution under the German Copyright Law.

Springer is a part of Springer Science+Business Media
springeronline.com
© Springer-Verlag Berlin Heidelberg 2005
Printed in Germany

The use of general descriptive names, registered names, trademarks, etc. in this publication does not imply, even in the absence of a specific statement, that such names are exempt from the relevant protective laws and regulations and therefore free for general use.

Product liability: The publisher cannot guarantee the accuracy of any information about dosage and application contained in this book. In every individual case the user must check such information by consulting the relevant literature.

The instructions given for the practical carrying-out of HPLC steps and preparatory investigations do not absolve the reader from being responsible for safety precautions. Liability is not accepted by the author.

Cover design: E. Kirchner, Springer-Verlag
Typesetting and Production: LE-TEX Jelonek, Schmidt & Vöckler GbR, Leipzig

Printed on acid-free paper   52/3141 YL – 5 4 3 2 1 0

Editor-in-Chief

Prof. em. Dr. Otto Hutzinger
Universität Bayreuth
c/o Bad Ischl Office
Grenzweg 22
5351 Aigen-Vogelhub, Austria
hutzinger-univ-bayreuth@aon.at

Advisory Board

Dr. T.A.Kassim
Department of Civil and
Environmental Engineering
College of Science and Engineering
Seattle University
901 12th Avenue P.O. Box 222000 Seattle,
WA 98122-1090, USA
tarek.kassim@oregonstate.edu

Prof. Dr. D. Barceló
Dept. of Environmental Chemistry
IIQAB-CSIC
JordiGirona, 18–26
08034 Barcelona, Spain
dbcqam@cid.csic.es

Prof. Dr. P. Fabian
Lehrstuhl für Bioklimatologie
und Immissionsforschung
der Universität München
Hohenbachernstraße 22
85354 Freising-Weihenstephan, Germany

Dr. H. Fiedler
Scientific Affairs Office
UNEP Chemicals
11-13, chemin des Anémones
1219 Châteleine (GE), Switzerland
hfiedler@unep.ch

Prof. Dr. H. Frank
Lehrstuhl fur Umwelttechnik
und Ökotoxikologie
Universitat Bayreuth
Postfach 10 12 51
95440 Bayreuth, Germany

Prof. Dr. J.P. Giesy
Department of Zoology
Michigan State University
East Lansing, MI 48824-1115, USA
Jgiesy@aol.com

Prof. Dr. R. Hites
Indiana University
SPEA 410 H
Bloomington 47405, USA
hitesr@indiana.edu

Prof. Dr. M.A.K. Khalil
Department of Physics
Portland State University
Science Building II, Room 410
P.O. Box 751 Portland, Oregon 97207-0751, USA
aslam@global.phy.pdx.edu

Prof. Dr. D. Mackay
Department of Chemical Engineering
and Applied Chemistry
University of Toronto
Toronto, Ontario, Canada M5S 1A4

Prof. Dr. A.H. Neilson
Swedish Environmental Research Institute
P.O. Box 21060
10031 Stockholm, Sweden
ahsdair@ivl.se

Prof. Dr. J. Paasivirta
Department of Chemistry
University of Jyväskylä
Survontie 9
P.O. Box 35
40351 Jyväskylä, Finland

Prof. Dr. Dr. H. Parlar
Institut für Lebensmitteltechnologie
und Analytische Chemie
Technische Universität München
85350 Freising-Weihenstephan, Germany

Prof. Dr. S.H. Safe
Department of Veterinary
Physiology and Pharmacology
College of Veterinary Medicine
Texas A & M University
College Station, TX 77843-4466, USA
ssafe@cvm.tamu.edu

Prof. P.J. Wangersky
University of Victoria
Centre for Earth and Ocean Research
P.O. Box 1700
Victoria, BC, V8W 3P6, Canada
wangers@telus. net

The Handbook of Environmental Chemistry
Also Available Electronically

For all customers who have a standing order to The Handbook of Environmental Chemistry, we offer the electronic version via SpringerLink free of charge. Please contact your librarian who can receive a password or free access to the full articles by registering at:

springerlink.com

If you do not have a subscription, you can still view the tables of contents of the volumes and the abstract of each article by going to the SpringerLink Homepage, clicking on "Browse by Online Libraries", then "Chemical Sciences", and finally choose The Handbook of Environmental Chemistry.

You will find information about the

– Editorial Board
– Aims and Scope
– Instructions for Authors
– Sample Contribution

at springeronline.com using the search function.

Preface

Environmental Chemistry is a relatively young science. Interest in this subject, however, is growing very rapidly and, although no agreement has been reached as yet about the exact content and limits of this interdisciplinary discipline, there appears to be increasing interest in seeing environmental topics which are based on chemistry embodied in this subject. One of the first objectives of Environmental Chemistry must be the study of the environment and of natural chemical processes which occur in the environment. A major purpose of this series on Environmental Chemistry, therefore, is to present a reasonably uniform view of various aspects of the chemistry of the environment and chemical reactions occurring in the environment.

The industrial activities of man have given a new dimension to Environmental Chemistry. We have now synthesized and described over five million chemical compounds and chemical industry produces about hundred and fifty million tons of synthetic chemicals annually. We ship billions of tons of oil per year and through mining operations and other geophysical modifications, large quantities of inorganic and organic materials are released from their natural deposits. Cities and metropolitan areas of up to 15 million inhabitants produce large quantities of waste in relatively small and confined areas. Much of the chemical products and waste products of modern society are released into the environment either during production, storage, transport, use or ultimate disposal. These released materials participate in natural cycles and reactions and frequently lead to interference and disturbance of natural systems.

Environmental Chemistry is concerned with reactions in the environment. It is about distribution and equilibria between environmental compartments. It is about reactions, pathways, thermodynamics and kinetics. An important purpose of this Handbook, is to aid understanding of the basic distribution and chemical reaction processes which occur in the environment.

Laws regulating toxic substances in various countries are designed to assess and control risk of chemicals to man and his environment. Science can contribute in two areas to this assessment; firstly in the area of toxicology and secondly in the area of chemical exposure. The available concentration ("environmental exposure concentration") depends on the fate of chemical compounds in the environment and thus their distribution and reaction behaviour in the environment. One very important contribution of Environmental Chemistry to the above mentioned toxic substances laws is to develop laboratory test methods, or mathematical correlations and models that predict the environmental fate of new chemical compounds. The third purpose of this Handbook is to help in the basic understanding and development of such test methods and models.

The last explicit purpose of the Handbook is to present, in concise form, the most important properties relating to environmental chemistry and hazard assessment for the most important series of chemical compounds.

At the moment three volumes of the Handbook are planned. Volume 1 deals with the natural environment and the biogeochemical cycles therein, including some background information such as energetics and ecology. Volume 2 is concerned with reactions and processes in the environment and deals with physical factors such as transport and adsorption, and chemical, photochemical and biochemical reactions in the environment, as well as some aspects of pharmacokinetics and metabolism within organisms. Volume 3 deals with anthropogenic compounds, their chemical backgrounds, production methods and information about their use, their environmental behaviour, analytical methodology and some important aspects of their toxic effects. The material for volume 1, 2 and 3 was each more than could easily be fitted into a single volume, and for this reason, as well as for the purpose of rapid publication of available manuscripts, all three volumes were divided in the parts A and B. Part A of all three volumes is now being published and the second part of each of these volumes should appear about six months thereafter. Publisher and editor hope to keep materials of the volumes one to three up to date and to extend coverage in the subject areas by publishing further parts in the future. Plans also exist for volumes dealing with different subject matter such as analysis, chemical technology and toxicology, and readers are encouraged to offer suggestions and advice as to future editions of "The Handbook of Environmental Chemistry".

Most chapters in the Handbook are written to a fairly advanced level and should be of interest to the graduate student and practising scientist. I also hope that the subject matter treated will be of interest to people outside chemistry and to scientists in industry as well as government and regulatory bodies. It would be very satisfying for me to see the books used as a basis for developing graduate courses in Environmental Chemistry.

Due to the breadth of the subject matter, it was not easy to edit this Handbook. Specialists had to be found in quite different areas of science who were willing to contribute a chapter within the prescribed schedule. It is with great satisfaction that I thank all 52 authors from 8 countries for their understanding and for devoting their time to this effort. Special thanks are due to Dr. F. Boschke of Springer for his advice and discussions throughout all stages of preparation of the Handbook. Mrs. A. Heinrich of Springer has significantly contributed to the technical development of the book through her conscientious and efficient work. Finally I like to thank my family, students and colleagues for being so patient with me during several critical phases of preparation for the Handbook, and to some colleagues and the secretaries for technical help.

I consider it a privilege to see my chosen subject grow. My interest in Environmental Chemistry dates back to my early college days in Vienna. I received significant impulses during my postdoctoral period at the University of California and my interest slowly developed during my time with the National Research Council of Canada, before I could devote my full time of Environmental Chemistry, here in Amsterdam. I hope this Handbook may help deepen the interest of other scientists in this subject.

Amsterdam, May 1980 *O. Hutzinger*

Twentyone years have now passed since the appearance of the first volumes of the Handbook. Although the basic concept has remained the same changes and adjustments were necessary.

Some years ago publishers and editors agreed to expand the Handbook by two new open-end volume series: Air Pollution and Water Pollution. These broad topics could not be fitted easily into the headings of the first three volumes. All five volume series are integrated through the choice of topics and by a system of cross referencing.

The outline of the Handbook is thus as follows:

1. The Natural Environment and the Biochemical Cycles,
2. Reaction and Processes,
3. Anthropogenic Compounds,
4. Air Pollution,
5. Water Pollution.

Rapid developments in Environmental Chemistry and the increasing breadth of the subject matter covered made it necessary to establish volume-editors. Each subject is now supervised by specialists in their respective fields.

A recent development is the accessibility of all new volumes of the Handbook from 1990 onwards, available via the Springer Homepage springeronline.com or springerlink.com.

During the last 5 to 10 years there was a growing tendency to include subject matters of societal relevance into a broad view of Environmental Chemistry. Topics include LCA (Life Cycle Analysis), Environmental Management, Sustainable Development and others. Whilst these topics are of great importance for the development and acceptance of Environmental Chemistry Publishers and Editors have decided to keep the Handbook essentially a source of information on "hard sciences".

With books in press and in preparation we have now well over 40 volumes available. Authors, volume-editors and editor-in-chief are rewarded by the broad acceptance of the "Handbook" in the scientific community.

Bayreuth, July 2001 *Otto Hutzinger*

Contents

Dissolved Oxygen and Salmon Cage Culture
in the Southwestern New Brunswick Portion of the Bay of Fundy
F. H. Page · R. Losier · P. McCurdy · D. Greenberg · J. Chaffey ·
B. Chang   1

Salmon Aquaculture, Nutrient Fluxes and Ecosystem Processes
in Southwestern New Brunswick
P. M. Strain · B. T. Hargrave 29

Ecosystem Indicators of Water Quality Part I.
Plankton Biomass, Primary Production and Nutrient Demand
W. G. Harrison · T. Perry · W. K. W. Li 59

Ecosystem Indicators of Water Quality Part II.
Oxygen Production and Oxygen Demand
P. E. Kepkay · W. G. Harrison · J. B. C. Bugden 83

Organic Enrichment at Cold Water Aquaculture Sites—
the Case of Coastal Newfoundland
M. R. Anderson · M. F. Tlusty · V. A. Pepper 99

Reconciling Aquaculture's Influence
on the Water Column and Benthos of an Estuarine Fjord –
a Case Study from Bay d'Espoir, Newfoundland
M. F. Tlusty · V. A. Pepper · M. R. Anderson 115

Modelling the Impacts of Finfish Aquaculture
C. J. Cromey · K. D. Black 129

Near-Field Depositional Model for Salmon Aquaculture Waste
D. Stucchi · T.-A. Sutherland · C. Levings · D. Higgs 157

Organic Enrichment from Marine Finfish Aquaculture
and Effects on Sediment Biogeochemical Processes
M. Holmer · D. Wildish · B. Hargrave 181

Lithium-Normalized Zinc and Copper Concentrations
in Sediments as Measures of Trace Metal Enrichment
due to Salmon Aquaculture
P. A. Yeats · T. G. Milligan · T. F. Sutherland · S. M. C. Robinson ·
J. A. Smith · P. Lawton · C. D. Levings 207

Sediment Geochronologies for Fish Farm Contaminants
in Lime Kiln Bay, Bay of Fundy
J. N. Smith · P. A. Yeats · T. G. Milligan 221

The Effect of Marine Aquaculture
on Fine Sediment Dynamics in Coastal Inlets
T. G. Milligan · B. A. Law 239

Far-Field Impacts of Eutrophication on the Intertidal Zone in the Bay of Fundy,
Canada with Emphasis on the Soft-Shell Clam, *Mya arenaria*
S. M. C. Robinson · L. M. Auffrey · M. A. Barbeau 253

Benthic Macrofaunal Changes Resulting from Finfish Mariculture
D. J. Wildish · G. W. Pohle 275

A Review and Assessment of Environmental Risk of Chemicals
Used for the Treatment of Sea Lice Infestations of Cultured Salmon
K. Haya · L. E. Burridge · I. M. Davies · A. Ervik 305

Antibiotic Use in Finfish Aquaculture: Modes of Action, Environmental Fate,
and Microbial Resistance
S. M. Armstrong · B. T. Hargrave · K. Haya 341

Assessing Nitrogen Carrying Capacity for Blue Hill Bay, Maine:
A Management Case History
J. W. Sowles 359

**The Suitability of Electrode Measurements for Assessment
of Benthic Organic Impact and Their use in a Management System
for Marine Fish Farms**
M. T. Schaanning · P. Kupka Hansen 381

Environmental Management and the Use of Sentinel Species
J. E. Stewart 409

**Assessing and Managing Environmental Risks Associated
with Marine Finfish Aquaculture**
B. T. Hargrave · W. Silvert · P. D. Keizer 433

Subject Index 463

Foreword

Mariculture of finfish in floating net pens is an expanding industry in many countries around the world. Although several species are cultured, Atlantic salmon (*Salmo salar*) is the species most intensively farmed in Australia, Canada, Chile, Ireland, Norway, Scotland, and the USA. However, potentially negative environmental effects and risks associated with large-scale marine finfish cage aquaculture have been identified and led to claims that the long-term sustainability of the industry is in doubt. High fish biomass held in pens over extended grow-out periods up to 24 months in cold water areas, often close to shore and in relatively shallow water, can lead to measurable changes in water and sediment variables both near and far from farm sites. Quantitative information is often not available to determine if these changes have more than local ecological consequence. Widespread environmental effects such as oxygen depletion and nutrient enrichment associated with release of waste feed, fish faeces and excretory products may occur in areas where hydrographic conditions prevent water exchange and result in sediment accumulation. The magnitude of these potential effects can be estimated using physiological models to predict the release of waste products by cultured fish.

Twenty chapters in this volume review some of the major environmental effects associated with marine finfish aquaculture. Chapters are arranged in four sections dealing with related issues: Eutrophication, Sedimentation and Benthic Impacts, Changes in Trophic Structure and Function, and Managing Environmental Risks. Models and observations are presented to show how knowledge of cultured fish biomass, temperature and local hydrographic conditions can be used to predict the magnitude and spatial scale of changes in some key marine environmental and ecological variables. Dispersion models, biogeochemical methods and chemical tracers sensitive to organic and nutrient enrichment from mariculture waste discharges are reviewed. At least one case study is included in each section to provide an example of how models and observations of changes in environmental variables can be used for regulatory purposes and to guide management decisions. Two chapters review toxicological and ecological effects of therapeutants used to treat parasites and microbial infections in cultured fish. The final section presents examples of management tools and models that can be used to quantify potentially adverse environmental risks. Although the epidemiology of disease and parasitic infections in cultured and natural fish stocks and effects of marine finfish aquaculture on wild fish populations through escapement are not specifically considered, the topics and possible linkages to environmental changes are described where appropriate.

In accordance with previous issues of The Handbook of Environmental Chemistry series chapters are written by experts with practical experience. The volume is intended for those working towards sustainable development inside and outside of the industry, graduate students, research scientists, environmental managers and decision-makers with regulatory responsibilities for finfish mariculture.

Owen Sound, Ontario, January 2005 *Barry Hargrave*

# Dissolved Oxygen and Salmon Cage Culture in the Southwestern New Brunswick Portion of the Bay of Fundy

F. H. Page[1] (✉) · R. Losier[1] · P. McCurdy[1] · D. Greenberg[2] · J. Chaffey[2] · B. Chang[1]

[1]Dept. of Fisheries and Oceans, Biological Station, 531 Brandy Cove Road, St. Andrews, New Brunswick, E5B 2L9, Canada
pagef@mar.dfo-mpo.gc.ca, losierr@mar.dfo-mpo.gc.ca, mccurdyp@mar.dfo-mpo.gc.ca, changb@mar.dfo-mpo.gc.ca

[2]Department of Fisheries and Oceans, Bedford Institute of Oceanography, P.O. Box 1006, Dartmouth, B2Y 4A2, Canada
greenbe@mar.dfo-mpo.gc.ca, chaffeyj@mar.dfo-mpo.gc.ca

1	Introduction	2
2	Ambient Concentrations of Dissolved Oxygen	4
3	Dissolved Oxygen in the Vicinity of Fish Farms	10
4	Dissolved Oxygen Within Fish Cages	11
5	Models of Dissolved Oxygen	13
6	Oxygen Depletion Index	16
7	Discussion and Conclusions	24
	Appendix	26
	References	27

Abstract Information on the spatial and temporal characteristics of ambient dissolved oxygen in the southwestern New Brunswick area of the Bay of Fundy is presented to help develop an understanding of dynamics of oxygen and salmon cage culture in the region. Some modelling efforts focussed on dissolved oxygen issues associated with fish farming in the area are also presented. A description and application of a simple oxygen depletion index is proposed to help identify the influence of salmon cage culture on the regional and farm-scale dissolved oxygen concentrations. The chapter concludes with a brief summary and discussion of observations and model development required to enhance the understanding of oxygen dynamics in the farms and bays of southwestern New Brunswick. The knowledge can be used by industry in their farm management practices and by environmental regulators in their efforts to define and sustain water quality standards.

1
Introduction

Commercial salmon cage culture in eastern Canada and the eastern United States began in 1978 in the Deer Island area of southwest New Brunswick. The first commercial harvest occurred in the autumn of 1979 [1]. The industry in southwest New Brunswick (SWNB) and eastern Maine has grown rapidly since its first commercial harvest. In 2003, annual production was about 33 100 tonnes and the value of the industry was $ 179 million [23].

The growth has manifested itself as increases in the number of farms, the average production per farm, the geographic domain utilized by the industry and the number of farms within each geographic area [2, 3]. The number of grow-out sites has steadily increased over the past 20 years from a single site in 1978 to 95 in 2003 (New Brunswick Department of Agriculture, Fisheries and Aquaculture personal communication; Fig. 1). The spatial distribution of the industry has also increased over the years. In 1980, one farm was located in the Deer Island area and one was in Grand Manan Island. By 1985, the industry had spread to other areas of Deer Island, two locations on Campobello Island and several locations within the Letang Inlet (Back Bay, Lime Kiln Bay and Bliss Harbour). Only one site remained in the Grand Manan area. By 1990, the number of sites in all of the above areas had increased, especially in the Fundy Isles and Letang Inlet areas, and new sites were established in Passamaquoddy Bay. Grand Manan continued to be under-developed with

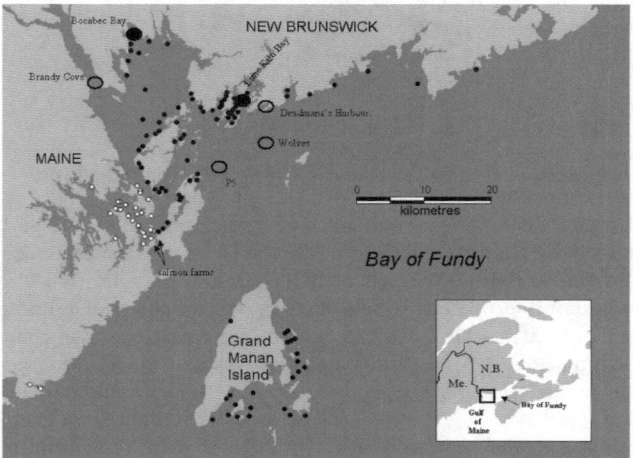

Fig. 1 Map of the southwestern New Brunswick area showing the location of salmon farms within Canada (*small black solid dots*) and the United States (*small white solid dots*) portion of the mouth of the Bay of Fundy in recent years

only two sites. By 1995, the number of sites had increased in all areas and by 2000, the continuing increase was confined mainly to the Passamaquoddy Bay and Grand Manan areas. Rapid development also occurred in the United States portion of the region so that by the year 2000, the density of sites within the Cobscook Bay area of eastern Maine was similar that of the Letang Inlet (Fig. 1).

As the industry developed and expanded in geographic extent, it became exposed to a wider range of environmental conditions and questions, and concerns related to dissolved oxygen (DO) increased. A few salmon farmers began to experience low oxygen concentrations on their farms and considered this to result in loss of fish production and hence a loss of income. They began to wonder whether it was a farm-scale, bay-scale or regional-scale phenomenon and whether it was related to natural events or to the presence of salmon farming.

The ambient concentration of DO in oceanic, shelf and coastal waters varies on a variety of time and space scales for natural reasons. The variability is caused by natural factors, including changes in temperature, exchange of oxygen with the atmosphere, production of oxygen by phytoplankton and other plants, utilization of oxygen by organism respiration processes, as well as chemical degradation processes. The dominant factor is temperature since, to a large extent, this controls the solubility of oxygen in water and therefore the saturation concentration for oxygen. Saturated warm water contains less oxygen than cool water and hence the annual cycle in the concentration of DO is out of phase with the annual cycle in water temperature. In the North Atlantic surface mixed layer, the concentration of DO is typically highest during the late winter and spring and lowest during the late summer and fall [13, 16]. Water temperatures are typically at a minimum during the winter and a maximum during the late summer or early fall. The biological production and respiration processes in the water column generate perturbations or departures from the temperature-controlled saturation concentrations. The oxygen production during plankton blooms often causes supersaturation in the spring when the water is warming, and oxygen utilization by decay processes causes under-saturation in the fall when the water is warmest and utilization rates are highest. The exact pattern varies somewhat with geographic location since there are regional variations in the timing and magnitude of the seasonal temperature cycle and the various oxygen production and utilization processes. For example, at middle to high latitudes, the amplitude of the annual oxygen cycle in the North Atlantic is greater in the west [16].

In addition to naturally induced variations, the concentration of DO can be altered by aspects of human activities that require oxygen, such as the oxygen demands associated with the chemical and organic loading in some industrial and municipal effluents. Fish farming also creates a demand for DO that is associated with the respiration needs of the fish being grown and the associated organic wastes, such as faeces and excess feed [28]. This demand is sometimes detectable as reductions in the concentration of oxygen near fish farms. For example, [13] towed a DO sensor along horizontal transects located near yellowtail farms in Japan. In addition to observing that the horizontal distributions varied seasonally, geographically and between farms, they noted reductions of 0–38% saturation between the upstream and downstream ends of farms. They found that in close proximity to sea bream and yellowtail farms, the concentrations of DO near the surface were lower than at depth. This was opposite to what they observed in areas away from fish farms where the highest concentrations of DO usually occurred near the surface and the concentration decreased with depth.

In the SWNB area of the Bay of Fundy the concentration of DO also varies seasonally [10]. It has also been suggested that in some localized areas the salmon farming industry has the potential to influence oxygen concentrations on the scale of the farm and, in some cases, on the scale of a bay [6, 19, 27]. Hence, both the salmon farming industry and environmental regulators would like to have an adequate understanding of the oxygen dynamics in the region so that informed farm management decisions can be made. This is particularly relevant to the generation of advice concerning the impact of production limits for salmon farms in the area and the potential for the multi-trophic culture of mussels, kelp and salmon farms to alter oxygen budgets within the host salmon farm.

In this chapter we attempt to contribute to the understanding of the dynamics of DO in the southwestern New Brunswick area and how the local fish farming industry interacts with these. Specifically we present information on the spatial and temporal characteristics of ambient DO of relevance to the SWNB area of the Bay of Fundy, and review some modelling efforts that have focussed on DO issues associated with fish farming in the area. We also describe a simple oxygen depletion index that helps to quantify the influence of salmon cage culture on the regional and farm scale concentrations of DO. The chapter is concluded with a brief summary and discussion.

2
Ambient Concentrations of Dissolved Oxygen

The concentration of DO in coastal waters undergoes spatial and temporal variations that are independent of the presence of aquaculture. The variations are driven by local processes and influxes of water from adjacent continental shelf and oceanic areas. The salmon aquaculture industry in SWNB is located within the Bay of Fundy, a part of the Gulf of Maine hydrographic system, which is on the eastern edge of the North Atlantic Ocean. To give a context to the SWNB situation, we briefly review the variation in the concentration of DO in these bodies of water.

In the North Atlantic Ocean the concentration of DO is characterized by a seasonal variation that is believed to be associated with the seasonal cycles in warming, water column stratification and productivity [14, 16]. The amplitude tends to be greater in the western Atlantic than in the eastern Atlantic, particularly at middle to high latitudes [14, 16]. The seasonal cycle in the oxygen concentration also varies with depth, with the highest concentrations usually occurring in the near-surface waters [16]. In the surface mixed layer of the western North Atlantic, the oxygen concentrations tend to be highest in the late spring to early summer, decrease throughout the summer and early fall and are lowest in the fall [16].

In the offshore regions of the Gulf of Maine, the concentration of DO has been summarized by [4] using the 1912 through 1970 data within the National Oceanographic Data Center (NODC) database. As in the North Atlantic, the concentration of DO varies seasonally and with depth. The DO concentration ranges from 5.2 to 14.2 mg L^{-1}. It is generally highest near the surface and decreases with depth (Fig. 2). In the upper 10 m the concentration ranges from 7.7–14.2 mg L^{-1}, whereas it ranges from 5.4–8.8 mg L^{-1} at 200 m. Within each depth level the concentration varies seasonally (Fig. 2) with the maximum concentration occurring in the spring (April through May) and the minimum in the fall (October through December). The amplitude of this variation is largest near the surface.

The concentration of DO in the offshore Bay of Fundy varies in a manner similar to that in the Gulf of Maine and North Atlantic. Early data from [10] shows that the concentration varies between about 80 and 120% saturation and that the saturation decreases with depth, particularly in August (Fig. 3). Their data also indicate the presence of spring (May) and summer (August) maxima in the concentration of DO in the upper 25 m of the water column.

More recent data from the Prince 5 hydrographic and Wolves plankton monitoring stations located in the offshore area of SWNB (Fig. 1) indicate that the concentration of DO is similar to the above observations made in the Gulf of Maine and outer Bay of Fundy in past decades. Recent concentrations typically range between about 7 and 13 mg L^{-1} (a saturation range of between about 80 and 130%; Fig. 4). The waters are generally supersaturated during the spring and under-saturated during the fall and winter, resulting in a single seasonal maximum that occurs in the spring (Fig. 4). This pattern is particularly evident at the Wolves (station 16), which does not receive the intensely mixed water that flows from the island passages to the northwest of Prince 5 (station 5) and influences the observations made at Prince 5.

In the nearshore/inshore areas of SWNB that support salmon farming, the general seasonal pattern in the concentration of DO is similar to that in the offshore [18] (Fig. 5). For example, the oxygen concentrations within Bocabec Bay and Lime Kiln Bay range between about 7 and 13 mg L^{-1} and 80 and

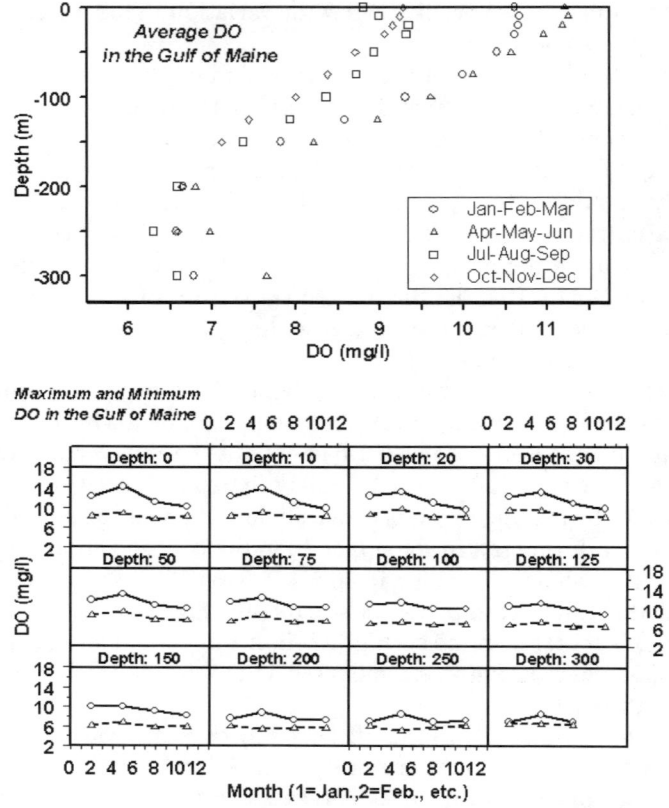

Fig. 2 A composite of the vertical (*top*) and seasonal (*bottom*) distributions of the average (1912–1970) dissolved oxygen (DO) in the Gulf of Maine. In the *bottom figure* each symbol represents an individual statistic (maximum or minimum) for a specific depth (units in metres) and seasonal period. The *solid* (*dashed*) lines join the monthly maximums (minimums) within a given depth level. The seasonal periods are January–March (month = 2), April–June (5), July–September (8) and October–December (11). The data are from [4], who calculated the statistics from the data archived in the NODC database

130%, with the seasonal maximum occurring in the spring and the seasonal minimum in the late summer and early fall. The few measurements that fall below these limits, particularly in Bocabec Bay, may be related to the greater number of measurements in the inshore in comparison to the offshore, or to localized processes such as phytoplankton blooms that are common in the Bocabec Bay area.

In contrast to the pattern of temporal variability described above, relatively little is known about the spatial variation in DO concentrations within the SWNB area. However, it is known that the concentration can be different in at least some areas. Wildish et al. [33] measured the concentration

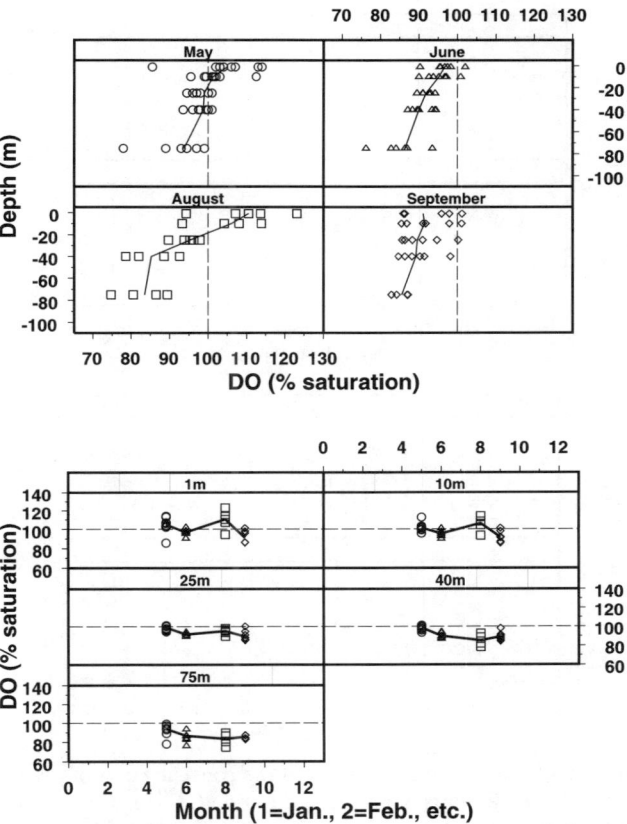

Fig. 3 A composite of the vertical (*top*) and seasonal (*bottom*) distribution of dissolved oxygen in the Bay of Fundy in 1932 as indicated by data contained within [10]. Each symbol represents an individual measurement from a specific depth and station location. The *solid lines* through the points join the median of the points at a given depth and month

of DO within Brandy Cove at approximately monthly intervals from January through October, and found the concentration to range between about 70 and 130% saturation. However, in contrast to the seasonal pattern observed within Bocabec and Lime Kiln Bays and at the offshore stations, the maximum saturation occurred in August. It is not known whether this is a consistent feature of the seasonal cycle in DO in the cove or whether it is indicative of an unusual year. A distinctly different DO environment exists in Blacks Harbour where a fish processing plant releases organic wastes into the harbour. This resulted in severe oxygen depletion within the harbour, especially at low tide [31]. This results in severe DO depletion limited to the scale of the harbour, especially at low tide [31]. No salmon farming occurs within either Brandy Cove or Blacks Harbour. Strain and Clement [26]

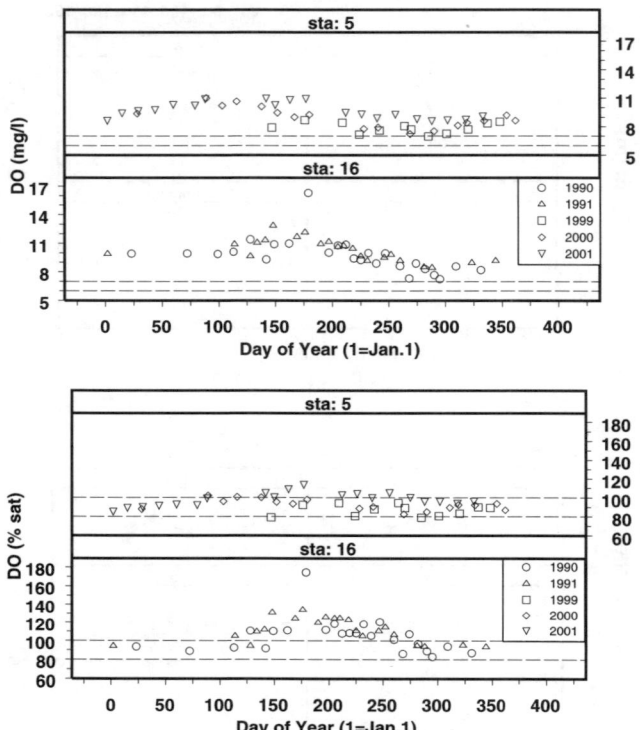

Fig. 4 A composite of the seasonal variation in the concentration of dissolved oxygen in units of mg L^{-1} (*top*) and in units of percent saturation (*lower*) at the Prince 5 (sta. 5) and Wolves (sta. 16) offshore monitoring stations (see Fig. 1). The Prince 5 data was collected with a SeaBird Electronics Model 25 CTD equipped with a YSI membrane-based dissolved oxygen sensor. The data for station 16 is based on Winkler titrations of water samples and is the data reported by [15]

measured DO at a series of stations within the Letang Inlet and Blacks Harbour during July, August and September of 1994. They found DO ranged from 61.7–116.1% saturation. The 61.7% value was from Blacks Harbour. The range for the Letang Inlet, excluding the upper L'Etang, which is influenced by pulp mill effluent, was 92.8–116.1%. This is an area of intense salmon farming.

In addition to the seasonal and spatial variations in DO, the data from both the offshore and inshore indicate a sub-seasonal variation in the concentration of DO that is about ±1 mg L^{-1} and ±10% saturation. This may be due to measurement errors or real variation in the DO concentration. The latter may be associated with interannual variability or high frequency variation on time scales of minutes to days or weeks that has not been resolved by the single sample measurement approaches referenced in the above.

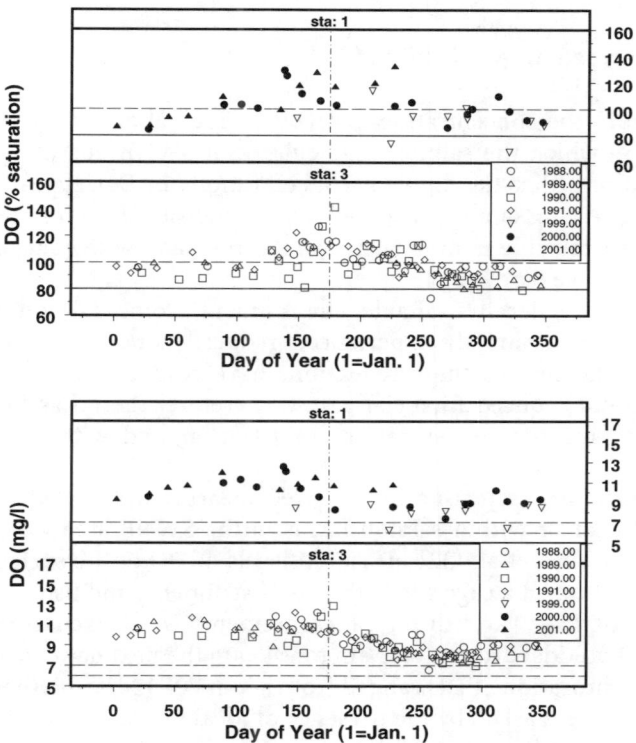

Fig. 5 A composite of the seasonal variation in the concentration of near-surface dissolved oxygen at the Bocabec Bay (sta. 1) and Lime Kiln Bay (sta. 3) inshore monitoring stations (see Fig. 1). The two panels express the same data using different units. The Bocabec Bay data was collected with a SeaBird Electronics Model 25 CTD equipped with a YSI membrane-based dissolved oxygen sensor. The data for Lime Kiln Bay is based on Winkler titrations of water samples and is the data reported by [15, 33]. The data are from depths of 0 and 1 m

The existence of such high frequency variation has been reported by [33], who sampled at hourly intervals in October at a location within Brandy Cove and found that the range in percent saturation over a 24-h period was about 15%.

There is relatively little information available to determine whether there have been decadal time scale variations in the concentration of DO. However, the saturation levels reported by [10] are similar to those reported above for more recent time periods, which suggests that large changes in the DO regime have not occurred. However, if areas of the coastal zone become severely eutrophied or nutrified, as suggested by [6, 27], the DO concentrations may change.

3
Dissolved Oxygen in the Vicinity of Fish Farms

The pattern in the concentration of DO described above is the regional environment in which the salmon cage culture in SWNB operates. The information indicates that the ambient concentrations of DO appear to be very suitable for growing salmon at some level of intensity since it is generally accepted that sustainable salmon farming requires water with concentrations of DO that are above 6 or 7 mg L^{-1} [34].

Despite this generally suitable environment, some salmon farmers in the area have occasionally experienced reduced concentrations of DO in their farms and believe these reductions have resulted in reduced salmon production and reduced financial gains. Therefore, there has been interest in characterizing the concentrations of DO near and within salmon cage sites.

Some of the first scientifically reported measurements of the concentration of DO near or within salmon farms in the SWNB area were published by [32, 33]. Wildish et al. [32] measured levels of DO at the surface, 5 m and 1 m off the bottom at a cage site with anoxic sediments and fish stocking densities of about 15–20 kg fish m^{-3}. Measurements were taken every hour for about 27 h (two tidal cycles; 10:40 am to 9:40 am the next day) during August. The percent saturation of DO ranged from about 70–120% and they noted the DO records were marked by the presence of local minima and maxima. They also made observations near a fish farm in Deadman's Harbour and which led them to suggest that the concentration of DO was often lowest during periods of slack tidal currents.

Wildish et al. [33] measured the seasonal variation in the concentration of DO immediately adjacent to a fish farm in Bliss Harbour. They made measurements at approximately monthly intervals from January through October and found the concentration of DO to range between about 80 and 120% saturation—a range that is very similar to that measured at locations distant from fish farms (see above section). Although the greatest percent saturation occurred in June (120%) and September (122%), the range (maximum – minimum concentrations) was greatest during August (27%) and September (27–38%) when the water temperatures were highest. At another farm site, located in Lime Kiln Bay, the range in percent saturation during October was about 13%, which was similar to that observed at a control station in Brandy Cove. This suggests the fish farm did not have a strong influence on the daily range in DO concentration.

In addition to the above, Wildish et al. [33] also sampled at hourly intervals over 24-h periods at locations immediately adjacent to fish cages in three separate areas within the southwestern Bay of Fundy. The percent saturation at these sites ranged from 65–122% over these time periods. The difference between the maximum and minimum concentrations over the 24-h period varied between 16 and 38%. The temporal pattern and trend in saturation was considerably different in each of the 24-h time series.

Although these initial investigations do not give strong evidence of oxygen depletion near the SWNB fish farms, it should be noted that the sampling was not spatially and temporally extensive and therefore may have missed a spatially varying and temporally transient signal. Extensive temporal (see next section) and spatial sampling using electronic oxygen meters has only just begun in the area and, if it continues, a better description of the horizontal and vertical distribution of the concentration of DO around fish farms should emerge.

4
Dissolved Oxygen Within Fish Cages

Although the measurements near fish cages have not detected a major influence of fish farms on the oxygen content of the adjacent waters, it is worth determining whether the concentration of DO within fish cages is different from ambient concentrations. Hence, preliminary measurements of DO within cages have been made by [32]. They measured levels of DO within the surface waters of a salmon cage with clean nets and a cage with heavily fouled nets. The cages contained an estimated 15–20 kg fish m^{-3}. The measurements were taken every half hour during eight consecutive daylight hours (10:00 am–18:00 pm) during September at a farm site with oxic sediments under the cages. Over the duration of the sampling period, the percent saturation of DO ranged from about 85–120% and the results suggested that DO levels in the cage with fouled nets were marginally lower, by 0.12 mg L^{-1}, than those in the cage with clean nets. The minimum DO concentration occurred at low tide in both cages and was suggested to be due to a lack of flushing associated with slack water [33].

More recently we have collected high frequency time series using internally recording oxygen sensors deployed within fish cages at a few selected fish farms in the SWNB area. In some cases the concentration of DO remained relatively constant for periods of days to weeks and, in other cases, the concentration varied on several different time scales.

Figure 6 shows a time series obtained from a cage in one of the SWNB salmon farms. A low-pass filtering of the time series shows a variation of about ±10% on a time scale of about one week. The deviations or residuals (i.e. the difference between the raw time series and the low pass filtered series) shows a high frequency variation with an amplitude of about 10–20% saturation (~ 1 mg L^{-1}) and a period of 12.42 h. This period corresponds to the dominant M_2 tidal cycle in the Bay of Fundy.

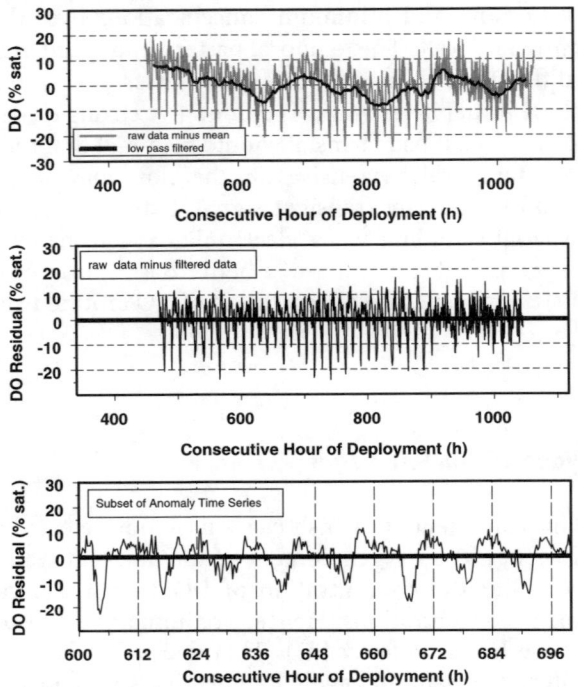

Fig. 6 Time series of the concentration of dissolved oxygen at a selected salmon farm in southwestern New Brunswick

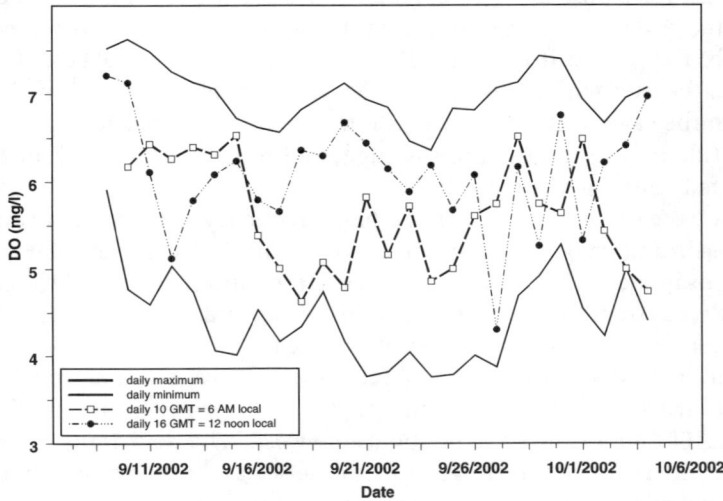

Fig. 7 The time series of the daily maximum (*upper solid line*) and minimum (*lower solid line*) in dissolved oxygen from the data shown in Fig. 6. The *open* and *solid* symbols indicate the concentration of dissolved oxygen at 6 am and 12 noon respectively

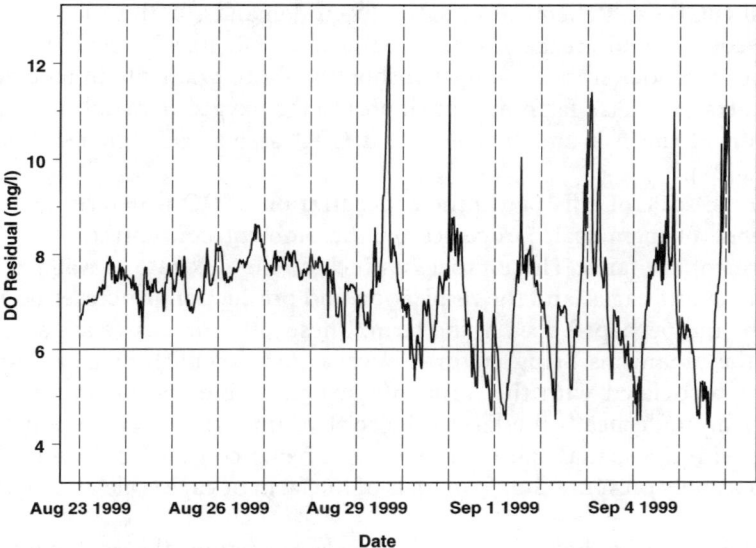

Fig. 8 Time series of dissolved oxygen within a salmon cage at a fish farm in southwestern New Brunswick. The data were collected during a phytoplankton bloom in the area

One of the consequences of variability such as the above is that the true characteristics of the oxygen environment within the fish cages is not detected by a sampling regime that consists of one measurement per day. The single sample per day sampling approach misses the tidal variation and tends to overestimate the concentration of DO in the cage (Fig. 7).

A second time series obtained from another farm in SWNB illustrates the temporal pattern in DO prior to and during a plankton bloom (Fig. 8). Prior to the bloom the average concentration of DO within the cage was consistent with expected values for the season, about 7–8 mg L^{-1}, with a variation around the mean of about ±1 mg L^{-1}. During the bloom the mean concentration decreased to between 6 and 7 mg L^{-1} and the amplitude of the daily variation increased to about ±3 mg L^{-1}, corresponding to a daily range from about 5 to 11–12 mg L^{-1}.

5
Models of Dissolved Oxygen

The concentration of DO in an area is controlled by the balance between the rates of local input and removal. DO is input into seawater by exchange with the atmosphere, by plant photosynthesis and by water mass exchange and mixing. DO is removed from the seawater by various sources of biological (BOD) and chemical (COD) oxygen demands. Both can occur in the water column and on the sea floor. BOD and COD may be generated in extensive intertidal areas leading to inshore-offshore gradients in ambient DO concentrations. Fish farming contributes to the oxygen demand through respiration of the fish and the BOD and COD associated with fish and feed wastes [27].

On the scale of a fish farm the concentration of DO is determined by the influence of farm scale processes on the ambient concentration of DO in the area of the farm. These processes include the flux rate of water through a farm and its cages, and the respiration and production processes occurring within, and perhaps beneath, the farm. These latter processes are associated with the organisms being cultured such as fish, shellfish or algae, the biofouling associated with the farm infrastructure and the benthic processes occurring underneath the farm. The relative importance of these processes varies with the spatial and temporal scales being considered. In this chapter we focus only on some simple models of the farm or cage scale and on the bay scale.

Several authors have developed models estimating the concentration of DO within fish farms. Silvert [21] presented some simple models exploring the role of current and respiration in determining oxygen concentrations in fish cages. A more detailed model such as the MOM model system includes an oxygen submodel that estimates the oxygen consumption by caged fish, the fish faeces and the uneaten fish food [25]. The objective of these models has been primarily for developing a quantitative understanding of the conditions under which the ambient concentration of DO may be reduced within or near fish cages, and of the relative importance of the various factors and processes influencing oxygen concentration in cages and bays.

Wildish et al. [33] applied a mass balance approach to the SWNB area to estimate the relative importance of the rates of fish respiration, anoxic sediment demand, re-aeration, phytoplankton production and tidal water exchange to DO concentrations on the spatial scale of a fish cage. They estimated that the rate of oxygen demand associated with the respiration of fish stocked at 18 kg m^{-3} was of the same order of magnitude as the rate of oxygen supply provided by tidal water exchange. The rates of anoxic sediment and re-aeration demand were, respectively, one and two orders of magnitude smaller than the fish demand. The rate of phytoplankton DO production was two to three orders of magnitude smaller than that provided by water transport. Hence, they concluded that on the scale of a cage, the balance between fish respiration demand and water exchange controls the DO levels at the scale of the farm. Silvert [22] also estimated that the DO demand by caged fish was much greater than that from heavily impacted sediment.

Calculations and oxygen budgets have also been made on the spatial scale of a bay in the SWNB area [32, 33]. Wildish et al. [32] estimated the total amount of DO within Lime Kiln Bay and the oxygen demands generated by microbial respiration, sedimentary oxygen demand, herring respiration, and salmon respiration. They estimated that the total oxygen demand could remove 5.6% of the available oxygen in one night. The demands of sediment, water column, herring and salmon respiration were of similar magnitude. Wildish et al. [33] also considered the relative importance of the rates of fish respiration, anoxic sediment demand, re-aeration, phytoplankton production and tidal water exchange to DO concentrations within Bliss Harbour. They estimated that the rate of oxygen demand associated with fish respiration was an order of magnitude less than the rate of oxygen supply associated with tidal water exchange. The rates of anoxic sediment and re-aeration demand were more than four orders of magnitude smaller than the fish demand. The rate of phytoplankton production was two orders of magnitude smaller than that provided by water transport. Hence, they concluded that on the scale of the bay, DO levels were also controlled by the balance between the demand by cultured fish respiration and the supply associated with tidal water exchange.

Strain, Wildish and Yeats [28] used a box model approach for oxygen in the Letang Inlet within SWNB. They concluded that the fish farms were cumulatively the largest anthropogenic source of oxygen demand in the inlet and that the fish farming, as it was practiced in the early 1990s, had the potential to reduce ambient concentrations of DO by a maximum of about 10%. These calculations have been updated recently by [27].

Trites and Petrie [29] describe a two-dimensional water quality model that estimated DO depletion due to fish farms in the Letang Inlet. The model included the respiration of salmon at the various fish farms within the area as spatially explicit sinks of DO. The model predicted localized reductions in DO in the vicinity of several of the farms that were located in relatively weak current areas, or in areas in which the residence time of the water was relatively long. Anecdotal information from the salmon farming industry suggests that, at times, some farms have experienced reduced concentrations of DO and some losses in salmon production consistent with the model predictions.

All of these models indicated a relative dominance of fish respiration and water circulation in controlling the oxygen concentration in a cage and in the fish farming bay areas that they considered. However, none of the above explicitly included the influence of atmospheric exchange on the oxygen budget. This may be a significant contributor to the oxygen budget on certain spatial scales since the air–sea flux of oxygen into the water is dependent on surface area. Therefore, a reconsideration of the importance of the various oxygen dynamic processes should include air–sea exchange and other potentially relevant processes.

6
Oxygen Depletion Index

As mentioned above, simple models estimating the concentration of DO within fish farms have been produced. We have used similar concepts to derive a simple index of the potential for caged fish to generate depressions of oxygen on scales of a cage, farm or bay. The index has proven to be a useful way to communicate aspects of farm-induced oxygen depletion to fish farmers and regulators. The simplest version of the index (I_{DO}) is given by the ratio of the time required for the fish biomass to reduce the ambient concentration of oxygen to a specified threshold level in the absence of flushing (τ_{thres}) to the time needed to flush the cage or farm (τ_{fl}):

$$I_{DO} = \frac{\tau_{thres}}{\tau_{fl}} \quad (1)$$

When the ratio is much less than 1, the respiration time scale is smaller than the flushing or exchange time scale. This means the fish are able to reduce the oxygen to threshold values in less time than the oxygen can be exchanged by flushing of the water within the cage, farm or bay. It therefore suggests the fish may be able to cause depressions in the DO regime; this may indicate the need for more careful consideration of the husbandry and ecological consequences of the depressions by fish farmers, regulatory agencies and other interested parties. When the ratio is much greater than 1, the respiration time scale is greater than the flushing or exchange time scale. This means it takes the fish longer to reduce the oxygen to threshold values than it takes to exchange the water in the cage or farm. It suggests that the fish are not likely to cause oxygen problems of husbandry, regulatory and/or ecological concern. When the ratio has a value of approximately 1, the respiration time scale is the same as the flushing or exchange time scale. This means the system is in the transition zone between being of concern or no concern.

The τ_{thres} is calculated as:

$$\tau_{thres} = \frac{C_0 - C_{thres}}{R} \quad (2)$$

where R is the respiration rate of the fish per unit volume of water within the cage or farm, C_0 is the ambient concentration of DO, i.e. the concentration of DO away from fish farms and C_{thres} is the minimum concentration of DO that is desired within the cage or farm. The ambient concentration of DO is assumed to be derived from empirical measurements of relevance to the area being considered. These measurements will therefore include the net effects of broader scale processes such as oxygen injection from wind-induced mixing and other turbulence processes, water mass exchange, the sources and sinks associated with ecosystem primary production and respiration and non-fish farming related sinks of DO.

Although the flow of water through a farm is very complex due to effects of cage structures, nets and farm layout on the local flow regime, it is often necessary and useful to assume the water velocity through the farm is spatially constant and consistent with locally ambient speeds. The τ_{fl} is therefore calculated as L/U where L is the length scale of the cage or bay (e.g. the diameter of a cage) and U is the typical water velocity in the area. This time scale can be thought of as either the time scale for a plug of water to flow through the cage, or the time scale for perfect mixing within the cage. Although neither of these flow patterns accurately represents the flow of water through a fish cage, they are useful and convenient approximations. In plug flow, a parcel of water passes through the cage and does not mix with the water around it. The substance within the water enters the cage as a plug, traverses across the cage and leaves the cage as a separate entity. In perfect mixing flow the water is instantaneously and completely mixed throughout the cage as soon as it enters. A substance within the entering water leaves the cage over time, such that the concentration of substance within the cage decreases exponentially with time. In this scenario τ_{fl} represents the e-folding time for flushing the cage.

As an example of the use of the above, we calculate the index for a representative salmon cage moored in SWNB. We adopt an arbitrary oxygen threshold of 6 mg L^{-1}, an ambient oxygen concentration of 7 mg L^{-1}, a cage diameter of 31 m, a pre-market fish weight of 4 kg, a swimming speed of 1 body length per second (bl s^{-1}), a stocking density of 4 fish m^{-3} and a water temperature of 12 °C. These conditions are somewhat representative of a pre-market situation in the late summer to early fall, when the water temperatures are near the annual maximum, the ambient concentration of DO is near the seasonal minimum and fish are growing rapidly.

The apparent swimming speeds of salmon in Bay of Fundy sea pens were observed by [20] and ranged from about 0.4–2.8 bl s^{-1}, with a mean of about 1 bl s^{-1}, or an overall relative mean speed of 43.9 cm s^{-1}. Although [20] did not measure absolute swimming speeds, the apparent speeds are probably indicative of the range of absolute swimming speeds. Water currents in the vicinity of the monitored farms ranged from 0–42 cm s^{-1} with the mean currents ranging from 2.6–17.6 cm s^{-1} [20] (Fig. 9). Hence, fish swimming at 44 cm s^{-1} against these currents would achieve apparent swimming speeds of about 0–3 bl s^{-1} for fish of approximate length > 44 to 15 cm, respectively. The respiration rate per non-feeding fish based on these assumptions is approximately 100 mg O$_2$ kg^{-1} of fish h^{-1}, based on the equations reported by [8, 12] and described more fully in the Appendix. The cage diameter corresponds to a circular cage with a circumference of 100 m.

The oxygen threshold concentration is based on scientific literature that suggests it is prudent to culture salmon at DO levels above about 6 mg L^{-1} to achieve optimal growth [34]. From an environmental protection perspective, Davis [5] suggests that salmonids may begin to show behavioural modifications at concentrations of DO less than about 6.75 mg L^{-1}. The United

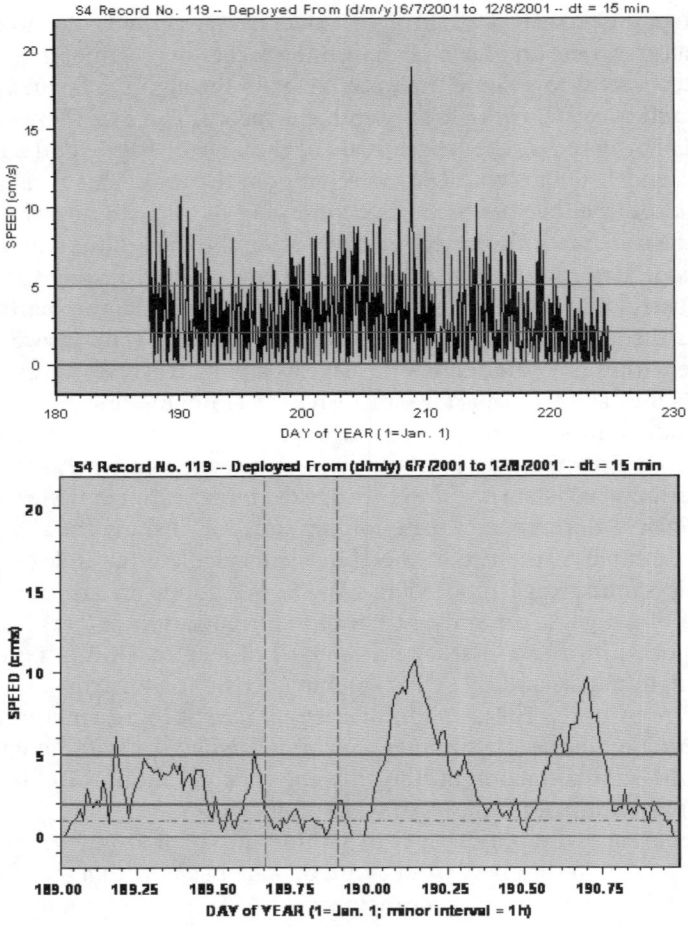

Fig. 9 Time series of the ambient current speed in the vicinity of a fish farm located in an area of southwestern New Brunswick. The *top panel* shows the entire data record. The *lower panel* shows a selected portion of the record that illustrates an event in which the current speed was less than 2 cm s^{-1} for about 5 h consecutively

States Environmental Protection Agency (USEPA) suggests that 5 mg L^{-1} is a reasonable environmental threshold for the middle Atlantic Bight since at concentrations above this level ecosystem integrity is largely maintained [30]. From a practical perspective, a threshold of 7 mg L^{-1} would perhaps be too high a target threshold if ambient DO concentrations approach this concentration, as is the case in the SWNB area. Hence, a threshold of 5–6 mg L^{-1} seems to be a reasonable threshold value that gives a high level of environmental protection and enables the local aquaculture industry to exist to some degree without reducing local levels of DO below the threshold. We therefore used a value of 6 mg L^{-1} for the illustrative calculations presented below.

Based on the above assumptions, the time scale for the fish in the cage to reduce the ambient oxygen concentration to the threshold concentration, a reduction of 1 mg L^{-1}, is $\tau_{thres} = 38$ min. A current flowing through the cage at a speed of 0.014 m s^{-1} (1.4 cm s^{-1}) results in a flushing time scale of $\tau_{fl} = 38$ min and an oxygen depletion index of $I_{DO} = 1$. If the fish are fed and we assume their respiration rate is doubled, the depletion index is 1 when the current speed is 0.028 m s^{-1} (2.8 cm s^{-1}). For comparison, salmon in their first summer in sea cages have a fish weight of about 0.5 kg and a $\tau_{thres} = 152$ min. This means that very weak currents are sufficient to flush cages containing smolts.

These calculations suggest that if currents are less than about 2 cm s^{-1} for more than about 40 min, the depletion index will be less than one for pre-market sized fish during the late summer and early fall, and the oxygen in a fish cage may be depleted below the desired threshold. An analysis of a current metre record collected from a location near a fish farm in the SWNB area, that has experienced reduced oxygen levels, illustrates that the currents in the area can often be less than 2 cm s^{-1} for periods greater than 40 min (Fig. 10).

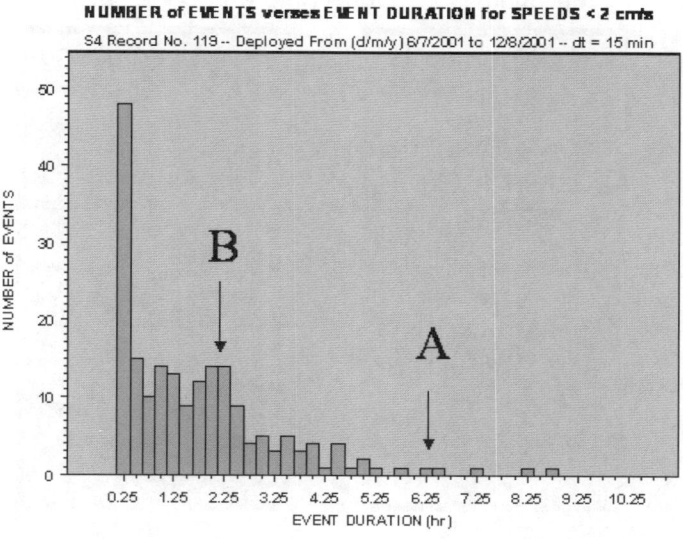

Fig. 10 Frequency distribution of the duration of events in which current speeds were less than 2 cm s^{-1}. The distribution is generated from an analysis of the current meter record shown in the top panel of Fig. 9. The *letters* are for illustration purposes only. They indicate that during the 35 days of the current meter deployment there was one event in which the current speed was less than 2 cm s^{-1} for about 6.25 h (*A*) and 14 events in which the current speed was less than 2 cm s^{-1} for about 2.25 h (*B*)

The geographic location of currents with these characteristics in SWNB is indicated by output from a three-dimensional, finite element tidal circulation model of the SWNB area described by [11]. Figure 11 shows a map of the model-derived contours of the longest consecutive period of time within a tidal cycle that the current speed is less than 2 cm s^{-1}. The heads of bays and inlets are predicted to experience current speeds less than 2 cm s^{-1} for periods of 1–2 h or longer. This is consistent with current metre records from the area and with anecdotal information that several of these areas are areas in which salmon farmers have experienced reduced oxygen concentrations in their farms at one time or another.

Although the index is best applied at spatial scales of a farm, the approach has the potential to be applied to larger spatial scales, such as that of the scale of a bay. In doing so, additional sources and sinks of oxygen, such as those that are area dependant should be taken into consideration. For example, the air–sea exchange of oxygen is dependent upon the surface area of the body of water being considered, as are processes such as sediment–water exchange and exchange associated with intertidal algae (see [27] for information on

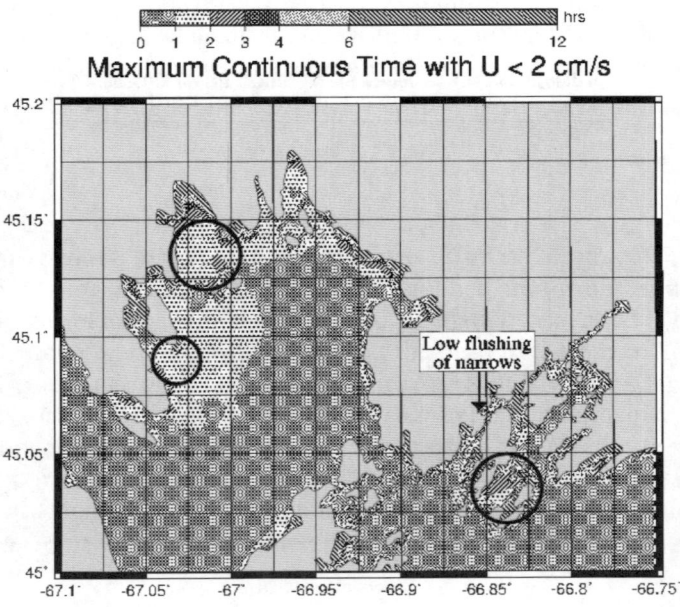

Fig. 11 Contour plot of the maximum consecutive number of hours within an M_2 tidal cycle that the modelled current speed is less than 2 cm s^{-1}. The *circles* indicate areas in which anecdotal information indicates salmon farms have experienced reduced concentrations of dissolved oxygen at one time or another. The *arrow* indicates an area in which salmon farms have experienced reduced concentrations of dissolved oxygen due to a low flushing rate of the channel rather than a low current speed at the farm site

the latter rates). The index could be generalized so the flushing time scale includes sources of oxygen in addition to flushing, and the threshold respiration time scale includes oxygen sinks in addition to fish respiration.

As an example of how an application of the simple depletion index approach to a bay scale situation might potentially be developed, we consider the Seal Cove area of southern Grand Manan within the Bay of Fundy (Fig. 1). Seal Cove is a semi-enclosed tidal embayment on Grand Manan Island in the mouth of the Bay of Fundy. The bay is approximately 4.8 km long and 2.1 km wide. The average depth at low tide is approximately 15 m. The tidal range in the bay is approximately 4 m. The volume of water within the bay at low tide is therefore approximately 1.6×10^8 m^3, the volume brought into the bay each tide is 4×10^7 m^3, and the volume of the bay at high tide is 2×10^8 m^3.

The rate of oxygen utilization by fish respiration was estimated from the approved fish production limits (APLs) for the farms within Seal Cove as established by the New Brunswick Department of Agriculture, Fisheries and Aquaculture (NBDAFA). As of the winter of 2003, the combined APL for salmon farms within Seal Cove was 1 269 000 fish and the cumulative estimated site potential (ESP) was 1 744 000 (Barry Hill, NBDAFA, personal communication). The oxygen consumption rate for salmon was assumed to be $\sim$100 mg O$_2$ kg^{-1} h^{-1}, assuming that each fish has a weight of 4 kg, the water temperature is 12 °C and the fish are swimming at 1 bl s^{-1}. The oxygen consumption rate for an entire 4-kg fish is therefore $\sim$400 mg O$_2$ fish^{-1} h^{-1} and the rate for the entire 1 269 000 fish is approximately 5×10^8 mg O$_2$ h^{-1} (Table 1) or 1.2×10^7 g O$_2$ d^{-1}. This consumption rate does not include the utilization of oxygen by fish faeces, which is about one third the rate of fish respiration [27] and does not include the enhancement of the fish respiration rate associated with fish feeding.

Calculations based on the above assumptions, and the assumption that water within the bay is instantaneously mixed throughout the bay, indicate it would take between approximately 45 or fewer days for the APL salmon in Seal Cove to use up the oxygen buffer in Seal Cove if there was no tidal exchange. It would take about 45 days if the buffer was 3 mg L^{-1}, 30 days for a 2 mg L^{-1} buffer and 15 days for a 1 mg L^{-1} buffer. These times would become smaller if the oxygen utilization rate within the bay was increased by other factors, such as an increased fish respiration rate due to feeding, a larger biomass of fish in the bay, an additional source of BOD such as that from sediments and fish faeces. These processes could easily double the respiration rate and would reduce the respiration time scales in half. For example, the oxygen demand by subtidal sediments located away from fish farms has been estimated to be 0.79 mmole m^{-2} h^{-1} [27] or 4.5×10^7 g O$_2$ day^{-1}. This is the same order of magnitude as the demand generated by the fish, and therefore increases the total demand for oxygen within the bay by a factor of about two and reduces the utilization time scales by about one half.

Table 1 Summary of flushing time, oxygen utilization time and utilization to flushing ratio estimates for Seal Cove, Grand Manan in the southwest New Brunswick area of the Bay of Fundy (from Page, 2003)

Seal Cove, Grand Manan	Primary scenario: –low temperature –average size
Physical characteristics	
Length (m)	5000
Width (m)	2100
Depth at low tide (m)	15
Tide height (m)	4
Low tide volume (m^3)	1.6×10^8
Tidal volume (m^3)	4.2×10^7
High tide volume (m^3)	2×10^8
Water temperature (°C)	12
e-Folding flushing times (tidal cycles, h, days)	4.25 tides, 53 h, 2.2 days
95% Flushing times (tidal cycles, h, days)	12.8 tides, 158 h, 6.6 days
Dissolved oxygen characteristics	
Ambient DO (mg L^{-1})	7–9
DO threshold (mg L^{-1})	6–7
Threshold DO deficits	0–3
Salmon farm characteristics	
Size of fish (kg)	4
Fish swimming speed (bl s^{-1})	1
Basic resp. rate/kg (mg DO kg^{-1} h^{-1}) at above temp.	99
Basic resp. rate/fish (mg DO/fish h^{-1})	398
Approved production limit (APL #fish)	1 269 000
Respiration rate for APL (mg DO h^{-1})	5×10^8
Time (d) for APL to utilize DO deficit of 1 mg L^{-1}	14.7
Estimated site potential (ESP #fish)	1 744 000
Respiration rate for ESP (mg DO h^{-1})	6.9×10^8
Time (d) for ESP to utilize DO deficit of 1 mg L^{-1}	10.7
*Ratio of utilization time/95% flushing time**	
0 mg L^{-1} deficit assuming APL and ESP	0
1 mg L^{-1} deficit assuming APL and ESP – basic resp.	2.2 (*1.1*); 1.6 (*0.8*)
– feeding respiration = 2 × basic resp.	1.1 (*0.6*); 0.8 (*0.4*)
2 mg L^{-1} deficit assuming APL and ESP – basic resp.	4.4 (*2.2*); 3.2 (*1.6*)
– feeding respiration = 2 × basic resp.	2.2 (*1.1*); 1.6 (*0.8*)
3 mg L^{-1} deficit assuming APL and ESP – basic resp.	6.7 (*3.3*); 4.8 (*2.4*)
– feeding respiration = 2 × basic resp.	3.3 (*1.6*); 2.4 (*1.2*)

* Ratio values are based on APL and 95% flushing (APL and 2 × 95% flushing); ESP and 95% flushing (ESP and 2 × 95% flushing), i.e. bracketed values are based on the assumption of the flushing rate being under-estimated by a factor of 2.

We assume the oxygen sources are provided mainly by the tidal flushing and air–sea exchanges of oxygen. We have not considered the net inputs associated with primary production by phytoplankton and macro-algae, although these have been considered to some extent by [27].

The residence time as estimated by the tidal prism method [7] (residence time = volume of bay at low tide divided by the high tide to low difference in the volume) is 4.25 tidal cycles which equals 53 h or 2.2 days. This time scale is usually interpreted as the time needed to exchange about 66% of the water in the bay. The time for 95% flushing is three times this or 6.6 days. This interpretation is based on the assumptions that the water entering the bay is instantaneously mixed with the water already in the bay, that water entering the bay has not been in the bay before and that none of the water leaving the bay each tide returns. None of these assumptions are completely correct and usually result in the residence time of a bay being underestimated, i.e. the actual residence time is longer or it takes more time to flush the bay than estimated by the tidal prism method. This underestimation may be by a factor of ~ 2 or more [9]. If it is by a factor of two, the flushing rate for Seal Cove is approximately 12 days. More accurate residence time estimates can be obtained via more extensive considerations such as circulation and drogue tracking models. Some preliminary explorations of particle tracks generated by our tidally driven drogue tracking model for southern Grand Manan [17] indicate that the assumption of no return of water into Seal Cove on each tide is not valid. This means the tidal prism flushing time estimate is actually an underestimate of the real residence time. However, the flushing time may be influenced by wind-driven water transport which we have not considered.

The rate of air–sea exchange of oxygen for the bay as a whole is 1.7×10^6 g day^{-1}, when estimated using the equations given in [24], a monthly average wind speed of 3.6 m s^{-1}, and a 1 mg L^{-1} difference between the ambient and saturated concentrations of oxygen. The observed monthly mean wind speed is actually 3.1 m s^{-1} for this area during the summer months (calculated from the 2002 through 2004 July through September wind records for the area). The higher value of 3.6 m s^{-1} was used in our calculations because the Stigebrandt [24] equations do not apply at wind speeds below this value. The time scale for air–sea exchange to replace a 1 mg L^{-1} (1 g m^{-3}) deficit in the concentration of oxygen within the low tide volume of the bay is therefore about 94 days (total oxygen deficit within the bay divided by the air–sea exchange rate). This is about eight times longer than the tidally generated exchange. This is not the case for most of the year when mean wind speeds in SWNB are consistently greater than the 3.6 m s^{-1}, and hence the replacement time is less. For example, during the winter, spring and fall typical mean wind speeds are 5 m s^{-1} and this generates a replacement time of about 12 days which is similar to the tidal flushing time scale.

Based on the above considerations, the oxygen budget within Seal Cove during the summer would appear to be dominated by fish respiration, sediment respiration and tidal exchange. At times, the wind-induced air–sea exchange of oxygen may be significant. If we ignore the sediment demand, the index calculations for Seal Cove, based on the cumulative APLs and ESPs, estimated DO utilization rates and estimated flushing times, indicate the depletion index is between 6.7 and 0.4 (Table 1). For the situation of a high ambient DO concentration of 9 mg L^{-1} and the corresponding buffer of 3 mg L^{-1}, the APL-based ratio varies between 6.7 and 1.6 and the ESP-based ratio varies between 4.8 and 1.2. If we assume the lower ambient DO concentration of 7 mg L^{-1} and a corresponding buffer of 1 mg L^{-1}, the APL-based ratio varies between 0.6 and 2.2 and the ESP-based ratio varies between 1.6 and 0.4. When the sediment demand is included, the ratios will be reduced by about one half and if the sediment demand and air–sea exchanges are included, the ratios remain about the same. The calculations would therefore suggest the APLs and ESPs for Seal Cove are near their limit for the specified threshold since they result in ratio values near to one. The fish farmers and environmental regulators are therefore giving more detailed attention to oxygen dynamics in the area to ensure their husbandry and environmental objectives are being met.

7
Discussion and Conclusions

In the SWNB region of the Bay of Fundy there is a period of several weeks to a few months during the late summer and early fall when the annual cycle in the concentration of ambient DO is at a minimum, water temperatures and fish respiration rates are at a maximum, and wind speed and air–sea oxygen exchange are at their annual minimum. During this time, the fish farms in areas of weak currents have the potential to reduce the concentration of DO below ambient values on localized spatial scales. The temporal nature of the reductions may vary between farms, with reduction events persisting for periods of several hours every tidal cycle (12.42 h) in some farms. There are also periods of time, such as during phytoplankton blooms, when the levels of DO cyclically falls to concentrations that are below the seasonal norm. Also, the oxygen depletion may be enhanced during times with extended periods of calm weather.

Index values for oxygen depletion that have been calculated for the area, although not calibrated and tuned for any particular farm, are qualitatively consistent with the observational data and salmon industry experience. They indicate that farms located in areas with weak water currents are susceptible to low oxygen events during the annual minimum in ambient DO. The index suggests that for typical fish stocking densities of approximately 4 fish per m^3, currents need to be greater than 2 cm s^{-1} if reductions in DO below a threshold of 6 mg L^{-1} are to be avoided. This current speed is likely to be an underestimate of the actual ambient current needed to flush a cage. It has been estimated using the conservative assumptions that the cage nets have no influence on the water flow and that fish respiration is the only source of oxygen utilization within the cage. The nets no doubt reduce the flow through the cage and hence reduce the flushing rate of the cage, and the oxygen demand by the farm is underestimated since we did not include the BOD associated with faeces [27, 28] and bio-fouling.

Although the significance of the periodic events of lower oxygen concentrations to fish production and the environmental water quality is not well understood, the experience of fish farmers in the region is that reductions could contribute significantly to losses in fish production. The high frequency variability in the oxygen concentrations also indicates that oxygen monitoring needs to be conducted regularly at short time intervals if the oxygen environment on a farm is to be sufficiently characterized.

The bay-scale calculations, although preliminary, indicate the importance of sediment respiration and winds on the oxygen budget on bay-wide scales. They also suggest that during periods of low winds bay-wide reductions in DO would be possible. This is consistent with anecdotal reports from the salmon farming industry indicating that, during summers with calm wind conditions for unusually long periods of time, the fish on many farms were showing symptoms of a reduced oxygen environment.

Finally, it should be noted that the above index is still being developed and the presented calculations are only a preliminary *indicator* of the potential influence of the fish farming component on the concentration of DO. If the index has a value that is near or below one, it is probably prudent to conduct a more detailed consideration of the processes influencing the DO budget. This should include detailed time series recordings of DOs in the area of interest, as well as enhanced modelling and scaling considerations. The spatial complexity of the water circulation should also be taken into considerations since water within a bay is unlikely to be homogenously mixed. This information will help define the appropriate size of box to use in the index calculations. Once this has been completed and the index refined, a desirable index value or values could perhaps be set that satisfies regulators and industry purposes. This value could then be used to estimate the number or biomass of fish that satisfies the index relationship.

In the future it is hoped that additional observational and modelling work will be conducted to enhance the understanding of oxygen dynamics in the farms and bays of SWNB, and that the knowledge gained can be used by industry in their farm management practices as well as by environmental regulators in their efforts to define and maintain water quality standards. Once confidence in the index approach is established, it can be used to estimate the number of caged salmon that will achieve a specified index value. This value may be one or greater or less than one depending upon the degree of protection the regulators wish to incorporate into the calculations.

Acknowledgements We sincerely thank the editor for the opportunity, encouragement and tolerance in the preparation of this chapter. We also thank the staff of the New Brunswick Department of Agriculture, Fisheries and Aquaculture and the staff of several fish farming companies, particularly Heritage Salmon Ltd. and Aqua Fish Farms for the positive interest they have shown and support they have provided for various aspects of our DO work over the past few years. The comments made by three anonymous reviewers were greatly appreciated and served to improve the manuscript.

Appendix

The rate at which salmon utilize oxygen is a function of several factors including water temperature, swimming speed, fish size, feeding activity and stress level. Many authors have reported measurements of oxygen consumption rates in salmon (see [8] and references therein). Unfortunately, few describe an equation that relates oxygen consumption to body size, water temperature, swimming speed and feeding rate and no such empirical relationship has been derived for fish reared in the Bay of Fundy. Hence, we have chosen to use the relationships provided by [8, 12].

Grottum and Sigholt [12] estimate the specific oxygen consumption rate by salmon (in units of mg of O_2 kg^{-1} of fish h^{-1}) as a function of fish body weight (M in kg), water temperature (T in °C) and fish swimming speed (U_{sw} in body-lengths per second, bl s^{-1}). Their relationship Eq. 3 estimates the specific oxygen consumption rate in fish that were starved for 24 h prior to being transferred into a respirometer:

$$DO_{\text{demand by fish}} = aM^b c^T d^{U_{sw}}. \tag{3}$$

The coefficients a, b, c and d have values of 61.6, 0.33, 1.03 and 1.79 mg L^{-1} h^{-1}, respectively. Coefficient "a" is the standard metabolic rate for a fish of 1 kg at 0 °C, "b" is the mass exponent, "c" is the temperature parameter with a value of 1.34, and "d" is the activity parameter [12]. When Eq. 3 is multiplied by the mass of the fish, b, the exponent for M becomes 0.67 and the relationship estimates the oxygen consumption rate for the whole fish (mg of O_2/fish h^{-1}).

Forsberg [8] estimates the incremental effect of feeding rate (F in percent body weight per day) on the specific oxygen consumption rate of salmon as:

$$DO_{\text{demand by fish}} = 1.0 + 1.74 F \quad \text{for} \quad 0 \leq F < 0.6 \tag{4}$$

and

$$DO_{\text{demand by fish}} = 2.0 \quad \text{for} \quad F \geq 0.6 \tag{5}$$

We combined the Forsberg [8] and Grottum and Sigholt [12] relationships to provide a working relationship that includes the effects of fish mass, temperature, swimming speed and feeding rate on the specific oxygen consumption rate. The complete relationship for specific oxygen demand is:

$$DO_{\text{demand by fish}} = aM^b c^T d^{U_{sw}}(1.0 + 1.74 F) \quad \text{for} \quad 0 \leq F < 0.6 \tag{6}$$

and

$$DO_{\text{demand by fish}} = 2.0\, aM^b c^T d^{U_{sw}} \quad \text{for} \quad F \geq 0.6 \tag{7}$$

The specific oxygen consumption rate in fasting salmonids is known to vary throughout the day by a factor of about 1.2–1.3 with peaks near sunrise and in the late evening [8]. Feeding activity increases the routine specific oxygen consumption rate by a factor of 1.8–3.6 when fish are fed at rates of about 0.60% or less of their body weight per day ([8] and references therein). The specific oxygen consumption rate does not continue to increase at feeding rates greater than about 0.60% body weight per day [8].

References

1. Aiken D (1989) World Aquacult 20:11
2. Chang BD (1998) Can Stock Assess Secretariat Res Doc 98/151
3. Chang BD (2003) The salmon aquaculture industry in New Brunswick: why go offshore? In: Bridger CJ, Costa-Pierce BA (eds) Open ocean aquaculture: from research to commercial reality. The World Aquaculture Society, Baton Rouge, Louisiana, United States, p 229
4. Churgin J, Halminski SJ (1974) Temperature, salinity, oxygen, and phosphate in waters off United States, vol I: Western North Atlantic. Key to oceanographic records documentation no 2. National Oceanographic Data Center, Washington, DC, US Department of Commerce, National Oceanic and Atmospheric Administration, Environmental Data Service
5. Davis JC (1975) J Fish Res Board Can 32:2295
6. Department of Fisheries and Oceans: Science Branch, Maritimes Region (2003) Can Tech Rep Fish Aquat Sci 2489:60
7. Dyer KR (1973) Estuaries: a physical introduction. Wiley, NY
8. Forsberg OI (1997) Aquacult Res 28:29
9. Gillibrand PA (2001) Est Coast Shelf Sci 53:437
10. Gran HH, Braarud T (1935) J Biol Bd Can 1:279
11. Greenberg DA, Shore JA, Page FH, Dowd M (2005) Modelling embayments with drying intertidal areas for application to the Quoddy region of the Bay of Fundy. Ocean Modelling 10:211
12. Grottum JA, Sigholt T (1998) Aquac Eng 17:241
13. Hirata H, Kadowaki S, Ishida S (1994) Bull Natl Res Inst Aquac Suppl 1:61
14. Louanchi F, Najjar RG (2001) Deep-Sea Res Part II 48:2155
15. Martin JL, Wildish DJ, LeGresley MM, Ringuette MM (1995) Can Manuscr Rep Fish Aquat Sci 2277:154
16. Najjar RG, Keeling RF (1997) J Mar Res 55:117
17. Page FH, Chang B, Greenberg D (2004) Can Tech Rep Fish Aquat Sci 2543:47

18. Page FH, Martin JL (2001) Seawater oxygen concentrations in the Quoddy Region and its relevance to salmon culture. In: Hargrave BT, Phillips GA (eds) Environmental studies for sustainable aquaculture (ESSA): 2002 Workshop Report. Can Tech Rep Fish Aquat Sci 2352:12
19. Page FH, Peterson R, Greenberg D (2002) Salmon aquaculture, dissolved oxygen and the coastal habitat: scaling arguments and simple models. In: Hargrave BT (ed) Environmental studies for sustainable aquaculture (ESSA): 2002 Workshop Report. Can Tech Rep Fish Aquat Sci 2411:16
20. Peterson RH, Page F, Steeves GD, Wildish DJ, Harmon P, Losier R (2001) Can Tech Rep Fish Aquat Sci 2337:117
21. Silvert W (1992) Aquaculture 107:67
22. Silvert W (1994) Modelling benthic deposition and impacts of organic matter loading. In: Hargrave BT (ed) Modelling benthic impacts of organic enrichment from marine aquaculture. Can Tech Rep Fish Aquat Sci 1949:1
23. Statistics Canada (2004) Aquaculture Statistics 2003. Statistics Canada, Catalogue no 23-222-XIE
24. Stigebrandt A (1991) Limnol Oceanogr 36:444
25. Stigebrandt A, Aure J, Ervik A, Kupka Hansen P (2004). Aquaculture 234:239
26. Strain PM, Clement PM (1996) Can Data Rep Fish Aquat Sci 1004:33
27. Strain PM, Hargrave BT (2005) Salmon aquaculture, nutrient fluxes and ecosystem processes in southwestern New Brunswick (in this volume). Springer, Berlin Heidelberg New York
28. Strain PM, Wildish DJ, Yeats PA (1995) Mar Pollut Bull 30:253
29. Trites RW, Petrie L (1995) Can Tech Rep Hydrogr Ocean Sci 163:55
30. United States Environmental Protection Agency (USEPA) (2000) Ambient aquatic life water quality criteria for dissolved oxygen (saltwater): Cape Cod to Cape Hatteras. United States Environmental Protection Agency EPA-822-R-00-012
31. Wildish DJ, Zitko V (1991) Can Manuscr Rep Fish Aquat Sci 2132:11
32. Wildish DJ, Zitko V, Akagi HM, Wilson AJ (1990) In: Saunders RL (ed). Proc. Canada-Norway Finfish Aquaculture Workshop, September 11–14, 1989, p 11
33. Wildish DJ, Keizer PD, Wilson AJ, Martin JL (1993) Can J Fish Aquat Sci 50:303
34. Willoughby S (1999) Manual of salmonid farming. Fishing News Books

Salmon Aquaculture, Nutrient Fluxes and Ecosystem Processes in Southwestern New Brunswick

P. M. Strain[1] (✉) · B. T. Hargrave[2]

[1]Institute of Ocean Sciences, Fisheries and Oceans Canada, 9860 West Saanich Road, Sidney, V8L 4B2, Canada
StrainP@pac.dfo-mpo.gc.ca

[2]Ecosystem Research Division, Fisheries and Oceans Canada, Canada
hargraveb@rogers.com

1	Introduction	30
2	**Aquaculture Discharges**	31
2.1	Salmon Growth Model	32
2.2	Mass Balances	35
2.3	Waste Fractions	39
2.4	Oxygen Demand	40
3	**Results**	41
3.1	Wastes Produced on Farm Scales	41
3.2	Wastes Produced on Inlet Scales	44
4	**Ecosystem Processes**	50
4.1	Primary and Macrophyte Production and Nitrogen Regeneration	51
4.2	Respiration	52
4.3	Comparisons with Fluxes from Salmon Aquaculture	52
4.4	Conclusions	55
	References	56

Abstract Salmon aquaculture discharges organic wastes into the marine environment. Salmon metabolism and the waste discharges add nutrients and organic matter to and remove oxygen from both the water column and sediments. The salmon industry in Southwestern New Brunswick (SWNB) is used to illustrate how waste discharges can be estimated in the absence of detailed information on farm operations. A fish growth model and mass balance calculations are used to estimate carbon, nitrogen and phosphorus wastes at the farm scale. Feed nutrition data and environmental measurements are used to partition wastes into fractions, and to estimate the oxygen demand. The predicted demand at the time of maximum discharge is 200 times greater than the oxygen uptake measured in surface sediments at cages, suggesting that most farm wastes are dispersed over wide areas and do not accumulate directly under cages. The number of farmed fish in an inlet is then used to predict total discharges to that inlet. In SWNB, salmon aquaculture is the largest anthropogenic source of organic input to the coastal zone. The significance of the wastes on inlet scales (2–25 km) is evaluated by comparing element fluxes through salmon farms with fluxes due to natural processes: primary production, nutrient regeneration, and community respiration. In intensively farmed bays, fluxes due to salmon farms reach values of 20, 330 and 160% of those due to natural processes for oxygen, nitrogen and carbon, respectively: significant changes to the ecosystem have occurred in these bays. Spatial scales are critical in describing such impacts: effects will be greater close to the farms, and smaller when averaged over larger areas.

Keywords Aquaculture · Salmon wastes · Ecosystem processes · Marine

1
Introduction

Sea-cage salmon aquaculture discharges a number of waste types into the marine environment, including nutrients (nitrogen and phosphorus) and organic matter. The organic wastes from excess feed and salmon faeces are partially decomposed by bacteria. This process consumes oxygen, adds nutrients and lowers the oxygen content in both the water column and surface sediments. The respiration of the farmed fish further reduces the oxygen levels. In areas of intensive aquaculture, these waste streams can have significant impacts on the ecosystem: organic wastes may add to suspended particulate matter (SPM) or enrich or smother the benthic habitat; nutrients can stimulate growth of phytoplankton and/or macroalgae; and reduced levels of dissolved oxygen can stress native organisms as well as the farmed fish. The increased SPM and resulting reduced water transparency, build-up of nutrients, reduction of dissolved oxygen levels and stimulation of algal growth are all aspects of eutrophication.

Growth of sea-cage salmon aquaculture in Southwestern New Brunswick (SWNB) has been quite rapid since its inception in 1978. In 2002, there were 93 operating farms (Fig. 1), with an estimated total of almost 20 million fish in sea cages based on the province of New Brunswick Department of Agriculture, Fisheries and Aquaculture (DAFA) Approved Production Limits (APLs). Production in 2002 was 39 450 tons worth $ 201 million (Can $).

This paper examines the environmental significance of the particulate and dissolved wastes discharged from salmon aquaculture in a coastal ecosystem, using the industry in SWNB in 2002 as an example, by comparing the discharges to natural levels and processing rates for carbon, oxygen and nutrients. Because our focus is on the environmental impacts in SWNB, we use data that we believe to be most representative of the industry there in 2002. The first step in this process is to estimate the magnitude of the discharges from the fish cages. The next step is to evaluate the impacts of the fluxes of organic matter, carbon, nutrients and oxygen demand from the farms by estimating how they might influence ambient conditions in the receiving waters and how they compare with other anthropogenic sources and the natural cycling of these elements elsewhere in the ecosystem, including both the water column and the sediments. Such comparisons measure the extent to which

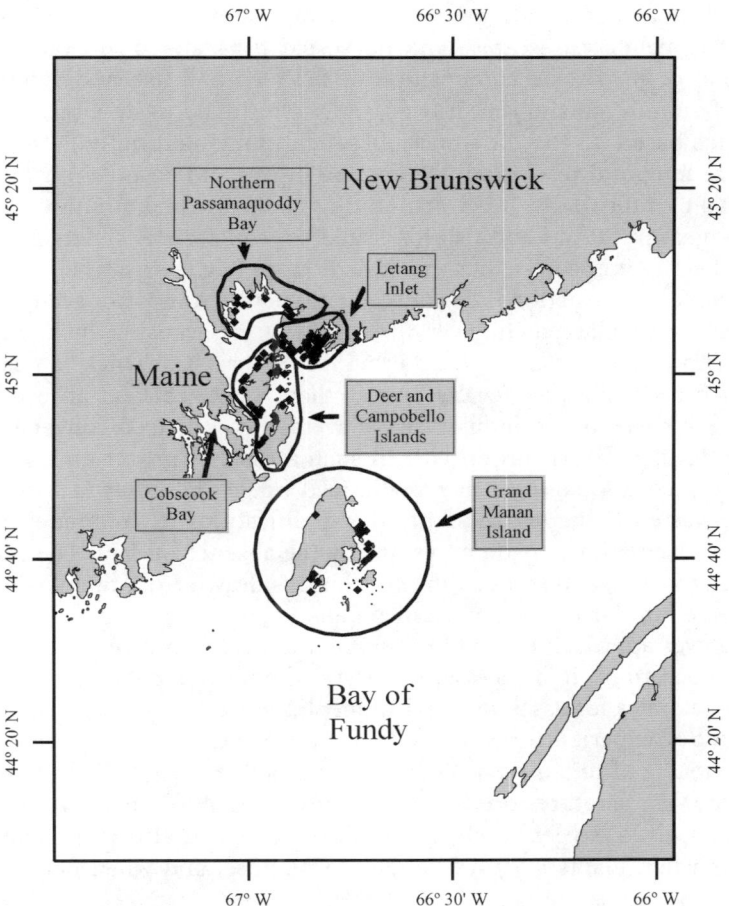

Fig. 1 Salmon farms in southwestern New Brunswick (diamond symbols) in 2002. Labelled areas in Canada are the CMRs referred to in the text. Some comparative calculations are also made for Cobscook Bay in Maine

aquaculture may be altering the ecosystem, and are indicators of the carrying capacity of the natural system.

2
Aquaculture Discharges

In land-based aquaculture, it is feasible to measure waste outputs directly by monitoring the composition of the waste stream. This is not possible in sea-cage aquaculture. The most rational alternative is to use a mass balance approach to compare the inputs to the farms (salmon smolts, feed) with the outputs (mortalities, escapes, and harvested fish). By consideration of the elemental or proximate composition of the feed and fish and some knowledge of salmon metabolism, it is possible to estimate waste discharges from the farms based on the differences between inputs and outputs. To achieve adequate temporal resolution, the data on inputs and outputs must be available on a monthly basis. It is worth noting that providing monthly reports of feed usage, mortalities and fish harvested to regulators is a statutory condition of licence for growers in some jurisdictions (e.g. Norway [1]). In SWNB, such detailed information is considered proprietary, and is not available to researchers or regulators. Only very general data are available, including APLs from farm licences issued by the Province of New Brunswick (actual numbers of fish on site may legally be less than the APLs) and an estimate of feed usage based on an industry-wide average for the feed conversion ratio (FCR = feed used/fish produced). In such a case, estimates and/or models must be used to quantify fish growth, feed usage and other factors that influence waste discharges. In addition, simplifications to the models or their inputs are sometimes made necessary by the absence of data. The predicted waste discharges, and some of the conclusions drawn from them, will be less constrained in the absence of detailed data.

Whatever approach is used to estimate the wastes discharged from finfish culture operations, it is necessary to assess the waste inputs on suitable time and space scales for assessing environmental impacts and risks. For temperate latitudes, experience has shown that coastal waters are most susceptible to acute impacts of organic wastes in late summer or early fall when relatively high water temperatures produce the highest rates of fish metabolism and the lowest dissolved oxygen levels. Macroalgal biomass is also maximum in late summer when plants start to die and decompose, and when natural nutrient concentrations are near the low point of their seasonal cycle. Discharge estimates on scales of no longer than a month are required to consider the worst-case scenarios arising from the seasonal cycle. The spatial scales of interest vary from the size of the cages, around which acute benthic impacts are localized (e.g. [2–4]), to inlet-wide scales that can be influenced by the dispersion of water-borne wastes.

2.1
Salmon Growth Model

Peterson et al. [5] surveyed 20 salmon farms in SWNB from 1995–1997, collecting data on smolt size, grow-out periods, growth rates and harvest weights. These are the most recent data available on growing conditions in the SWNB industry. Based on average values, 92 g smolt are placed in sea cages on 1 May and are grown for 21 months before harvesting at a weight of 4.9 kg at the end of January in the second calendar year following their introduction.

Therefore the salmon are in the cages during 3 calendar years: first-year fish are present from May to December; second-year fish for the entire calendar year; and third-year fish for 1 month.

Silvert [6] developed a model for salmon aquaculture in SWNB that was used to estimate nutrient release and oxygen demand from salmon farms in the Letang Inlet for 1992 [7]. This model is based on a temperature-corrected allometric expression for growth of the form:

$$G \propto W^a e^{(bT)}$$

where:
G = Absolute growth rate (i.e. g d^{-1})
W = Body weight (g wet weight)
a = Allometric exponent
T = Temperature (°C)
b = Exponent related to the Q_{10} value ($Q_{10} = G(T+10)/G(T)$)

Silvert fitted this model to fish cultured in SWNB using growth data available at the time and the seasonal temperature cycle for SWNB, deriving values of $a = 0.64$ and $Q_{10} = 6.4$. In the formulation of this model, growth was calculated relative to the growth rate of a 50 g smolt at 6 °C, determined from a doubling time for 50 g smolt that was provided in feed specifications from some local feed manufacturers at the time. These base growth rates are no longer routinely provided by the feed suppliers. For the update of the model, we continued to use the growth parameters selected by Silvert, ran the model using the SWNB seasonal temperatures, smolt size and grow-out period [5], and tuned the doubling time to match the harvest weight reported for farms in SWNB [5]. This empirical model has now been fitted using three parameters, and will be referred to as the Q_{10} model.

Although the Q_{10} model has a history of application to the SWNB salmon industry, it is not widely used. A model based on a thermal-unit growth coefficient [8, 9] has been much more widely used to predict the growth of salmonids. We will refer to this model as the TGC model. It has the form:

$$W_f = [W_i^{1/3} + \sum (TGC * T * t)]^3$$

where:
W_i, W_f = Initial, final body weights (g wet weight)
TGC = Thermal-unit growth coefficient
t = Time (days)

As we did for the Q_{10} model, we ran the TGC model using SWNB temperature data, smolt size and grow-out period and adjusted the TGC to a value of 0.00265 g$^{1/3}$ °C^{-1} d^{-1} to correctly predict the harvest weight of salmon in SWNB. This TGC value can be compared with other values reported for farm strains of Atlantic salmon grown in seawater. Sveier et al. [10] reported a value of 0.0022 g$^{1/3}$ °C^{-1} d^{-1} for salmon grown from a mean size of ~ 630 g to ~ 1000 g.; Thodesen et al. [11] reported a mean value of 0.0030 g$^{1/3}$ °C^{-1} d^{-1}

for the range 810–1460 g; Refstie et al. [12] reported values of 0.0039 (salmon from ∼ 560 to ∼ 1780 g) and 0.0020 g$^{1/3}$ °C^{-1} d^{-1} (∼ 1780 to ∼ 2750 g). The tuned value of 0.00265 g$^{1/3}$ °C^{-1} d^{-1} falls within the range of these measured values.

Figure 2 compares the daily growth rates predicted by the Q_{10} and TGC models for the full growout period using the growing conditions in SWNB. The two models produce very similar ranges of fish growth rates, from ∼ 0.2 to 1.5% d^{-1}, and have very similar seasonal cycles. The mean difference between the two models is 15% of the growth rate. Peterson et al. [5] presented some statistical models for growth derived from their observations of fish weights at 20 farms in SWNB from 1995 to 1997. These models predict fish weight for the first 5 months of growth, and for 430 days after smolt introduction. Since these models do not explicitly include temperature, and do not cover the entire grow-out period, they are not useful for a growth model that will also be used to scale fish metabolism to calculate total waste production. They can, however, be used as to verify the Q_{10} and TGC model predictions. For 92 g smolts, Peterson et al. predicted weights of 513 g on 17 September of the first year (based on their preferred model, model 3) and 1818 g in early July of the second year. The Q_{10} model predicts weights of 526 and 1764 g at those times; the TGC model predicts weights of 493 and 1788 g. Both the Q_{10}

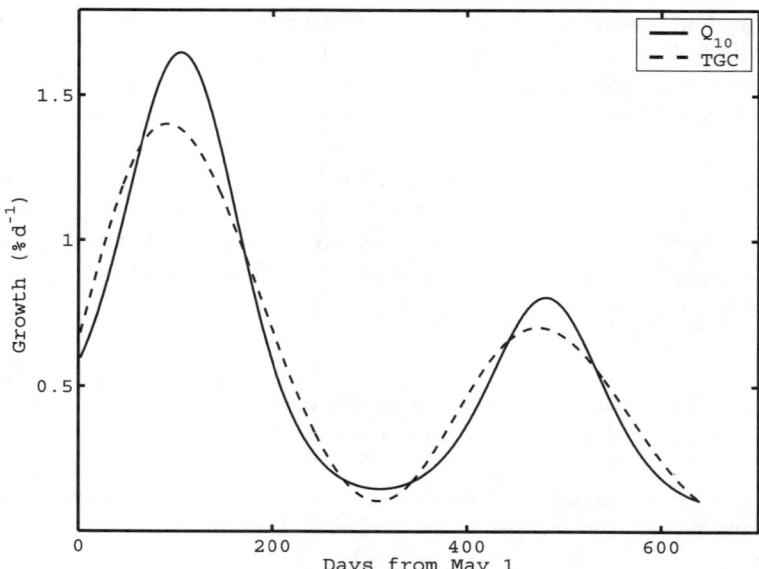

Fig. 2 Daily growth rates for 92 g smolts as predicted by the Q_{10} and TGC models. Both models use a sinusoidal temperature cycle whose standard deviation, maximum temperature and date of T_{max} match data for the area. Both models were tuned to produce harvest weight fish of the correct size (see text)

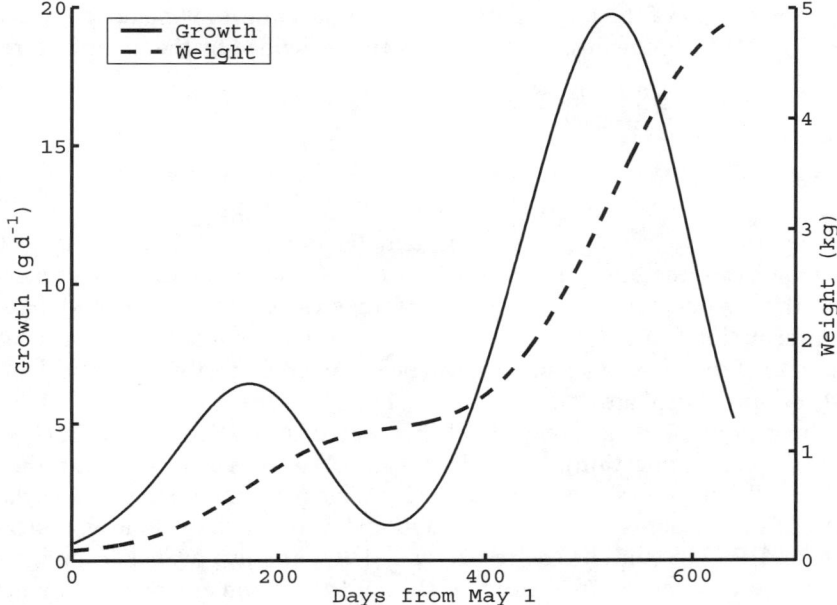

Fig. 3 Daily weight gain and body weights for 92 g smolts predicted by the *TGC* model

and *TGC* models do reasonably well in predicting fish size. All further waste calculations are based on the *TGC* model.

Daily biomass gain (g wet weight added per day) will be used to calculate the waste discharges. Fig. 3 show the biomass gain and the fish size predicted for SWNB by the *TGC* model.

2.2
Mass Balances

Carbon, nitrogen and phosphorus waste discharges can now be calculated by subtracting the C, N and P content stored in the fish (estimated from the growth model) from the C, N and P contents of the feed. The amount of feed used is simply related to fish growth by the *FCR*:

feed = growth × *FCR*

This is the economical *FCR*, defined as the ratio of feed used (wet weight) to fish produced (wet weight, whole fish). As stated earlier, only an industry average value of *FCR* is available for SWNB in 2002: 1.1 (B. Glebe, Department of Fisheries and Oceans, St Andrews, New Brunswick, personal communication). Since *FCR*'s relate total feed weight to total growth, the feed amount must then be adjusted for its water content (SWNB industry average: 10%, B. Glebe, personal communication). The C and N contents may be estimated from the weight of the feed using standard biochemical formulae that calculate C and N from the proximate composition provided by the manufacturer:

$$C = 0.5 \times f_{prot} + 0.7 \times f_{fat} + 0.4 \times f_{carbohydrate}$$

$$N = 0.16 \times f_{prot}$$

where

C, N = fraction (by mass) of carbon, nitrogen in feed

$f_{prot}, f_{fat}, f_{carbohydrate}$ = fractions of protein, fat and carbohydrate in the feed

In some cases, the phosphorus content of the feed may be directly available from the manufacturer (for SWNB, an average value of 1.2% was used, based on information from feed suppliers), or it can also be estimated if the ash content of the feed is known, since ash can be considered mostly a mix of calcium and sodium phosphate.

The average water content of salmon reared in SWNB is 68% (B. Glebe, personal communication). A similar approach was used to calculate the C in the salmon as was used for the feed. Using proximate composition data from [11], C accounts for 54% of the salmon dry weight. Direct measurements of N and P (2.88 and 0.45%, respectively [13]) are also available for whole salmon: these were used instead of the indirect estimates from proximate composition data.

The waste C discharge includes the carbon consumed by fish respiration as well as the organic carbon in fish faeces and uneaten feed. The N and P discharges include the N and P excreted by the salmon as dissolved wastes and faeces, and the nitrogen and phosphorus in the uneaten feed. The mass balance for N may be written:

$$N_{feed} = N_{retained} + N_{waste}$$

where

$N_{retained}$ is the nitrogen retained in the fish tissues

N_{waste} is the total of the nitrogen wastes: dissolved wastes, waste feed and faeces

Abbreviating the subscripts and rearranging yields:

$$N_w = N_{fd} - N_r$$

which may be rewritten explicitly in terms of the model inputs available to calculate the waste outputs:

$$N_w = 0.16 \times f_{prot} \times DM_{fd} \times FCR \times G - f_{Nfish} \times G$$

where

DM_{fd} = fraction of dry matter in feed

G = absolute growth (g d^{-1})

f_{Nfish} = fraction of nitrogen in whole salmon

The total N waste, N_w, is directly proportional to growth. Such a direct link between waste and growth is partially an artefact of determining the feed used from growth and FCR in SWNB. If independent estimates for feed use were available, then the variables would not be coupled so closely.

Dividing by the growth:

$$N_w/G = 0.16 \times f_{prot} \times DM_{fd} \times FCR - f_{Nfish}$$

Thus, the N waste produced per unit of biomass gain may be calculated from the four model inputs on the right hand side of this equation. It is also straightforward to evaluate the partial derivatives of N_w/G with respect to each of those four inputs. Multiplying such a partial derivative by an estimated uncertainty for the corresponding parameter makes it possible to assess and compare the sensitivity of N_w/G to how well the different parameters are known. Table 1 lists the model input values used for the SWNB model and estimates of their uncertainties. N_w/G is most sensitive to errors in the protein content of the feed (f_{prot}) and the feed conversion ratio (FCR) which translate to uncertainties of $\sim$ 19% and 15%, respectively, in the nitrogen discharge. The waste discharge is also quite sensitive to the nitrogen (or protein) content of the fish (f_{Nfish}) but relatively insensitive to the water content of the feed (DM_{fd}). Fig. 4 shows the daily discharges of nitrogen for a single salmon over its entire grow-out cycle, and their sensitivity to the FCR.

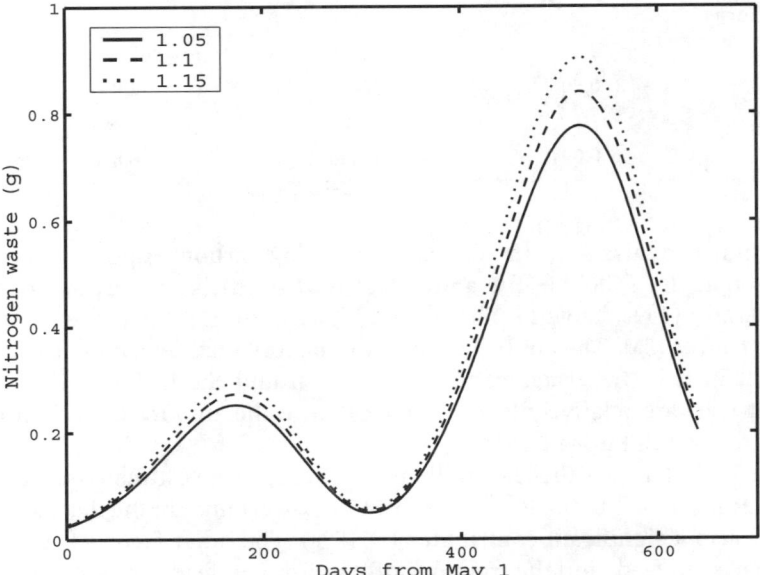

Fig. 4 Total nitrogen waste (waste feed + fish faeces + dissolved waste) generated by the growth of a single salmon through its entire grow-out cycle. Waste discharges are shown for three FCR values: 1.05, 1.1 (the industry average for SWNB) and 1.15

Table 1 Sensitivity of waste estimates to model inputs. The right hand column shows the percentage uncertainty in the waste due to the uncertainty of each model input, expressed as a percentage of the waste per unit growth. W is the nitrogen (N_w), carbon (C_w), or phosphorus waste (P_w)

Parameter (X)	Value	Uncertainty (ΔX)	$100 \times \frac{\partial(W/G)}{\partial X} \times \frac{\Delta X}{(W/G)}$
Nitrogen			
f_{prot}	0.45	0.05	19%
DM_{fd}	0.90	0.05	9.3%
FCR	1.1	0.10	15%
f_{Nfish}	0.029	0.005	−12%
Carbon			
f_{prot}	0.45	0.05	8.1%
f_{fat}	0.26	0.05	11%
$f_{carbohydrate}$	0.19	0.04	5.2%
DM_{fd}	0.90	0.05	8.7%
FCR	1.1	0.10	14%
DM_{fish}	0.32	0.05	−5.2%
f_{Cfish}	0.54	0.03	−5.3%
Phosphorus			
f_P	0.012	0.005	67%
DM_{fd}	0.90	0.05	8.9%
FCR	1.1	0.10	15%
f_{Pfish}	0.0045	0.0005	−6.8%

A similar analysis for the C waste (including carbon respired by salmon) shows that the *FCR* (i.e. the amount of feed used) is the largest source of uncertainty (14%, Table 1). The second highest contributor is the fat content of the feed (11%). The contributions from factors that determine the carbon content in fish (the water content of the fish and the fraction of C in dry weight fish) are relatively low ($\sim$ 5%), because the amount of carbon in the fish is fairly well known.

The P discharge is the least well known, because the estimated uncertainty in the amount of P in the feed leads to a 67% uncertainty in the discharge. The *FCR* is also a significant contributor ($\sim$ 15%). The other factors (the percent dry matter in feed, and the fractions of P in fish and feed) are less important, translating to 7–9% uncertainty in the P discharge.

2.3
Waste Fractions

The organic matter, carbon, nitrogen and phosphorus in the wastes are partitioned into a number of different fractions. The wastes (defined as anything other than the material retained in the fish tissues) are consumed by metabolic processes, released in uneaten feed, or released in salmon faeces, urine or through the gills. The waste types are important, because they determine the type of impact each waste stream will have. Salmon metabolism/respiration creates an immediate oxygen demand in the water and releases N and P into the water at the farm site. Waste feed includes many large particles that settle rapidly to the seabed and impact the benthic community by accumulating close to the farm, again consuming oxygen and releasing nutrients, on time-scales of days to months. Salmon faeces, and the smaller particles from waste feed, settle more slowly and therefore have the potential to be more widely dispersed. These wastes are degraded in the water column, creating oxygen demand and nutrient releases distributed over broad areas, on a time-scale of several days. Some of the waste feed and faeces will be decomposed by bacteria; some will be eaten by higher organisms (zooplankton, benthic organisms, and organisms in the fouling communities on the cages) to fuel their respiration and growth.

The partitioning of C_w can be written:

$$C_w = C_{respiration} + C_{wastefeed} + C_{faeces}$$

where the subscript respiration indicates respiration and other metabolic losses.

Feed digestibility and retention have been extensively studied for salmonids. These data have been reported in many ways, but are often described in terms of apparent digestibility coefficients (*ADCs*) or retention efficiencies. For example, the *ADC* for energy (or carbon) is:

$$ADC = 1 - \frac{E_{faeces}}{E_{intake}} \approx 1 - \frac{C_{faeces}}{C_{intake}}$$

where
ADC is expressed as a fraction
E (*C*) = the energy (carbon) in the faeces or the feed ingested by the fish.
The energy (carbon) retention efficiency can be expressed as a fraction of the digestible energy (carbon) that is retained in the fish tissues:

$$ERE = \frac{E_{retained}}{DE} = \frac{E_{retained}}{E_{respiration} + E_{retained}} \approx \frac{C_{retained}}{C_{respiration} + C_{retained}}$$

where *ERE* = energy retention efficiency
DE = digestible energy intake = that portion of the energy ingested that is not lost to the faeces

Rearranging the ADC and ERE equations leads to the following expressions:

$$\frac{C_{\text{respiration}}}{C_{\text{retained}}} = \frac{1 - ERE}{ERE}$$

$$\frac{C_{\text{faeces}}}{C_{\text{retained}}} = \frac{1 - ADC}{ADC \times ERE}$$

$C_{\text{respiration}}/C_{\text{retained}}$ and $C_{\text{faeces}}/C_{\text{retained}}$ can be used to calculate $C_{\text{respiration}}$ and $C_{\text{faeces}} \cdot C_{\text{wastefeed}}$ can then be calculated by difference. A number of studies have reported ADCs for energy and EREs or their equivalents for Atlantic salmon grown in seawater on fish meal diets [11, 12, 14–16]. Based on average values from these studies, $C_{\text{respiration}}/C_{\text{retained}} = 0.87$ and $C_{\text{faeces}}/C_{\text{retained}} = 0.43$.

The nitrogen wastes are partitioned as follows:

$$N_w = N_{\text{diss}} + N_{\text{wastefeed}} + N_{\text{faeces}}$$

where N_{diss} is the nitrogen lost by metabolic processes in the urine and through the gills, almost all of which will be in dissolved form.

Separation of the nitrogen wastes into fractions is exactly analogous to the carbon case, except that $ADCs$ and retention efficiencies for nitrogen (or protein) are used. From average values in the literature [10–12, 14–17], $N_{\text{diss}}/N_{\text{retained}} = 0.84$ and $N_{\text{faeces}}/N_{\text{retained}} = 0.28$. Phosphorus is partitioned in exactly the same way as nitrogen. Since data on P digestibility and retention is less available than for C or N for Atlantic salmon in seawater, we have supplemented the salmon data [17] with data for rainbow trout in freshwater [18] and calculated values of $P_{\text{diss}}/P_{\text{retained}} = 0.60$ and $P_{\text{faeces}}/P_{\text{retained}} = 0.89$.

As stated above, it is possible to calculate the waste feed component of the C, N and P wastes by difference after assessing the respiration, dissolved and faecal components based on the $ADCs$. The waste feed component is greater than the faecal component for both carbon and nitrogen ($C_{\text{wastefeed}}/C_{\text{faeces}} = 1.09$; $N_{\text{wastefeed}}/N_{\text{faeces}} = 1.27$), showing that waste feed dominated the solid wastes at the FCR achieved by the SWNB salmon industry in 2002. In other terms, the waste feed is 17% of the total feed used based on carbon or 14% of the total feed used based on nitrogen.

2.4
Oxygen Demand

The reduction of dissolved oxygen levels, both in the water column and in surface sediments, will have a potentially significant impact on salmon aquaculture. Fish respiration consumes oxygen in the fish pen, and more is consumed by the decomposition of waste feed and faeces. It is possible to estimate the respiration demand of O_2 from $C_{\text{respiration}}$. To quantify the demand from the decomposition of waste feed and faeces, the fraction of those waste streams susceptible to rapid bacterial decomposition must be estimated. For wastes on the seabed, some fraction of the organic matter will be permanently buried in the sediments without decomposing. For waterborne organic waste, a fraction will decompose quickly enough to impact conditions in the local (i.e. inlet-scale) area. We have measured O_2 uptake rates over a period of 20 days by particles that were resuspended from heavily biofouled nets, and in surface sediments containing high concentrations of feed pellets and faeces. Fitting both linear and exponential models produced similar results for the oxygen consumed after 5 days: 216 and 203 mg O_2 (g sediment)$^{-1}$. By combining these estimates with the water and organic carbon content of the sediments, these rates are equivalent to oxidizing 49% of the available carbon in 5 days. This is consistent with the 1:2 ratio of biological oxygen demand (BOD) and COD found in fresh fish processing wastes [7]. We will assume that 50% of the waste feed and faeces decomposes quickly enough to contribute BOD, whether the material is decomposed in the water column or in newly deposited sediments that are in close contact with the water. The remaining 50% of the carbon in the waste feed and faeces is assumed to be buried in the sediments, although not necessarily in the footprint of the cages. We will assume that 50% of the nitrogen and phosphorus in these wastes will be converted to dissolved forms during remineralization. Some transport of faeces further from the sites undoubtedly occurs [4, 19]. This buried organic material will be subject to further decomposition in the sediments but over longer time scales.

3
Results

3.1
Wastes Produced on Farm Scales

Table 2 summarizes the outputs of the fish growth model, listing total O_2 consumed, and the total discharges of carbon, nitrogen and phosphorus to the water column and sediments. These discharges are expressed as the total waste per ton of fish produced (i.e. wastes are summed over the entire grow-out cycle) and as maximum discharges for each of the 3 calendar years that the fish are in the water during their grow-out cycle (expressed per 1000 fish). Third year sites are those with mature fish still on site in January at the end of the grow-out cycle. Wastes are partitioned between the water column and the sediments as described above. Water temperatures and fish metabolism combine to produce maximum daily discharges in October for first and second year sites. Average daily discharges over the entire calendar for first, second and third year sites may be determined by dividing the values in Table 2 by

Table 2 Wastes from salmon aquaculture operations in Southwest New Brunswick (as weight of O_2, C, N or P)

Substance	kg waste/ tonne of fish produced[1]	Maximum rate of discharge[2]: Year1 kg d^{-1} (1000 fish)$^{-1}$	Maximum rate of discharge[3]: Year2 kg d^{-1} (1000 fish)$^{-1}$	Maximum rate of discharge[4]: Year3 kg d^{-1} (1000 fish)$^{-1}$
Water column impacts				
O_2 consumed (salmon respiration)	390	2.6	7.9	3.7
O_2 consumed (decomposition of wastes)	200	1.3	4.1	1.9
Carbon respired by fish	150	0.97	3.0	1.4
Carbon respired during decomposition of wastes	76	0.50	1.5	0.71
Nitrogen discharges (dissolved)	33	0.21	0.66	0.30
Phosphorus discharges (dissolved)	4.9	0.032	0.10	0.046
Sediment impacts				
Carbon buried	76	0.50	1.5	0.71
Nitrogen buried	9.0	0.059	0.18	0.084
Phosphorus buried	2.3	0.015	0.046	0.021

[1] Amount of waste produced during the entire grow-out cycle, per metric ton of fish produced.

[2] Maximum rate of discharge found during the first calendar year of growth, for each 1000 fish. This occurs on 18 October. Average daily discharges (averaged over the entire calendar year) are 2.37 times less than the values in this column.

[3] Maximum rate of discharge found during the second calendar year of growth, for each 1000 fish. This occurs on 2 October. Average daily discharges (averaged over the entire calendar year) are 2.00 times less than the values in this column.

[4] Maximum rate of discharge found during the third calendar year of growth, for each 1000 fish. This occurs on 1 January. Average daily discharges (averaged over the entire calendar year) are 16.4 times less than the values in this column.

2.37, 2.00 and 16.4, respectively; total annual discharges can then be determined by multiplying the average daily discharges by 365. Since all of the results in Table 2 depend on the seasonal temperature cycle and farming practices (feed use, type of feed, grow-out period), they are specific to SWNB, but they are unlikely to change quickly, since farming practices evolve relatively slowly. Since such farm-scale quantities are not dependent on the distribution of farms or the bay management strategies which may change from year to year, they are useful starting points for estimating wider-area impacts. (These estimates are also used in the calculations of aquaculture discharges into Blue Hill Bay, Maine [20].)

Despite their dependence on local conditions, the waste estimates in Table 2 are similar to those predicted for modern farm operations in other areas. For example, the mass balance model used in the Scottish locational guidelines for marine salmon farms predicts 48.2 kg total nitrogen waste for each ton of fish produced [21], compared to a total of 42 kg per ton (= 33 kg dissolved discharges + 9 kg buried in the sediments) for SWNB (the estimated *FCR*ss are similar for the two areas: 1.17 for Scotland; 1.1 for SWNB). Adjusting the Scottish estimates of the particulate and dissolved wastes to account for the rapidly decomposed fraction as we did in SWNB makes it possible to compare the water column and sediment impacts predicted for the two locations. The water column discharges are 41.9 kg N (Scotland) and 33 kg N (SWNB) per ton of fish produced; the sediment impacts are 6.3 kg N (Scotland) and 9.0 kg N (SWNB) per ton of fish produced. The higher sediment impacts calculated for SWNB result from the differences between the estimates of waste feed in the two areas. The Scottish model specifies waste feed at 5% of feed used, although reliable data are rare [21]. As noted above, the feed nutrition information in the SWNB model makes it possible to calculate the rate of feed wastage at $\sim 15\%$.

The use of *FCR*s to determine feed use and the way in which the partitioning of wastes into different fractions has been calculated means that the total wastes are directly proportional to growth: i.e. curves showing all the waste discharges have the same shape as the growth curve in Fig. 3. Peak discharges occur in September/October for years 1 and 2: there is a lag of a few weeks between the maximum water temperatures and maximum discharges because the biomass of the fish is increasing fast enough at this time of year to more than compensate for the slight decline in water temperature. Maximum rates of discharge are about 3 times higher in the second year of growth than in the first and are a consequence of the greater biomass on site in the second year. Peak discharges occur on 1 January for the third year because these fish are harvested early in the year.

Determining what fraction of the wastes from salmon aquaculture are confined to the highly visible debris piles under salmon cages versus those that are more widely dispersed is an important question for the environmental management of aquaculture. Since the early 1990s, we have made extensive measurements of the fluxes of oxygen, carbon and nutrients into or out of surface sediments in SWNB, both in the footprint of cages and at reference sites (e.g. [2, 22]). By applying a typical stocking density for the fish in the cages, and assuming that the "footprint" of the cage is twice the cage area, it is possible to compare waste discharges from the farms with the measured sediment/water fluxes. For example, the maximum oxygen consumption due to waste feed and faeces during the second year of growth is 4.1 kg d^{-1} (Table 2), which is equivalent to approximately 320 mmol m^2 h^{-1} at a stocking density of 18 kg m^{-3} in a cage that is 10 m deep. This consumption is $\sim$ 50 times greater than the maximum consumption of O_2 in the sediments measured at cage sites (6.2 mmol m^{-2} h^{-1}) and is more than 200 times greater than the median consumption observed at all cage sites (1.55 mmol m^{-2} h^{-1}, n = 51). Clearly, most of the wastes are not being confined to the area of the cages. Comparisons between the waste fluxes of carbon and the nutrients produce similar results: sediments under farms that are highly disturbed by waste inputs can only account for a small fraction of the total discharges.

It is also possible to compare the accumulation of wastes in sediments under cages with the flux of refractory carbon from the farms. Using the same assumptions above about stocking density in the farms, averaging the C_{burial} flux to sediments under cages over odd and even years corresponds to a deposition rate of $\sim$ 20 g organic matter m^{-2} d^{-1}. Using data from Smith et al. [23] for sedimentation rates measured at farm sites (0.25 cm y^{-1} for stations 1–3 in Lime Kiln Bay), sediment density (0.6 g cm^{-3}), the maximum percent organic matter ($\sim$ 10%) and the sedimentation increase due to the inputs from the farm (x 2), the flux of organic matter to the sediments is $\sim$ 0.41 g m^{-2}d^{-1}, approximately 50 times smaller than the discharge of refractory material from the farm. Although the debris piles found under some cages may be the most visible fate of farm wastes at some locations, most of the wastes are being transported further from the farms even in depositional conditions that promote the accumulation of wastes under the cages.

3.2
Wastes Produced on Inlet Scales

It is possible to assess the wider area impacts of the discharges of nutrients and oxygen demand from salmon aquaculture by examining the total waste loads for larger areas. Several years ago, the Canadian Department of Fisheries and Oceans proposed four Coastal Management Regions (CMRs) for management of salmon aquaculture in SWNB (Fig. 1 [24, 25]). These were considered to be hydrologically separate regions, based on three-dimensional tidal models for the area where mixing of water within each region was greater than that between regions. More than 90% of the SWNB salmon industry was located within these CMRs in 2002. They are used here to examine the significance of discharges from salmon aquaculture by combining the estimates of the wastes from the fish growth model with the APLs for odd and even year classes for each of these areas. At present, CMRs are not used for isolating year classes of salmon: this is done on the basis of much smaller bay management areas. Hence, both year classes are present in each of the CMRs in any given year.

Table 3 Impacts of fish farm wastes on proposed CMRs in SWNB. The approved production limits are the number of smolt permitted for introduction in each CMR in odd and even years as of 2002. However, discharges are the total discharges due to all odd and even year class fish in each CMR during odd and even years

Substance	Pass,[1] odd/even[6]	Letang[2] odd/even	Deer Isl.[3] odd/even	Grand Manan[4] odd/even
Approved production Limits (2002 APL×10^6)	1.08/0.27	4.1/0.34	0.86/5.5	3.2/2.7
Total annual discharges (metric tons (calendar year)$^{-1}$)				
O$_2$ consumed	1400/2600	3700/9300	13 000/5900	8300/9000
Carbon respired by salmon	340/640	920/2300	3100/1450	2000/2200
Carbon (waste feed + faeces)	350/660	950/2400	3200/1500	2100/2300
Nitrogen (total)	96/180	260/650	880/410	580/630
Phosphorus (total)	17/31	45/110	150/71	100/110
Water column impacts, maximum daily discharges[5] (metric tons d^{-1})				
O$_2$ Consumed (salmon respiration)	4.9/9.2	16/33	45/21	30/32
O$_2$ consumed (decay of faeces and waste feed)	2.5/4.8	8.0/17	23/11	15/17
Carbon respired by fish	1.8/3.5	5.8/12	17/8.0	11/12
Carbon respired during breakdown of wastes	0.95/1.8	3.0/6.5	8.8/4.1	5.7/6.2
Nitrogen discharges	0.41/0.77	1.3/2.8	3.8/1.8	2.5/2.7
Phosphorus discharges	0.062/0.12	0.20/0.42	0.57/0.27	0.37/0.40

Table 3 (continued)

Substance	Pass.[1] odd/even[6]	Letang[2] odd/even	Deer Isl.[3] odd/even	Grand Manan[4] odd/even
Sediment impacts, maximum daily discharges[5] (metric tons d^{-1})				
Carbon buried	0.95/18	3.0/6.5	8.8/4.1	5.7/6.2
Nitrogen buried	0.11/0.21	0.35/0.76	1.0/0.48	0.67/0.73
Phosphorus buried	0.028/0.054	0.091/0.20	0.27/0.12	0.17/0.19
Worst-case changes in ambient water column concentrations (μM)				
Dissolved O_2	– 1.4	– 43	– 1.4	–
Nitrogen	0.18	5.5	0.18	–
Phosphorus	0.012	0.38	0.012	–

[1] Northern Passamaquoddy Bay.
[2] Letete Passage, Back Bay, Bliss Harbour, Lime Kiln Bay (collectively called Letang).
[3] Deer and Campobello Islands.
[4] Grand Manan Island.
[5] Maximum inputs occur from 3–12 October, except for odd years in the Letang and even years in the Deer Island CMRs. In the latter two cases, the maximum discharges occur on 1 January.
[6] Odd/even refers to whether the year is an odd or even number (different numbers of smolts are introduced into these CMRs in odd and even years).

Table 3 lists the APLs and calculated waste discharges into each of the CMRs in a number of different ways. Note that the APLs in Table 3 are the number of smolts that growers are permitted to put on their sites in odd or even years. On the other hand, the calculated discharges are the totals for odd and even year class fish in a CMR in an odd or even year. For example, in an even year, Northern Passamaquoddy Bay would be expected to have a total 0.27 million year first year fish, 1.08×10^6 second year fish and 0.27×10^6 third year fish.

The total discharge information for the Letang Inlet CMR, which consists of Letete Passage, Back Bay, Bliss Harbour, Lime Kiln Bay and Letang Harbour (Fig. 5) can be compared with the aquaculture discharges presented in Fig. 2 of Strain et al. [7], which included farms in Back Bay, Bliss Harbour, Lime Kiln Bay and Letang Harbour, but not Letete Passage (Table 4). Discharges predicted for this CMR for odd years are similar to those predicted for the Letang Inlet in 1992 prior to introduction of single year class management in the area. However, discharges predicted for even years are approximately 2.4 times higher than they were in 1992. This switch to single year class management has exacerbated potential environmental impacts in those areas with the highest stocking density, because most smolts are introduced to this CMR in odd years (Table 3). Hence mature fish occur at most sites during the same year, instead of being spread out over 2 years. The situation has been made even worse by a marked increase in APLs in recent years. In 1992, the aquaculture industry was clearly the largest anthropogenic contributor of oxygen demand and nutrients to the Letang Inlet. Other human sources included two fish processing plants (one of which is no longer operating), a pulp mill, and a sewage treatment plant. The dominance of salmon aquaculture sources has continued: in odd years salmon aquaculture discharges are now comparable to those from the second largest anthropogenic source (the fish plant); in even years, discharges from salmon aquaculture are estimated to be 1.6–3.5 times greater than the second largest anthropogenic source (Table 4).

Table 3 also lists the predicted maximum daily discharges for each waste into each CMR. Once again, the maximum daily discharges occur in September when water temperatures and fish biomass are high. These maximum daily discharges are 2.0 times higher than the corresponding daily discharges averaged over the entire calendar year, except for the Letang CMR in odd years, for which the multiplier is 2.3. Note that there are some substantial differences for some CMRs between the total APLs for odd and even year sites, with corresponding differences in the discharges.

Table 3 also includes estimates for changes in ambient oxygen and nutrient concentrations that would be caused by the fish farm wastes released into three of the four CMRs. These are worst-case estimates because they are based on the maximum daily waste discharges and the year of the odd/even cycle with the higher discharges, but they do predict the severity of environmental impacts of the farms in the early fall. These calculations assume

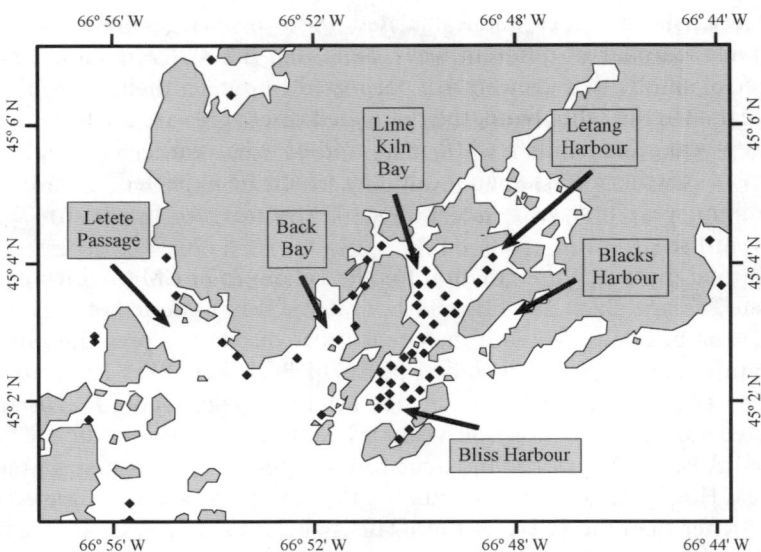

Fig. 5 The Letang Inlet Coastal Management Region. Diamonds show positions of fish farms in 2002

Table 4 Inputs (MT d^{-1}) of BOD, dissolved and particulate carbon, nitrogen and phosphorus from some natural and anthropogenic sources to the Letang Inlet. Data for anthropogenic sources for 1991–92 from Strain et al. (1995) with values as daily discharge rates (annual totals ÷365) for sewage, pulp mill and fish plant inputs increased by 25% to reflect industrial and population growth since 1992. Fish processing plant wastes enter Blacks Harbour and pulp mill effluents enter the head of Letang Harbour with river discharge (see Fig. 5). Daily inputs from salmon aquaculture were derived from total annual discharges ÷365 for 2002 for odd and even years in the Letang/Letete CMR (from Table 4). Carbon inputs from aquaculture include the carbon in faeces and waste feed, but not the carbon consumed by salmon respiration

Source	BOD	Carbon	Nitrogen	Phosphorus
Runoff	0.35	0.82	0.03	0.0018
Precipitation	–	–	0.05	0
Sewage treatment	0.24	0.14	0.013	0.0024
Pulp mill	0.65	0.38	0.011	?
Fish plant	7.2	4.18	0.75	0.10
Salmon aquaculture				
1991–1992	12.0	2.3	0.79	0.12
2002 odd years	10	2.6	0.71	0.12
2002 even years	25	6.5	1.8	0.31

no biological alteration of the farm wastes: their purpose is to compare the magnitude of the potential impacts of fish farm wastes with natural concentrations. The other inputs to the calculations are the volumes of water in the CMRs (from [26], or estimated from hydrographic charts) and estimates of the residence times for the CMRs [24]. The residence times for the three CMRs (15, 9 and 3 days for the Passamaquoddy, Letang and Deer Island CMRs, respectively) are significant because they are long enough that water-borne wastes from farms will have time to decompose before they are transported offshore into the wider Bay of Fundy. Data to do the same calculations for the Grand Manan Island CMR are not readily available, but residence times are likely shorter, and volumes equal or greater than, the ones for the other three CMRs. If so, the expected changes in ambient concentration for this CMR would be less than the smallest values in Table 3. It must be emphasized that the scale on which these calculations are done is very important. Severe localized impacts could occur within a larger CMR for which predicted average changes in ambient levels are relatively small.

The predicted changes in ambient oxygen and nutrient concentrations for the Northern Passamaquoddy Bay and Deer/Campobello Island CMRs are very small. A decrease of $1.4\,\mu M$ for dissolved oxygen is $\sim 0.5\%$ of oxygen saturation (275 μM) at the ambient temperature (12 °C) and salinity (33.5). Changes of $0.18\,\mu M$ in nitrogen and $0.012\,\mu M$ in phosphorus are much smaller than the concentrations of these nutrients in offshore waters in the Bay of Fundy (e.g. [27]). However, the predicted changes for the Letang CMR are very significant. A 43 μM decrease in oxygen is equivalent to a drop of 16% in oxygen saturation or a decrease of $1.4\,mg\,L^{-1}$. An increase in dissolved inorganic nitrogen of 5.5 μM is comparable to or greater than observed background levels.

There are some field data confirming that very high concentrations of nitrogen do occur in these waters. Bugden et al. [28] reported nitrate + ammonia concentrations as high as 9.7 μM in Back Bay in September 1999. The increased availability of inorganic nitrogen could be promoting increased biomass of micro- and macroalgae as fouling communities on net-pens and in intertidal areas. Macroalgae have a large capacity for nutrient uptake and may form extensive beds in response to nutrient enrichment [29]. While increased production of commercial macroalgal species such as *Ascophyllum* has an economic advantage, other species such as the green algal *Enteromorpha* may cause adverse ecological effects on soft-shell clams in intertidal areas [30, 31]. The increase in phosphorus is not as great as that for nitrogen (since excess phosphorus is generally present in these waters), but is still within the range expected for background levels. All of these predicted changes apply to an area much larger than that of the farm sites themselves. Changes in dissolved oxygen and nutrients close to the farm sites will undoubtedly be greater.

It is noteworthy that intertidal sediments at sampling sites in SWNB that are not close to individual farms but are in inlets with aquaculture are more enriched in organic matter than elsewhere in the Bay of Fundy. Average (± SD of the mean) organic carbon concentrations in fine grained sediment (modal size < 100 μm) from intertidal areas in Cumberland Basin (0.88 ± 0.25, $n = 22$) [32] and Pocologan (0.88 ± 0.17, $n = 3$) (B. Hargrave unpublished data), two areas in the Bay of Fundy where salmon aquaculture has not been developed, are significantly ($p < 0.05$) lower than mean values measured in Hinds Bay (inner end of Lime Kiln Bay) (1.91 ± 0.08, $n = 6$) and Mascarene Shore (Passamaquoddy Bay) (1.24 ± 0.12, $n = 6$) (B. Hargrave, unpublished data) where salmon aquaculture sites have been active for up to 2 decades. The O_2 fluxes in these intertidal areas in SWNB (1.60 ± 0.15 mmol m^{-2} h^{-1}, $n = 24$) are not significantly different (at $p = 0.1$) from those in sub-tidal sediments near cages (1.94 ± 0.15, $n = 51$) but both these groups are significantly different ($p < 0.001$) from those in sub-tidal sediments at reference sites away from cages (0.79 ± 0.038, $n = 125$). The conclusion that some intertidal zones in SWNB may be experiencing enrichment from aquaculture wastes is supported by observations on effects of eutrophication on clam populations discussed in [31].

4
Ecosystem Processes

Fluxes of carbon, oxygen and nutrients are important aspects of ecosystem function which are measures of the overall ecosystem activity or "metabolism". In an open system like the SWNB aquaculture region, rates of carbon uptake by primary production, nutrient regeneration or community respiration are measures of the cycling of these elements in the ecosystem which are driven by a combination of internal processes, inputs from freshwater and other land-based sources, and exchanges with offshore waters in the Bay of Fundy. Practically speaking, it is very difficult to separate out the externally driven processes in this area from internal ones: measuring net marine exchanges is very difficult because of the large tidal flows; tools such as salinity-driven box models are unsuitable because of the low salinity gradients. More sophisticated models can be used to parameterize exchanges and to estimate residence times (e.g. [24]), but the available data for oxygen or nutrients are not detailed enough to allow the calculation of net exchanges. Nevertheless, measurements of the rates of basic ecosystem processes are useful indicators of overall ecosystem metabolism, and can be used to gauge the significance of additional processing of carbon, oxygen and nutrients by anthropogenic influences like salmon aquaculture.

This section will compare the relative magnitude of oxygen, carbon and nitrogen fluxes due to salmon in various CMRs with fluxes due to the primary natural sources or sinks for these elements. If BOD loadings and nutrient discharges from salmon aquaculture are within the range of natural variability when all natural processes in an area are considered, then it might be concluded that an ecological capacity exists to assimilate increased inputs of nutrients and organic matter. On the other hand, if calculated releases of waste products by salmon are a significant fraction of total fluxes from natural sources, then significant changes to both the structure and functioning of the ecosystem might be expected.

For scaling up measurements of ecosystem processes to CMRs, calculations were based on mid-tide surface area and volumes. No attempt was made to account for advection and tidal mixing since the CMRs were originally selected on the basis of a circulation model that predicted more thorough mixing within than between CMRs. The Letang Inlet CMR was sub-divided into Lime Kiln Bay, Bliss Harbour, Back Bay, and Letete Passage because this region is potentially the most seriously impacted (Table 3, Fig. 5). Separate calculations were made for Cobscook Bay, Maine (current APLs provided by John Sowles, Maine Department of Marine Resources, Boothbay, Maine, personal communication), as another example of a semi-enclosed bay in the region, which could also interact with the adjacent the Campobello/Deer Island CMR (Fig. 1).

Releases of waste products by any organism are allometrically scaled to body size so that with maturity feeding, growth, respiration and excretion increase rapidly. Metabolic rates also increase with temperature. As noted above, the combined effects of increasing body size and temperature lead to maxima in all of these metabolic processes during the second year of salmon culture in September. Estimates of primary and macroalgal production, pelagic and benthic respiration, nitrogen, oxygen and carbon fluxes are presented for September, when many of these processes are at maximum rates as well (e.g. [2]), so that they may be compared with the maximum releases from salmon aquaculture.

4.1
Primary and Macrophyte Production and Nitrogen Regeneration

Phytoplankton production has been measured at several locations in SWNB and offshore areas in the lower Bay of Fundy [33–36]. The range of daily phytoplankton production during late summer and fall months reported from these previous studies (0.6 to 1.7 g C m^{-2} d^{-1}) is similar to values summarized in [37] (0.7 to 1.3 g C m^{-2} d^{-1}) assuming a 15 m euphotic depth. Phytoplankton may be light limited during summer periods of highest growth. In many areas, particularly nearshore, in the Bay of Fundy the depth of light penetration is relatively shallow (5–15 m) due to high SPM and dissolved/colloidal coloured substances [35, 38]. The production system alternates between net autotrophic during summer (production-driven) to net heterotrophic during winter when respiration increases markedly [36, 39]. For the purposes of our calculations the photic zone is assumed to have a maximum depth of 15 m with an estimate of average phytoplankton production during later summer/fall months of $1 \, g \, C \, m^{-2} \, d^{-1}$.

Studies of macrophyte distribution in intertidal and sub-littoral zones throughout SWNB provide estimates of biomass and production [38, 40, 41]. The average biomass of the predominant commercial species (rockweed, *Ascophyllum nodosum*) in intertidal areas in SWNB in 1991 varied from 4.6 to $10.4 \, kg \, m^{-2}$. Since annual production:biomass ratios are 0.4–0.5 depending on the method of calculation [41], values for annual production would range from 1.8 to 5.2 kg dry weight (equivalent to 2.5 and $7.1 \, g \, C \, m^{-2} \, y^{-1}$ assuming 50% of dry weight as organic carbon). Intertidal areas represent from 15 to 40% of the total area in various CMRs, so representative values for macrophyte production on an inlet-wide basis would vary from 0.4 to $2.8 \, g \, C \, m^{-2} \, y^{-1}$. Since this range is the same as that observed for phytoplankton production, for the purposes of these calculations the daily rate of phytoplankton production was doubled to estimate total autotrophic carbon supply from marine primary production. A C : N ratio of 6 was used to calculate nitrogen requirements of phytoplankton/macroalgae. 50% could reasonably be supplied by ammonia regeneration in the water column. A mean value for benthic ammonia release was derived from measurements using sediment cores collected in central areas of Lime Kiln Bay and Bliss Harbour [2]. Based on these calculations, the water column (pelagic) regeneration is 1–2 orders of magnitude greater than the benthic regeneration.

4.2
Respiration

Pelagic and benthic oxygen consumption was estimated from data presented in [36] assuming no depth variation and applied uniformly in all CMRs. Measurements of oxygen uptake by particulate matter suspended in flood tide water in Cumberland and Minas Basins at the head of the Bay of Fundy [42] fall within the range of values measured in Lime Kiln Bay, Bliss Harbour and at the offshore Wolves station during late summer/fall months [36, 39]. Subtidal sediment oxygen uptake was calculated using the mean rate measured in September in central Bliss Harbour [2] and applied equally across the total mid-tide area in each CMR. As with nitrogen, the pelagic contribution dominates: it is 1–3 orders of magnitude greater than the benthic O_2 consumption.

4.3
Comparisons with Fluxes from Salmon Aquaculture

Table 5 compares these estimates of the natural processing of oxygen, nitrogen and carbon with those from salmon aquaculture. The numbers in Table 5 are the ratios of the total fluxes due to salmon aquaculture in odd and even years to the total fluxes from natural processes, expressed as percentages. The totals for aquaculture include: for O_2, the O_2 consumed by both salmon respiration and the decomposition of waste feed and faeces; for N, the N released in dissolved form, during the decomposition of waste feed and faeces, and the organic N buried in the sediments, but not the N in the salmon biomass, which is removed from the ecosystem at harvest time; for C, the C respired by salmon, the C respired during the decomposition of waste feed and faeces and the C buried in the sediments, but not the C in the salmon biomass. Natural oxygen fluxes include water column respiration (including macrophytes) and sediment oxygen consumption; natural nitrogen fluxes include ammonia regeneration (phytoplankton + macrophytes) and sediment regeneration; natural carbon fluxes include primary production by phytoplankton and macrophytes.

The amount of O_2 consumed by salmon respiration and the breakdown of wastes from salmon farms varies from less than 0.1 (northern Passamaquoddy Bay) to 20% (Lime Kiln Bay in even years) of the natural ecosystem metabolism, a range which probably corresponds to minimal to significant impacts on the ecosystem. The extra O_2 demand from farms in three of the four sub-regions of the Letang Inlet CMR are $\geq$ 9% of the natural O_2 cycle. Salmon respiration in the semi-enclosed Cobscook Bay is < 1% of

Table 5 Comparisons between fluxes of oxygen, nitrogen and carbon from salmon aquaculture with those from natural processes, expressed as the ratio of salmon fluxes/natural (in percent). For the purposes of this calculation, the Letang Inlet CMR has been further sub-divided into four sub-regions (see Fig. 5)

CMR Year	Oxygen		Nitrogen		Carbon	
	Odd	Even	Odd	Even	Odd	Even
Passamaquoddy Bay	0.04	0.08	1.7	3.2	1.1	2.1
Letang Inlet	1.1	2.6	22	70	14	34
Lime Kiln Bay	8.8	20	93	330	58	160
Bliss Harbour	3.6	9.1	58	210	37	100
Back Bay	5.5	13	47	170	29	84
Letete Passage	0.26	0.34	6.2	11	3.9	5.2
Campobello / Deer Islands	0.40	0.18	16	7.3	10	4.6
Grand Manan Island	0.26	0.28	10	11	6.5	7.1
Cobscook Bay	0.77	0.76	8.1	8.1	4.5	4.5

natural respiration, as are the values in each of the three other CMRs outside of the Letang Inlet.

The division of the Letang Inlet CMR into sub-regions clearly shows the importance of considering scales when doing this kind of analysis. In even years, O_2 consumption due to salmon farms is 2.6% of natural metabolism in the Letang Inlet as a whole, but 20% in Lime Kiln Bay within that CMR: local impacts in smaller sub-regions can be much greater than in the CMR as a whole. Impacts at individual farm sites will be greater still. Such calculations show why conditions of sub-optimal dissolved oxygen concentrations requiring re-aeration in late summer months have occurred at some farms with restricted circulation and relatively long water residence times in the Letang Inlet CMR.

On the basis of the comparisons in Table 5, the cycling of nitrogen is more perturbed by salmon aquaculture than oxygen or carbon. In Lime Kiln Bay, nitrogen flux attributable to salmon in even years is 330% higher than natural nitrogen fluxes. This means that more nitrogen is introduced to the ecosystem through salmon farming than is cycled naturally in the water column and sediments. Cultured fish are now a major biogeochemical pathway for nitrogen in both the bays that make up the Letang Inlet CMR and some of the larger CMRs. As stated above, these nutrient fluxes are evident in unusually high dissolved inorganic nitrogen concentrations [28], and may be responsible for promoting growth of intertidal macrophytes and other algae [31].

O_2 demand and nitrogen release discussed above can be considered net fluxes attributable to aquaculture (e.g. the nitrogen in the salmon biomass is not included in these estimates, since it is removed from the ecosystem at harvest time). The processes either use resources from the ecosystem or may induce changes in its functioning. The situation for carbon is slightly different. Since the impact of excess CO_2 from salmon respiration is likely to be small in seawater that has high natural levels of dissolved CO_2, the carbon that is respired has little impact on the receiving environment: its addition does not pose a threat, and its production does not rely on natural processes in the ecosystem. The carbon processed by fish respiration accounts for $\sim$ 49% of the carbon in aquaculture fluxes (Table 5). In Lime Kiln Bay in even years, 1.64 times as much carbon is cycled through aquaculture than is processed naturally by the ecosystem if the carbon respired by the salmon is included. But even if it is not included, the fluxes of carbon due to aquaculture are still more than 80% of those due to natural processes. Salmon aquaculture also plays a very significant role in carbon cycling in Bliss Harbour and Back Bay, but a relatively minor role in Letete Passage and the other CMRs.

It is possible to compare the discharge of refractory carbon from the fish farms with sedimentation rates on a bay scale in a manner analogous to the earlier comparison on farm scales. Bay-wide discharges for refractory carbon (C_{burial}) based on the data in Table 3 averaged over odd and even years are equivalent to organic matter burial rates from 0.09 to 0.17 g m^{-2} d^{-1} for the bays of the Letang Inlet CMR. Sedimentation rates in the Letang CMR (Smith et al., Table 1 [23]), range from 0.16 to 0.55 cm y^{-1}. Using the average sediment density (0.6 g cm^{-3}) and organic matter content of 5% for sediments away from cage sites [23] shows that this sedimentation rate range is equivalent to organic matter burial rates from 0.13 to 0.45 g m^{-2} d^{-1}. Thus the refractory organic wastes from the fish farms in the Letang CMR could be responsible for substantial portions of the total organic matter buried in the sediments on an inlet-wide scale. This is another indication of the significance of the fish farm wastes on natural biogeochemical cycles in the area.

4.4
Conclusions

Quantifying the overall impacts from marine salmon aquaculture is a task with many steps. Measurements or models of the wastes generated at individual farms, the partitioning of those wastes into different fractions, and the assessment of their significance in comparison with other anthropogenic sources and natural ecosystem processes are required both close to the farms and at larger bay scales. In the ideal case, such calculations should be data driven as much as possible, using detailed information on the operations at individual farms to quantify feed types and amounts used, numbers of fish and fish growth. In the absence of such data more general models on fish growth and industry wide averages for feed application must be used. Predictions made in this way are both more complex and less reliable than those based on specific data. Such calculations also treat the entire industry uniformly: there are no rewards for efficient farm operation, and no penalties for inefficient operations.

The outputs of the models presented here were analyzed for their sensitivity to changes in inputs. The principal uncertainties in predicting waste discharges from farms are in the amount of feed used, its protein content, and, to a lesser extent, the nitrogen and phosphorus content of the fish. In SWNB in 2002, solid wastes were still dominated by waste feed: further improvements in feeding strategies could significantly reduce environmental impacts. It is difficult to predict the reliability of the waste estimates, but the sensitivity of the models suggests that the overall precision may be in the range of 30–50%. However, because impacts in some areas are so large the overall conclusions are robust to the uncertainty in the waste estimates. In the Letang Inlet, salmon aquaculture contributes 1.6–3.5 times more waste to the marine environment than fish plants, the second largest anthropogenic source. The cycling of nitrogen and carbon through salmon farms is up to 3.3 and 1.6 times larger, respectively, than the processing of those elements by natural processes in the ecosystem: in other words, substantial changes to the functioning of the ecosystem have occurred due to the presence of the salmon farms. Even in a larger, less intensively farmed area like the Campobello/Deer

Island CMR, fluxes of carbon and nitrogen from salmon aquaculture are 10 and 16%, respectively, of those due to natural processes.

The impacts we have estimated using mass balance calculations affect areas much larger than the farms themselves, with size scales of many kilometres. Oxygen consumption due to wastes released from salmon farms can reach 20% of natural respiration. Although all of these impacts will be more severe close to the farms, comparisons of sediment processes with the magnitudes of discharges suggest that most of the wastes are distributed over wide areas. Time and space scales are critical in evaluating the potential impacts.

The fact that salmon aquaculture can dominate natural processes over quite large scales shows the need for establishing integrated management goals for marine finfish aquaculture. Judgements must be made on what levels of exploitation of the environment are acceptable, and these judgements must recognize that wastes from net-pen aquaculture may significantly impact areas much larger than the farms themselves.

References

1. Schaanning MT, Kupka Hansen P (2005) The suitability of electrode measurements for assessment of benthic organic impact and their use in a management system for marine fish farms (in this volume). Springer, Berlin Heidelberg New York
2. Hargrave BT, Duplisea DE, Pfeiffer E, Wildish DJ (1993) Mar Ecol Prog Ser 96:249
3. Tlusty MF, Pepper VA, Anderson MR (2005) Reconciling aquaculture's influence on the water column and benthos of an estuarine fjord – a case study from Bay d'Espoir, Newfoundland (in this volume). Springer, Berlin Heidelberg New York
4. Cromey CJ, Black KD (2005) Modelling the impacts of finfish aquaculture (in this volume). Springer, Berlin Heidelberg New York
5. Peterson RH, Page F, Steeves GD, Wildish DJ, Harmon P, Losier R (2001) Can Tech Rep Fish Aquat Sci 2337
6. Silvert W (1994) J Appl Icthyol 10:349
7. Strain PM, Wildish DJ, Yeats PA (1995) Mar Poll Bull 30:253
8. Iwama GK, Tautz AF (1981) Can J Fish Aquat Sci 38:649
9. Cho CY (1992) Aquaculture 100:107
10. Sveier H, Wathne E, Lied E (1999) Aquaculture 180:265
11. Thodesen JB, Grisdale-Helland B, Helland SJ, Gjerde B (1999) Aquaculture 180:237
12. Refstie S, Storebakken T, Baeverfjord G, Roem AJ (2001) Aquaculture 193:91
13. Einen O, Holmefjord I, Asgard T, Talbot C (1995) Aquaculture Res 26:701
14. Refstie S, Storebakken T, Roem AJ (1998) Aquaculture 162:301
15. Sveier H, Lied E (1998) Aquaculture 165:333
16. Storebakken T, Shearer KD, Baeverfjord G, Nielsen BG, Asgard T, Scott T, De Laporte A (2000) Aquaculture 184:115
17. Storebakken T, Shearer KD, Roem AJ (1998) Aquaculture 161:365
18. Bureau D, Cho CY (1999) Aquaculture 179:127
19. Stucchi D, Sutherland T-A, Levings C, Higgs D (2005) Near-field depositional model for salmon aquaculture waste (in this volume). Springer, Berlin Heidelberg New York

20. Sowles JW (2005) Assessing nitrogen carrying capacity for Blue Hill Bay; a management case history (in this volume). Springer, Berlin Heidelberg New York
21. Gillibrand PA, Gubbins MJ, Greathead C, Davies IM (2002) Scottish Fisheries Research Report Number 63/2002 Fisheries Research Services, Marine Laboratory, Aberdeen
22. Hargrave BT, Phillips GA, Doucette LI, White MJ, Milligan TG, Wildish DJ, Cranston RE (1997) Water Air Soil Pollut 99:641
23. Smith JN, Yeats PA, Milligan TG (2005) Sediment geochronologies for fish farm contaminants in Lime Kiln Bay, Bay of Fundy. Springer, Berlin Heidelberg New York
24. Thompson KR, Dowd M, Shen Y, Greenberg D (2002) Cont Shelf Res 22:1603
25. Department of Fisheries and Oceans (2003) Can Tech Rep Fish Aquat Sci 2489
26. Gregory D, Petrie B, Jordan F, Langille P (1993) Can Tech Rep Hydrog Ocean Sci 143
27. Martin JL, LeGresley MM, Strain PM (2001) Can Tech Rep Fish Aquat Sci 2349
28. Bugden JBC, Hargrave BT, Strain PM, Stewart ARJ (2001) Can Tech Rep Fish Aquat Sci 2356
29. Chopin TD, Welles D, Belyea E (2000) Huntsman Mar Sci Centre Occ Rep 2000/1:39
30. Auffrey LM, Robinson SMC, Barbeau MA (2002) Can Tech Rep Fish Aquat Sci 2411:84
31. Robinson SMC, Auffrey LM, Barbeau MA (2005) Far-field impacts of eutrophication on the intertidal zone in the Bay of Fundy, Canada, with emphasis on the soft-shell clam, *Mya arenaria* (in this volume). Springer, Berlin Heidelberg New York
32. Hargrave BT, Phillips GA, Neame PA, Prouse NJ (1982) Can Data Rep Fish Aquat Sci 354
33. Prouse NJ, Gordon DC Jr, Hargrave BT, Bird CJ, McLachlan J, Lakshminarayan JSS, Sita Devi J, Thomas MLH (1984) Can Tech Rep Fish Aquat Sci 1256:65
34. Emerson CW, Roff JC, Wildish DJ (1986) Ophelia 26:165
35. Harrison WG, Perry T (2001) Can Tech Rep Fish Aquat Sci 2352
36. Kepkay P, Harrison G, Bugden J, Perry T (2002) Can Tech Rep Fish Aquat Sci 2352:51
37. Harrison WG, Perry T, Li WKW (2005) Ecosystem indicators of water quality. Part I. Plankton biomass, primary production and nutrient demand (in this volume). Springer, Berlin Heidelberg New York
38. Logan A, Page FH, Thomas MLH (1984) Est Coast Shelf Sci 18:571
39. Kepkay PE, Harrison WG, Bugden JBC (2005) Ecosystem indicators of water quality. Part II. Oxygen production and oxygen demand (in this volume). Springer, Berlin Heidelberg New York
40. Sharp G, Semple R (1992) Department of Fisheries and Oceans Can Atlantic Fisheries Advisory Comm Working Paper 92/218, Dartmouth NS
41. Sharp G, Ugarte R, MacEachron T, Semple R, Black G (1999) Dept. of Fisheries and Oceans Regional Advisory Process Working Paper 99/29:18 p, Dartmouth, NS
42. Hargrave BT, Prouse NJ, Philips GA, Neame PA (1983) Can J Fish Aquat Sci 40:229

Hdb Env Chem Vol. 5, Part M (2005): 59–82
DOI 10.1007/b136004

Published online: 15 July 2005

Ecosystem Indicators of Water Quality Part I. Plankton Biomass, Primary Production and Nutrient Demand

W. Glen Harrison (✉) · Tim Perry · William K. W. Li

Biological Oceanography Section, Ocean Sciences Division, Fisheries and Oceans Canada, Bedford Institute of Oceanography, P.O. Box 1006,
Dartmouth, Nova Scotia, B2Y 4A2, Canada
HarrisonG@mar.dfo-mpo.gc.ca, PerryT@mar.dfo-mpo-gc.ca, LiB@mar.dfo-mpo.gc.ca

1 **Introduction** 60

2 **Methods** 61
2.1 Sampling 61
2.2 Analytical 63

3 **Results** 63
3.1 Phytoplankton Biomass, Bacterial Abundance and Primary Production . . 63
3.2 Nutrient Concentrations, Demand and Turnover 68

4 **Discussion** 72
4.1 Phytoplankton Biomass, Bacterial Abundance and Particulate Organic Matter 72
4.2 Primary Production and Nutrient Demand 73
4.3 Light vs. Nutrient Limitation of Primary Production 75
4.4 Bacteria 78
4.5 Water-Quality Indicators 79

5 **Summary and Conclusions** 80

References 81

Abstract Seasonal measurements of plankton (phytoplankton and bacteria) biomass and abundance, primary production, and nutrient demand were conducted in the coastal waters of southwestern New Brunswick (SWNB) in 2000–2002 to investigate the far-field effects of finfish (salmon) aquaculture on the pelagic ecosystem. Plankton biomass and production varied seasonally with peak concentrations and activity in summer–fall and lows in winter. Nutrient demand followed a similar pattern with nitrogen (nitrate and ammonium) turnover times ranging from greater than a week in winter to less than a few days in summer. Ammonium concentrations were elevated at the aquaculture sites relative to control sites, however, effects on other nutrients, phytoplankton biomass, bacterial abundance, and primary production were not discernible despite the significant flux of nutrients into the system from finfish farming. Several lines of evidence point to the conclusion that primary production in SWNB is under light rather than nutrient control and that phytoplankton there have limited capacity to process additional nutrients produced as aquaculture in the region expands. The ratio of bacterial abundance to phytoplankton biomass (B/P ratio) is proposed as an easily measured water-quality indicator for assessing the trophic balance (autotrophy vs. heterotrophy) of the pelagic ecosystem in coastal waters.

Keywords Aquaculture · Phytoplankton · Bacteria · Primary production · Nitrate · Ammonium · Nutrient demand · Light-limitation · Water quality

1
Introduction

Finfish aquaculture is a fast growing industry worldwide. In southwestern New Brunswick (SWNB), Canada, the farming of Atlantic salmon has grown dramatically in the last 30 years with the number of active farms doubling and the harvest increasing fourfold over the last 10 years [1]. The growth of this industry has been accompanied by mounting concern about its environmental impacts on coastal ecosystems. There exists now an extensive literature and several comprehensive reviews on the documented and potential environmental effects of aquaculture in the coastal zone (e.g. [2]).

In reviewing existing knowledge and research needs on environmental effects of finfish aquaculture, Hargrave [3] noted that surprisingly little is known about far-field effects (ecosystem-scale), particularly on the pelagic ecosystem, when compared with knowledge of near-field effects (within or adjacent to fish farms), especially on the benthos. Aquaculture impacts on the water-column are mainly concerned with farm effluents/wastes, i.e. inorganic and organic nutrient enrichment and their effects on plankton growth dynamics and community structure, i.e. eutrophication (e.g. [4]). Eutrophication is the consequence of nutrient enrichment [5, 6] and manifests itself in the pelagic zone as increases in plankton (phytoplankton and bacteria) biomass and production (including increased frequency and intensity of benign and Harmful Algal Blooms, HABs [7]) and changes in community structure and trophic state [8]. Among the latter are concerns about the effects of aquaculture on the balance between autotrophic and heterotrophic biomass/production that determine the pelagic oxygen balance of coastal ecosystems [9]. A fundamental question, therefore, is, "Can ecosystems maintain their natural state under the influence of aquaculture activity and for how long?". More specifically for the pelagic zone, "What is the capacity of plankton to process the effluents of aquaculture?"

This question was at the core of a recently completed multidisciplinary project, Environmental Studies for Sustainable Aquaculture (ESSA), aimed at

1. Evaluating current far-field environmental effects of salmon aquaculture on three contrasting Canadian coastal ecosystems;
2. Constructing models to predict future ecosystem changes and;
3. Developing standard methodologies for effective management of the Canadian finfish aquaculture industry for sustainability [10, 11].

This and the accompanying chapter [9] report on results of the research carried out during the ESSA project to evaluate the effects of salmon aquaculture activity on the pelagic component of the SWNB coastal ecosystem. Properties of the pelagic ecosystem addressed in this paper include water-column transparency, inorganic nutrient concentrations (specifically nitrogen), particulate organic matter, microbial biomass (phytoplankton) or abundance (bacteria), and phytoplankton production (inorganic carbon and nitrogen utilisation rates). Emphasis is placed on nitrogen because of its established role in the regulation of coastal primary production [6, 12, 13]. The chapter that follows [9] looks at dissolved organic matter, microbial respiration and the production/respiration (P : R) balance. Both papers offer simple indices that may be useful in assessing the capacity of pelagic ecosystem in SWNB to process aquaculture wastes now and under future projections.

2
Methods

2.1
Sampling

The SWNB study area, which includes Passamaquoddy Bay and adjacent waters at the western mouth of the Bay of Fundy (Fig. 1), is a shallow (< 100 m at the mouth) semi-enclosed system with irregular bathymetry and coastline inundated with numerous islands and channels that lead to a strong and highly complex and structured tidal circulation that dominates the flow fields in the region [14, 15]. Fresh water input from the St. Croix River in the west and seawater exchange through the Western and Letete Passages in the south and east set up a cyclonic circulation in Passamaquoddy Bay; the residence time of bay waters is on the order of 15 days [14, 16]. The ESSA aquaculture sites were concentrated principally in the shallow (< 30 m) northeastern, Letang/Letete region where water exchange occurs predominantly through the Letete Passage and open Bay of Fundy and residence times are on the order of 9 days [16].

Observations were conducted from fall 2000 to spring 2002. Three process study sites were chosen in close proximity to fish farms (i.e. within a few 100 meters), two in Lime Kiln Bay and one in Bliss Harbour (Fig. 1). Samples were collected for nutrients, biomass (particulate organic matter, chlorophyll a) and phytoplankton productivity over four seasons: fall, 19–21 September, 2000; summer, 24–26 July, 2001; winter, 4–6 December, 2001; and spring, 15–17 May, 2002. The inner and outer Lime Kiln Bay sites were only about 0.5 km apart and the Bliss Harbour site about 3 km from them. An

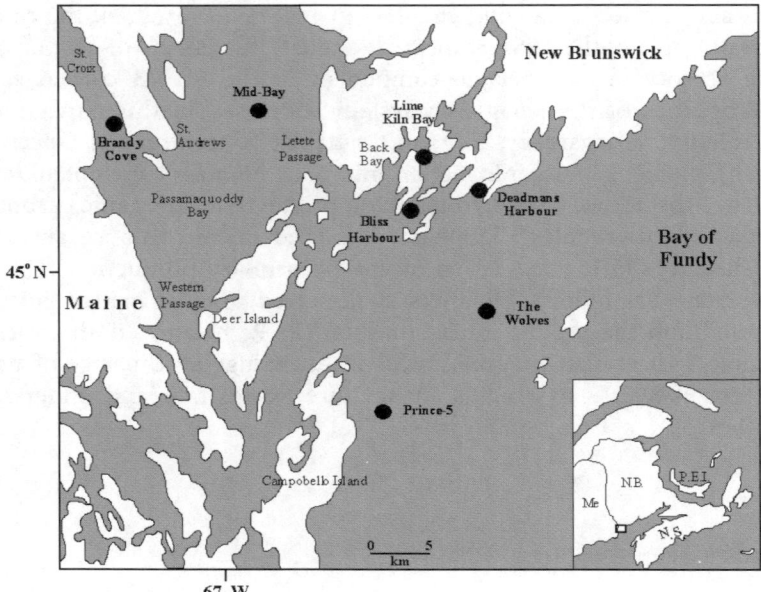

Fig. 1 Locations of aquaculture (Bliss Harbour, Deadman's Harbour, Lime Kiln Bay) and control (Brandy Cove, Mid-Passamaquoddy Bay, Prince-5, The Wolves) sampling sites in southwestern New Brunswick (SWNB)

offshore control site (~ 10 km to the southeast of the aquaculture sites) at The Wolves was also selected for study but due to logistical constraints and weather, it was only sampled in fall. In addition to the seasonal process studies, additional samples for chlorophyll a, water transparency and nutrient concentrations were taken approximately bi-weekly at the Lime Kiln and The Wolves sites in 2000 (chlorophyll measurements started mid-2000) and continued throughout 2001. During the later portion of this time-series (May–Nov, 2001), samples were also collected for bacterial abundance at these two sites, and two additional aquaculture sites (Brandy Cove, Deadman's Harbour) and a control site, mid-Passamaquoddy Bay, far removed from fish farms [17]. Due to unforeseeable problems (weather, logistics constraints), comparative measurements of some properties at aquaculture and control sites were made only during the fall period. Relevant data (chlorophyll a, water transparency, nutrients) collected at another offshore (control) site during 2001, Prince-5, located ~ 15 km south of the aquaculture sites and southwest of The Wolves but not part of the ESSA program, were also used in this study. The Prince-5 data are collected as part of Canada's eastern Atlantic Zone Monitoring Program, AZMP [18, 19].

2.2
Analytical

Water samples for nutrients, biomass and productivity measurements were collected with Niskin bottles at up to 7 "light depths" (approx. 90, 50, 25, 15, 10, 5, 1% surface light) determined from a Secchi disc. Nutrient samples were frozen (polyethylene bottles) and later analysed in the laboratory for nitrate, nitrite, ammonium, phosphate, silicate using standard automated methods [20]. Chlorophyll a was measured fluorometrically [21] and for particulate organic carbon (POC) and nitrogen (PON) by high temperature combustion [22]. Bacteria were fixed in paraformaldehyde (1%) and stored at $-80\,°C$. In the laboratory, bacteria were stained with the DNA-binding fluorochrome SYBR Green-1 and detected by green fluorescence in flow cytometric analysis [23]. Productivity measurements were made using stable isotopes (^{13}C-bicarbonate, ^{15}N-nitrate, ^{15}N-ammonium) at tracer concentrations, i.e. $0.2\,\mathrm{mol\,m^{-3}}$ for ^{13}C, and $0.1\,\mathrm{mmol\,m^{-3}}$ for ^{15}N. Water samples collected for productivity measurements were transported to a nearby shore facility, dispensed into 500 ml polycarbonate bottles, inoculated with the isotope tracers and incubated for $\sim 3\,h$ under natural light conditions in attenuated incubator boxes. At the end of incubation, the particulate material was collected on pre-combusted ($475\,°C$, 12 h) glass-fibre filters and stored for later mass spectrometric analysis (Europa Scientific) in the laboratory. Nutrient utilisation rates were calculated using standard equations [24]. No corrections were made for "isotope dilution" (due to biological production of ammonium in particular) during the incubation.

Non-parametric statistics (Wilcoxon-Mann-Whitney Rank Sum test) were employed where appropriate to assess differences in properties among aquaculture sites and between them and control sites.

3
Results

3.1
Phytoplankton Biomass, Bacterial Abundance and Primary Production

The seasonal cycle of phytoplankton biomass (chlorophyll a) in waters close to fish farms (Lime Kiln Bay) and at the control site (The Wolves), revealed from the time-series measurements in 2000 and 2001, were remarkably similar in temporal pattern and magnitude (2a). Surface chlorophyll a concentrations varied from a low of $< 1\,\mathrm{mg\,m^{-3}}$ in winter (January–February) to a high of $\sim 10\,\mathrm{mg\,m^{-3}}$ during the pronounced spring (May) and late summer/fall (August–October) blooms. A similar seasonal cycle in chlorophyll a concentration was seen at the Prince-5 station in 2001 although the spring maximum was not as pronounced. Overall, surface chlorophyll concentrations at the aquaculture site were not statistically different from concentrations at the control sites. Other oceanographic properties observed during the time-series studies (surface temperature and salinity) were also similar between the aquaculture and control sites, however, marked difference were seen in water transparency (2b). Secchi depths at The Wolves varied from 4–8 m (summer to winter) but were only 2–5 m in Lime Kiln Bay. Secchi depths did not show much seasonal variability at Prince-5, remaining in the range of 7 ± 1 m and similar in magnitude to Secchi depths at The Wolves. Differences in Secchi depths between the aquaculture site and control sites were statistically significant.

Due in part to their close proximity to each other, particulate biomass (chlorophyll a, POC and PON concentrations) and compositional ratios (C : N) measured during the process studies showed similar patterns among the three aquaculture sites, both within and among seasons (Tables 1 and 2). Particulates were highest in fall and lowest in winter. The spring peak in chlorophyll a seen in 2001 in the time-series observations (2a) was not seen during the process studies in May, 2002 where concentrations were

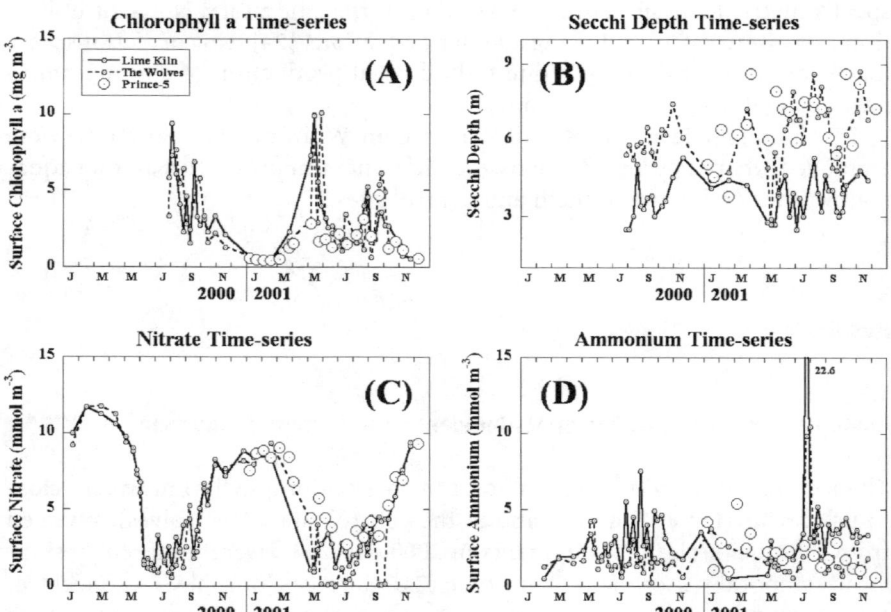

Fig. 2 Seasonal variability in surface chlorophyll a concentrations (**a**), Secchi depths (**b**), surface nitrate concentrations (**c**), and surface ammonium concentrations at the Lime Kiln Bay aquaculture site and The Wolves and Prince-5 control sites in 2000 and 2001

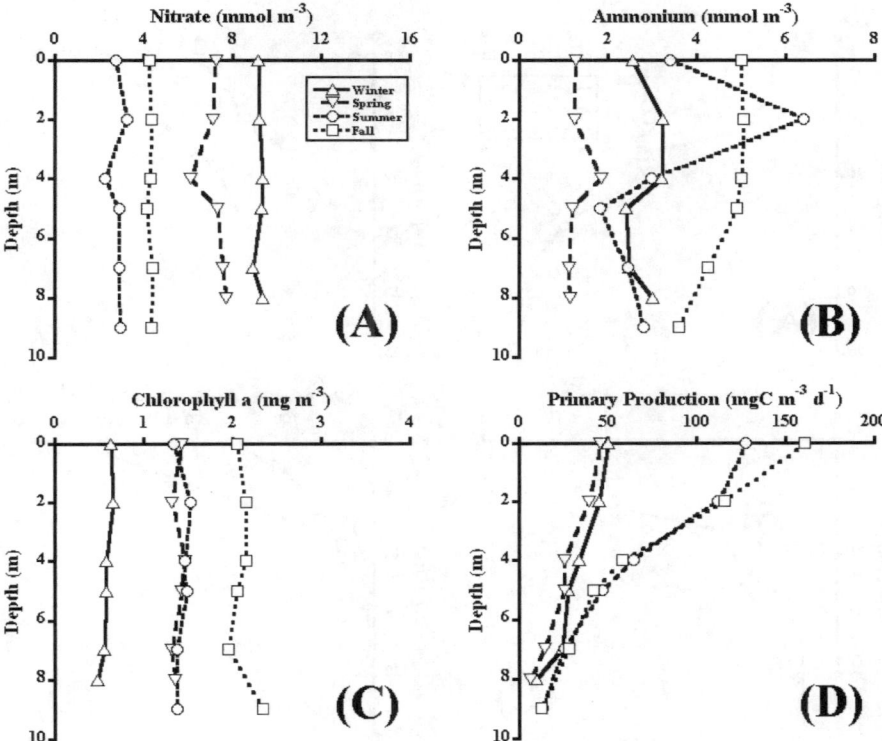

Fig. 3 Seasonal variability and vertical structure of nitrate (a) and ammonium (b) concentrations, chlorophyll a (c) and primary production (d) at the (Inner) Lime Kiln Bay aquaculture site

< 1.5 mg m^{-3}. Compositional ratios (C:N) varied little among stations or seasons and were higher (8–10) than the so-called "Redfield Ratio" [25] of 7. Analysis of variance revealed that season was the most important factor accounting for the variability of particulates while site was of minor significance. Chlorophyll a, POC, PON concentrations and compositional ratios at The Wolves were statistically indistinguishable from concentrations and ratios at Lime Kiln Bay or Bliss Harbour in fall, the only season in which the complete suite of variables were measured at both aquaculture and control sites. The shallow waters of SWNB are "macro-tidal" (mean tidal range > 2 m) with strong tidal mixing which explains the uniform vertical distribution of particulates (chlorophyll a, POC, PON) seen at the aquaculture sites in all seasons (3c). Likewise, chlorophyll a concentrations are generally uniform with depth for most of the year at the deeper Prince-5 control site where tidal mixing is also important [19]. At The Wolves, in contrast, stratification of the upper water-column was evident in fall; highest chlorophyll a, POC and PON concentrations were observed in near-surface (< 10 m) waters (4c).

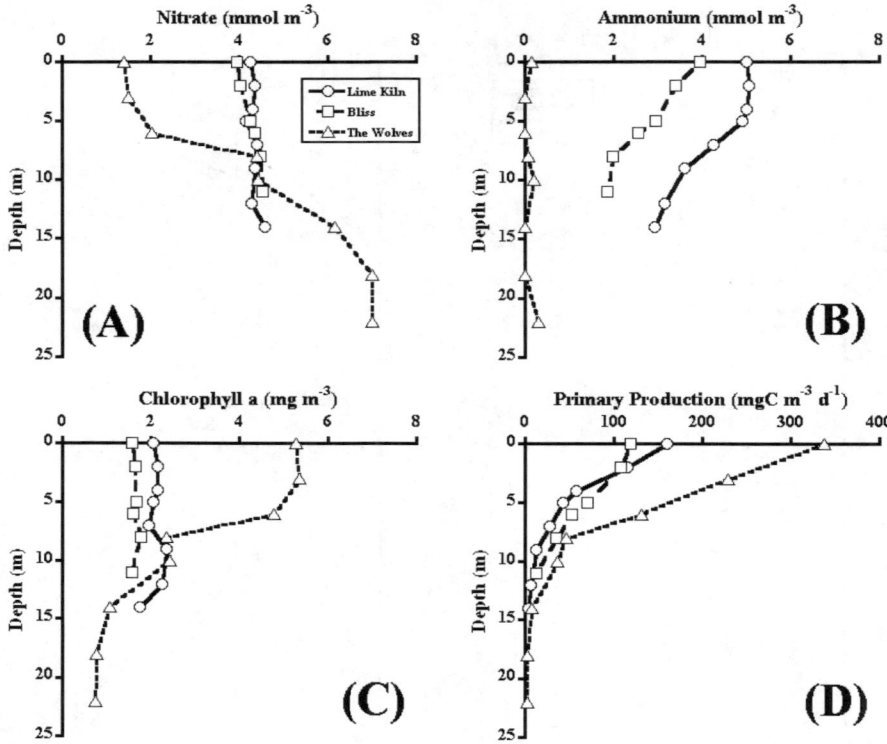

Fig. 4 Vertical structure of nitrate (a) and ammonium (b) concentrations, chlorophyll a (c) and primary production (d) at the aquaculture (Lime Kiln Bay and Bliss Harbour) and control (The Wolves) sites in fall (19–21 September), 2000

From May to November 2001, there was a general coherence in the seasonal development of bacteria in SWNB, which ranged from 0.4 to 2 million cells mL^{-1}. Throughout the entire sampled region, cell abundance was low in spring, increased from the summer solstice to the fall equinox, and then decreased with the approach of winter (**5f**). In surface waters, the difference in bacterial abundance between Lime Kiln Bay (**5d**) and The Wolves (**5e**) was not statistically significant. However, these two sites together with Deadman's Harbour (**5c**) were, on average, less abundant in bacteria (but not statistically different) than the two innermost sites of Brandy Cove (**5a**) and Mid Passamaquoddy Bay (**5b**).

Primary production rates were also similar among the three aquaculture sites. Highest rates were observed in summer (60–100 mg C m^{-3} d^{-1}) and lowest rates in winter (10–30 mg C m^{-3} d^{-1}) (Tables 1 and 2; (**3d**)). Rates were somewhat higher in Bliss Harbour than in Lime Kiln Bay but the difference was not statistically significant. During the fall sampling, primary production rates were significantly higher at the control site (The Wolves) than at

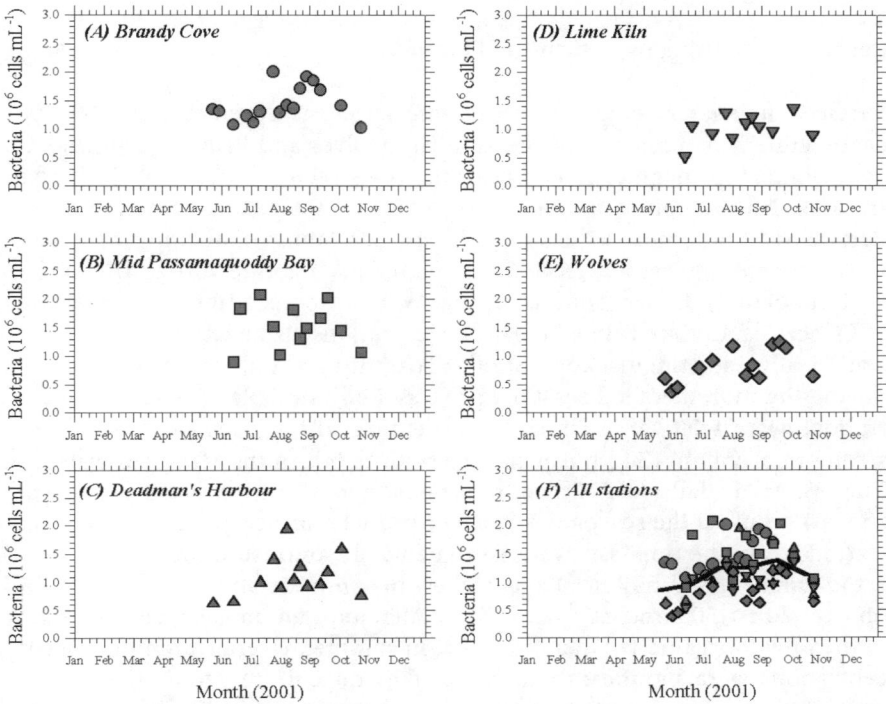

Fig. 5 Bacterial abundance (cells mL^{-1}) in the surface waters (1 m) at 5 locations in SWNB from May to November, 2001. The *solid trend line* in panel (**f**) is the monthly average for all stations combined

the aquaculture sites, whether calculated on a column-average or areal basis, (**4d**), however, when normalised to chlorophyll biomass (P : B, Table 1), the differences were not significant. Stratification and higher water transparency likely resulted in a more favourable light environment for phytoplankton at the control site and contributed to the higher absolute production there.

Vertical profiles of primary production and light penetration measurements, i.e. from Secchi depths, showed that euphotic depths ($\sim 3\times$ Secchi depth) or compensation depths, D_c (depth of zero primary production) were on the order of 10 m or less at the aquaculture sites and 20 m at the control site (Tables 1, 2; (**3d**), (**4d**)). In all cases, euphotic depths were shallower than estimated (from vertical temperature profiles) mixing depths: 10–17 m at the aquaculture sites, 43 m at the control site.

3.2
Nutrient Concentrations, Demand and Turnover

Surface nutrient concentrations from time-series measurements in 2000–2001 at the Lime Kiln aquaculture site and The Wolves and Prince-5 control sites were similar in magnitude and seasonality as seen for chlorophyll concentrations. Nitrate concentrations were high in winter (8–12 mmol m^{-3}) and decreased to minimum values in summer (**2c**). Summertime nitrate concentrations were highest at Prince-5 ($\sim$ 4 mmol m^{-3}) and lowest at The Wolves (< 1 mmol m^{-3}). Concentrations at The Wolves were significantly lower than at Prince-5 and Lime Kiln whereas concentrations at the latter two sites were statistically indistinguishable. Similar magnitudes and patterns were seen during the more detailed seasonal process studies (Tables 1 and 2). Concentrations were relatively uniform with depth at the aquaculture sites due to strong tidal mixing (**3a**) but increased with depth in the stratified waters of The Wolves in fall (**4a**). Nitrite, phosphate and silicate concentrations (not shown) followed the seasonal pattern of nitrate. Surface ammonium concentrations from the time-series measurements, in contrast to nitrate, were lowest in winter (< 1 mmol m^{-3}) and highest in summer/fall (> 4 mmol m^{-3} and on occasion > 20 mmol m^{-3}) (**2d**). In winter, ammonium represented $\sim$ 25% of the total inorganic-N but almost 50% in summer. Overall, ammonium concentrations were significantly higher at the aquaculture site than at the two control sites. This was also evident from the process studies where ammonium concentrations at The Wolves in fall were < 0.1 mmol m^{-3} compared with > 2 mmol m^{-3} at the two aquaculture sites (Table 1). Note, however, that these concentrations at The Wolves were considerably lower than observed in fall during the time-series study (**2d**). In contrast to the vertical distribution of nitrate, ammonium concentrations were not uniform with depth at the aquaculture sites but were somewhat higher in near surface waters, particularly in summer and fall ((**3b**), (**4b**)). Concentrations at The Wolves, however, were uniform (and low) despite the highly stratified water column apparent at that site. Analysis of nutrient ratios showed that the aquaculture sites were enriched in ammonium in summer and fall (relative to nitrate) compared with ratios at the control sites. Nitrate: ammonium ratios at the aquaculture sites in fall, for example, averaged 1.2 compared with a ratios 12–66 at control sites. N : P and N : Si ratios at the aquaculture sites, however, were statistically indistinguishable from ratios at the control sites. N : P ratios at all sites in fall (i.e. 6–8) were significantly lower than requirements for phytoplankton growth, i.e. 16, the Redfield Ratio [25].

Regional and seasonal variations in nitrogen utilisation rates at the aquaculture sites were similar to concentration variations, i.e. nitrate utilisation was highest in spring and winter (40–120 µmol m^{-3}h^{-1}) when nitrate concentrations were highest and ammonium utilisation was highest in summer and fall (40–100 µmol m^{-3}h^{-1}) when concentrations peaked (Tables 1 and 2). Pat-

Table 1 Water-column properties (euphotic depth average) at the SWNB aquaculture and control sites

	Spring		Summer		Fall			Winter	
	Lime Kiln	Bliss	Lime Kiln	Bliss	Lime Kiln	Bliss	The Wolves	Lime Kiln	Bliss
Euphotic depth (m)	9	11	8	11	9	11	22	10	11
Mixing depth (m)	10	12	10	12	10	12	43	10	12
Nutrients									
NH_4 (mmol m^{-3})	1.38	0.91	3.44	2.36	4.14	2.73	0.07	2.88	1.45
NO_3 (mmol m^{-3})	7.13	8.0	2.89	2.73	4.36	4.27	4.55	9.25	9.64
Biomass									
CHL (mg m^{-3})	1.38	1.36	1.44	1.45	2.14	1.64	2.59	0.59	0.58
PON (mmol m^{-3})	2.5	2.1	3.0	2.6	3.4	3.4	3.3	1.9	1.3
POC (mmol m^{-3})	22.6	20.9	24.0	22.7	25.8	29.1	27.7	17.5	13.3
C : N (molar)	9.0	10.0	8.0	8.7	7.6	8.6	8.5	9.4	10.1
Productivity									
NH_4 Util (μmol m^{-3} h^{-1})	13	13	74	51	38	40	8.1	17	9.1
NO_3 Util (μmol m^{-3} h^{-1})	53	118	55	72	2.1	4.5	32	53	57
Prim prod (mgC m^{-3} d^{-1})	28	38	65	89	46	65	80	34	16
f-ratio	0.81	0.90	0.43	0.58	0.05	0.10	0.80	0.76	0.86
P : B (mg C mg CHL^{-1} h^{-1})	1.7	2.3	3.7	5.1	1.8	3.3	2.6	4.9	2.4
NH_4 turnover (d)	4.6	2.9	2.0	1.9	4.6	2.9	0.35	7.3	6.7
NO_3 turnover (d)	5.6	2.8	2.2	1.6	88	39	5.8	7.2	7.1

f-ratio = [NO_3 Util/(NO_3 Util + NH_4 Util)]

terns and magnitudes were similar among the aquaculture sites; an analysis of variance showed that season was the most significant factor contributing to variability. Nitrate utilisation exceeded ammonium utilisation much of the year, i.e. nitrate fuelled 80–90% of the primary production in spring and winter, 40–50% in summer and < 10% in fall (see f-ratios, Tables 1 and 2). Nitrate-based production at The Wolves control site in fall was on the same order as winter/spring values at the aquaculture sites, i.e. f-ratio ~ 0.8, and significantly higher than nitrate-based production at the aquaculture sites in fall (f-ratio: 0.05–0.10). The anomalously low ammonium concentrations measured at The Wolves in fall, however, may have accounted for the low ammonium utilisation and high f-ratios observed. Short-term carbon (primary production): nitrogen (nitrate plus ammonium uptake) utilisation ratios (molar) at the aquaculture sites were well below Redfield ratios in winter (C : N Util: 2–3), but increased through spring and summer (C : N Util: 3–5) to ratios above Redfield in fall (C : N Util: 8–10); highest ratios (C : N Util: 13) were observed at the offshore control site where surface inorganic nitrogen concentrations were lowest.

Table 2 Comparison of water-column properties (euphotic depth average) between inner and outer Lime Kiln Bay, SWNB

	Spring		Summer		Winter	
	Inner	Outer	Inner	Outer	Inner	Outer
Euphotic depth (m)	9	11	8	10	10	12
Mixing depth (m)	10	17	10	17	10	17
Nutrients						
NH_4 (mmol m^{-3})	1.38	1.64	3.44	5.43	2.88	2.07
NO_3 (mmol m^{-3})	7.13	7.50	2.89	2.29	9.25	9.21
Biomass						
CHL (mg m^{-3})	1.38	1.64	1.44	1.86	0.59	0.57
PON (mmol m^{-3})	2.5	2.4	3.0	3.8	1.9	1.3
POC (mmol m^{-3})	22.6	33.3	24.0	29.4	17.5	12.2
C : N (molar)	9.0	14.1	8.0	7.7	9.4	9.3
Productivity						
NH_4 Util (μmol m^{-3} h^{-1})	13	12	74	101	17	9
NO_3 Util (μmol m^{-3} h^{-1})	53	39	55	56	53	36
Prim prod (mg C m^{-3} d^{-1})	28	26	65	104	34	10
f-ratio	0.81	0.77	0.43	0.36	0.76	0.80
P : B (mg C mg CH L^{-1} h^{-1})	1.7	1.3	3.8	4.7	4.8	1.4
NH_4 turnover (d)	4.6	5.7	2.0	2.2	7.3	9.7
NO_3 turnover (d)	5.6	8.0	2.2	1.7	7.2	10.8

f-ratio = [NO_3 Util/(NO_3 Util + NH_4 Util)]

Table 3 Comparison of mean inorganic nutrient concentrations and nutrient ratios in Bliss Harbour, SWNB: 1990–2002. DIN(dissolved inorganic nitrogen) = $NO_3 + NH_4$

	SiO_3 (mmol m^{-3})	PO_4 (mmol m^{-3})	NO_3 (mmol m^{-3})	NH_4 (mmol m^{-3})	DIN:SiO_3 (molar)	DIN:PO_4 (molar)	NO_3:NH_4 (molar)	Ref
1990								[33]
Spring	10.6	0.7	9.0	1.3	1.0	14.4	6.8	
Summer	1.4	0.6	1.8	3.0	3.4	8.4	0.6	
Fall	3.9	1.0	3.6	5.3	2.3	9.0	0.7	
Winter	10.1	1.0	9.4	1.1	1.0	10.6	8.9	
1999								[34]
Spring	–	–	–	–	–	–	–	
Summer	–	–	–	–	–	–	–	
Fall	1.8	0.6	1.0	2.5	1.9	5.8	0.4	
Winter	–	–	–	–	–	–	–	
2000–2002								[Present study]
Spring	8.5	0.8	8.0	0.9	1.1	11.5	8.8	
Summer	4.0	0.6	2.7	2.4	1.3	8.0	1.2	
Fall	5.3	0.9	4.3	2.7	1.3	7.7	1.6	
Winter	9.6	1.0	9.6	1.5	1.2	11.1	6.6	

Nitrate and ammonium turnover (concentration/utilisation rate) was comparable at the aquaculture sites for most seasons, i.e. turnover times were on the order of a week in spring and winter and ∼ 2 days in summer. During the fall, ammonium turnover times were still relatively short (3–5 days) but nitrate turnover was 40–90 days, due to the low utilisation rates then, and only ∼ 10% of the summer rates. An explanation for the anomalously low nitrate utilisation rates observed in fall at both Lime Kiln Bay and Bliss Harbour is unclear. In contrast, nitrate and ammonium turnover times at The Wolves in fall were short; < 1 day for ammonium and less than a week for nitrate.

4
Discussion

One of the principal environment concerns surrounding the aquaculture industry is eutrophication: increased (above natural levels) biomass and/or production of autotrophs (phytoplankton and macrophytes) and microheterotrophs (bacteria in the sediments or water-column) from the addition of inorganic and organic nutrients to the ecosystem from the industry's wastes. In addition to increased biomass and metabolism, alterations in community structure and disruption of the natural balance between autotrophic and heterotrophic processes are also known consequences [2, 26]. The ability to quantify environment impacts of aquaculture is considerably more difficult when dealing with far-field, ecosystem-level effects compared with near-field, localised effects [3], due in large part to the often dominant influence of other anthropogenic sources of nutrients (domestic sewage, industry, agriculture) in coastal waters [27, 28]. Understanding far-field effects of finfish aquaculture on the pelagic ecosystem is a great challenge as evidenced by the lack of published information on the subject [3]. Detailed mass-balance calculations of fish farm wastes in SWNB [16, 29] have shown that nitrogen fluxes from farms, for example, dominate anthropogenic nitrogen inputs to the region and can exceed natural nitrogen fluxes by as much as threefold in some locations. The question is: how are these excess nutrients affecting the pelagic (and benthic) ecosystems of SWNB?

4.1
Phytoplankton Biomass, Bacterial Abundance and Particulate Organic Matter

Similarities in the seasonal cycles of phytoplankton (chlorophyll a concentration) in SWNB at sites close to salmon pens and sites far removed lead one to the conclusion there are no discernible effects of the aquaculture activity on this ecosystem property. Differences in the vertical distribution of phytoplankton between the aquaculture sites and The Wolves control site were observed in fall but can be explained by local hydrography. The upper watercolumn at The Wolves is highly stratified in summer and fall but well-mixed at the aquaculture sites year-round due to strong tidal forcing. A more appropriate comparison, i.e. between the well-mixed aquaculture sites and the well-mixed Prince-5 control site, showed that the magnitudes and seasonal cycles of phytoplankton biomass were similar. In a like manner, the concentrations and elemental composition (C : N ratio) of particulate organic matter were similar at the aquaculture and control sites. Although these properties were measured at both sites only in fall, the impacts of aquaculture wastes on the pelagic ecosystem may be most significant then [16, 30]. In an earlier study of finfish aquaculture effects on water chemistry and plankton in SWNB, Wildish et al. [31] came to similar conclusions about phytoplankton biomass, i.e. no measurable differences were observed in chlorophyll a concentrations in the proximity of and away from fish farms. Evidence in the literature of significant effects of finfish aquaculture wastes on phytoplankton biomass and associated particulates from other regions is lacking [3, 26].

Seasonal measurements of bacterial abundance showed no clear differences between the aquaculture and control sites either. Bacterial abundances relative to phytoplankton biomass (B/P ratios) were slightly higher at the Lime Kiln Bay aquaculture site compared with The Wolves control site but well within the range of ratios seen in offshore shelf and oceanic waters far from the influence of aquaculture and other human activities. We found one study in the literature that has investigated near- and far-field effects of finfish aquaculture on bacteria, in Gokasho Bay, Japan (34.3N 136.6E) in late summer [32]. This is a shallow system (15–30 m) but with much higher temperatures (> 25 °C compared with late summer maximum temperatures in SWNB of < 15 °C), nutrients and plankton biomass than seen in SWNB. They observed sharp spatial gradients in nutrients, chlorophyll a and bacterial abundance and production rates with highest levels closest to the fish farms and lowest levels at the mouth of the Bay, some 4 km away. Bacterial biomass and production rates were an order of magnitude higher at the farms compared with the mouth of the Bay. Chlorophyll a and nutrients were higher at the farms by a factor of five and dissolved organic matter by a factor of two. Unfortunately, the data were not presented in a way that B/P ratios could be calculated and compared with those observed in SWNB.

4.2
Primary Production and Nutrient Demand

Similar to the results for phytoplankton biomass, primary production rates varied seasonally at the aquaculture sites but differences among the sites were insignificant overall. Comparative data at The Wolves control site were collected only during fall and productivity (normalised to biomass, P : B) there was not significantly different from P : B ratios at the aquaculture sites;

again, impacts of aquaculture activities on ecosystem properties in the water-column may be most significant during the summer-fall seasons [16, 30]. At the aquaculture sites, nitrate was the dominate nitrogen source fuelling primary production in winter and spring at all sites and ammonium was relatively more important in summer and fall. Nitrate, however, was the principal source for production at the offshore control site in fall but the anomalously low ammonium concentrations observed then may have contributed. We are not aware of any published studies of the effects of aquaculture on nutrient utilisation by phytoplankton, however, our observations are similar to those seen in other coastal waters where aquaculture is not an issue. L'Helguen et al. [33], for example, investigated the seasonal cycles of nitrate and ammonium in permanently well-mixed coastal waters in the western English Channel off the coast of France and found magnitudes and patterns of nitrate and ammonium utilisation remarkably similar to what we observed in SWNB. Our estimates of nitrate and ammonium turnover times were similar among the experiment sites, ranging from the order of a week in winter and spring to 2 days in summer. The anomalously long nitrate turnover times at the aquaculture sites in fall (40–80 days) are difficult to explain but resulted from very low measured utilisation rates. At the offshore control site, extremely low ammonium concentrations in fall resulted in very short turnover times, < 1 day. From the time-series study, it would appear that much higher ammonium concentrations are typical at that site in fall and thus turnover times may generally be longer.

Because the sources of nutrients in SWNB vary [16, 29], it is difficult to attribute observed levels to aquaculture activity specifically. Analysis of absolute nutrient levels and nutrient ratios, however, suggests that the aquaculture sites were enriched in recycled nitrogen and ammonium (relative to nitrate) compared with the control sites. Wildish et al. [31] carried out an earlier (1989–1990) seasonal study of the influence of salmon aquaculture activity on phytoplankton biomass and nutrients in SWNB. Similar to our results, they found no evidence of elevated concentrations of nitrate and silicate in the proximity of fish farms compared with open water control sites. However, ammonium concentrations, and phosphate to a lesser extent, were high near the farms. A more recent fall survey in SWNB almost 10 years later, 1999, [34] showed, again, elevated ammonium concentrations and low N : Si and N : P ratios at the aquaculture sites but nutrient concentrations and ratios were not substantially different from those seen in 1989–1990 (Table 3). Gowen and Bradbury [26] observed elevated ammonium concentrations in the vicinity of finfish farms in Scottish coastal waters but no apparent effects on phytoplankton biomass. Over the long term, changes in nutrient concentration ratios in coastal waters as a result of nitrogen enrichment from a variety of sources, including aquaculture, have been implicated in dramatic alterations in phytoplankton community composition, and increases in the frequency and intensity of phytoplankton blooms, including harmful algal blooms (HABs) world-wide (e.g. [7]). In other cases, the alterations are more subtle and can only be discerned from a long record of measurements made over many years. For example, in Bedford Basin, it appears that annual average ratios of N : Si and P : Si are higher now than they were 30 years ago. Concomitantly, there has been an increase in the abundance of smaller phytoplankters (pico- and nanoplankton), which do not generally use silicate as a macro-nutrient [35]. Potentially, this change in the size structure of the primary producers may have an influence on ecosystem dynamics. Phytoplankton community structure in SWNB observed over the past decade, including the incidence of HABs, has not changed appreciably with the growth of the aquaculture industry [36, 37]. Absence of evidence at present of significant effects of aquaculture activity in SWNB on nutrients or phytoplankton may not continue indefinitely into the future, however. Salmon rearing and fish processing are the dominant anthropogenic sources of nutrients and oxygen demand in this region [16, 29] and continued expansion will translate to increasing nutrient loads and stress on the oxygen balance of the system (see also [9, 30]). How much longer the pelagic ecosystem of SWNB can absorb these inputs without negative consequences is a critical unknown.

4.3
Light vs. Nutrient Limitation of Primary Production

Observations made during this study strongly suggest that primary production rates at both the aquaculture and control sites were not under the control of nutrient supply. Seasonal variability in nutrient concentrations indicated that at no time were nutrients completely depleted in surface waters, even at the control sites. Moreover, during summer–fall when winter nutrient stores (e.g. nitrate) are reduced due to biological utilisation, internally recycled nutrients (e.g. ammonium) reached their maximum levels at the aquaculture sites and control sites. Both nitrate and ammonium turnover times exceeded the normal doubling time of coastal phytoplankton ($\sim$ 1 doubling d^{-1}), even in summer/fall when concentrations were lowest, suggesting further that nutrient supply was not limiting in this region. Nutrient concentration ratios indicated that phosphorus and silicate were sufficient for or in excess of metabolic requirements even during summer. Taken together, these observations provide convincing evidence that factors other than nutrients are controlling primary production (and setting an upper limit on nutrient demand) in SWNB.

Primary production profiles and measurements of water transparency during this study showed that euphotic depths were exceeded at all stations by mixing depths, a strong indicator of light-limitation. In offshore coastal and oceanic waters, the attenuation of light is largely due to the presence of phytoplankton and phyto-detritus [38]. However, this is not the case for strongly mixed, shallow inshore waters, including at least one of the control sites in this study, where other dissolved and particulate constituents dominate the optical attenuation of water. At Prince-5, for example, extinction coefficients (K_d) calculated from phytoplankton biomass [39] were significantly less (0.06–0.25 m^{-1}) than those determined from Secchi depths (0.17–0.38 m^{-1}) which is a measure of total light extinction. The indication is that in these waters, phytoplankton account for half or less of the light attenuation. In a similar way, phytoplankton contributed < 25% to light attenuation at the aquaculture sites compared with the Wolves control site (2b).

In shallow macro-tidal coastal waters such as SWNB, turbidity is an important factor limiting primary production and evaluating the response of phytoplankton in these environments to nutrient over-enrichment from human activities such as aquaculture is not straight forward [8, 40]. In an attempt to address this problem, Cloern [41] developed a simple index of the relative sensitivity of natural phytoplankton populations to light and nutrients based on knowledge of the resource (i.e. submarine light conditions and nutrient concentrations) and physiological properties of the phytoplankton (i.e. growth response to light and nutrients). Cloern determined light sensitivity from estimates of the mean euphotic zone light field (from incident light and extinction coefficient) scaled by the half-saturation growth constant for coastal phytoplankton grown under light-limited conditions. Similarly, nutrient sensitivity was determined from the local euphotic zone nutrient concentration scaled by the half-saturation growth constant for coastal phytoplankton grown under nutrient-limited conditions. The ratio of these sensitivity estimates (R) is then the basis for evaluating light versus nutrient limitation, i.e. $R = 1$ is the partition between light and nutrient limitation where $R > 1$ is the light-limited domain ($R > 10$ represents strong light limitation) and $R < 1$ is the nutrient limited domain ($R < 0.1$ represents strong nutrient limitation).

We applied this procedure in a similar way to our data from SWNB using incident light, extinction coefficients (calculated from Secchi depths) and observed inorganic nitrogen concentrations. A single growth half-saturation constant for light (derived from photosynthesis versus light curves in local waters, i.e. 50 W m^{-2} or $\sim$ 235 µE m^{-2}s^{-1}) was applied to all data. Similarly, a single half-saturation constant for nutrient limited growth (1.5 mmol m^{-3}), typical for coastal phytoplankton [41] was used. Physiological response to both light and nutrients can change with resource conditions (i.e. half-saturation constants are not truly constant), however, Cloern's sensitivity analysis suggests that this limitation index is robust, i.e. fairly insensitive to changes in the growth constants. It is evident from this analysis that phytoplankton at both the aquaculture and control sites in SWNB are strongly light-limited, even in summer when incident light is at its maximum (Fig. 6). Severity of light-limitation is highest in winter when incident light is at its minimum and day-length shortest. Interestingly, light-limitation was more acute at the Prince-5 control site than at the aquaculture sites despite the

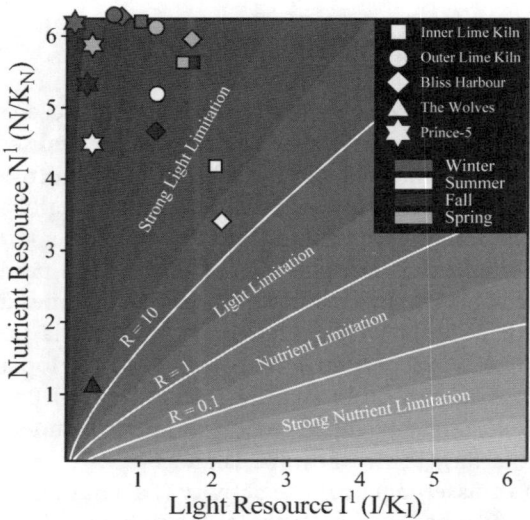

Fig. 6 Index (ratio, R) of the relative strength of light and nutrient limitation on phytoplankton growth from [41]: seasonal data from SWNB aquaculture and control sites superimposed

fact that turbidity was greater inshore, by at least a factor of two. Because mixing depths are much greater (~ 90 m) at Prince-5 compared with the aquaculture sites (10–20 m), however, the "average" light levels that Prince-5 phytoplankton would be exposed to were far less than average levels at the shallow aquaculture sites, resulting in stronger light-limitation at the former site.

Because aquaculture activity has not measurably affected the pelagic primary producers in SWNB, one might conclude that their capacity to process nutrients has not been exceeded. However, an important implication of the results described here is that the phytoplankton in SWNB may, in fact, have little excess capacity for increasing growth and production in response to increased nutrient loading. It is apparent that light has basically set the upper limit on pelagic primary production in this system. Moreover, light-limitation will likely follow a course of greater severity as suspended particulates and turbidity increase with increasing aquaculture activity. In general, therefore, it might not be unreasonable to suggest that the autotrophic component of the pelagic ecosystem in SWNB is probably at or close to its capacity and that any further response to aquaculture activity (and nutrient loading) may only be manifest in the micro-heterotrophs. The same may not be true for the macro-flora, however, where there is some evidence that nutrient enrichment from fish farms in SWNB may have influenced the growth of inter-tidal green macroalgal mats (*Entermorpha* and *Ulva* species) [42, 43].

4.4
Bacteria

In offshore coastal and oceanic waters which do not receive significant allochthonous input of organic matter, the maximum realised abundance of bacteria depends on an interplay between bottom-up control by phytoplankton and top-down control by bacterivores and viruses. A global macroecology of the bacteria-phytoplankton relationship in oceanic regions [44] indicates a well-defined domain within which almost all of 13 974 paired observations have been recorded (**7a**). In this context, the SWNB measurements occur close to the median trend (**7b**), with Lime Kiln bacteria just slightly more abundant than The Wolves bacteria when scaled to phytoplankton biomass. This is in contrast to two other marine inlets: Bedford Basin, Nova Scotia ($44°41'$N, $63°38'$W), heavily impacted by municipal and industrial wastes (**7c**), and Tracadie Bay, Prince Edward Island ($46°24'$N, $63°00'$W), a site of intense shellfish (mussel) aquaculture activity (**7d**). In these latter cases, bacteria occur at abundances (scaled to phytoplankton abundance) far exceeding the maximum levels realised in natural coastal and oceanic regions remote from anthropogenic influence. More important, however, is the observation that at present, bacterial abundance at aquaculture sites in SWNB does not appear to be significantly higher in most cases than levels seen at sites far removed from salmon farms. Bacterial abundance in SWNB, similar to phytoplankton biomass, shows little effect from aquaculture activity. Moreover, bacterial abundance relative to phytoplankton biomass, i.e. B/P ratios, in these waters are statistically indistinguishable from ratios observed in offshore and oceanic waters far-removed from anthropogenic inputs. In contrast to the expected response of phytoplankton to increasing aquaculture activity, the micro-heterotrophs are not constrained by light and would be expected to respond positively to increased nutrient loading (inorganic and organic; particulate and dissolved) relatively unchecked. Low short-term C : N utilisation ratios compared with particulate elemental ratios observed in this study during most seasons (below Redfield ratios) suggest that bacteria may be competing with phytoplankton for inorganic nitrogen [45] and that microheterotrophs are already playing a significant role in the nitrogen dynamics in SWNB. With the production side of oxygen balance in the water-column effectively in check through light-limitation, increasing bacteria abundance and metabolism will tip the balance in the water-column to excess respiration and subsequent oxygen depletion [9]. Not only does this have implications for the ecosystem but for the aquaculture industry as well [30]. The shift from a balanced towards a heterotroph-dominated system, although not evident yet in SWNB, is nonetheless a distinct possibility and has already been observed in coastal waters close by impacted heavily by municipal and industrial wastes (Bedford Basin, Nova Scotia) and waters supporting extensive shellfish aquaculture (Tracadie Bay, Prince Edward Island). It would be interesting to know

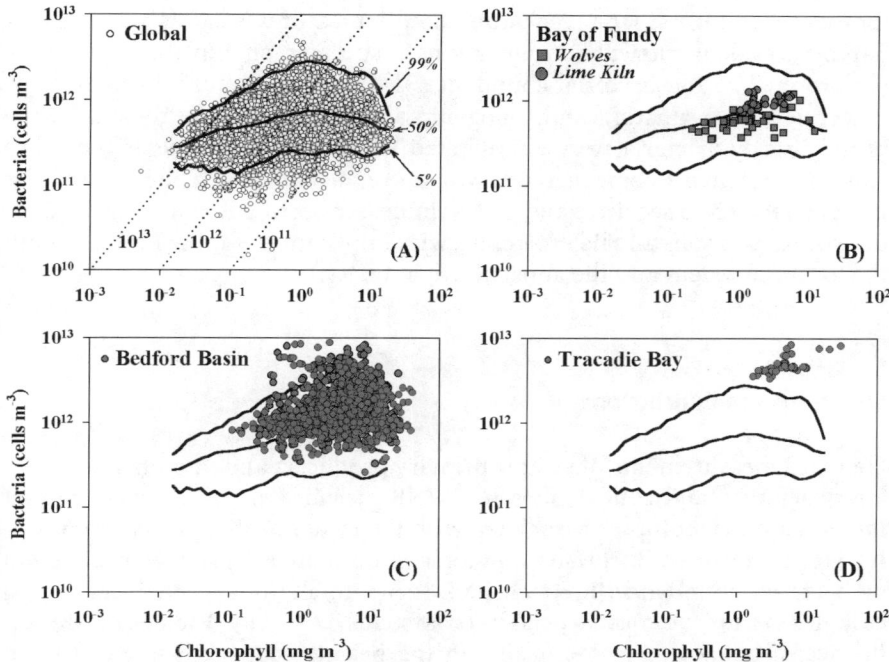

Fig. 7 Macroecological analysis of the relationship between bacterial abundance (Cells m^{-3}) and chlorophyll concentration (mg m^{-3}). (**a**) A global dataset of 13 974 paired measurements in coastal offshore and oceanic waters [44]. Quantile lines are constructed from bacterial distributions within successive binned chlorophyll intervals of 0.1 logarithmic unit (99 percentile = *upper*; 50 percentile = *middle*; 5 percentile = *lower*). *Straight dashed lines* indicate bacteria to chlorophyll ratios of 10^{13}, 10^{12} and 10^{11} cells mg^{-1}. (**b**) SWNB – finfish aquaculture and control sites. (**c**) Bedford Basin, Nova Scotia – major metropolitan/industrial area. (**d**) Tracadie Bay, Prince Edward Island – major shellfish acquaculture site

to what extent the different B/P ratios seen in SWNB and Tracadie Bay are due to differences in nutrient loading or more a consequence of the fact that phytoplankton are the major food source for shellfish (mussel in the case of Tracadie) as opposed to finfish culture [46] and the industry may be a major phytoplankton loss term in that ecosystem.

4.5
Water-Quality Indicators

Several properties of the water-column and plankton were evaluated as indicators of far-field effects of finfish aquaculture on the pelagic ecosystem of SWNB. Relatively complicated measures of plankton production provide information on nutrient demand and turnover and data useful for determining oxygen budgets. However, more simple measures such as water transparency, residual nutrient concentrations in summer, phytoplankton biomass (chlorophyll a) and bacterial abundance also provide useful information for detecting and evaluating anthropogenic disturbances in the region. In the present study, bacteria were enumerated using a flow cytometer but more conventional microscopic methods would work as well [23]. The B/P ratio, for example, may be a sensitive and robust indicator of trophic status, i.e. balance between autotrophs and heterotrophs, with implications for the health of both the local ecosystem and the aquaculture industry.

5
Summary and Conclusions

Standard measurements of pelagic primary producers and microheterotrophs at aquaculture and control sites in SWNB provided no clear evidence that increased nutrient inputs associated with the local finfish aquaculture industry [16, 29] resulted in elevated phytoplankton biomass, primary production or bacterial abundance, however, this is not totally unexpected due to the complex nature of coastal pelagic ecosystems [2–4, 26]. Evaluating the influence of aquaculture specifically on the pelagic ecosystem in SWNB will require a longer time-series of observational data (i.e. monitoring on the time scale of industry development), robust mass-balance calculations [16, 29] and sophisticated bio-physical models. [30] with the capability of predicting system response to aquaculture inputs against a large and highly variable suite of inputs and exchanges from natural and other anthropogenic sources [3].

Several lines of evidence point to strong light-limitation of phytoplankton growth and suggest that the pelagic autotrophs in SWNB have little capacity to process the increased nutrient loads that will accompany further growth in the aquaculture industry in the region. Although there is no evidence at present of a serious imbalance in the autotrophic and heterotrophic components of the water-column in SWNB, the constraints imposed on phytoplankton by light do not affect bacteria and pelagic microheterotrophs which may continue to increase in abundance and metabolism as aquaculture grows. The result will likely be negative consequences for the system's oxygen balance and water quality affecting both the ecosystem and the industry. Simple measures such as chlorophyll a concentration and bacterial abundance (and derived B/P ratios) may prove to be useful indicators of pelagic trophic status and water quality in coastal system stressed by human activities. More complex indices of productivity and nutrient demand will be useful for mass-balance calculations and modelling and provide insight into the physiological mechanisms by which the pelagic ecosystem responds to anthropogenic stresses.

Acknowledgements The authors wish to thank High Akagi, Alex Hanke, Muerielle LeGresley, Randy Losier, Jennifer Martin, Fred Page, Michelle Ringuette, Peter Strain, Dave Wildish and the New Brunswick Salmon Growers Association for assistance, provision of facilities to carry out the field work of this project and additional data. In addition, staff in the Biological Oceanography Section of the Ocean Sciences Division at Bedford Institute of Oceanography provided analytical support. Funding for ESSA was provided by Fisheries and Oceans' Environmental Science Strategic Research Fund (ESSRF). Helpful suggestions for improving this manuscript were provided by four anonymous reviewers.

References

1. Fisheries and Oceans Canada (2003) Can Tech Rep Fish Aquat Sci 2489
2. Black KD (2001) Environmental Impacts of Aquaculture. CRC Press, Boca Raton, Florida
3. Fisheries and Oceans Canada (2003) Can Tech Rep Fish Aquat Sci 2450
4. Rosenthal H, Scarratt DJ, McInerney-Northcott M (1995) In: Boghen (ed) Cold-Water Aquaculture in Atlantic Canada, 2nd ed. Can Inst Res Reg Dev, Univ. Moncton. Tribune Press, Sackville, p 451
5. Nixon SW (1995) Ophelia 41:199
6. Rabalais NN, Nixon SW (2002) Estuaries 25:639
7. Smayda TJ (1989) In: Cosper EM, Bricelj VM, Carpenter EJ (eds) Novel Phytoplankton Blooms. Springer, Berlin Heidelberg New York, p 449
8. Cloern JE (2001) Mar Ecol Prog Ser 210:223
9. Kepkay PE, Harrison WG, Bugden JBC (2005) Ecosystem indicators of water quality part I. Oxygen product and oxygen demand (this volume)
10. Fisheries and Oceans Canada (2003) Can Tech Rep Fish Aquat Sci 2352
11. Fisheries and Oceans Canada (2003) Can Tech Rep Fish Aquat Sci 2411
12. Hecky RE, Kilham K (1988) Limnol Oceanogr 33:796
13. Howarth RW (1988) Ann Rev Ecol Syst 19:89
14. Trites R, Garrett C (1983) In: Thomas M (ed) Marine and coastal systems of the Quoddy Region. Can Spec Pub Fish Aq Sci 64:9
15. Greenberg DA, Shore JA, Page FH, Dowd M (2004) Ocean Modelling (In Press)
16. Strain PM, Hargrave BT (2005) Salmon aquaculture, nutrient fluxes and ecosystem processes in southwestern New Brunswick (this volume)
17. Li WKW, Martin JL, Ringuette MM (2002) In: Hargrave BT (ed) Environmental Studies for Sustainable Aquaculture (ESSA): 2002 Workshop Report. Can Tech Rep Fish Aquat Sci 2411:49
18. Therriault J-C, Petrie B, Pepin P, Gagnon J, Gregory D, Helbig J, Herman A, Lafaivre D, Mitchell M, Pelchat B, Runge J, Sameoto D (1998) Can Tech Rep Hydrogr Ocean Sci 194
19. Harrison G, Sameoto D, Spry J, Pauley K, Maass H, Soukhovtsev V (2002) Can Sci Advis Sec Res Doc 2002/056
20. Grasshoff K (1983) Methods of Seawater Analysis, 2nd Ed. Verlag Chemie, Weinheim, FRG
21. Holm-Hansen O, Lorenzen CJ, Holmes RW, Strickland JDH (1965) J Cons Perm Int Explor Mer 30:3
22. Sharp JH (1974) Limnol Oceanogr 19:984
23. Li WKW, Dickie PM (2001) Cytometry 44:236

24. Dugdale RC, Wilkerson FP (1986) Limnol Oceanogr 31:673
25. Redfield AC, Ketchum BH, Richards FA (1963) The influence of organisms on the composition of seawater. In: Hill MN (ed) The Sea, Vol. 2. Wiley Interscience, New York, p 26
26. Gowen RJ, Bradbury NB (1987) Oceanogr Mar Biol Ann Rev 25:563
27. Ackefors H, Enell M (1994) J Appl Ichthyol 10:225
28. Seitzinger SP, Kroeze C, Bouwman AF, Caraco N, Dentener F, Styles RV (2002) Estuaries 25:640
29. Strain PM, Wildish DJ, Yeats PA (1995) Mar Poll Bull 30:253
30. Page FH, Losier R, McCurdy P, Greenberg D, Chaffey J, Chang B (2005) Dissolved oxygen and salmon cage culture in the southwestern New Brunswick portion of the Bay of Fundy (this volume)
31. Wildish DJ, Keizer PD, Wilson AJ, Martin JL (1993) Can J Fish Aquat Sci 50:303
32. Sakami T, Abo K, Takayanagi K, Toda S (2003) Est Coast Shelf Sci 56:111
33. L'Helguen S, Madec C, Le Corre P (1996) Est Coast Shelf Sci 42:803
34. Fisheries and Oceans Canada (2003) Can Tech Rep Fish Aquat Sci 2356
35. Li WKW (2003) Bedford Basin plankton monitoring program. Workshop on Repairing Environmental Damage to the Halifax Harbour Watershed, Halifax Harbour Watershed Restoration Coalition, Dalhousie University, April 2003
36. Martin JL, LeGresley MM, Strain PM, Clement P (1999) Can Tech Rep Fish Aq Sci 2265
37. Martin JL, LeGresley MM, Strain PM (2001) Can Tech Rep Fish Aq Sci 2349
38. Jerlov NG (1976) Marine optics. Elsevier, Amsterdam
39. Platt T, Sathendranath S, Caverhill CM, Lewis MR (1988) Deep-Sea Res I 35:885
40. Monbet Y (1992) Estuaries 15:563
41. Cloern JE (1999) Aquat Ecol 33:3
42. Auffrey LM, Robinson SMC, Barbeau MA (2002) In: Hargrave BT (ed) Environmental Studies for Sustainable Aquaculture (ESSA): 2002 Workshop Report. Can Tech Rep Fish Aquat Sci 2411:84
43. Robinson SMC, Auffrey LM, Barbeau MA (2005) Far-field impacts of eutrophication on the intertidal zone in the Bay of Fundy with emphasis on the soft-shell clam, *Mya arenaria* (this volume)
    In: Hargrave BT (ed) Environmental Effects of Marine Finfish Aquaculture. The Handbook of Environmental Chemistry, Vol 5. Springer, Berlin Heidelberg New York
44. Li WKW, Head EJH, Harrison WG (2004) Deep-Sea Res (In Press)
45. Kirchman DL (2000) In: Kirchman DL (ed) Microbial Ecology of the Oceans. Wiley-Liss, New York, p 261
46. Silvert W, Sowles JW (1996) J Appl Ichthyol 12:75

Ecosystem Indicators of Water Quality Part II. Oxygen Production and Oxygen Demand

Paul E. Kepkay (✉) · W. Glen Harrison · John B. C. Bugden

Biological Oceanography Section, Ocean Sciences Division, Fisheries and Oceans Canada, Bedford Institute of Oceanography, P.O. Box 1006, Dartmouth, B2Y 4A2, Canada
KepkayP@mar.dfo-mpo.gc.ca, HarrisonG@mar.dfo-mpo.gc.ca, BugdenJ@mar.dfo-mpo.gc.ca

1	Introduction	84
2	**Oxygen Production and Demand**	85
2.1	Measurement of Oxygen Production and Demand	85
2.1.1	Productivity (P) Measurements	85
2.1.2	Respiration (R) Measurements	86
2.1.3	Interpretation of P and R Measurements	88
2.2	Oxygen Production, Oxygen Demand and PNDB Systems	89
3	**Production, Respiration and P/R**	90
4	**Carbon Load, Phytoplankton and Bacteria**	92
4.1	Measurement of Total, Particulate and Dissolved Organic Carbon	92
4.2	Bioreactivity of the Coastal Carbon Load	93
5	**Photochemistry - A Seasonal Link in the Coastal Carbon Cycle?**	94
6	**Summary and Conclusions**	95
	References	96

Abstract Seasonal transitions from oxygen production to oxygen demand at coastal aquaculture sites in southwestern New Brunswick (SWNB) can be defined in terms of the production-respiration (P/R) ratio. During the summer, when P/R is greater than 1, an autotrophic ecosystem is in place, and dissolved oxygen (DO) in surface waters remains above thresholds for optimal fish growth. During the fall and winter, when P/R is less than 1, a heterotrophic system is in effect, and DO can decrease to below threshold. Photochemical decomposition may act as a seasonal link, contributing to the onset of net oxygen demand by facilitating the breakdown of terrestrial and marine organic carbon in the fall. The overall carbon load is not a useful index of the bioreactive material that creates oxygen demand. Instead, bacterial number and chlorophyll concentration (expressed in terms of a bacteria:chlorophyll ratio) may be better indicators of the seasonal regulation of oxygen dynamics by the ecosystem. Low DO is the cumulative effect of sustained oxygen demand; a seasonal change in P/R from greater to less than 1 is an early warning of this demand. The application of both indices (P/R and DO) in tandem can be used to develop ecosystem-sensitive plans for the management of water quality at aquaculture sites.

Keywords Aquaculture · Oxygen production and demand · P/R ratio · Oxygen dynamics · Water quality

1
Introduction

The rapid growth of coastal finfish culture has forced the aquaculture industry to face a number of problems related to water quality. For example, caged salmon production in the surface waters of the southwestern New Brunswick (SWNB) coastal region has increased by a factor of about 40 between 1983 and 2003 [1]. Given that aquaculture will continue to expand in SWNB as the culture of other species is perfected, there is an ongoing requirement for clear indices of water quality - not only as affected in the near field by caged fish, but also as affected in the far field by the seasonal changes encountered in this temperate coastal ecosystem. As it continues to grow, the aquaculture industry in SWNB will deal with many of the issues faced by finfish aquaculture as a whole. This means that lessons learned about environmental quality within the region may be applicable worldwide.

One of the main environmental factors associated with inshore aquaculture is the oxygen demand generated by respiration in water, sediment and fish. At both aquaculture and non-cultured sites in SWNB, the results from mass balance calculations [2] indicate that the majority of respiration is located in the water column. In addition, these results suggest that salmon aquaculture wastes may be distributed over wide areas, and that accurate estimates of the time and space scales of waste discharge are required to determine impacts of the discharge. Hargrave [3] has also suggested that if salmon respiration is greater than 20% of total oxygen demand at a particular cage site, caution should be exercised to ensure that dissolved oxygen (DO) does not become critically depleted in the water column.

There is an annual inshore and offshore cycle of ambient oxygen in SWNB surface waters [4], as high (saturated/supersaturated) DO in the summer gives way to lower (undersaturated) DO during late summer/early fall when respiration tends to dominate the surface water ecosystem [5]. Page et al. [4] have shown that caged fish can amplify this seasonal low in DO, but only during the second year of growth. During the first year of fish grow-out, DO is not reduced below a subsaturation threshold of 6.75 mg O_2 L^{-1}. During the second year, DO is reduced further (by larger fish and their higher respiration rates) to less than threshold. The potential for this amplification of a seasonally-low DO to be maintained over the length of time required to cause stress in caged fish then becomes dependant on ambient DO, current speed and their combined effect on the delivery of oxygen from outside of a cage.

Even though it is reasonable to assume that oxygen demand by respiration in the water column will be a prime regulator of the ambient DO available for optimal fish growth, the production of oxygen by the phytoplankton is equally important. In addition, the transition from oxygen production during summer months to oxygen demand in the late summer/early fall [5]

Ecosystem Indicators of Water Quality Part II. 85

and the association of this transition with a seasonal decrease in DO [4] are both important in any useful monitoring of oxygen as a water quality index.

2
Oxygen Production and Demand

A number of results have been reported for oxygen production by the phytoplankton and oxygen consumption by respiration in coastal ecosystems. Often the results are from the same embayment (e.g. Cloern's [6] and Smith and Hollibaugh's [7] data from San Francisco and Tomales Bays). Very rarely can the measurements be compared to each other because they have not been made in samples taken at the same time, location and depth. In situ measurements of gross (autotrophic) O_2 production and dark (heterotrophic) O_2 consumption by respiration have been compared in offshore waters [8, 9], but equivalent data from coastal sites are rare and often contradictory. For example, the Gulf of Gdansk coastal zone has been advanced as either a year-round autotrophic [10] or heterotrophic ecosystem [11]. This means that the interpretation of oxygen production and respiration measurements from SWNB or any coastal ecosystem requires a clear definition of how the measurements can be applied in tandem.

2.1
Measurement of Oxygen Production and Demand

Water samples for productivity, respiration and associated measurements (e.g. of chlorophyll, nutrients and bacterial number) were collected from SWNB surface waters as described by Harrison et al. [12]. In order to provide the information most relevant to salmon aquaculture in surface cages, coincident measurements of oxygen production (by phytoplankton productivity) and oxygen demand (by plankton respiration) were carried out using water taken from a depth of 1 m. The water samples were transported to a nearby shore facility and subdivided for productivity and respirometer incubations.

2.1.1
Productivity (P) Measurements

Productivity measurements were carried out using stable isotopes (^{13}C-bicarbonate, ^{15}N-nitrate and ^{15}N-ammonium) at tracer concentrations (0.2 mol m^{-3} for ^{13}C and 0.1 mmol m^{-3} for ^{15}N) and the tracer incubation protocol of Harrison et al. [12]. Duplicate water samples were dispensed into 500 ml polycarbonate bottles, inoculated with the tracers and typically incubated for 3 h under natural light conditions in attenuated incubator boxes. At the end of the incubations, particulate material was collected on precombusted (475 °C, 12 h) 0.4 μm glass-fiber filters (MFS Advantec GF75) and stored frozen for later mass spectrometer (Europa Scientific) analysis in the laboratory.

Nutrient utilization rates were calculated using standard equations [13]. No corrections were made for isotope dilution (due to biological production of ammonium in particular) during the incubation or for isotope fractionation when converting carbon or nitrogen fixed to oxygen production. In addition, given the focus of the work on surface waters associated with the deployment of salmon cages, the measured rates of oxygen production were not integrated with depth. The standard deviations (1 σ) between rates of production calculated from ^{13}C and ^{15}N-tracer incorporation into particulates in duplicate samples were less than 7.6%.

2.1.2
Respiration (R) Measurements

A pulsed oxygen electrode system (Endeco T.1125) adapted to a 16-electrode multiplexer was used to measure short-term respiration in unfiltered seawater [14]. Each respiration measurement included the combined oxygen consumption of the phytoplankton and microbial heterotrophs.

Unfiltered seawater from each 1 m sampling depth was added to a glass, 1-l, 4-electrode, water-jacketed respirometer and maintained at in situ temperature in the dark. Once the electrode signals had stabilized over an initial 50 min, oxygen consumption in each respirometer was monitored as decreasing signals obtained from 4 electrodes over an additional 60 min. In order to correct for electrode drift, the sample was replaced with ultrafiltered seawater (that had passed through a 1000 KDalton Millipore PrepScale cartridge) and the blank water allowed to reach in situ temperature over 15–30 min.. Electrode drift was then measured for a further 60 min and subtracted from the respiration measurement to obtain a final corrected value for respiration. The total incubation time required to obtain measurable linear decreases of electrode signals in unfiltered and blank water was 3 h or less.

Electrode signals were converted to oxygen concentration and respiration (the rate of oxygen decrease) using linear calibration curves obtained from respirometers containing aged, filter-(0.2 μm)-sterilized seawater sparged at in situ temperature with three different analyzed gas mixtures of 5%, 10% and 20% oxygen in nitrogen. The electrode calibrations were carried out in the same manner as the respiration measurements and the concentration of oxygen in the seawater in each of the three calibration respirom-

Ecosystem Indicators of Water Quality Part II. 87

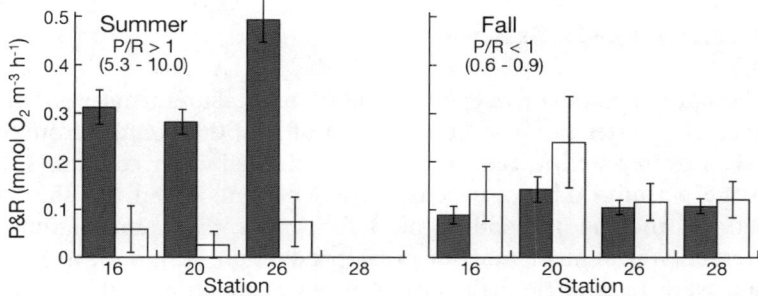

Fig. 1 Phytoplankton production (P, *gray columns*) and microbial community respiration (R, *white columns*) during the summer and fall in surface (1 m depth) water from four stations at approximately 59°N and 51°W in the Labrador Sea [16]. The values for P are means calculated from replicate ^{13}C and ^{15}N incubations on the basis of 1 mole oxygen evolved per mole carbon fixed. The values for R are the means calculated from four individual electrode signals in each respirometer. Note that P/R is greater than 1 when oxygen production predominates during the summer and less than 1 when net oxygen demand rules during the fall

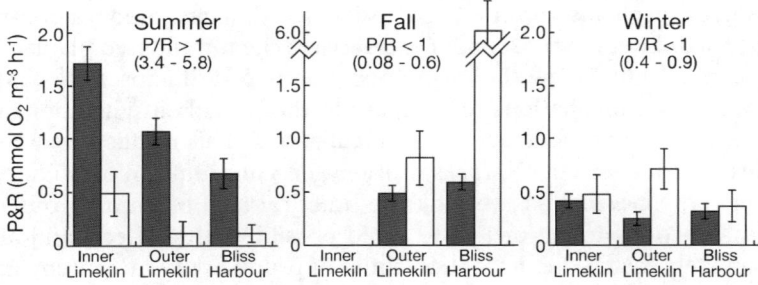

Fig. 2 Phytoplankton production (P, *gray columns*) and microbial community respiration (R, *white columns*) during the summer, fall and winter in surface waters at three coastal aquaculture sites in the southwestern New Brunswick region [5]. The values for P are means calculated from replicate ^{13}C and ^{15}N incubations and the values for R are the means calculated from four individual electrode signals in each respirometer. Note that P/R is greater than 1 when oxygen production rules in the summer and less than 1 when oxygen demand predominates during the fall and winter

eters was determined by Winkler titration. In order to account for variations produced by the differing response of electrodes to changing oxygen concentration, final respiration values (Figs. 1 and 2) were expressed as the means of four values generated by the four individual electrodes in each respirometer. The standard deviations (1 σ) between respirations determined from each 4-electrode array in a 1-l respirometer chamber were less than 6.8%.

2.1.3
Interpretation of P and R Measurements

Even though our productivity and respiration measurements were applied only to surface water samples from a depth of 1 m, the samples from SWNB were taken from a well-mixed, severely light-limited water column [12] subject to the periodic tidal resuspension of sediment. Given the short incubation times and the probability that light-limited tracer incubations yield measurements at or near rates of gross production [15], we assumed that our data were reasonable estimates of gross production, but not of gross photosynthesis.

We also assumed that the results obtained using the two incubation techniques (Figs. 1 and 2) were directly comparable because they were taken from the same water samples incubated for short times (< 3 h) with no discernible bottle effects. During previous applications of our respirometry techniques to coastal [5] and offshore [16, 17] surface waters, we found that short-term linear decreases of oxygen concentration in unfiltered seawater gave way to steeper non-linear decreases when dark incubations were extended beyond 2–3 h to times approaching 24 h. We have attributed the longer-term increases in oxygen consumption to bacterial growth in enclosed water samples because we did not observe similar increases in the ultrafiltered blanks. However, we could not follow the recommendations of Williams et al. [15] and carry out in situ incubations to compare to those made in vitro nor did we measure bacterial number during the incubations. This means that we could only describe longer-term increases in oxygen consumption as an apparent bottle effect. Nevertheless, we took the practical step of limiting our respiration incubations to 3 h or less to avoid possible bottle effects and also to approximately match the incubation times of productivity measurements.

Our results can be used to establish a preliminary relationship between P and R in the surface waters associated with salmon aquaculture, but they cannot be used to infer an overall relationship between P and R over a 24 h (diurnal) period. In order to achieve more complete (long-term) estimates of P and R, a diurnal correction (based on results obtained from back-to-back P and R incubations of surface water over a 24 h period) would need to be applied to the results. In addition, no attempt was made to estimate P and R integrated with depth. Given the large tidal excursions encountered in SWNB, integrated values would be of limited utility in the shallow coastal waters of the region. As a result, we restricted our P and R measurements to the surface waters most relevant to the environments associated with salmon pens.

Systematic offsets are introduced when oxygen production is calculated from productivity determined via isotope incubation. For example, in the Baltic (Gdansk) coastal zone, the seasonal variations in oxygen production and demand in surface waters are small [10, 11]. This means that only a 5% change in productivity (to account for isotope fractionation when isotope fixed is converted to oxygen evolution) results in the same ecosystem interpreted as year-round autotrophic [10] rather than heterotrophic [11]. The stronger seasonal variations observed in surface waters at more northerly latitudes in the Labrador Sea [16], eastern Scotian Shelf [17] and SWNB [5] would be less affected by systematic offsets in the calculation of oxygen production from isotope incubations. In particular, the overall pattern of seasonal transition from oxygen production in summer to sustained oxygen consumption in the late summer/early fall (Figs. 1 and 2) would remain unchanged.

2.2
Oxygen Production, Oxygen Demand and PNDB Systems

Some of the major biological processes contributing to oxygen demand and production in well-mixed coastal inlets are summarized as the phytoplankton-nutrients-detritus-bacteria (PNDB) system in Fig. 3. Within this simplified model, both the oxygen production associated with phytoplankton growth (in surface waters) and the demand associated primarily with bacterial growth (in water and sediment) are fuelled by nutrients. This is slightly different from some models of coastal eutrophication where nutrients are assumed only to supply phytoplankton growth.

The C:N ratio of total organic matter (TOM) in coastal marine systems is often far higher than the Redfield ratio of 6:1 encountered in particulate organic matter [14]. The C:N of TOM is high enough (20–30:1) to make it more worthwhile for the bacteria to obtain the nitrogen required for pro-

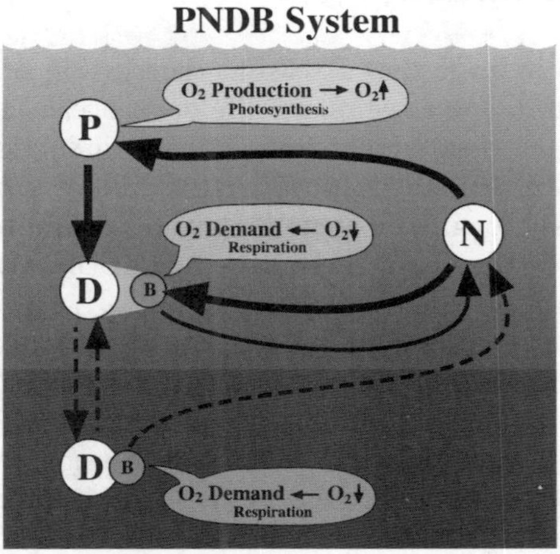

Fig. 3 Schematic of a phytoplankton-nutrient-detritus-bacteria (PNDB) ecosystem in coastal water and sediment. Note that nutrients supply the growth of bacteria and their oxygen demand as well as the growth and oxygen production of the phytoplankton

tein synthesis and growth from dissolved nutrients [18, 19]. This is especially the case in surface waters from SWNB, where high (non-limiting) concentrations of ammonium and nitrate are routinely encountered [12]. In addition, adaptive models of the dynamics of organic matter cycling in seawater [20] highlight the ability of bacteria to adapt their metabolism and take advantage of above-threshold concentrations of both organic carbon and nutrients in surface waters.

The control exerted by the bacteria (Fig. 3) on the decomposition of both dissolved and particulate components of organic detritus requires that nutrients be diverted from phytoplankton to bacterial growth. This means there can be a high oxygen demand associated with a eutrophic nutrient load that is not related to just the bacterial breakdown of particulate phytoplankton detritus. Respiratory oxygen demand can also be regulated by the bacterial utilization of nutrients and dissolved bioreactive detritus from a number of sources (Fig. 3).

At this point, it is important to remember that oxygen demand in a PNDB system cannot be examined in isolation from oxygen production (Fig. 3). The production of oxygen by light-driven phytoplankton communities will act to counterbalance demand, especially during the summer growing season. As a result, it should be possible to utilize measurements of production and demand to calculate production-respiration (P/R) ratios and locate when autotrophy gives way to heterotrophy in a surface-water ecosystem.

3
Production, Respiration and P/R

The data in Fig. 1 indicate that a seasonal succession of plankton physiology is in operation in some regions of the open ocean [16]. During a short and intense summer season of plankton growth, production (P) near the surface of the Labrador Sea is driven by phytoplankton growth. In the fall, P decreases as the day length shortens [21], and respiration (R) increases as bacteria break down the dissolved organic matter (DOM) accumulated over the summer growing season.

When production and respiration are expressed as P/R (Fig. 1), oxygen production rules in the summer and P/R is greater than 1. In the fall, with P/R less than 1, the consumption of oxygen predominates. This simple seasonal relationship between P and R is evident in waters associated with the Gully marine protected area on the Scotian Shelf [17] and is also evident in SWNB, where oxygen production during the summer gives way to a strong demand during the fall (Fig. 2) and a lower demand during the winter [5]. If P/R is utilized as an ecosystem-based index of water quality then, with an autotrophic system in place and a P/R of greater than 1, there is little chance of DO decreasing below a threshold for optimal fish growth. With P/R at less than 1, a heterotrophic system is in place and there is a far greater potential for DO to decrease below threshold.

Our results are very different from those of Blight et al. [22], who found that respiration in a temperate coastal ecosystem exceeded gross production for only short periods (of 1-2 weeks) at the end of spring blooms. They also pointed out that the coincidence of cell-specific respiration with phytoplankton blooms in this environment was "consistent with a very close coupling between bacterial respiration and phytoplankton production and biomass". The close coupling of bacterial respiration to phytoplankton production is probably not the case in SWNB. Instead, the bacteria in this particular PNDB ecosystem would be fuelled by an admixture of dissolved and particulate organic carbon derived from phytoplankton decomposition, the resuspension of sediment (including aquaculture waste) and river discharge [5]. This would result in the uncoupling of respiration from phytoplankton production and its linkage to other processes, such as the recycling of DOC from a number of allochthonous sources [23] with a range of reactivities.

To date, the transition from an autotrophic to heterotrophic ecosystem in SWNB has been defined on the basis of only a few seasonal snapshots of data (Fig. 2). Even though the snapshots can give an approximate idea of when autotrophy gives way to heterotrophy, a more-continuous monitoring of oxygen production versus demand is required. Ideally the monitoring would be based on in situ measurements over time scales ranging from tidal-to-diurnal-to-seasonal to provide the early warning required by salmon growers to time the grow-out of caged fish. The growers would then be able to avoid the periods of high oxygen consumption associated with DO at below the threshold for optimal fish growth [4].

A seasonal change of P/R from greater to less than 1 is an early warning of the onset of net oxygen demand; low DO is the cumulative effect. By careful monitoring of both P/R and DO, the onset and maintenance of oxygen demand can be tracked. This could then provide advance warning of, for example, the early (summer season) onset of heterotrophic conditions and sustained low DO fuelled by the widespread distribution of aquaculture wastes [2]. In this way, the definition of oxygen dynamics in terms of two simple indices (P/R and threshold DO) could be utilized as a monitoring tool by both industry and government for the management of water quality in the coastal zone.

4
Carbon Load, Phytoplankton and Bacteria

4.1
Measurement of Total, Particulate and Dissolved Organic Carbon

A 30-ml sub-sample of water from each Niskin cast was run into an acid-washed 125 ml fluorinated-polyethylene bottle (Nalgene). Ninety (90) µl of 40% phosphoric acid was mixed with a 10-ml sub-sample and 3.1 ml of the acidified (pH 2.5) seawater was dispensed into each of two acid-washed 4-ml amber glass vials. The vials were sealed with black phenolic caps equipped with red Teflon/white silicone septa inserts, and stored in a refrigerator at 4 °C until analyzed.

Analysis of the contents of the duplicate vials was performed using an MQ1001 total carbon analyzer (MQ Scientific, Ada, Oklahoma) equipped with a LICOR CO_2 detector and a computer running PeakSimple for Windows (version 1.83). The combustion column and auto-injection upgrades of Peterson et al. [24] were applied to the stock analyzer to improve analytical precision and increase column life under the salt loads generated by the high temperature combustion of repeated injections of seawater.

As part of the automated analysis sequence, the acidified sample was first sparged with ultra-pure oxygen for 5 minutes before each injection to remove dissolved inorganic carbon and a sequence of three injections was drawn from each vial. The area under each CO_2 peak generated by the combustion of 6 individual injections was determined with the PeakSimple software; the concentration of total organic carbon (TOC) was calculated from mean peak area and standard curves generated by the analysis of known standards at the beginning and end of a sample run. To ensure uniform instrument function and results that could be compared to those from other laboratories, quality control samples in the form of Certified Reference Material (obtained from W. Chen and D.A. Hansell at the Rosenstiel School of Marine and Atmospheric Science, Miami Florida USA) were also included with known standards at the beginning and end of each sample run. Typical analytical precisions (1 σ) were < 2%.

Particulate organic carbon (POC) was analyzed by the high temperature combustion [25] of material collected on pre-combusted (475 °C, 12 h) 0.4 µm (MFS Advantec GF75) glass-fiber filters. Dissolved organic carbon (DOC) was determined by either subtracting POC from TOC or by prefiltering a sample through a pre-combusted (475 °C, 12 h) 0.4 µm glass-fiber filter and collecting the filtrate to run through the MQ1001 analysis sequence.

4.2
Bioreactivity of the Coastal Carbon Load

As with nearly all marine ecosystems, the majority of the carbon load in SWNB waters is carried as dissolved organic carbon (DOC). In Fig. 4, DOC is about 90% of the total organic carbon (TOC) at two inshore aquaculture sites; the remaining 10% is particulate organic carbon (POC).

The bioreactivity of DOC in seawater is not easy to define because most of the carbon is regarded as semi-labile or refractory [26] and cannot be directly utilized by the bacteria. There are also wide variations in the organic carbon load carried by surface waters in SWNB. The concentration of DOC in surface water can be as high as 6800 μM C and as low as 100 μM C during the summer [5]. In addition, the concentration of organic carbon cannot be related in any simple way to oxygen demand. Instead, carbon load is probably regulated by a complex interaction between the sedimentation of phytoplankton detritus, the tidal resuspension of sediment (including aquaculture waste) and the supply of terrestrial carbon by coloured-water rivers rich in dissolved humic material [5].

Given the complex relationship that exists between carbon load and oxygen dynamics, there may be simpler ways of monitoring the main elements of the ecosystem responsible for ambient oxygen. Chlorophyll (an indicator of the phytoplankton biomass responsible for oxygen production) and bacterial number (an indicator of the microbial biomass responsible for oxygen demand) are two of the simplest variables that can be measured with con-

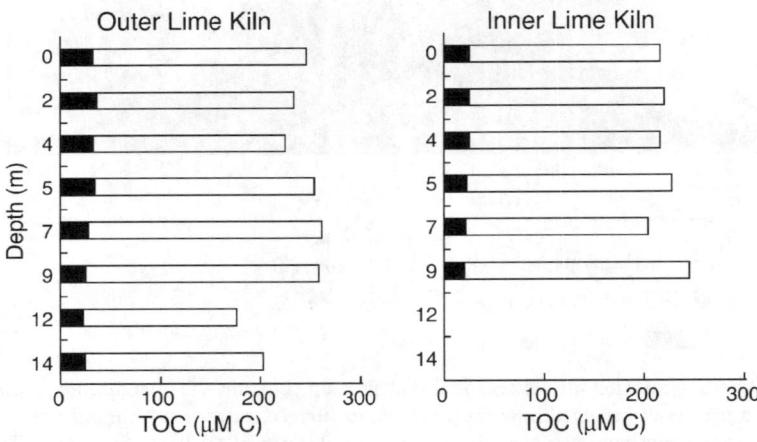

Fig. 4 Distribution of total organic carbon (TOC) load between particulate organic carbon (*black bar sections*) and dissolved organic carbon (*white bar sections*) in samples taken during September 2002 at two coastal aquaculture sites in Lime Kiln Bay, southwestern New Brunswick [5]

fidence. As Harrison et al. [12] have pointed out, the combination of these two variables into a bacteria/chlorophyll ratio may make them the best candidates for defining the overall response of an ecosystem moving seasonally from oxygen production to demand.

5
Photochemistry - A Seasonal Link in the Coastal Carbon Cycle?

A distinct seasonal succession of carbon inputs and outputs (Fig. 5) is characteristic of the Atlantic Canada (including the Bay of Fundy) coastal zone [27].

The input of terrestrial carbon by river discharge peaks as freshwater runoff reaches a maximum in spring and fall. The input of marine carbon by photosynthesis peaks in June and July when longest day length and maximum solar intensity are evident [12]. As a light-driven process, the consumption (output) of carbon by photo-oxidation follows a similar pattern and again peaks during June and July [27].

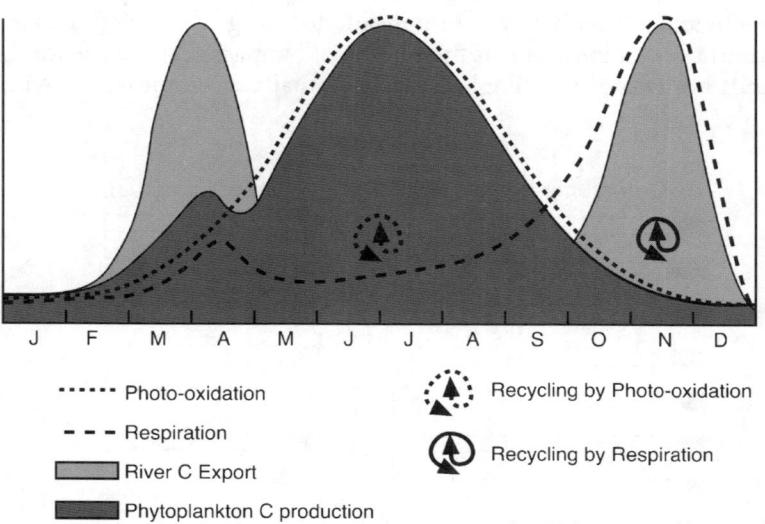

Fig. 5 Schematic of the annual seasonal distribution (but not the absolute magnitude) of the main processes involved in carbon cycling in surface waters from the Atlantic Canada (including the southwestern New Brunswick) coastal zone [27]. Note that phytoplankton production and photo-oxidation peak in association with maximum day length in the summer. Respiration peaks in the fall, associated with a maximum in river (terrestrial) carbon export and the seasonally-delayed accumulation of marine organic carbon in late summer/early fall

As day length and solar intensity decrease in the late summer/early fall (Fig. 5), respiration becomes the primary carbon-consuming component of the carbon cycle. The onset of respiration is stimulated by the seasonally delayed accumulation of relatively large amounts of dissolved organic carbon (DOC) in late summer [26, 28, 29]. This seasonal accumulation can be explained by simply assuming that DOC is not reactive [30]. However, a good deal of evidence on the photo- and bioreactivity of DOC argues against a simple explanation. For example, diurnal photochemical degradation will generate a complex mixture of more and less bioreactive carbon [31, 32], and possibly enhance the recycling and consumption of carbon by respiration. In this way, photochemical degradation may act as a seasonal link, preparing the organic carbon accumulated during the summer for degradation in the fall and (to a lesser extent) in the winter.

The idea of photochemistry as a connecting link in the seasonal transition from oxygen production to demand (Figs. 1 and 2) may be most important when river discharge peaks in the fall (Fig. 5). River carbon in the Atlantic Canada coastal zone is often discharged as coloured organic material [33] – an additional source of large amounts of terrestrial carbon that can be photo-decomposed to material suitable for respiration, even at the lower light levels encountered later in the year [12]. As a result, knowledge of the photo-oxidation in operation during the shoulder season between summer and fall in SWNB would be useful backup information when P/R is used as an early warning of low cumulative DO.

6
Summary and Conclusions

The individual importance of phytoplankton production (P) or respiration (R) in the coastal zone has been the focus of a number of individual reviews, e.g. by Cloern [6] and Smith and Hollibaugh [7]. Yet the combined importance of P and R has often been overlooked. Given the strong seasonal cycle of oxygen production and demand associated with P and R in surface waters from SWNB (Fig. 2), this knowledge gap should be addressed. The following conclusions are based on preliminary data collected to fill the gap:

1. Seasonal transitions from production to oxygen demand in surface waters can be expressed in terms of the production-respiration (P/R) ratio (Figs. 1 and 2). With P/R greater than 1 and an autotrophic system in place during the summer, there is a relatively small chance of dissolved oxygen (DO) decreasing below the threshold for optimal fish growth. With P/R at less than 1 and a heterotrophic system in place during the fall [5], DO can decrease below threshold [4].

2. Carbon load is variable in terms of source and lability over time [5] and is not a useful measure of the bioreactive material available to create oxygen demand. Instead measurements of chlorophyll, bacterial number and the bacteria/chlorophyll ratio may be better indicators of ecosystem response [12] and its regulation of oxygen dynamics.
3. Photochemical degradation may act as a seasonal link (Fig. 5), inducing a sustained oxygen demand by preparing marine and terrestrial organic carbon for the respiration of degradation products in the fall.
4. The seasonal change of P/R from greater to less than 1 is an early warning of the onset of net oxygen demand; low DO is the cumulative effect. By careful monitoring of both P/R and DO, the onset and maintenance of oxygen demand can be tracked. This would then provide advance warning of the early onset of heterotrophic conditions and the sustained low DO associated with less than optimal fish growth.

Even though these four conclusions suggest that oxygen responds in a reasonably straightforward manner to seasonal changes in surface waters, the question still remains – Can the monitoring of oxygen in the water column be used as a coastal management tool? This question is especially important given that phytoplankton are severely light-limited in the SWNB ecosystem [12] and that the long-term development of the system is biased towards an increase in the utilization of nutrients and oxygen by the heterotrophs. When applied together, P/R (the early warning of oxygen depletion) and DO (the cumulative result of this depletion) could indeed become useful tools for designing ecosystem-sensitive plans to monitor oxygen demand as a first index of water quality at aquaculture sites.

Acknowledgements We would like to thank Hugh Akagi, Alex Hanke, Murielle LeGresley, Randy Losier, Jennifer Martin, Fred Page, Michelle Ringuette, Dave Wildish and the New Brunswick Salmon Growers Association for assistance, ship time and the shore-based facilities to carry out field work in southwestern New Brunswick. Staff in the Biological Oceanography Section of the Ocean Sciences Division at Bedford Institute of Oceanography provided analytical support, and funding was provided by Fisheries and Oceans' Environmental Science Strategic Research Fund (ESSRF).

References

1. Chang B (2003) Fisheries and Oceans Can Tech Rep Fish Aquat Sci 2489:3
2. Strain PM, Hargrave BT (2005) Salmon aquaculture, nutrient fluxes and ecosystem processes in southwestern New Brunswick (in this volume). Springer, Berlin Heidelberg New York
3. Hargrave BT (2003) Fisheries and Oceans Canada Tech Rep Fish Aquat Sci 2489:30
4. Page FH, Losier R, McCurdy P, Greenberg D, Chaffey J, Chang B (2005) Dissolved oxygen and salmon cage culture in the southwestern New Brunswick portion of the Bay of Fundy (in this volume). Springer, Berlin Heidelberg New York

5. Kepkay PE, Harrison WG, Bugden JBC, Perry T (2002) Fisheries and Oceans Canada Tech Rep Fish Aquat Sci 2411:51
6. Cloern JE (2001) Mar Ecol Prog Ser 210:223
7. Smith SV, Hollibaugh JT (1993) Rev Geophys 31:75
8. Williams PJLeB, Purdie DA (1991) Deep-Sea Res 38:891
9. Langdon C, Marra J, Knudsen C (1995) J Geophys Res 100:6645
10. York JK, Witek Z, Labudda S, Ochicki S (2001) Oceanologia 43:365
11. Witek Z, Ochocki S, Maciejowska M, Pastuszak M, Nakonieczny J, Podgorska B, Kownacka JM, Mackiewicz T, Wrzesinska-Kwiecien M (1997) Mar Ecol Prog Ser 148:169
12. Harrison WG, Perry T, Li WKW (2005) Ecosystem indicators of water quality part I. Plankton biomass, primary production and nutrient demand (in this volume). Springer, Berlin Heidelberg New York
13. Dugdale RC, Wilkerson FP (1986) Limnol Oceanogr 31:673
14. Kepkay PE, Jellett JF, Niven SE (1997) Mar Ecol Prog Ser 150:249
15. Williams PJLeB, Morris PJ, Karl DM (2004) Deep-Sea Res I 51:1563
16. Kepkay PE, Bugden JBC (2002) Fisheries and Oceans Can Tech Rep Fish Aquat Sci 2403:80
17. Kepkay PE, Harrison WG, Bugden JBC, Porter CJ (2002) Fisheries and Oceans Can Tech Rep Fish Aquat Sci 2377:61
18. Goldman JC, Dennett MR (1991) Mar Biol 109:369
19. Kirchman DL (2000) In: Kirchman DL (ed) Microbial Ecology of the Oceans. Wiley-Liss, New York, p 261
20. Pahlow M, Vezina AF (2003) J Mar Res 61:127
21. Li WKW, Harrison WG, Head EJH, Kepkay PE, Platt T, Vezina AF (2002) Fisheries and Oceans Can Tech Rep Fish Aquat Sci 2403:85
22. Blight SP, Bentley TL, Lefevre D, Robinson C, Rodrigues R, Rowlands J, Williams PJLeB (1995) Mar Ecol Prog Series 128:61
23. Hopkinson CS (1985) Mar Biol 87:19
24. Peterson ML, Lang SQ, Aufdenkampe AK, Hedges JI (2003) Mar Chem 81:89
25. Sharp JH (1974) Limnol Oceanogr 19:984
26. Carlson CS, Ducklow HW, Michaels AF (1994) Nature 371:405
27. Kepkay PE (2001) Fisheries and Oceans Can Sci Advisory Secretariat 2001/027:37
28. Copin-Montegut G, Avril B (1993) Deep Sea Res 40:1963
29. Williams PJLeB (1995) Mar Chem 51:17
30. Legendre L, LeFevre J (1995) Aquat Microb Ecol 9:69
31. Kaiser E, Herndl GJ (1997) Appl Environ Microbiol 63:4026
32. Kepkay PE (2000) In: Wangersky PJ (ed) Marine Chemistry. The Handbook of Environmental Chemistry, Vol 5. Springer, Berlin, p 35
33. Clair TA, Ehrman JM (1996) Limnol Oceanogr 41:921

Hdb Env Chem Vol. 5, Part M (2005): 99–113
DOI 10.1007/b136006
© Springer-Verlag Berlin Heidelberg 2005
Published online: 15 July 2005

Organic Enrichment at Cold Water Aquaculture Sites— the Case of Coastal Newfoundland

M. Robin Anderson[1] (✉) · Michael F. Tlusty[2] · Vern A. Pepper[1]

[1]Science Branch, Fisheries and Oceans Canada, PO Box 5667, St. John's, NL A1C 5X1, Canada
andersonro@dfo-mpo.gc.ca, pepperv@dfo-mpo.gc.ca

[2]New England Aquarium, Central Wharf, Boston, MA 02110, USA
mtlusty@neaq.org

1	Introduction	100
2	Aquaculture in Newfoundland	101
3	Hydrography of the Newfoundland Coast and Its Influence on Organic Matter Deposition	102
4	Quantification of Organic Matter Deposition in Newfoundland— Using Redox Potential and Total Sulfides as Indicators	104
4.1	Study Sites	104
4.2	Materials and Methods	105
4.3	Sediment Characteristics in Newfoundland Coastal Waters	105
4.4	Sediment Redox—Sulfide Relationships in Newfoundland Coastal Waters	106
5	Sulfate Reduction in Cold Sediments	108
6	Conclusion— Monitoring Organic Enrichment at Cold Water Aquaculture Sites	111
References		112

Abstract Benthic organic matter (OM) enrichment is a frequent environmental effect of coastal aquaculture. There is a need for simple, general methods of predicting and monitoring such effects. One approach uses a geochemical relationship between total sulfides and redox potential to define organic enrichment. This empirical relationship developed for shallow, macrotidal aquaculture sites in Southwest New Brunswick (SWNB), Canada, is used to determine organic matter (OM) loading and to monitor shifts in benthic quality. A similar relationship is seen for salmonid farm sites in coastal Newfoundland but not for sediments under mussel farms or at nearby reference sites. Overall, redox potentials and total sulfides are not correlated and 70% of observations fall below the relationship documented for SWNB. Sediments in Newfoundland coastal waters are often rich in OM, [1.5 to > 30%, median 8% loss on ignition (LOI)]. In addition, they are seasonally ($< 0\,°C$) or always cold (-1.8 to $< 5\,°C$). Much of the coast is exposed to high-energy conditions often with an effective fetch exceeding 700 km. Inner basins of some bays and fjords are protected from the waves associated with such exposure and may naturally experience seasonal anoxia that will significantly influence sediment-water exchange processes. All these factors will moderate sedimentation, remobilization and eventual remineralization of organic matter. The application of simple geochemical indices of organic enrichment as monitoring and assessment tools must be tempered with understanding of the environmental processes that regulate these factors under diverse geomorphological conditions.

Keywords Environmental effects of aquaculture · Organic enrichment · Redox potentials · Total sulfides · Temperature

Abbreviations
Eh   redox potential
HDPE   high density polyethylene
LOI   loss on ignition
NHE   normal hydrogen electrode
OM   organic matter
PBS   performance-based standards
SRR   sulfate reduction rate
SWNB   South West New Brunswick, Canada

1
Introduction

This chapter presents a case study to examine variables used as indicators of sediment organic enrichment with respect to the development of finfish and shellfish aquaculture in coastal areas of Newfoundland. Regulators within Fisheries and Oceans Canada are proposing the use of performance-based standards (PBS) to assist in both preliminary site assessments and ongoing site monitoring of aquaculture leases [1]. A monitoring method, based on an empirical relationship between total sulfides and redox potential [2, 3], has been proposed and tested as a method for monitoring salmon aquaculture sites in SWNB. In near-shore, depositional environments, increased organic loading with its effects on benthic organisms and chemistry has been shown to be closely related to a decrease in redox potentials (Eh), and increases in total sulfides and organic matter in surface sediments [4]. These easily measured variables are correlated and can be used independently to determine if organic loading from a farm is causing sediments to become anoxic and thus impact benthic community structure and functioning.

We undertook a study of the utility of measurements of sediment organic matter, redox potential and total sulfides as a means of monitoring surface sediments at finfish and mussel aquaculture sites in Newfoundland coastal waters. In this chapter, we relate our findings to those of Wildish et al. [2] and discuss applicability of these methods to environmental assessment and monitoring of aquaculture sites in Newfoundland.

2
Aquaculture in Newfoundland

Although small compared to other regions of Canada, marine aquaculture in coastal Newfoundland has grown significantly since its experimental beginnings in the early 1980s. Commercial production ($MT\,yr^{-1}$) of blue mussels (1700), steelhead trout (1600), Atlantic salmon (1270) and Atlantic cod (227) had a total value of 21 330 000 $CDN in 2002 [5].

Cage culture of salmonids started in the late 1980s in Bay d'Espoir, a large complex fjord on the south coast of the Island, and is now expanding into other south coast bays and fjords (Fig. 1). Long-line mussel aquaculture was originally sited on the Northeast coast in Notre Dame Bay and has since spread along the Northeast coast and to some sites on the west and south

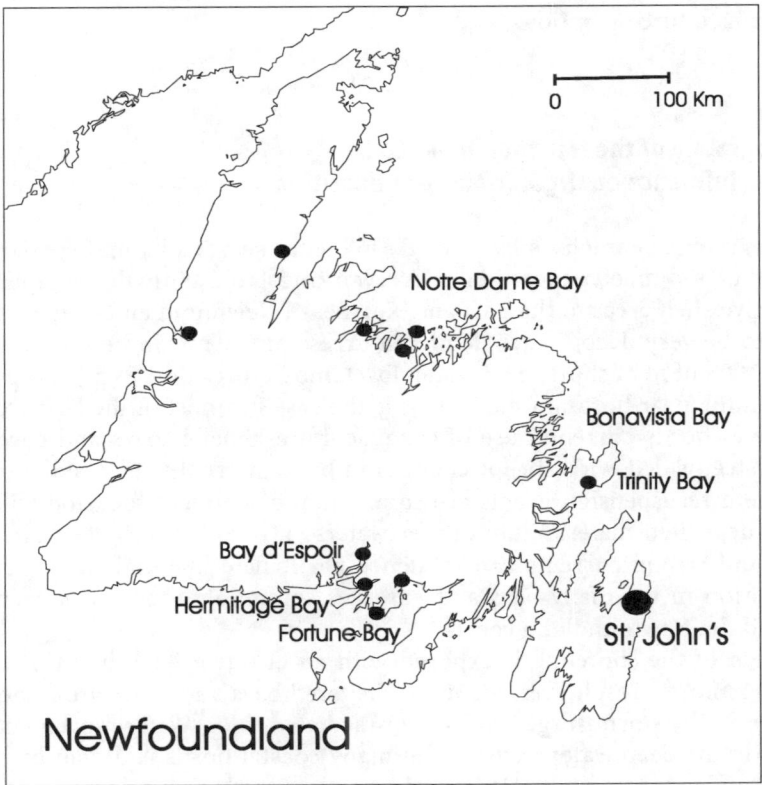

Fig. 1 Map of the Island of Newfoundland showing the location of sediment sample sites (•). Each site consists of a finfish or mussel lease and one or more nearby reference sites

coasts. Cod aquaculture started primarily as a grow-out operation of wild-caught fish and is now shifting to hatchery-raised stock.

Throughout the history of development of the aquaculture industry in Newfoundland, growers and regulators have recognized the importance of understanding the environmental consequences of aquaculture operations on Newfoundland coastal waters and of avoiding or minimizing deleterious environmental impacts [6]. Research has been ongoing to determine how farms affect their environment and the environmental limits for sustainable aquaculture production. A study of the potential assimilative capacity of Bay d'Espoir for salmonid aquaculture [7] concluded that there was little or no evidence for measurable effects of cages on the water column [8]. However, localized organic enrichment of the benthos was observed close to the cages at some sites [8, 9]. The focus of more recent work has been on the potential effects of aquaculture operations on the benthic environment. In addition, the inclusion of aquaculture as an "undertaking" under the Canadian Environmental Assessment Act has reinforced the need for scientifically defensible tools to predict and measure environmental effects of both finfish and shellfish aquaculture operations.

3
Hydrography of the Newfoundland Coast and Its Influence on Organic Matter Deposition

Hydrographic characteristics strongly influence seasonality and spatial variability of sediment deposition and reworking [10]. With the exception of extensive shelf areas to the East and Northeast, Newfoundland coastal waters tend to be very deep. Water depths in excess of 500 m can be found within a few 100's of m of shore. As a result, low temperatures ($-1.8-5\,°C$) are often seen in the near bottom waters. This is the case for most of the large coastal bays and fjords currently used for aquaculture. Smaller bays and coves are often steep sided with abrupt changes in bathymetry. In steep sided fjords, sediment resuspension events are common and sediment focusing will also move deposited material into deeper waters [11–15]. Shoal waters are common and strong currents can influence particulate material transport and deposition in the many bays and passages commonly used for aquaculture around the Newfoundland coast.

Much of the shoreline is exposed with an effective fetch often in excess of 700 km. As a result even the more sheltered bays and coves are frequently influenced by storm surges and storm-mediated tides. Water movement particularly the deep water exchange in many coastal fjords is driven by storm events and/or spring tides [16]. In addition to directly influencing mixing and upwelling, strong winds and tides can also combine to modify estuarine circulation. For example at one cage holding site near the head of Bay d'Espoir, salinity can vary by as much as 20 ppt over a 24-hour period [8].

Water column structure is strongly influenced by the hydrographic characteristics of the coast. In summer many of the deeper coastal bays are stratified with a two or three-layered system (Fig. 2). Bay d'Espoir for example, typically has three layers in the water column of the main stem of the Bay [17]. As a result of inputs from a large hydroelectric project in its head waters, surface waters are brackish and temperature varies between 0 to ~ 20 °C. Surface waters freeze in the winter [17, 18]. Mid-depth waters are typical of the Gulf of St. Lawrence with seasonally variable temperatures ranging from − 1.8 to ~ 17 °C. The deepest waters are from the deep water in the Hermitage Channel and remain relatively constant at ~ 5 °C year round. Bottom water exchange occurs at infrequent intervals primarily during winter storm events and/or spring tides [16, 17]. Shallower bays (< 20 m) may stratify for shorter periods but are often mixed to the bottom. Interannual and interdecadal variation in deepwater exchange has also been observed in coastal fjords [12, 13]

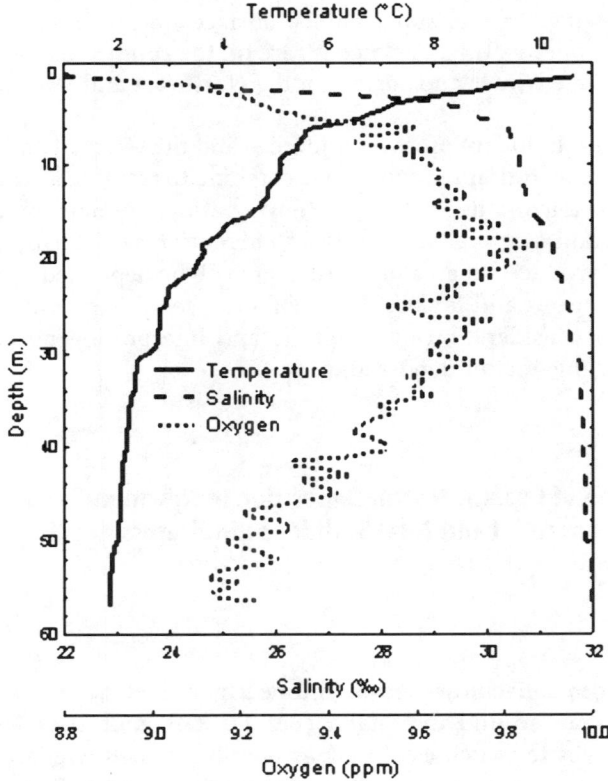

Fig. 2 Water column profile for mid Bay d'Espoir

and may be related to long-term climatic variability [19]. Such processes can significantly influence the oxygen content of the underlying sediments resulting in large shifts in benthic fauna in response to oxygen depletion [12, 13].

Bottom substrates in Newfoundland coastal waters are dominated by rock and cobble with depositional areas in protected pockets determined by depth and bathymetry. The long effective fetch results in frequent large waves particularly in winter. There is significant reworking of mobile sediments and little accumulation of fine-grained deposits. Winter storms can resuspend fine material at depths of up to 100 m and even cobble can be reworked to depths > 20 m in shallow areas [20]. As a result, even if fine particulate material is deposited near a farm site during the summer it will often be transported away to long-term depositional areas in winter months.

Winter ice conditions also affect the local environment by limiting the direct effect of wind on water column mixing and temperature. Sheltered estuarine environments such as Bay d'Espoir and many of the smaller protected bays and coves are characterized by the presence of land fast ice during the coldest winter months. This characteristic is used to advantage by the salmonid growers in Bay d'Espoir where cages frozen into the ice are protected from winter kill in super chilled surface waters and protected from crushing due to ice rafting. Pack ice is rare on the south coast but commonly occurs in the late winter and spring on the Northeast and West coasts of the Island.

The constant influence of waves and tides and the effect of bathymetry due to site-specific hydrographic conditions combine to result in a strong seasonal pattern of erosion and deposition at many locations suitable for aquaculture. In addition, non-seasonal variability in currents may determine where organic matter produced at an aquaculture site will be deposited. Thus local hydrographic features and temporal variability of sedimentation regimes must be taken into consideration in designing and interpreting monitoring programs for organic matter deposition.

4
Quantification of Organic Matter Deposition in Newfoundland—Using Redox Potential and Total Sulfides as Indicators

4.1
Study Sites

Sediments from aquaculture leases and nearby reference sites were sampled by Eckman grab around the Island (Fig. 1). Reference sites were selected for all farm sites to match as closely as possible farm site characteristics of bathymetry, exposure and circulation. Salmonid farms in Bay d'Espoir and more recently, Fortune and Hermitage Bays were sampled (Fig. 1). Farm samples were collected from around the outside of the salmonid cages. Most mussel farm sites sampled were located on the Northeast coast. Farm samples were taken under the mussel lines. Data included in this analysis came from samples collected in 1997–98 (finfish—Bay d'Espoir), 1999–2002 (finfish—Bay d'Espoir and mussels around the Island) and 2003 (finfish—Bay d'Espoir, Hermitage Bay, Fortune Bay).

4.2
Materials and Methods

Surface sediment samples were collected in triplicate using a (15 × 15 × 22.5 cm) Eckman grab. Sample processing followed the method of Wildish et al. [2, 3]. Samples were rejected if the sediment water interface was not intact and covered with water. Redox potentials were measured directly in the grab using an Orion platinum electrode inserted to a depth of 2–3 cm. The probe was calibrated using Zobell's solution, prior to each field trip. The reading in mV was corrected to normal hydrogen electrode (Eh_{NHE}) potential using the temperature-corrected values supplied by the manufacturer [21]. Temperature was measured directly in the Eckman grab using a thermometer or the thermistor associated with the redox probe. Two 50-ml surface (to a depth of 2–3 cm) sediment samples were collected from the Eckman grab using a cutoff 60-ml syringe. These samples were placed in individually labelled 150-ml high-density polyethylene (HDPE) bottles and stored on blue ice for transport to the field laboratory.

One sample was analyzed for total sulfide content within 6 hours of collection using an Orion silver/sulfide probe [2]. The probe was calibrated daily using standards between 1 and 100 000 μM. Readings were recorded in mV and converted to μM using the appropriate calibration curve. The calibration was not linear below 10 μM therefore, values below this were rejected. The other sample was frozen and returned to the laboratory for organic matter analysis. Organic matter content was determined as loss on ignition (LOI) by ashing at 500 °C for 6 hours [22].

4.3
Sediment Characteristics in Newfoundland Coastal Waters

Depositional areas of the Newfoundland coast are characterized by fine organic-rich muds. Sediments under finfish aquaculture farms are of a noticeably different biogeochemical signature than naturally occurring sediments. They tend to be flocculent with a higher water content [9, 23], and a higher organic matter content [24, 25] than naturally occurring sediments of a similar grain size in the same area. Organic matter content from pristine sites around the Island ranged from 2 to 35% with a median of 8% (Table 1). At

Table 1 Characteristics of coastal Newfoundland sediments found at aquaculture sites or nearby reference sites. Shown is the median, range and sample size (in parentheses) for mussel farms, finfish farms and nearby reference sites

	Reference sites	Mussel farms	Finfish farms
% Organic matter	8	30	12
Range	2–34	13–36	2–23
(n)	(47)	(42)	(19)
Temperature (°C)	− 1.7–17.8	− 1.7–12	2.5–10
(n)	(6)	(46)	(40)
Redox potential (Eh)	− 35	− 109	61
Range	− 175–227	− 232–77	− 170–303
(n)	(66)	(54)	(40)
Total sulfides (µM)	705	276	4914
Range	42–10913	26–1367	< 10–16784
(n)	(63)	(52)	(27)

mussel farms situated in relatively sheltered environments, organic matter ranged from 13 to 36% while at finfish farms which tend to be situated in slightly more dispersive environments, organic matter ranged from 2 to 23%. The ranges of redox and sulfides observed at mussel and finfish farms overlapped those found at pristine sites (Table 1). Redox potential varied from anoxic (− 232) to oxic (303) and total sulfides from below detection (< 10 µM) to 16.8 mM. Sediment temperature with a range between − 1.7 and 17 °C (Table 1) varied as a function of season, depth and water circulation patterns.

4.4
Sediment Redox—Sulfide Relationships in Newfoundland Coastal Waters

Overall there was no significant relationship between sediment redox potentials and total sulfide levels for sediments from coastal Newfoundland waters (Fig. 4, $r^2 = 0.004$, $P = 0.439$, $n = 139$). Similarly, there was no significant relationship for mussel farm sites ($r^2 = 0.013$, $P = 0.426$, $n = 52$) or pristine sites when analyzed separately. Total sulfides were negatively related to redox potential at salmonid farm sites ($r^2 = 0.173$, $P = 0.039$, $n = 25$). In general, sediments from both the salmonid farms and their nearby reference sites tended to have significantly higher redox potentials than did mussel farm sites ($t = − 8.505$, $P < 0.001$) while total sulfides did not differ significantly ($t = − 1.959$, $P = 0.061$), they did show a greater range (Table 1) at the salmonid sites.

At finfish aquaculture sites, sediment redox potentials were generally lower and sulfide content higher than at nearby reference sites (Fig. 3). This sug-

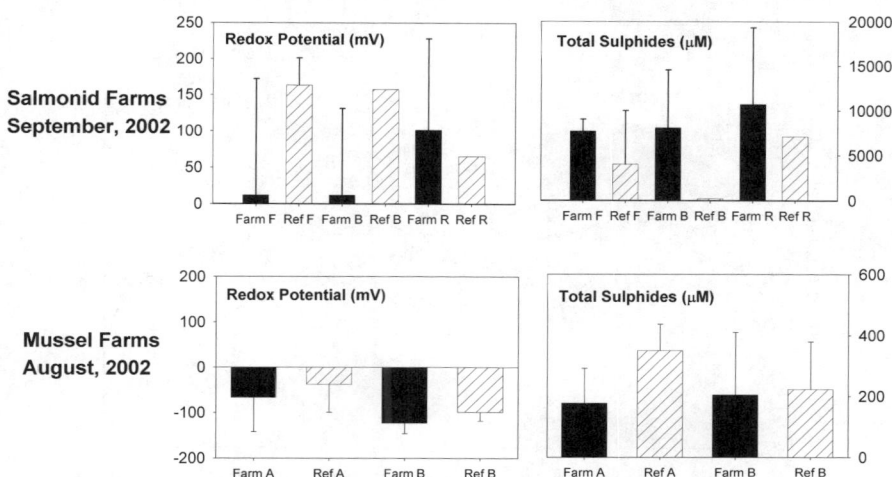

Fig. 3 Sediment redox potentials and total sulfides at Newfoundland aquaculture sites compared to nearby reference sites. For mussel farms (*top panels*) redox potentials and total sulfides differed between farms but not between each farm and its reference sites. Sediments at salmonid farms generally had lower redox and higher sulfides than those found at nearby reference sites. Data are for a mussel farm in Notre Dame Bay and salmonid farms in Bay d'Espoir and Fortune Bay. *Error bars* are standard deviations

gests that redox potential and total sulfide levels can be used to monitor the response of sediments under fish cages to increases in organic matter loading. In contrast, sediments under mussel farms did not have significantly different redox potentials or total sulfide levels when compared to nearby reference sites (Fig. 3). At these sites, organic matter is naturally elevated (average 11.6%) and the presence of a mussel farm does not significantly alter either the organic matter content, redox potentials or total sulfides in the sediments.

The relationship between total sulfides and redox potentials for Newfoundland salmonid aquaculture sites is similar to that of the Bay of Fundy (SWNB) [2] while mussel farms and pristine sites tend to fall below the relationship. Overall, 70% of the Newfoundland observations are below the regression line describing an inverse relationship between Eh and total sulfides for SWNB data (Fig. 4). This means that a majority of the sites have lower sulfide levels than similar sites in the Bay of Fundy. These differences may relate to the type of site selected for the two types of aquaculture and result from the combined influence of low temperature and quantity and quality of organic matter on the processes that generate sulfides [26]. Thus for Newfoundland, although site specific comparisons with appropriate reference sites do show significant differences for finfish farms (Fig. 3) the wide variability in sediment characteristics, in particular organic matter content (Table 1), mask any general trends among farm types and reference sites (Fig. 4). If however, sed-

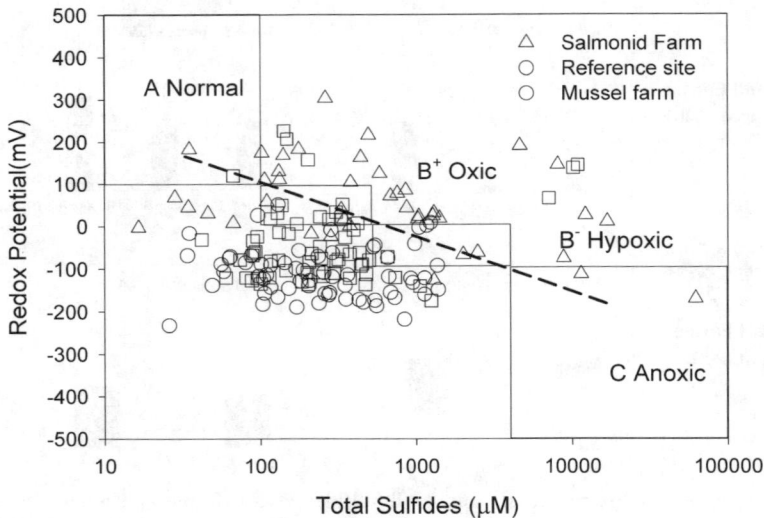

Fig. 4 Redox potentials and total sulfides at aquaculture sites and nearby reference sites around the Island of Newfoundland compared to the regression line for similar data from the Bay of Fundy, NB and Tracadie Bay, PEI (Eh = 473 − 66 ln($S^=$), [2]). *Boxes* represent the site classification categories based on the ecological responses to geochemical conditions related to organic matter loading, as applied to aquaculture sites in New Brunswick [2]

iment organic matter is included as an independent variable with sulfide in a step-wise linear regression, the resulting model explains more than 70% of the variation in redox potential (adjusted r^2 = 0.704, n = 54, P < 0.001). This suggests that natural organic matter content should also be taken into consideration in the development of monitoring and assessment methodologies for aquaculture sites in coastal Newfoundland.

5
Sulfate Reduction in Cold Sediments

Sulfate reduction is the major pathway of organic matter breakdown in most coastal sediments [27, 28]. The rate and magnitude of sulfate reduction are influenced by the quantity and quality of organic matter in the sediments and by temperature (Fig. 5). For a given temperature and substrate type, sulfate reduction rates (SRR) appear to be a concave (non-linear) function of sediment organic matter content with highest values found at intermediate organic matter levels. SRR increased significantly in sediments near fish farms amended with varying quantities of fish feed [28, 29]. The greatest increases in SRR were observed with low and moderate additions. High fish

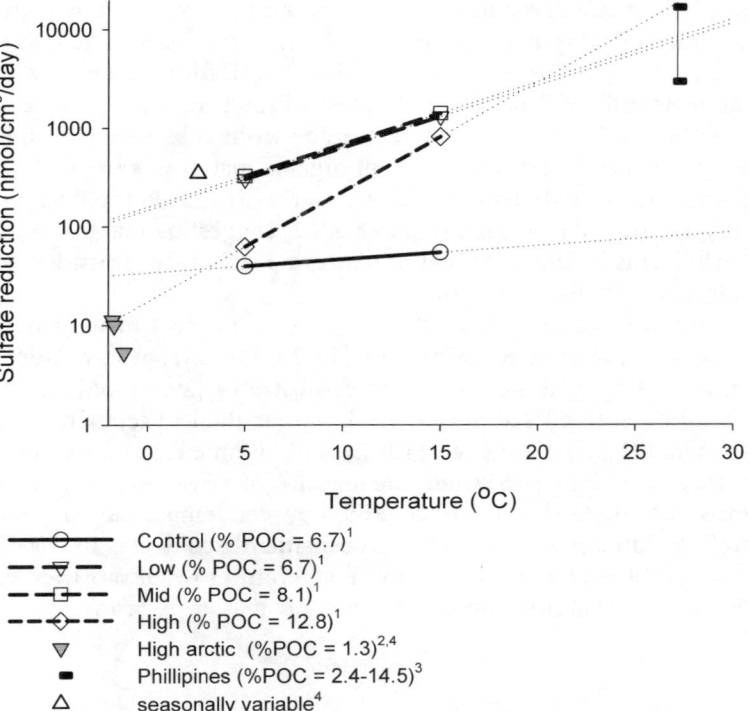

Fig. 5 Sulfate reduction rates as a function of sediment temperature and organic matter content. Data summarized from: [1]Holmer and Kristensen, 1994 [29], [2]Rysgaard et al., 1998 [36], [3]Holmer et al., 2003 [37], [4]Knoblauch et al., 1999 [27]

feed additions (doubling sediment POC) resulted in lower SRR and this effect was greatest at 5 °C [29]. Holmer and Kristensen [30] suggest that SRR in organic rich sediments can be inhibited by metabolite build-up or shifts in sediment pH as discussed in Chapter 9 [28].

SRR is also influenced by the quality of organic substrates. Coastal sediments receive organic matter inputs from both allochthonous and autochthonous sources. Terrigenous organic matter input tends to be refractory as does some of the autochthonous input since it has already been subject to microbial degradation [31–33]. Refractory organic matter is less amenable to bacterial decomposition and thus in coastal environments naturally rich in such material SRR may be lower than at sites enriched with high quality organic material like fish feed or sedimented phytoplankton [29, 34–36].

As with most microbial processes, temperature plays a significant role in determining SRR (Fig. 5). The Q_{10} of SRR in unamended temperate sediments is low (1.38) relative to those receiving moderate amounts of fish feed (Fig. 5). A low Q_{10} suggests that the activity is not greatly affected by temperature and will be relatively high even at low temperatures. However, sediments from the Arctic and the Philippines representing the extremes of natural temperature variability do not fit this Q_{10} response (Fig. 5). SRR for permanently cold arctic sediments falls well below the line established for temperate sediments. This may be the result of the interaction between substrate availability and temperature since the percent sediment organic matter was low (1.3%) at that study site. SRR in sediments under fish farms in the Philippines were the highest yet observed in natural sediments [37] suggesting that the Q_{10} for SRR under fish farms is similar to that of temperate sediments amended with fish feed (Fig. 5) which have a Q_{10} of 4.3.

Temperature and organic matter supply also interact to influence bacterial metabolism in cold environments [24, 30, 38]. SRR at low temperatures are stimulated by additions of organic substrates [29, 34–36]. As a result, SRR in cold (ca. 0 °C) sediments will be lower than in temperate or tropical sediments (Fig. 5). However, SRR in cold organic rich sediments may be higher than at organic poor sites. The low SRR observed in the Arctic (Fig. 5) may thus be a function of the interaction between temperature and substrate availability. Similarly at seasonally or permanently cold sites in coastal Newfoundland, SRR may be reduced at low temperatures when substrate supply is limited by the refractory nature of the organic matter present.

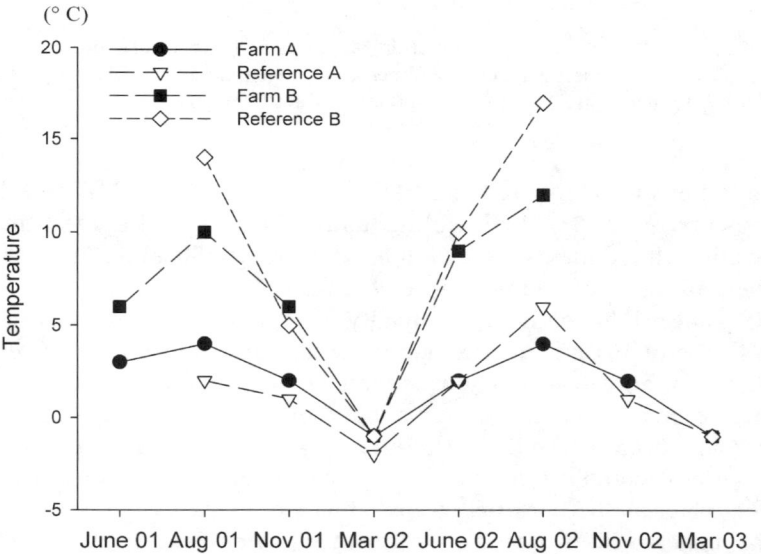

Fig. 6 Annual variation in sediment temperatures at two mussel farms and nearby reference sites in Notre Dame Bay, Newfoundland. One site is relatively shallow (< 20 m) and the other deeper (ca. 40 m). Sediments at the deeper sites remain cold (< 5 °C) all year while those at the shallow site vary seasonally

Bacterial community structure may also affect the rate of sulfate reduction (SRR). Optimal temperatures for sulfate reduction have been shown to be above 25 °C even for bacteria from permanently cold sediments [39, 40]. SRR for mesophilic bacteria are significantly reduced at low temperatures (0–5 °C), while SRR of psychrophilic and psychrotrophic bacteria are similar to those observed for mesophiles under more temperate conditions [27]. Bacteria isolated from permanently cold sediments show higher SRR at low temperatures than do bacteria from seasonally variable sites [27]. Sediment temperatures in coastal Newfoundland vary depending on season, depth and water exchange. For example, sediment temperatures in two shallow bays (Z_{max} = 14 m) varies between – 1.0 and 17 °C while sediment temperatures at two deeper sites (Z_{max} = 40 m) where the bottom waters exchange with the outer bay only a few times a year, varied only between – 1.7 and 6.3 °C (Fig. 6). Thus the bacterial communities and hence SRR may differ significantly between such sites.

6
Conclusion—
Monitoring Organic Enrichment at Cold Water Aquaculture Sites

Depositional areas around the coast of Newfoundland show wide variation in sediment characteristics. Although the accumulation of organic matter at individual finfish farm sites can be followed by measuring sediment redox and sulfide levels, the intersite variability precludes broad application of a single relationship between redox potential and total sulfide concentration as a standard for site classification and selection as has been proposed for other parts of Canada. Site characteristics of bathymetry, hydrography and sediment organic matter content (which varies between 2 and 35% at reference sites around the Island) and quality as well as seasonal variability in temperature regimes and bottom water renewal, must be taken into account in developing appropriate performance-based standards for aquaculture sites.

Organic matter decomposition, as with all bacterial processes, is influenced by temperature. Furthermore, in seasonally or permanently cold sediments the rate of decomposition and sulfate reduction are also dependant upon substrate supply and the quality of the available substrate. Thus sediment temperature and organic matter content will have a greater influence on the redox sulfide continuum than they would under more temperate conditions. In consequence, aquaculture site evaluation and monitoring programs must include measurement of both organic matter content and temperature to provide context for the interpretation of redox and sulfide measurements. For interannual comparisons it is also critical to sample at the same time of year to avoid the confounding influences of temperature and site hydrodynamics on organic matter decomposition.

Acknowledgements Support for this research was provided by the Canadian Department of Fisheries and Oceans Environmental Science Strategic Research Fund, AquaNet Canadian Network of Centres of Excellence for Aquaculture and the Aquaculture component of the Canada-Newfoundland Economic Renewal Agreement. This research could not have been carried out without the help and support of many finfish and shellfish growers. In particular, the authors would like to acknowledge the Salmonid Growers of Newfoundland, Black Gold Inc. and Atlantic Ocean Farms for site access and assistance on the farms. The authors are grateful to the editor and three anonymous reviewers for thoughtful and helpful advice and comments on the manuscript.

References

1. Department of Fisheries and Oceans Canada (DFO) (2003) Interim Guide to Information Requirements for Environmental Assessment of Marine Finfish Aquaculture Projects. http://www.dfo-mpo.gc.ca/aquaculture/ref/AAPceaafin_e.htm
2. Wildish DJ, Akagi HM, Hamilton N, Hargrave BT (1999) Can Tech Rep Fish Aquat Sci 2286
3. Wildish DJ, Hargrave BT, Pohle G (2001) ICES J Mar Sci 58:469
4. Hargrave BT, Phillips GA, Doucette LI, White M, Milligan TM, Wildish DJ, Cranston RE (1994) Can Tech Rep Fish Aquat Sci 2062
5. Newfoundland and Labrador Department of Fisheries and Aquaculture (NLDFA) (2003) Seafood Industry 2003 Year in Review. Government of Newfoundland and Labrador
6. Tlusty MF, Pepper VA, Anderson MR (2000b) World Aquaculture 31:50
7. Tlusty MF, Pepper VA, Anderson MR (1999) Can Tech Rep Fish Aquat Sci 2273. Vi + 32pp
8. Tlusty MF, Pepper VA, Anderson MR (2005) Reconciling aquaculture's influence on the water column and benthos of an estuarine fjord—a case study from Bay d'Espoir, Newfoundland (in this volume). Springer, Berlin Heidelberg New York
9. Tlusty MF, Hughes Clark JE, Shaw J, Pepper VA, Anderson MR (2000a) Mar Tech Soc J 34:59
10. Heiskanen A-S, Tallberg P (1999) Hydrobiologia 393:127
11. Hapberg J, Tunberg BG (2000) Estuarine Coastal Shelf Sci 50:373
12. Keck A, Wassmann P (1996) Sarsia 80:259
13. Reigstad M, Wassmann P (1996) Sarsia 80:245
14. Rosenberg R, Agrenius S, Hellman B, Nilsson HC, Norberg K (2002) Mar Ecol Prog Ser 234:43
15. Wassmann P (1984) Mar Biol 83:83
16. Hay AE, de Young B (1989) J Geophys Res 94:843
17. Richard JM, Hay AE (1984) The physical oceanography of Bay d'Espoir, Newfoundland. Institute of Cold Ocean Sciences. Memorial University of Newfoundland, St. John's, NL, 30pp
18. Marine Science Research Laboratory (MSRL) (1980) In: Bay d'Espoir Feasibility Study, Section 10, pp 1–37
19. Nordberg K, Filipsson HL, Gustafsson M, Harland R, Roos P (2001) J Sea Research 46:187
20. Schwinghamer P, Hawryluk M, Powell C, MacKenzie CH (1994) Aquaculture 122:171

21. Thermo Orion (1997) Platinum Redox Electrodes—Instruction Manual. Orion Research Inc., Beverley, MA
22. Kristensen E, Andersen FØ (1987) J Exp Mar Biol 109:15
23. Holmer M (1991) J European Aquaculture Society Special Publication 16:155–175
24. Weibe WJ, Sheldon WM, Pomeroy LR (1993) Microb Ecol 25:151
25. Tlusty MF, Anderson MR, Pepper VA (1998) Bull Aquacult Assoc Can 98:35
26. Kostka JE, Thamdrup B, Glud RN, Canfield DE (1999) Mar Ecol Prog Ser 180:7
27. Knoblauch C, Jørgensen BB, Harder J (1999) Appl Environ Microbiol 65:4230
28. Holmer M, Wildish DJ, Hargrave BT (2005) Organic enrichment from marine finfish aquaculture and effects on sediment processes (in this volume). Springer, Berlin Heidelberg New York
29. Holmer M, Kristensen K (1994) FEMS Microbiology Ecology 14:33
30. Holmer M, Kristensen E (1996) Biogeochemistry 32:15
31. Pusceddu A, Dell'Anno A, Danovaro R, Manini E, Sara G, Fabiano M (2003) Estuaries 26:641
32. Raymond PA, Bauer JE (2001) Nature 409:497
33. Tlusty MF, Snook K, Pepper VA, Anderson MR (2000c) Aquaculture Research 31:745
34. Flindt MR, Nielsen JB (1992) Hydrobiologia 235:283
35. Glud RN, Holby O, Hoffmann F, Canfield DE (1998) Mar Ecol Prog Ser 173:237–251
36. Rysgaard S, Thamdrup B, Risgaard-Petersen N, Fossing H, Berg P, Bondo Christensen P, Rosenberg R, Agrenius S, Hellman B, Nilsson HC, Norling K (2002) Mar Ecol Prog Ser 234:43
37. Holmer M, Duarte CM, Heilsov A, Olesen B, Terrados J (2003) Mar Pollut Bull 46:1470
38. Rivkin RB, Anderson MR, Lajzerowicz C (1996) Aquat Microb Ecol 10:243
39. Arnosti C, Joergensen BB, Sagermann T, Thramdrup B (1998) Mar Ecol Prog Ser 165:59
40. Isaksen MF, Jørgensen BB (1996) Appl Environ Microbiol 62:408

Reconciling Aquaculture's Influence on the Water Column and Benthos of an Estuarine Fjord – a Case Study from Bay d'Espoir, Newfoundland

Michael F. Tlusty[1,3] (✉) · Vern A. Pepper[2] · M. Robin Anderson[2]

[1] Newfoundland Salmonid Growers Association, St. Alban's, NL, A0H 2E0, Canada
mtlusty@neaq.org

[2] Science Branch, Department of Fisheries and Oceans, P.O. Box 5667, St. John's, NL, A1C 5X1, Canada
nobull@jetstream.net, AndersonRo@dfo-mpo.gc.ca

[3] *Present address:*
New England Aquarium, Central Wharf, Boston, MA, 02110, USA
mtlusty@neaq.org

1	Introduction	116
2	Bay d'Espoir, Newfoundland Salmonid Aquaculture	118
3	Reconciling Water Column and Benthic Impacts	121
	References	127

Abstract One unifying principle proposed for the environmental influence of aquaculture is that when flushing is poor (> 2 d.), the maximal biomass produced in an area will be constrained by accumulation of waste products in the water column. In the Bay d'Espoir estuarine fjord on the south coast of Newfoundland, under-ice salmonid culture in cages in protected bays with low flushing rates (5 to 20 d.) is a challenging component of the annual production cycle. However, in two years of environmental monitoring of such protected bays, no significant change to water quality was observed. A measurable influence on the benthos was more frequently detected, but localized. Thus the inconsistency of Bay d'Espoir; it has a low flushing rate, yet there was no observable change to the water column. Possible reasons for this are discussed, and include: the sheer amount of water (i.e., potential for within-basin mixing/dilution and biodegradation) in this estuarine fjord; increased surface transport of nutrients; the benefit of fallowing; and, diminished relative loadings to the water column and benthos in winter conditions for an industry in its early stages of development. Further refinement of assimilative capacity estimates for this and other similar suboptimal areas will have to resolve this apparent contradiction prior to espousing "unifying principles".

Keywords Assimilative capacity · Benthos · Environmental impact · Modeling · Water column

Abbreviations
H holding or assimilative capacity
ΔC allowable change in nutrient level (μmol N l^{-1})

R   rate of nutrient release by fish (mol N d^{-1} kg^{-1})
T   flushing rate (m^3 d^{-1})
V   volume (m^3)
mt metric tons (1 mt = 1000 kg)

1
Introduction

Monitoring and simulation efforts have been put to use in several countries to understand and control the impacts of net-pen aquaculture on the environment [1]. These efforts have been directed at investigating both water column [2–6] and benthic impacts [7–13]. However, few studies have compared simultaneous effects on the water column and benthos [10, 12–16]. This growing body of work suggests that aquaculture operations will cause greater change to the water column surrounding the net pens than to the benthos beneath the net pens. An estimated 47.6% to 63.8% of the total carbon [9, 17], 65.0 to 65.1% of the total nitrogen [9, 18], and 31.4% of the total phosphorus [19] input as feed ultimately enter the water column as soluble waste. Furthermore, approximately 50% of the particulate nitrogen and phosphorus reaching the sediment layer translocates back to the water column [20]. This would increase the aforementioned losses to 82.5% to 83.5% of nitrogen, and 68.3% of phosphorus input as feed that ultimately enters the water column as soluble waste. Thus using a mass balance approach to investigate nutrients lost to the environment, relatively more enter the water column than the benthic food chain.

While the mass balance approach suggests a theoretical proclivity towards increased water column impacts, degradation of benthic habitats, particularly for salmonid mariculture, is more frequently observed [1, 8, 10]. It is this benthic degradation rather than that of the water column that is likely to cause negative impacts and subsequent fish disease issues for the farmer [21]. Comparing those studies that have demonstrated a direct impact of aquaculture on water column variables, it was often observed that changes are most pronounced in eutrophic, shallow, or very confined areas that have long flushing times [11, 15, 22, 23], and included the Baltic [24], Hong Kong [15], and areas of Norway [14]. As an example, prolonged flushing times are associated with a greater incidence of change to the water column [25].

For any given body of water, the production capacity can be calculated using either a benthic model developed from carbon burial rates (B) [12], or from water quality variables (W) [6, 12]. A B/W production ratio can be calculated as the benthic production estimate divided by the water column production estimate, and can be used to determine which milieu is limiting. Values less than 1.0 indicate that the water column estimate is larger, and thus changes to the benthic environment are limiting. Conversely, values greater than 1.0 indicate that the benthic estimate is greater, and changes to the water column environment are limiting. Using data published elsewhere [12], a comparison of benthic and water column derived production estimates shows that the B/W production ratio is greater than 1.0 for flushing times longer than 2 d (Fig. 1). This suggests that a flushing time of 2 d appears to be the point at which water quality degradation may become problematic as opposed to benthic deterioration [25]. Production capacity estimates based on sediment carbon burial rates [12] are typically more than threefold larger than estimates using water quality parameters [6, 12] when the flushing time is > 2 d (see Fig. 1), and imply that changes to the water column will be limiting at long flush times.

The lack of observed impacts of aquaculture on the water column likely is a result of the process of locating aquaculture operations in areas that are biophysically appropriate, and abandoning those that are not [21]. However, as the demand for aquaculture product increases globally, the optimum growing sites will reach their full capacity or be removed from produc-

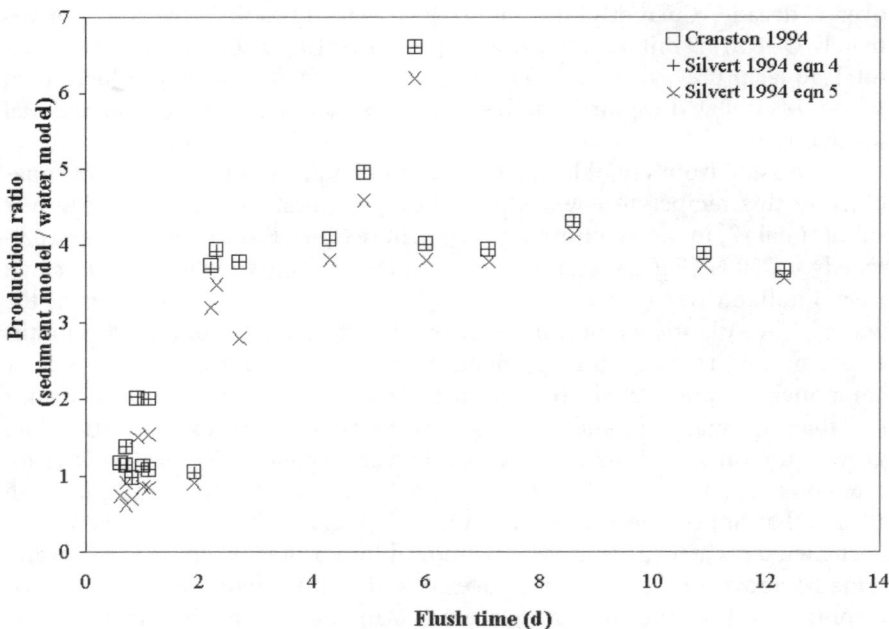

Fig. 1 The relationship between flushing time and assimilative capacity for Salmonid aquaculture operations. The B/W production ratio is the ratio of the benthic derived production estimate [12] to that of the water column derived estimate (see legend). B/W production ratios > 1.0 indicate that the benthic estimate is greater, and changes to the water column will limit the holding capacity. Site characteristics are given in [12]

tion through societal conflict [26] and suboptimal production areas will be developed. Bay d'Espoir, Newfoundland, with its winter ice cover, estuarine-fjord hydrographic dynamics and confounding influence of freshwater input from hydroelectric development, is one such area of development in a suboptimal environment [2, 7, 27]. The water column of Bay d'Espoir has long flushing times, but there is no evidence of increased nutrient loading in this domain. On the other hand, there is localized evidence of benthic accumulation beyond what can be considered the norm for an estuarine fjord. Thus the question of why this area runs counter to current concepts needs to be considered.

2
Bay d'Espoir, Newfoundland Salmonid Aquaculture

Bay d'Espoir aquaculture businesses have demonstrated their ability to produce high quality Atlantic salmon (*Salmo salar*), rainbow trout (*Oncorhynchus mykiss*), and brook charr (*Salvelinus fontinalis*) for the international marketplace. These businesses have been operating in the bay for over a decade. Currently, the industry in Bay d'Espoir is worth approximately CDN$14.9 million annually [28], representing > 70% of the province's total aquaculture value. In a national context, total annual production of salmonids in Bay d'Espoir is modest at approximately 2% of the total national production.

Quite aside from tangible demonstration of aquaculture production capability in this temperate environment, in a comparative sense this region is suboptimal [7] for aquaculture for two main reasons. First, located at approximately $47°50'N$ (Fig. 2), and with the largest freshwater inflow of any small Newfoundland Bay (2.0×10^6 m^3 d^{-1}) [2], the fjord freezes over during the winter. The Atlantic salmon production cycle spans 32 months (20 months estuarine on-growing), thereby making under-ice cage culture a necessary component of the annual production cycle within the fjord. This makes for less than optimal salmonid farming conditions since fish exhibit little to no growth for three to four months of the year. Secondly, Bay d'Espoir is located on the south coast of Newfoundland. It is a complex estuarine fjord with 12 sills that impede deep-water flushing [2] (Fig. 2). This is particularly true in the winter when cages need to be moved into protected coves where water remains above $-0.7\,°C$, and ice cover is sufficiently stable to prevent catastrophic cage loss due to shifting ice pans. In addition to the impact of the basin/sill topography on flushing time during ice-free seasons, the average current speed is further constrained by the formation of ice cover (Fig. 3). To assure sufficient water quality within a cage, currents should remain above 10 cm sec^{-1} for a major proportion of the tidal cycle [29]. In consideration

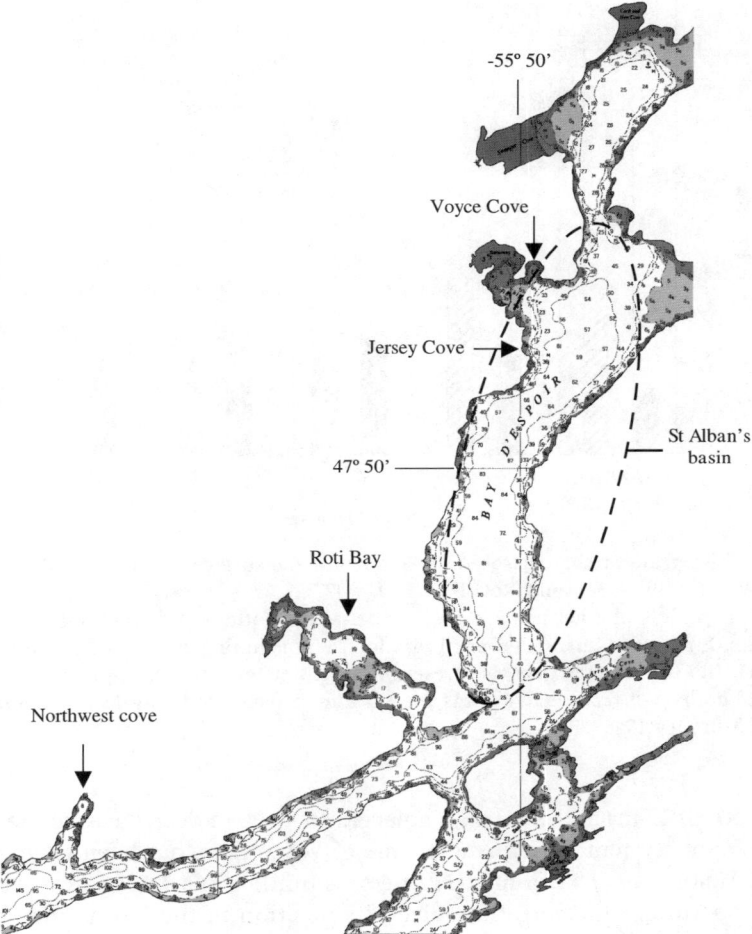

Fig. 2 A chart of Bay d'Espoir. Salmonid production sites are identified. Depths are in fathoms. The chart image was supplied by NDI/CHS and is not to be used for navigation

of the Newfoundland industry's reliance on seasonally-limited areas for holding salmonid stocks, a monitoring program for Bay d'Espoir over-wintering sites was undertaken to determine if conditions within the spatially-confined over-wintering sites of the estuarine fjord are satisfactory for long-term sustainability of aquaculture development.

Although the entire Bay d'Espoir fjord encompasses some 250 km^2, historically only four geographically-confined areas have been used for over-wintering of the salmonid inventory: Voyce Cove, holds the market fish; Jersey Cove; Northwest Cove; and, Roti Bay (Fig. 2), which hold all other fish. Voyce Cove ($250\,000 \text{ m}^2$, average depth of 10 m) and Jersey Cove ($150\,000 \text{ m}^2$, aver-

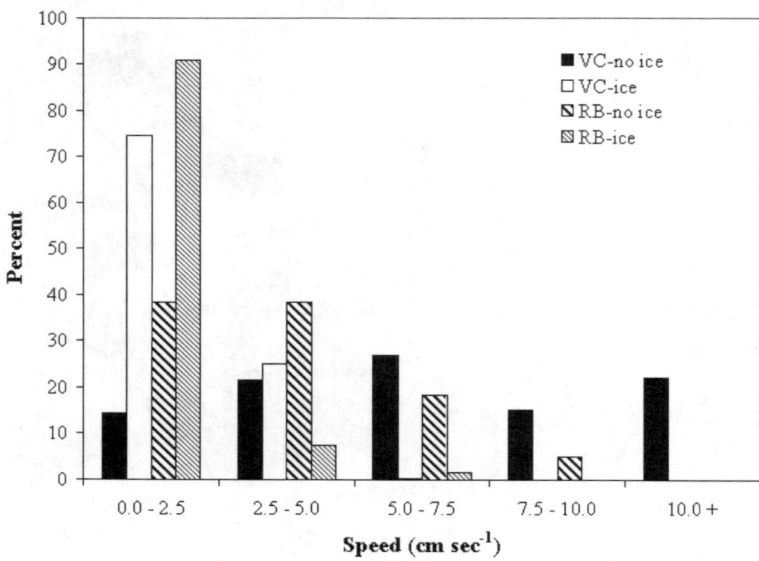

Fig. 3 A histogram of current speeds at 5 m below the surface in Voyce Cove (N 47° 52′ 022″, W − 55° 49′ 763″), and Roti Bay (N 47° 47.22′, W − 55° 51.72′). The meter in VC was in the middle of a net pen system, 2 m below the bottom of the nets and 5 m off the substrate. It recorded currents every 1 min for 24 h beginning 30 November 1998 (no ice present), and 16 February 1999 (ice present). The RB meter was in the mouth of the bay in water 43 m deep. It recorded currents every 1 h for 5 days beginning 25 November 1997, and 15 February 1998

age depth of 12 m) are shoal bays adjacent to a 100 m deep basin of the upper Bay d'Espoir system. The flushing time of Voyce Cove has been estimated at 5 d (J. Helbig, DFO St. John's, NL pers. comm.) while Jersey Cove is likely to have a similar flushing time given its position in the basin, and the observed pattern of ice pan movement through the area during breakup. Voyce Cove was been used for winter production of fish for over a decade, and in 1997 carried 850 mt. Jersey Cove was used continuously for a 2.5 year period (1996–1999), and in 1997 carried 15 mt [2].

Roti Bay (2 662 000 m^2) and Northwest Cove (600 000 m^2) are more enclosed bays with a length : width ratio of 4 : 1 (Fig. 2). Roti Bay is the major overwinter site as approximately 85% of all premarket fish have been held here. It has two useable 30 m deep basins for aquaculture, which are separated by a 10 m deep sill. If considered as a single unit, Roti Bay has an estimated flushing time of 20 d (J. Helbig, DFO St. John's, NL pers. comm). Northwest Cove has a simple bottom topography in which there is no defined sill, and an average depth of 12 m (Fig. 2). At the time of the present study, Roti Bay had been used continuously for winter production for seven years, while Northwest was used continuously for 2.5 years (1995 to 1998, [2]). These areas carried 595 and 86 mt respectively in the winter of 1997.

Placed in perspective, the relative effort directed at the two most important over-wintering areas, (Voyce Cove and Roti Bay), is considerable. In British Columbia, a year-round production level of $1.3\,\mathrm{t\,ha^{-1}}$ has been suggested as maximal to prevent significant environmental change [30]. In comparison, the over-wintering biomass of Voyce Cove in 1997 was approximately $34.0\,\mathrm{t\,ha^{-1}}$, while that of Roti Bay was $2.2\,\mathrm{t\,ha^{-1}}$. Although the Newfoundland industry is still quite small, at least for the winter interval of the production cycle, it is a very concentrated effort.

Due to concern about assimilative capacity for aquaculture in these areas, industry wanted an evaluation of the tonnage of salmonids that could be safely overwintered in the four disparate geographic locations. The science steering committee for the undertaking determined that construction/adoption of a numerical model would be appropriate. The first step was collection of background data. Beginning in January of 1997, monthly water quality samples were obtained from 30 stations throughout the bay. The over-winter sites were the main focus of the sampling, with appropriate reference samples well removed from aquaculture sites being taken for comparison [2]. In the summer of 1997, a program of equivalent effort was initiated for the benthos [7]. Based on this sampling effort (> 25 000 samples), water quality showed no increased nutrification [2]. In contrast, benthic impacts were detected but were localized under the aquaculture cages and were highly variable with time and between sites [7]. In an estuary-wide analysis, any apparent decrease in water quality or benthic status was more prevalent in areas removed from the aquaculture sites [2]. Erosion of environmental quality was associated with the geomorphology of the estuarine fjord (e.g., points of stream entry well removed from aquaculture cages) rather than the aquaculture production sites.

3
Reconciling Water Column and Benthic Impacts

A popular water-based model states that the assimilative or holding capacity for finfish of an area (H) can be determined by:

$$H = \left(\frac{\Delta C}{R}\right) \times \left(\frac{V}{T}\right) \tag{1}$$

where ΔC is the allowable change in nutrient level, R is the rate of nutrient release by fish, V is the volume of water, and T is the flushing rate [6], see also [31–33]. This model produces an estimate of holding capacity based solely on water quality, and is the basis for much of the ongoing work to assess environmental change [15, 31–33]. We used this model as a base from which we produced the first estimate of the holding capacity tonnage for Bay d'Espoir over-wintering areas as well as using it to determine where additional information was necessary. The original plan was to create an individualized model for Bay d'Espoir. We used this existing model as a starting point for an interim analysis in the hope of directing future research and modeling efforts.

In the Bay d'Espoir situation, application of the data to this model produced estimates compromised by considerable uncertainty [7] (Fig. 4). This uncertainty is derived from ambiguity within the assessment of each of the four variables within the model. Beginning with volume (V), the assumption is that the defined areas of the coves and bays are a functionally discrete unit. However, water transfer over a sill is often not a distinct event, but is the gradual mixing of a parcel of water that is being transferred back and forth over the sill (B. Silvert, pers. comm.). Within the Bay d'Espoir system, several of the sites are coves that do not have a sill defining an outer boundary (Voyce Cove, Jersey Cove, and Northwest Cove, Fig. 2). Given the open configuration of these shoals, they may operate as part of the larger volume of water to which they are attached. If these shoals are functionally part of the larger, 100 m deep basins, then the immense volume of the water mass to which the calculation is being applied may be a confounding factor. While Roti Bay is

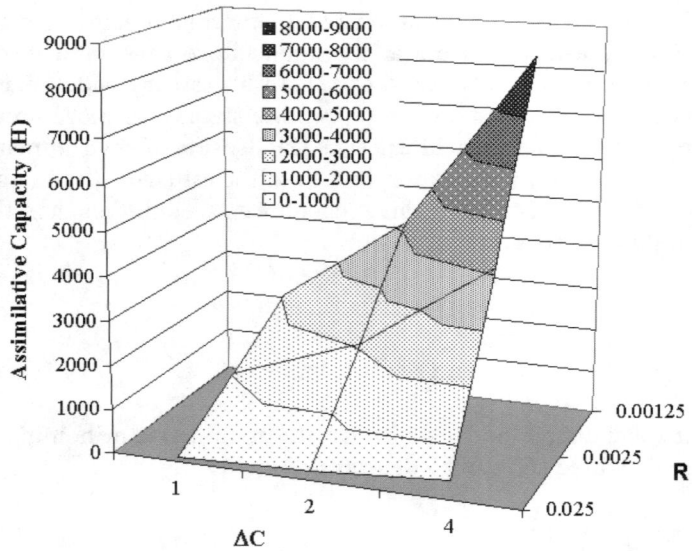

Fig. 4 The assimilative capacity of Roti Bay based on a volume of 51 636 794 m³, and a flushing time of 20 days, assuming it functions as a single basin. The rate of nutrient release (R, mol N d^{-1} kg^{-1}) ranges from the value fish typically excrete [10] to a value 1/20th of that based on the reduced amount of food being fed to the fish in winter. The allowable concentration increase (ΔC, μmol N l^{-1}) ranges from the typical allowable increase [10] to increased values because of fallowing

a well defined water body (Fig. 2), it is unknown if the water within the bay is a single parcel, or operates as two separate basins. The effort to determine assimilative capacity (Fig. 4) used a single basin assumption. Because of connectivity between the two inner basins, a two-basin model would likely yield a lower estimate of assimilative capacity than the single basin model (Fig. 4).

Finally, the production level has to be placed in context of the sheer volume of water of the production areas, and the entire fjord. It may be that the water volume of the over-winter areas relative to the biomass being produced may be sufficient enough that the soluble nutrients added within the basin are largely dispersed via diffusion, advection, and biodegradation of the wastes as they are transported away from the farm sites. The relatively small total biomass of fish produced per year (1000 t) by the present Newfoundland industry may not represent a measurable nutrient loading influence to the water column. In either case, waste products would not accumulate in any part of the system, and the background levels of ammonia, nitrite, nitrate or phosphorus in the water column may not be altered to a substantial degree. As the biomass carried by the Bay d'Espoir salmonid aquaculture industry increases, there may be a greater disposition for increased water column degradation in these over-wintering areas. Continued monitoring as the industry grows will support or refute this hypothesis.

Flushing time (T) also are poorly known and selected values may be questioned. Estimates of this variable have been made based solely on the larger basins [2], and exist for Roti Bay as a single unit instead of two basins separated by a sill, and Voyce cove, as part of the larger St. Alban's basin (Fig. 2). The present estimates also do not account for the substantial vertical stratification of the water column. Estuarine fjords are characterized by marked vertical stratification in the salinity and temperature of the water column. In Bay d'Espoir, the large hydroelectric input of freshwater overlies the denser tidally-influenced water, resulting in pronounced water-column stratification [2, 7]. This freshwater layer is highly influenced by wind as, at times of northerly gale ($\sim$ 35 knots) winds, the freshwater layer throughout the bay diminishes. Surface salinity has been seen to vary from < 5 ppt to > 25 ppt in less than one day (Fig. 5).

While the freshwater lens can be removed in less than a day, the surface currents are significantly faster than those below the halocline (no ice conditions, $\chi^2 = 24.87$, df = 4, p < 0.001; ice conditions, $\chi^2 = 188.48$, df = 4, p < 0.0001, Fig. 3). In addition, it appears that little mixing occurs between the surface layer and higher-density lower layer as the upper layer is still only 1/3 the salinity of the tidal layer after traveling 10 km through the estuary [2]. In a cage, the fish, particularly rainbow trout and brook char, spend a significant amount of time in the upper low-salinity layer (C. Collier, Markland Aquaculture Inc., Conne River, NL, pers. comm.). This is in contrast to both Atlantic salmon and steelhead (anadromous *O. mykiss*) that demonstrate preference for marine salinities. If caged aquaculture salmonids excrete solu-

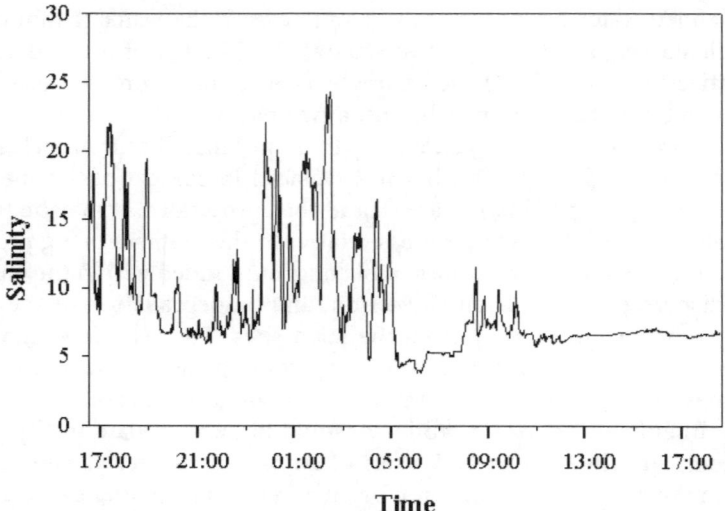

Fig. 5 Salinity at a depth of 1.5 m in Voyce Cove (N 47 51.987, W – 55 49.694) beginning on 30 November 1998. These readings occurred at the tail end of a storm that ended approximately 1100 h on 1 December. Measurements were taken with an Interoceans S4 current meter every min continuously for 24 h. Salinity fluctuation did not correlate to tides as high tides were at 21:56 on 30 November and 10:26 on 1 December, with the low tide at 3:59 on 1 December

ble nutrients into the upper layer, and there is little entrainment of the denser marine layer, then the soluble nutrients may be transported away from the site at a greater rate than that produced by tidal flushing. The increased rate of transport of nutrients in the surficial waters will decrease the flushing time, and thus substantially increase the theoretical biomass of fish that an area can assimilate based on a water column model [6] (Fig. 1). A similar explanation for the lack of differences between coastal and fjord waters, coincident with a decrease in fjord water quality below the sill, was postulated for Norwegian secondary embayments and archipelagos [3]. Refinements to T in the holding capacity model necessitate development of a meaningful hydrographic algorithm for Bay d'Espoir.

The rate at which nutrients are released (R) has been examined for salmonids under controlled conditions, and at temperatures generally > 5 °C [34–36]. In Bay d'Espoir, fish can spend up to four months at temperatures < 5 °C, with a significant proportion of this period at temperatures from 0 to 2 °C. Preliminary field estimates suggested that salmon digestion becomes very inefficient at low temperatures, based on the observation that fecal organic content increased considerably [27] during the winter. This, coupled with a decrease in metabolic rate at low temperatures [37] would partition a smaller proportion of nutrients to an end as soluble wastes, with a concomitant increase in particulate waste production. Therefore, more of the nutrients introduced to the system as food will end up being deposited as fecal waste. This increase in organic loading to the benthos is moderated by a smaller input of feed, as the overall amount of food offered is decreased at low temperatures. In Bay d'Espoir, winter feeding rates can be 1/20th those of summer.

The question of quality vs. quantity of particulate waste and the respective impacts on the benthos has not been explored. However, if fish feeding intensity decreases faster than the amount being fed, the amount of feed wasted will increase. Any increase in feed wastage will have a disproportionate additional negative impact on the benthos. Feed was shown to have a greater environmental persistence time relative to feces as only 5% of the feed organic matter was lost over a period of 12 d, while 50% of fecal organic matter was lost during the same time interval [27].

Finally, the allowable change in nutrients (ΔC) in areas with a production cycle similar to that of the Bay d'Espoir system, needs to be examined further. In Silvert's [6] original model, the value for ΔC considered year-round production at the site in question. The management plan for Bay d'Espoir [38] calls for a six-month mandatory fallow period each year for all winter sites. Since sites are being fallowed, they may be less prone to exhibit time-integrated (yearly) water quality impacts. It has yet to be determined if there is an inverse relationship between the proportion of a year that the site is used and the allowable ΔC. Recent research has examined the full benefit of fallowing [39]. In Bay d'Espoir, the importance of fallowing is apparent at the Northwest Cove site. This site was operated continuously for 2.5 years and had a substantial increase in organic matter (to a substrate depth of 10 cm, Tlusty, unpublished data) relative to fallowed sites (Voyce Cove and Roti Bay). This site also did not conform to the increase in nematode diversity during a fallowing period observed at the other sites (Fig. 6) [40]. The recovery observed at Roti Bay was much faster than the 0.5 to 1.5 years typically cited in the literature [41–43].

Summarizing each of the variables, we assume that the volumes of each area will stand. This assumption is valid for Roti Bay since it is an enclosed bay. We await detailed analysis of drifter-drogue work to determine the linkage between Voyce Cove and the rest of St. Alban's basin. As for the other variables, in winter conditions, R will be 1/20th that of summer values based on feed delivery alone. Any additional reduction in feed utilization by the fish will lower this value. ΔC will also change because of bay specific management practices. Assuming linearity, ΔC for this area may be increased twofold since the sites are fallowed for half a year. Finally, it is difficult to know the accuracy of the estimated value for T, as the overlying freshwater layer increases the transport of soluble wastes away from farm sites thus reducing flushing time. However, the overall change in flushing time is currently unknown. If freshwater-layer dynamics are assumed to decrease flushing time by half, then these three "winter" model inputs together will give an assimilative capacity

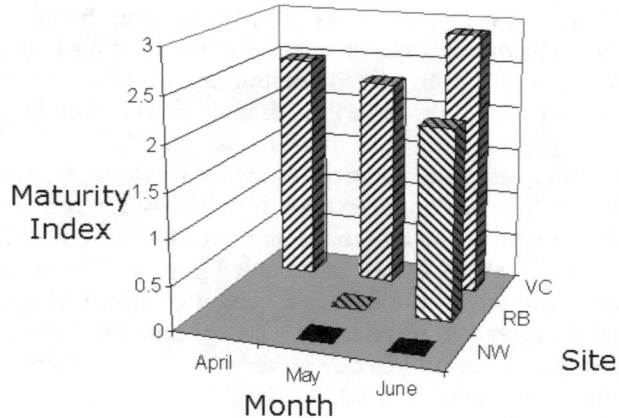

Fig. 6 Nematode diversity in three aquaculture sites (Northwest Cove – NW, Roti Bay – RB, and Voyce Cove – VC) during the first three months of a six month fallow period in 1998. The maturity index assess if each nematode species present in the sample had a tendency to colonize (1) or persist (5), and was calculated as the weighted average of the C-P index. The closer M is to 5.0, the more mature or stable the nematode community [45]

estimate nearly two orders of magnitude greater than that calculated using "summer" values for the model inputs (Fig. 4).

In 1997 the 850 mt in Voyce Cove was approximately 50% of its theoretical maximum of 1600 mt based on modeling the production of nitrogenous wastes, while Roti Bay was at 7% of its theoretical maximum. Since the industry goal for Roti Bay of eight 1000 mt farms is near the theoretical maximum calculated here (Fig. 4), we urge extreme caution and additional monitoring as the industry expands. For example, if freshwater transport is important for flushing, and with Roti Bay being enclosed, additional cages may significantly block (and reduce) this upper-layer flushing.

Newfoundland is one of few areas where an assimilative-capacity estimate has been attempted prior to significant growth or geochemically-significant changes to environmental quality and function. At this stage in our evaluations, we are uncertain how the assimilative capacity estimates may change with increased industry effort. Meaningful application of Silvert's [6] water-based model of assimilative capacity requires accurate and precise estimates of the four model variables and clarification of upper- and lower-layer dynamics. In short, we were directed to provide a quick and simple answer to a complex biological issue. Refinement of model variable estimates is the next logical step along with creation of a new model specific to fjord dynamics of Bay d'Espoir. Such a model would have to account for a stratified water layer, presence of ice cover, wind events, disparate bottom topography and variable aquaculture industry practices.

The uncertainty associated with our present estimates of these four variables for Bay d'Espoir becomes extremely problematic for immediate use of this model as a management tool. The range of potential solutions is too great at this stage in the analysis to specify a sound industry-operations protocol on which to base industry expansion, and effort must be made to decrease the uncertainty of these variables for areas where the production cycle operates differently compared to the major world centers of salmonid aquaculture production. In addition, it is critical to assure that this uncertainty is emphasized and properly conveyed with any estimate of production potential, so that regulators can make a more informed decision, as well as understand the consequences of setting goals too high or low [44]. Performance-based standards, properly tuned to the geomorphology of a given area, are important prerequisites to meaningful licensing and adaptive management of aquaculture for any jurisdiction.

The key to further understanding aquaculture production in estuarine fjords is to identify the functional geomorphologic unit in which each aquaculture facility operates. As part of this, the relationship between the freshwater surficial flow and deep water tidal exchange need to be further elucidated. This will not only lead to a better estimate of flushing time, but will also determine if soluble and particulate wastes are being similarly dispersed by the water. It is likely that in estuarine fjords with greater surficial flow, the more disparate water column and benthic production estimates will be. The Bay d'Espoir example illuminates many questions that need to be answered before salmonid aquaculture can flourish over the long term as other suboptimal environments become exploited.

Acknowledgements This program was funded under the Aquaculture Component of the Canada-Newfoundland Economic Renewal Agreement. J.A. Helbig provided access to current meter data for Roti Bay. C. Diamond, R. Mercer, L. Snook, and K. Snook, were invaluable field and laboratory assistants. T. Nicholls and S. Kenny (DFO) assisted with collection and data archiving. B. Hargrave, B. Silvert, and two anonymous reviewers provided valuable comments on a draft of this manuscript.

References

1. Ervik A, Hansen PK, Aure J, Stigebrandt A, Johannessen P, Jahnsen T (1997) Aquaculture 158:85
2. Tlusty MF, Pepper VA, Anderson MR (1999) Can Tech Rep Fish Aquat Sci 2273:1
3. Aure J, Stigebrandt A (1990) Aquaculture 90:135
4. Håkanson L, Wallin M (1991) Envirometrics 2:49
5. Wallin M (1991) PhD Thesis, Uppsala University, Sweden
6. Silvert W (1994) Can Tech Rep Fish Aquat Sci 1949:1
7. Tlusty MF, Pepper VA, Anderson MR (2000) World Aquaculture 31:50
8. Beveridge M (1987) Cage aquaculture. Fishing News Books, Oxford
9. Gowen RJ, Bradbury NB (1987) Oceanogr Mar Biol Ann Rev 25:563

10. Gowen RJ (1990) An assessment of the impact of fish farming on the water column and sediment ecosystem of Irish coastal water including a review of current monitoring programmes. Department of the Marine, Leeson Lane, Dublin
11. Gowen RJ, Brown J, Bradbury NB, McLusky DS (1988) PhD Thesis, University of Stirling, Scotland
12. Cranston R (1994) Can Tech Rep Fish Aquat Sci 1949:92
13. Wu RSS (1995) Mar Poll Bull 31:159
14. Abdullah MI, Danielsen M (1992) Hydrobiologica 235/236:711
15. Wu RSS, Lam KS, MacKay DW, Lau TC, Yam V (1994) Mar Env Res 38:115
16. Silvert W, Sowles JW (1996) J Appl Ichthyology 12:75
17. Hall POJ, Anderson LG, Holby O, Kolberg S, Samuelson MO (1990) Mar Ecol Prog Ser 61:61
18. Hall POJ, Holby O, Kolberg S, Samuelson MO (1992) Mar Ecol Prog Ser 89:81
19. Holby O, Hall POJ (1991) Mar Ecol Prog Ser 70:263
20. Enell M, Ackefors H (1991) ICES CM 1991/F:56
21. O'Connor B, Hartnett M, Costelloe J (1991) J Europ Aquacult Soc Spec Pub 16:191
22. Muller-Haeckel A (1986) Vatten 42:205
23. Iwama GK (1991) Crit Rev Environ Contr 2:177
24. Persson G (1991) Proc First Int Symp Nutrit Strat Mgmt Aquacult Wastes. University of Guelph, Guelph, Ontario, 1990:163
25. Frid CLJ, Mercer TS (1989) Mar Pollut Bull 20:379
26. Black EA (1991) J Europ Aquacult Soc Spec Pub 16:441
27. Tlusty MF, Anderson MR, Pepper VA (1998) Bull Aquacult Assoc Can 98:35
28. Department of Fisheries and Oceans (2004) www.dfo-mpo.gc.ca/communic/statistics/aqua/aqua02_e.htm; last accessed 4/7/2004
29. Weston DP (1986) PhD Thesis, University of Washington, Seattle, WA
30. Levings CD (1994) J Applied Aquaculture 4:65
31. Gillibrand PA, Turrell WR (1997) Aquaculture 159:33
32. Strain PM, Hargrave BT (2005) Salmon aquaculture, nutrient fluxes and ecosystem processes in southwestern New Brunswick (in this volume). Springer, Berlin Heidelberg New York
33. Sowles JW (2005) Assessing nitrogen carrying capacity for Blue Hill Bay, Maine – a management case history (in this volume). Springer, Berlin Heidelberg New York
34. Windell JT, Foltz JW, Sarokon JA (1978) Trans Ma Fish Soc 107:613
35. Hudon B, de la Noue J (1985) J World Maricul Soc 16:101
36. Mäkinen T (1993) Aquacult Fish Mgmt 24:213
37. Brett JR (1979) In: Hoar WS, Randall DJ, Brett JR (eds) Fish physiology Volume 8: Bioenergetics and growth. Academic Press, New York, p 599
38. Newfoundland and Labrador Department of Fisheries and Aquaculture (1998) Roti Bay management plan. PO Box 8700, St John's, NF A1B 4J6
39. Stewart JE (1998) Can Tech Rep Fish Aquat Sci 2218:1
40. Powers VA, Finney-Crawley J, Tlusty MF (1999) Proc 38th Ann Mtg Can Soc Env Biol 1998:103
41. O'Connor B, Costelloe J, Dinneen P, Faull J (1993) ICES CM 1993/F:19:1
42. Johannessen PJ, Botnen HB, Tvedten ØF (1994) Aquacult Fish Mgmt 25:55
43. British Columbia Environmental Assessment Office (1997) Salmon Aquaculture Review http://www.eao.gov.bc.ca/ report 6 January 1998
44. Hilborn R, Pikitch EK, Francis RC (1993) Can J Fish Aquat Sci 50:874
45. Bongers T, Alkemade R, Yeates GW (1991) Mar Ecol Prog Ser 79:135

Modelling the Impacts of Finfish Aquaculture

C. J. Cromey (✉) · K. D. Black

Scottish Association for Marine Science, Oban, Argyll PA34 4AD, UK
chjc@sams.ac.uk

1	Introduction	130
2	Primary Considerations of Aquaculture Impact Models	131
2.1	Bathymetry	131
2.2	Cage Layout and Positioning	132
2.3	Husbandry Data	132
2.4	Hydrodynamic Data	133
2.5	Dispersion Coefficient Data	134
2.6	Faecal and Feed Properties	135
2.7	Wild Fish Populations	135
2.8	Biochemical Components	136
2.9	Resuspension	137
2.10	Standardisation of Data	139
2.11	Complexity of Scenarios	140
2.12	Model Output	141
3	Detailed Investigations	142
3.1	Drifter Data and Dispersion Aspects	142
3.2	Hydrographic Data	145
3.3	Cage Movement and Shadowing Effects	148
3.4	Feeding Systems and Waste Particle Release Times	151
4	Conclusion	153
	References	153

Abstract The main components of lagrangian (particle tracking) models used in the assessment of aquaculture impact have been reviewed. Trends in input data requirements have also been examined, with some emphasis on general modelling issues such as standardisation of data and scenario complexity. Specific data aspects were tested in a flux and benthic effects (impact) model and their importance assessed. Dispersion coefficients resolved from drifting buoy data were found to influence model predictions widely depending on the criteria used in the analysis. Different lengths of hydrographic data subsampled from a 206 day record measured at a Scottish fish farm also resulted in different model predictions. This highlighted the importance of taking representative hydrographic measurements for regulatory modelling of maximum farm biomass. Both cage movement and the timing of feeding and defecation events were also tested using observed data and found to be of less importance.

Keywords Aquaculture · Benthic impact · Dispersion · Model

1
Introduction

Models for predicting the dispersion and subsequent deposition of particulate aquaculture wastes have been in common use for a decade or more [1]. Given information on site depth, hydrodynamic conditions, settling characteristics of wastes and discharge rate of waste faecal and food particles, vertical particulate flux (g solids deposited m^{-2} bed yr^{-1}) at the sea bed can be predicted [2]. Through the years, these models have been developed using increasingly complex aspects of these fundamentals such as spatially varying flow fields and detailed bathymetry [3]. These developments have been partially driven by the recognition of certain important processes, but also by the increased availability and accuracy of instrumentation used for data collection. For instance, acoustic profiling instruments sample flow fields more accurately and frequently compared to the rotary instruments used in early studies. Settling velocity experiments are also increasing in accuracy and sample size. Until recently, emphasis was placed on feed pellets [4] as these were regularly wasted and readily available for experimentation. Now, as feed wastage is being reduced, research emphasis is switching to faecal material as this is the main component of the discharged waste. Physical and biochemical properties of the faecal material vary between fish species, size and diet and these are increasingly being measured and modelled.

Although some models which predict vertical flux alone are of interest, these lack processes such as resuspension which are fundamentally important in some environments. Redistribution of waste material and associated chemical components by this process requires quantification if realistic impact predictions are sought. Some advances have been made in validation of aquaculture resuspension models [5, 6], but problems exist in finding tracers representative of the farm waste, or in distinguishing farm derived from natural sediment in the erosion process. Having predicted flux before or after resuspension, many researchers use this information subjectively to assess farm impact. Some models take this a step further and predict benthic effects from relationships between predicted flux and a suite of benthic indices.

Clearly, the main modelling developments in the coming years are likely to be in biochemical components which link flux and benthic effects, as well as in the benthic effects relationships themselves. A model which can predict a biochemical component measurable in the field will be useful for monitoring of existing sites. Key research areas are also likely to be bioturbation and sediment consolidation processes, particularly important as sediment characteristics and fauna change widely along the organic gradient. As newer species are being farmed and different environments with improved husbandry characteristics utilised, refined models will become available.

The main data input requirements, trends and limitations of certain data representations are assessed here. Later, some specific data aspects are investigated in a flux and benthic effects model to examine their importance.

2
Primary Considerations of Aquaculture Impact Models

Lagrangian (particle tracking) models used in the assessment of aquaculture impact are commonly made up of some or all of the following components: grid generation, fish bioenergetics, particle tracking, resuspension, biochemical and benthic faunal response. A grid containing information on depth, cage and sampling station positions for the area of interest is required. Given wastage rates of fish food and faeces from a bioenergetics model, hydrodynamic data and settling velocity of wastes, initial deposition of particles on the sea bed can be predicted with a particle tracking component. A resuspension model then redistributes particles according to near-bed current flow fields to predict net solids accumulated on the sea bed. At this stage, some biochemical modelling of the particulate material in the sediments may be included. Finally, some impact assessment is required and this can be undertaken by correlating predictions of a physical (e.g. flux) or biochemical aspect (e.g. oxygen demand) of the waste with benthic faunal response.

2.1
Bathymetry

Local topography in the area of interest will determine how much depth detail is required in the model. At high biomass depositional sites, detail will be required near the cages particularly for closely spaced sampling stations. Less detail is required at increasing distance from the farm as fewer particles are advected to these areas. Model grid resolution will depend on the expected dispersion and the spatial sampling frequency of the survey data, with grid resolution between 5 and 25 m being typical.

Bathymetry is commonly measured using an echo sounder interfaced with a GPS. Other sources of data include discrete soundings using a hand held echo sounder, SCUBA diver measurements and tape or lead line soundings. Charts can provide useful information but accuracy is limited in dated charts or areas along coasts away from major shipping channels. Positional information usually requires some modification prior to implementation into the model as coordinate systems need to be compatible. In addition, bathymetric data collected from different sources may be referenced to different datums. Once corrected for tidal height, irregular spaced data can then be gridded in a contouring package and exported as a grid with equal spacing between nodes. For the wider picture, local topography along with general circulation patterns are useful for assessment of advection and subsequent fate of resuspended material. This is especially important if the residual current is towards an area of restricted exchange or an area containing sensitive species [7, 8].

2.2
Cage Layout and Positioning

With information on cage layout and dimensions, point sources can be defined in the model with initial particle distributions within the cage volume. Cage maximum depth and shape are required as predictions from lagrangian models are sensitive to these variables [9]. Cage positions in a datum compatible with the bathymetry data must also be known and for new sites, an accurate proposed location of the cages is required. For a large leased area containing steep bathymetry, deposition and benthic effects will vary depending on the final farm location and cage rotation within the leased area. Accurate sampling station positions are of paramount importance if the transect is located along a high deposition gradient or steep bathymetry. Relatively small errors in station location in the model grid can change flux predictions by an order of magnitude.

2.3
Husbandry Data

The detail of husbandry data required depends on the length and complexity of the modelling study. Feed input data (kg cage^{-1} d^{-1}) are most commonly used, but information on fish species, numbers, average weight, total biomass and feed diameter should be obtained where available. The study objectives will also determine whether these data are required for the whole farm, each cage group or individual cages. Monthly husbandry data summaries are sufficient for long term predictions of benthic effects but for short term studies (e.g. 24 hours), daily data are necessary for each cage. This is particularly important in polyculture operations where different sized species are contained in adjacent cages, as is often the case for gilthead sea bream (*Sparus aurata* L.) and sea bass (*Dicentrarchus labrax* L.) culture in the Mediterranean Sea.

Obtaining detailed husbandry data is nearly always problematic and reluctance to providing this information is commonplace, irrespective of whether a confidentiality agreement exists. Although information obtained direct from the farmer at the time of survey will be accurate, data from the fish farm regional offices often have the advantage of being in an electronic format covering an extended period. It is crucial to obtain the cage numbering system the farmer uses, as detailed cage by cage husbandry data will refer to this numbering system. This numbering system put in place locally by the farmer may be completely unknown to the office where husbandry data are collated.

Specific timing of feeding and defecation events can be important aspects which require modelling, this is tested later. The number of daily feeding events as well as the feeding method (i.e. hand or automatic) are increasingly being modelled and information on this is best obtained during the site survey. Little information exists in the literature on the evacuation of faecal material by farmed fish in relation to feeding times, so defecation events may be typically modelled over the whole day rather than at specific times. This is particularly the case for longer term studies.

Few studies exist in the scientific literature on feed wastage as generalisations are difficult given the variation in husbandry feeding practices. Modelling studies currently use uneaten feed as 1 to 5% of feed input, with UK regulatory modelling studies using 5% to represent a worse case scenario [10]. Feed digestibility and water content will vary with feed pellet type, ambient water temperature and fish size and species. Modelling scenarios are increasingly varying feed pellet size between cages and over time in the farm growing cycle. Although these data show some sensitivity in model predictions, the accuracy of hydrodynamic data and faecal settling data are of higher importance.

2.4
Hydrodynamic Data

Sensitivity analyses show hydrodynamic data to be an important driver of particle tracking model predictions. Early models used summary statistics of current or flow fields which did not vary spatially within the grid or with depth [11]. More recently, the most common approach uses depth varying, horizontally homogeneous flow fields [9]. This approach provides sufficient detail close to the cages and current meter mooring location, but accuracy decreases at increasing distances from the farm where circulation patterns change. To overcome this, aquaculture models can be coupled with hydrodynamic models allowing flow fields to vary horizontally around the grid [3], particularly important for the assessment of solids export via erosion. For areas where hydrodynamic models exist such as in coastal Maine, US, aquaculture models have been applied in this way for a while [12]. However, coarse grid resolution is a problem for the few existing hydrographic models in Scottish and Norwegian fjordic areas. These models are also normally pseudo-3D as generalised vertical profiles are fitted to the flow fields.

Typical aquaculture modelling studies use several current data records to represent change in flow fields with depth. With discrete depth sampling current meters, two or three instruments can achieve this but a single profiling instrument such as an Acoustic Doppler Current Profiler (ADCP) can sample most of the water column. Ideally, these instruments should be used in combination with a low threshold electromagnetic current meter deployed in the surface layer not sampled by the ADCP. Data used in modelling vary between 5 minute intervals to hourly averages, with individual particle trajectory steps being resolved on a smaller timescale than the current data time step (of the order of tens of seconds). Current record length depends on whether modelled or observed current data are being used as well as the study objectives. For example, short term sediment trap validation studies of a few days require detailed current measurements during the deployment period. Conversely, for modelling of benthic effects longer hydrographic records of several months around the sampling date will increase confidence in predictions.

Where limited numbers of instruments are available, priority should be given to obtaining surface and near-bed measurements. Where only one instrument is available, mid-water may be the best deployment depth particularly if the site is deep. In the case of a deep site, surface measurements only in the model would tend to over-predict dispersion as the higher surface flows are used at all depths. A further advantage in deploying a mid-water current meter is that this is commonly at a similar depth to the net bottom. Measurement of the current at this depth is desirable as sediment trap flux measurements have demonstrated that the exit point of particles at some Mediterranean sites was the net bottom and not the net sides (Cromey, unpublished). In addition, measurements at mid-water depth are less likely to be influenced by the shadowing effect of cages. Fitting a theoretical vertical profile between data sampled at discrete depths is generally inaccurate and unnecessarily elaborate, especially in stratified waters.

2.5
Dispersion Coefficient Data

It is convenient for particle tracking models to consider flow fields as the sum of two components: a slowly varying mean current which can be represented suitably by current meter data and a more rapidly varying, random turbulent component whose mean value is zero. The mean current initially transports the particles away from their release point with turbulence dispersing the particles more widely. Random excursions introduced into the particle trajectories can represent turbulence and be related numerically to horizontal and vertical dispersion coefficients [13]. Drifting buoy and dye studies used to measure dispersion are more commonly associated with long sea outfalls or marine dumping grounds than fish farms. The MERAMED project (www.meramed.com) undertook a number of drifter surveys using six DGPS drifting buoys at Mediterranean fish farms (fix interval 30 s; accuracy 57% ± 1 m, 99% ± 4 m; sock depth 6 m; Cromey et al., unpublished). The main limitation of such studies is that only a snap shot of conditions are obtained during the survey period and generalisations are often required when modelling several sites. For example, in Scotland regulatory models apply a standardised horizontal dispersion coefficient (k_x, k_y)

of $0.1 \text{ m}^2 \text{ s}^{-1}$ [10] unless site-specific data are provided. k_x is resolved for the model x-axis (090°–270° true axis) and k_y for the model y-axis (000°–180° axis). These regulatory models use $0.001 \text{ m}^2 \text{ s}^{-1}$ for the vertical dispersion coefficient (k_z).

2.6
Faecal and Feed Properties

A recent review of faecal settling velocity data available in the literature showed salmonids as the most studied farmed species [14]. Detailed experiments undertaken by these authors on the settling rates of bass and bream faeces found significant differences between species. The faecal mass associated with particles of different settling velocities was also shown to be important as well as the inadequacies of using averaged data. As more detailed experiments are being undertaken with faecal material [15, 16], differences between species farmed in polyculture operations are being included in modelling studies. In these cases, as adjacent cages contain different species with different food conversion ratio and faecal settling rates, cage by cage detail is necessary in the model. Where long term simulations are being undertaken a combined distribution representative of both species is usually more practical.

Information on feed pellets is more commonly available in the literature as these are easier to obtain for experimentation than faeces [4, 17–19]. Feed digestibility, water content and wastage are used to determine faecal outputs of the fish in a bioenergetics model. Although this output will vary at different stages of the growing cycle as a result of changes in the fish digestive system and diet, changes in faecal output are not commonly modelled. Relationships between feed pellet diameter, pellet type (i.e. pelletised and extruded) and settling velocity allow detailed modelling of the settling behaviour of feed particles. Where husbandry data do not include information on pellet diameter, averaged values are acceptable and model sensitivity is generally low particularly where removal of waste pellets by wild fish is simulated. In Mediterranean farms where different size classes are farmed in the same group, there is an increasing need to specify feed properties for each cage.

2.7
Wild Fish Populations

Wild fish populations significantly effect the fate pathway of waste particles from fish farms [20] and experiments in the Eastern Mediterranean have confirmed that these populations may completely remove waste feed pellets (H. Thetmeyer, pers. comm.). Theoretically the removal of each waste type can be modelled via wild fish feeding on the bed or in the water column. The actual mechanism however, will vary on a site by site basis and depend on the species present, mode of feeding and season. Although estimation of wild fish feeding on farm derived wastes is possible, the effect of wild fish defecation is less easily determined. Adequate justification of input data are required on a site by site basis usually by wild fish population surveys and examination of behaviour.

2.8
Biochemical Components

The complex biochemical processes in impacted fish farm sediments are well studied [21, 22]. Bacterial decomposition of organic material has been shown to decompose via first-order kinetics [23] as the rate of sulphate reduction was in direct proportion to the amount of organic material added. Evidence suggested that this material can be divided into two fractions of considerably different reactivity, with a third non-metabolizable fraction (G_{NR}). The G-model is a simple multi-first-order rate law for organic decomposition in marine sediments:

$$G_T(t) = G_{01}[e^{-k_1 t}] + G_{02}[e^{-k_2 t}] + G_{NR} \qquad (1)$$

where G_T is the concentration of organic carbon at time t, G_{01} and G_{02} are initial concentrations of highly reactive and less reactive carbon respectively and k_1 and k_2 are rate constants for the different components. This useful model allows carbon with different reactivities to be modelled separately. Where links are sought between some prediction of a physical aspect of the waste (e.g. deposition rates) and an observed benthic response, use of the G-model may improve these relationships. Some research has been undertaken on sulphate reduction of fish farm wastes, concentrating mainly on feed [24]. This focus on feed is not ideal as both waste types (i.e. food and faeces) require definition of G_1, G_2 and G_{NR} fractions in the G-model. However, some simplification may be possible by using a single rate constant for each waste type. Further research is required to determine the rate constants of each of these wastes, particularly at different temperatures.

Modelling different metabolic fractions of carbon is particularly useful when adopting an approach which depends on the balance between oxygen supply and demand as a predictor of benthic response. These models use the ratio between oxygen supply and demand in sediments and have been successfully developed as an indicator for benthic response to salmon aquaculture operations [25, 26]. Oxygen supply to the benthos was calculated based on Fickian diffusion and 2 hour averaged near-bed flow velocities. A relationship between carbon sedimentation rates and rates of benthic metabolism allowed an estimation of oxygen demand for predicted sedimentation rates around the farm. The ratio between oxygen supply and demand could then be correlated with the degree of benthic effect for different sites. The impact index (I) is:

$$I = \frac{O_{2\,supply}}{O_{2\,demand}} \qquad (2)$$

where $I > 1.0$ indicates minimum impact, for $I \sim 1.0$ impacts are moderate and if $I < 1.0$ then impacts are potentially severe. This model would benefit from having a temperature dependent term for oxygen as well as building in some variation in ambient seawater oxygen concentration. The relationship used in the model between carbon flux to the sediment and benthic metabolism measured by O_2 consumption could be improved by considering different waste types. The degradability of carbon is likely to vary spatially across the deposition footprint area, with more highly degradable carbon arising from waste feed proximal to the cages. The proportion of highly degradable carbon will decrease at increasing distances from the cages and further compartmentalisation of these different types of carbon in this model would be beneficial. In addition, rather than using a 2 hour averaged velocity term, the model could be tested to see how the index varies over short time periods when oxygen stress is high during periods of slack water at neap tides.

2.9
Resuspension

Resuspension has been a large component of estuarine models for many years [27–31], usually compartmentalised into several different processes which can be studied individually. Erosion is typically assumed to take place above a certain threshold of bed shear stress. Although this threshold is dependent on sediment characteristics such as depth of the so called "fluff" layer, sediment grain size, density and cohesiveness [32–34], a constant threshold in space and time is more commonly modelled. However, some advances have been made in varying critical thresholds throughout the tidal cycle in estuarine system modelling [31]. Once material is resuspended, it is then subject to advection and dispersion in near-bed flows and will redeposit if bed shear stress reduces to a level below a critical threshold for deposition. Some research has concluded modelling of redeposition to be less important than erosion in determining resuspension effects as a whole [35]. Consolidation in the model may also take place depending on the concentration of bed material or as a function of the time the material has been on the bed [36, 37].

Far fewer advances have been made in modelling resuspension of aquaculture wastes. Researchers using the AWATS modelling package published some of the first detailed observational and model validation studies of fish farm waste resuspension in the US [5, 12]. Prior to this, most published models had concentrated primarily on details of particle transport from source to sea bed [1, 11]. Although resuspension is likely to be more important in macrotidal areas, the usefulness of a model is not necessarily reduced if it lacks a resuspension component. For example, in primarily wind-driven environments such as the Mediterranean Sea, resuspension of aquaculture wastes is likely to be less important than accurate hydrographic or faecal settling measurements. A recent integrated hydrodynamic and lagrangian model for a Western Mediterranean fish farm excluded resuspension [3].

Although these resuspension models achieve a level of validation [6, 12], the values measured and subsequently used for the critical threshold for erosion are quite different. Critical thresholds used in some models vary between 33 and 66 cm s^{-1} [5] and this is in stark contrast to the value of 10 cm s^{-1} used in other models [6]. As this threshold is such a sensitive parameter, AWATS and DEPOMOD would give quite different predictions of resuspension effects despite both of these models being validated in macro-tidal areas containing salmonid farms. Not only would differences in predicted accumulation of solids underneath the cages be different but so would predictions of net export of solids from the area. As well as being highly organic, this resuspended material may also contain chemotheraputants [38, 39] and heavy metals [40, 41]. Some user interpretation as to the fate of this material in adjacent waters is required and where hydrodynamic models or observations are absent, this will be subjective.

Several studies support a low erosion threshold (< 15 cm s^{-1} near bed speed [42–47]) others a higher one (> 15 cm s^{-1} [5, 12, 27, 48]) although the type of particulate material, environment and method employed varies between these studies. One study deployed annular flumes on the sea bed and undertook concurrent turbidity measurements, water sampling and videos at different current magnitudes [5]. The measured turbidity was then correlated with shear velocity to provide the critical erosion threshold for use in modelling. The authors conclude that these experiments are likely to measure the erosion properties of both the fish farm material and the natural sediment, including fine cohesive and coarse material. This then requires modelling of the bulk sediment, as the threshold of the farm derived and natural material cannot be separated. By comparison, another validation study used a UV fluorescent particle tracer which had a similar diameter, specific gravity and settling velocity to faecal particles [6]. This was deployed on the sea bed and its redistribution measured by intensive grab sampling over time. The resuspension model was validated by varying parameters within reasonable bounds determined from literature studies. A value of 9.5 cm s^{-1} resulted in the best agreement between observed and predicted mass budgets, agreeing with studies focussed on freshly deposited material. The use of a higher threshold (e.g. 33 cm s^{-1}) determined for the AWATS model would have resulted in zero resuspension, clearly not the case from observations. For this tracer study [6], the effectiveness of sampling discrete tracer particles is an uncertainty, even though up to thirty nine 0.1 m^2 van Veen grabs were taken per sampling event. In addition, manufactured tracer particles will also lack some of the cohesive properties of farm derived material.

There are several other methods which potentially could be used to validate a fish farm resuspension model. Sediment traps have been used to assess particle resuspension in a sea grass (*Posidonia oceanica*) meadow [7]. This method examines sediment trap flux at various heights and considers material yet to deposit on the bed (primary flux) and material which has already deposited and subsequently been eroded (resuspended flux). These two fluxes can be decoupled assuming that the resuspended flux declines exponentially with height above the bed. This approach worked where the primary flux was a diffuse source of natural particulate material, but it may not work if there are high concentrations of farm-derived material in the water column. Trap deployment outside the main deposition zone or use of a model to predict primary flux at each trap location may be required. Another study has successfully used labelled sea grass leaves to assess resuspension in a sea grass meadow [49].

Any signature in the farm waste which can be accurately determined could in principle be used to determine the deposition and subsequent resuspension. Isotope analysis has been used for this purpose [50] but other tracers such as Titanium dioxide in feed have been tested (H. Thetmeyer, pers. comm.). Other methods such as CHN analyses tend to be consistently unreliable. Any such tracer needs to have a reasonable half-life, a minimal background concentration and be exclusively associated with the farm waste material. Fundamentally though, the tracer will need to be acceptable to the fish farmer and perhaps already a component of the feed, thus limiting the number of suitable tracers.

2.10
Standardisation of Data

Aquaculture model input data generally falls into three categories made up of site-specific survey data, site-specific data obtained from the fish farmer and standardised (default) data. Hydrographic data, bathymetry, cage and sampling station layout are necessary for modelling a site and should be given priority during survey planning. Occasionally, dispersion coefficients are available from a specifically designed survey. For validation of model predictions, benthic faunal data, sediment trap data and tracer concentrations from the site can be used to test model predictions if these are available. The wild fish population and its effect on the fate of wastes is also site-specific. The second category of data is mostly comprised of husbandry data obtained from the farmer and these are required for accurate modelling. The final category is comprised of feed wastage estimates, feed digestibility and water content, feed and faecal settling velocities. Standardising these data means that differences in predicted flux and benthic effect between scenarios will be primarily a result of differences in feed input, hydrography and bathymetry.

2.11
Complexity of Scenarios

Scenario complexity and recommendations for setting up deposition and benthic effect models for salmonid and sea bass/bream farms are made here (DEPOMOD [9] and MERAMOD—Cromey et al., unpublished). It is good modelling practice to decide on modelling study objectives prior to a site survey so that appropriate data can be collected. During the modelling exercise it is useful to begin with a simple scenario, followed by an increase in the level of complexity in stages with appropriate checks on model output. In addition to building confidence in the model, the effect on model predictions of increasing scenario complexity can be assessed. Highly complex scenarios can make little improvement to model predictions and a simple robust model that performs reliably is more desirable than one requiring extensive data input. Despite this, a simple model should still use good quality input data and where the reliability of these data are uncertain, sensitivity should be tested.

The lowest level of complexity should allow a quick assessment of the site to check that measured hydrodynamic data are accurate. The second level of complexity should use accurate bathymetry, feed input and species data for individual cages (if available). Faecal settling velocities appropriate to the species being farmed should be used and feed settling velocity according to pellet diameter. Where the species for each cage is unknown, a combined faecal settling velocity distribution representative of both species should be used. A wild fish module with justified variables for the site is appropriate. Site-specific data for dispersion coefficients should also be used, where available. The third and final level of complexity includes modelling the timing of feeding and defecation events and modifying the release position of particles within the cage volume. Appropriate justification is required if the latter variable is modified from default as this is highly sensitive.

After the model has been run for the first level of complexity, a deposition footprint contour plot should exhibit similar features to a scatter plot of current velocity vector components. Any features such as a strong residual current should be apparent in both, although if the footprint has been derived using several current velocity data sets this will be less apparent. For every level of complexity and for the final scenarios, the user should check model output with a simple mass balance calculation. This enables the user to verify wastage rates are being represented accurately in the model and this check should be undertaken as necessary throughout scenario development. In addition, the mass exported as a result of erosion processes should be calculated. As the level of complexity is increased, comparisons can be made between the predicted flux and benthic effects and observations.

When undertaking predictions of flux and benthic effect, the detail of husbandry data available from the farmer is usually lower than the level of detail which the model can use. Typically monthly summaries for the whole farm are provided with biomass, fish numbers and weight and feed input and as a result, mean feed input and settling velocity data have to be used for all cages. If benthic faunal sampling was undertaken underneath cages containing large fish, the real feed input may be greater than the averaged value used in the model. In this case, the model is likely to underpredict flux and benthic effects at the benthic sampling location and overpredict effects at shallower locations where cages containing smaller fish are located. It is typical to use an average feed input of the three months prior to benthic sampling, but some modifications to this approach can be used if these data are largely different between months.

Planning scenarios of new sites will typically consist of modelling a lengthy period of the farm growing cycle. Such scenarios are crucially dependent on using representative hydrodynamic data ideally measured during the same season when maximum farm biomass is expected. Hourly averaged hydrographic data (ten or twenty minute sampling interval) of at least one month are typical. Mean and maximum feed inputs for a growing cycle and different cage locations and configurations within the study area are useful scenarios. Modelling existing farms usually has the advantage that a monitoring fieldwork survey exists and these data are available for model validation.

Model validation of solids or carbon flux can be undertaken by comparing predictions with observed sediment trap data. Current meter data measured over the complete study duration are essential and for short term validation studies (e.g. a few days or less), unaveraged hydrographic data (five minute sampling interval) are ideal. For such detailed studies, feed input and species data for individual cages for the experimental period are required. Experimental field studies using tracers or chemicals associated with the particulate wastes can be used to validate model predictions and can be several months in length. Concentrations of the chemical on the feed and faeces are required, or the concentration on the feed and its retention in the species. These studies require hourly averaged ten or twenty minute sampled hydrographic data collected over the complete study duration, even if this is for several months. Decay of chemical or carbon mass via a first order kinetics decay model such as the G-model is usually necessary [23]. For all the complexity levels and scenarios described, meteorological data should be collected concurrently as these are useful for interpretation of hydrographic data and harmonic analyses.

2.12
Model Output

A contouring package is useful for visualisation of output in contour plots of flux and benthic effect. Attention should be paid to the minimum contour level used in the contour maps and this should not be unnecessarily low (e.g.

50 g m^{-2} yr^{-1} minimum). In addition, artefacts of contouring algorithms and indeed the model are often highlighted in these maps in initial scenarios. Action should be taken to investigate and eliminate these or at least highlight them by attaching caveats.

3
Detailed Investigations

The effect on model predictions of two important data types (drifter and hydrographic data) and two less important processes (cage movement and waste release times) are explored. Drifter and hydrographic data are a fundamental component of a detailed site study and can be analysed and implemented into the model in many different ways. Cage movement exists at every site but few authors have tested the effect this has on model predictions with real positional data. Timing of feeding and defecation events appear to be an unnecessary detail in the majority of studies, with the exception of short-term sediment trap validation studies.

In these investigations, both predictions of flux (g m^{-2} yr^{-1}) and the benthic faunal descriptor Infaunal Trophic Index (ITI) [51] obtained from the DEPOMOD model [9] were used to test model sensitivity. Flux predictions in this model were validated with sediment trap observations and are generally satisfactory ($\pm 20\%$ dispersive site, $\pm 13\%$ depositional site). Despite shortcomings of sediment traps such as resuspension of contents and collection of non-farm derived material, these effects can generally be minimised by deploying well designed traps (height : diameter = 5 : 1) for short periods in the main deposition zone. The ITI describes the feeding behaviour of soft bottom benthic communities in terms of a single parameter. This much debated index [52] classifies animals according to their mode of feeding (i.e. suspension, surface detritus, surface deposit, sub-surface deposit). It has the range 0 (grossly impacted) to 100 (pristine), but 0 to 60 is more commonly observed. During validation studies, natural variability of the benthic community resulted in an ITI of 60 ± 18 ITI for unimpacted stations and 5 ± 5 ITI for impacted stations.

3.1
Drifter Data and Dispersion Aspects

Drifter technology is commonly used for measuring dispersion in the marine environment. As technology has developed over the last few decades, most of the application has been in the global ocean and shelf seas [53–56]. Methods employed in early studies varied widely from the simplest of studies involving the release of objects (i.e. oranges) to the use of more complex technology such as aerial photography and radar [57]. Now as selective availability has been switched off, GPS technology is commonly used. The spatial and temporal information provided by these different studies varies considerably and the most appropriate method will depend on the environment and the scale of the process being studied. For oceanic applications where the study area is potentially massive, the frequency and accuracy of the positional information can be reduced so that a long survey period is achieved through preservation of battery life. There are very few examples of inshore and freshwater lake drifter studies [58–60]. Generally, greater sampling frequency and accuracy is required as the main aims are to measure circulation patterns and dispersion properties of a comparatively small area (e.g. a few km^2). In the case of the oceanic survey where the frequency of each position fix is small, the reliability in obtaining and subsequently storing or relaying the positional information is important. The use of drifter technology in the coastal environment has been mainly used for assessment of domestic sewage or industrial discharges with studies around aquaculture operations not so common in the literature. The use of drifters in this environment will typically involve GPS, but DGPS systems do exist. Given the low cost and availability of GPS, materials are commonly available in the market place to manufacture low cost drifters using this technology [61, 62].

One such DGPS system developed at SAMS has been deployed in coastal areas in Scotland and the Eastern Mediterranean to measure dispersion and circulation patterns around fish farms (Cromey et al., unpublished). The main advantage of this system is that it provides DGPS information without the need for reliance on differential corrections from external sources. The base station can be set up in any location and is not affected by steep sided topography which can interfere with corrections from external commercial sources. The frequency and accuracy of the positional information is useful for inshore studies but line of sight is required with the base station. A simplification of this system which involved attachment of hand-held waterproof GPS units with internal logging to drogues has been found to be useful. Line of sight is not required with a base station so that a topographically complex area can be surveyed with this system.

The main objective of a fish farm drifter study is to provide information on dispersion and circulation patterns so that this can be incorporated into modelling studies of the discharge. The variance of positional data (x, y) at time t for N drifters [57] is:

$$\sigma_x^2 \equiv \frac{1}{N-1} \sum_{i=1}^{N} [x_i(t) - \bar{x}(t)]^2 \quad \sigma_y^2 \equiv \frac{1}{N-1} \sum_{i=1}^{N} [y_i(t) - \bar{y}(t)]^2 , \tag{3}$$

$$\sigma_{xy}^2 \equiv \frac{1}{N-1} \sum_{i=1}^{N} [x_i(t) - \bar{x}(t)][y_i(t) - \bar{y}(t)]$$

Dispersion coefficients (k_x and k_y) are then calculated as:

$$k_x(t) = \frac{1}{2}\frac{\partial \sigma_x^2}{\partial t}, \quad k_y(t) = \frac{1}{2}\frac{\partial \sigma_y^2}{\partial t} \tag{4}$$

Typically in aquaculture modelling studies, dispersion coefficients are resolved for east-west and north-south axes. As a result k_x is representative of dispersion along the east-west axis and k_y representative of the north-south axis. Although convenient to resolve in this manner, more accurate representation of the data may be obtained if the dispersion coefficients are resolved according to major and minor axes of flow [57]. In the case of the example shown in Table 1, dispersion coefficients were resolved for east-west (x) and north-south (y) axes respectively using site-specific data for a North Atlantic salmon site ($k_x = 0.278$ m^2 s^{-1}, $k_y = 0.108$ m^2 s^{-1}). Dispersion coefficients were then recalculated by varying the direction of the major and minor axes until covariance (σ_{xy}^2) was minimised. This resulted in major and minor axes of 058–238° and 148–328° respectively with values of k_ζ and k_η of 0.386 and 0.001 m^2 s^{-1} respectively. As variations on the method of data analysis give different answers, it is useful to test the effect on model pre-

Table 1 The axis used for evaluation of the dispersion coefficient (k) from drifter data results in different flux predictions for both macro-tidal (Scotland) and micro-tidal (Eastern Mediterranean) environments

Environment	Scotland			Eastern Mediterranean	
Species	Salmon	Salmon	Salmon	Sea bream	Sea bream
Source of k data	Measured	Measured	Theoretical	Measured	Measured
Major – k_ζ (m^{2}s^{-1})	0.108	0.386	0.1	0.108	0.446
Minor – k_η (m^{2}s^{-1})	0.278	0.001	0.1	0.353	0.015
Major axis (°)	0	58	0	0	118
Predicted vertical flux at sea bed[a] (g m^{-2} yr^{-1})					
0 m	9381	10855	10396	8048	9806
5 m	–	–	–	5591	7412
10 m	–	–	–	4168	5539
25 m	6572	8126	6848	1305	1829
50 m	2240	3009	2294	326	469
100 m	733	934	655	–	–

[a] An increase in flux at stations between scenarios is associated with a reduction in the mean flux directly underneath the cages so that mass balance is preserved.

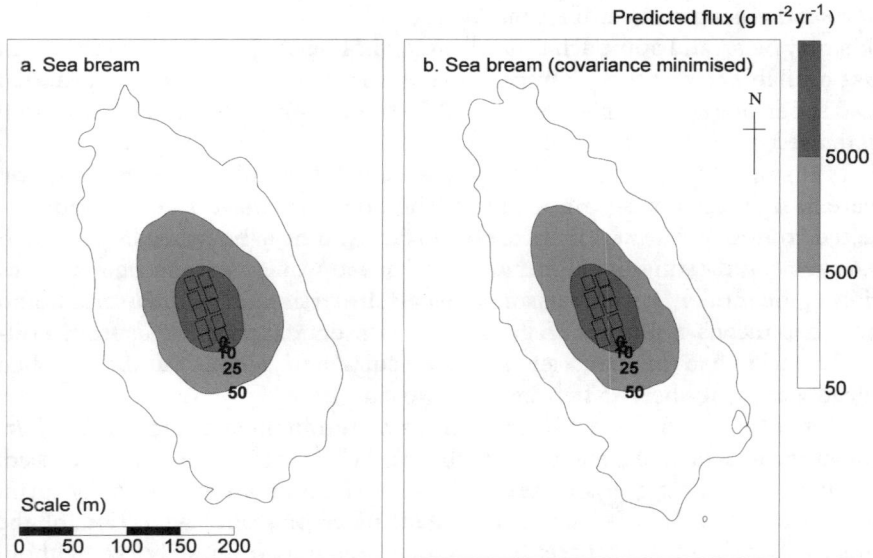

Fig. 1 Predicted flux arising from a Sea bream farm using different representations of drifting buoy data. In scenario (a), dispersion coefficients were resolved using true north as the major axis ($k_\zeta = 0.108 \text{ m}^2 \text{ s}^{-1}$, $k_\eta = 0.353 \text{ m}^2 \text{ s}^{-1}$, $\zeta = 0\,°C$). For scenario (b), covariance minimisation resulted in dispersion coefficients more representative of the plume elongated along the main current axis ($k_\zeta = 0.446 \text{ m}^2 \text{ s}^{-1}$, $k_\eta = 0.015 \text{ m}^2 \text{ s}^{-1}$, $\zeta = 118\,°C$).

dictions. Table 1 also shows the difference between data resolved for North Atlantic and Mediterranean sites and the effect on predicted flux at varying distances from the cages. For both examples, the predicted flux at each station is increased. This is caused by enhanced dispersion of material from directly underneath the cages in the direction of the major axis (and residual current for both examples). Fig. 1 showing the Mediterranean example demonstrates the difference in deposition footprint shape. Model predictions of discharges containing fine, slow settling sea bream faecal material will be influenced the most by this effect due to the long time period in which particles are being subject to random walk processes. Correlations between the current velocity component and k resolved for the major axis may well exist, as well as correlations between the major axis and the residual current direction.

3.2
Hydrographic Data

The importance of the record length in determination of farm biomass in regulatory procedures was investigated. In Scotland, farmers are required to measure current for a minimum of 15 days and this is taken as representative of a spring-neap cycle at the site. For sites which are primarily tidally driven this may be a valid approach but a 206 day time series for a Scottish fish farm was available to test this. Summary hydrographic data for the most different 15 day periods sub-sampled from the 206 days are presented in Table 2 as well data used in regulation of the site.

Mean surface and near-bed current speeds for the whole 206 day data set were 10.6 and 5.9 cm s^{-1} respectively. The most dispersive 15 day period selected from this data set had a mean surface and near-bed speed of 14.8 and 9.6 cm s^{-1}, indicating the higher near-bed speed would cause more resuspension in the model. By comparison, the least dispersive period had surface and near-bed means of 6.4 and 2.0 cm s^{-1}. The consent data which was from a different period had the lowest surface mean current of all data, but the near-bed speed was somewhere in between the two extremes.

Marked differences in ITI predictions were obtained using these different current data representations (Table 3). Where dispersive data are used, the model predicts a less severe (16.9 ITI), small area deposition footprint (0.75 × 10^5 m^2). This is due to the effect of erosion and advection of the wastes away from the source. If the model was inappropriately run without resuspension, the scenario using the most dispersive data would result in the least severe, largest area footprint. These results suggest that where modelling is used to consent the size of farms, the method should include prediction of resuspension effects.

Table 2 Summary hydrographic data statistics for a macro-tidal Scottish fish farm. The most and least dispersive 15 day periods were selected from the 206 day record. The consent record was used in regulatory procedures for the site and was from a different period

Portion of data set	Meter location	Period	Speed (cm s^{-1}) Mean	Max.	Resid.	Resid. direction (° true)	Length (d)
Whole	Surface	23/07/01–14/02/02	10.6	48.8	3.2	219	206
	Near bed		5.9	47.0	0.6	207	
Most dispersive	Surface	07/10/01–21/10/01	14.8	48.8	5.4	223	15
	Near bed		9.6	47.0	1.0	241	
Least dispersive	Surface	29/12/01–12/01/02	6.4	19.6	1.8	209	15
	Near bed		2.0	7.1	0.1	74	
Consent record	Surface	06/08/97–21/08/97	2.7	21.8	1.7	72	15
	Near bed		3.8	16.7	0.6	279	

Table 3 Predicted Infaunal Trophic Index (ITI) and Envelope of Acceptable Precision (EAP) at increasing distance from the cages using different periods of hydrographic data (with resuspension modelled). The severity and extent of the Allowable Zone of Effect (AZE) assist in comparing predicted effects between scenarios. More dispersive data result in higher predicted resuspension and thus a shorter distance to the AZE boundary. These results imply farm biomass could be unwisely increased if these dispersive data were used

		Predicted ITI (± EAP)		
Portion of data set	All data	Most dispersive	Least dispersive	Regulatory data set
0 m	4±4	15± 5	1±4	4± 4
25 m	5±4	22± 5	4±4	21± 5
50 m	16±5	59±18	15±5	59±12
100 m	34±8	59±18	36±9	59±18
AZE severity – Average predicted ITI under cages	1.9	16.9	1.1	1.6
AZE extent – Area enclosed by 37 ITI contour ($\times 10^5 \text{m}^2$)	3.95	0.75	4.27	2.36
Distance from cages to AZE boundary: SW[a]	100 m	25 m	100 m	25 m
NE[a]	100 m	0 m	100 m	50 m

[a] The boundary of the AZE is the 37 ITI contour for this site

For a quiescent site, the deposition footprint may be concentrated in a small area and have a severe effect underneath the cages. Conversely, dispersive areas favoured in recent years by the industry and regulator will tend to dilute the discharge resulting in less severe effects over a larger area. The Area or Allowable Zone of Effect (AZE) is an area around the farm which may be impacted by the farms activities. For this exercise, the boundary of the AZE is taken as an ITI of 37, which is 80% of the ITI value for the reference station at the site. The contour plots in Fig. 2 show that use of the regulatory data resulted in a predicted AZE boundary somewhere in between the two extremes of most and least dispersive conditions. This is satisfactory in a sense but in reality was fortuitous as the ideal scenario uses the whole data set (Fig. 2a), as this is the most representative. An iterative procedure was undertaken by increasing farm biomass until the extent of the AZE using the dispersive data equalled the area of $3.95 \times 10^5 \text{ m}^2$ or severity of 1.9 ITI (values for the whole data set). Using this method, the severity criteria were met first. Approximately three times more farm biomass using the dispersive data resulted in a similar AZE to the scenario using the whole data set. Thus,

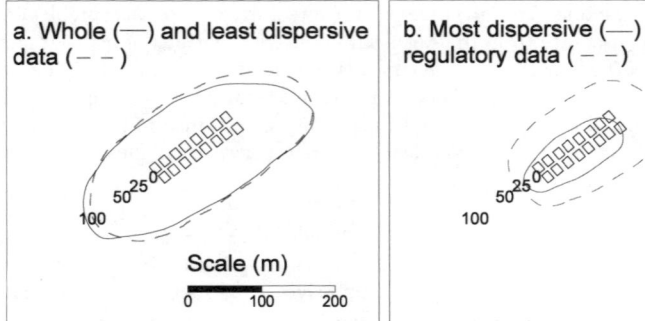

Fig. 2 The predicted boundary of the AZE using the ITI contour of 37 for different periods of current data collection. Predictions using the (a) whole data set and the least dispersive data resulted in a large AZE with a boundary at 100 m. The most dispersive data used in (b) results in a small AZE due to high predictions of high resuspension. Data used in regulation of the site resulted in an AZE somewhere in between the two extremes

any regulatory method using the dispersive data only would have resulted in a significantly higher consented biomass. As modelling is evolved within regulatory procedures, the importance of these hydrographic data considerations cannot be ignored.

In conclusion, short 15 day records on the whole are unlikely to be representative of a site and therefore biomass consented using modelling will be completely dependent on the conditions during the measurement period. To this end, the farmer takes some risk in undertaking the shortest possible deployment as little information is provided by a short data set. In practice, it is impractical for the regulator to require the farmer to undertake lengthy measurements such as 206 days but for planning scenarios, the less emphasis placed on short term fluctuations the better. Reliability of predictions will be increased with longer current records.

3.3
Cage Movement and Shadowing Effects

Physical cage structure and implications for particle exchange have been assessed using a finite element circulation model of flow for Sungo Bay, China, a site of dense cultivation of both marine bivalves and kelp [63]. Model output indicated that suspended aquaculture resulted in a 20% reduction in current speed in the main navigation channel and a 54% reduction in speed in the midst of a culture area. The research concluded that as a result, carrying capacity was over estimated, but in practice, these effects are rarely considered in measurement or modelling. Cage and mooring design and the effect on deposition is particularly important at quiescent sites as cage movement is large in relation to the deposition footprint area. Single point moorings have been found to cause a two to seventy-fold reduction in deposition due to the wastes being dispersed over a much larger area [64]. Submerged cages can reduce net drag and mooring tension by 60% [65] and mooring arrangements can be optimised to minimise the effects of wave action [66]. Such complex mooring arrangements are not commonly included in aquaculture impact models, most probably due to their infrequent occurrence.

The limit of cage group movement will depend primarily on mooring design but the movement within these limits will depend on the various forces acting on the cage group and its resistance to these forces (Fig. 3). The balance of tidal and wind forces on the group is a primary factor and the resistance to these forces will depend on the degree of net biofouling, stocking density, frequency and location of empty cages. All these factors make it impossible to generalise a set of variables which could easily be used in deposition modelling. Although such models do exist for the testing of mooring designs, the coupling of such a model with a flux and benthic effects model is impractical but not impossible. Such a model would require hydrographic data of current and wind, with some additional information on the factors influencing resistance outlined above. Validation of a cage movement model could be undertaken by comparing predicted cage movements with DGPS fixes of cage corners over a period of time.

To test the effect of cage movement on the predictions of a benthic effects model, the model was run with both static and moving cage groups. At this North Atlantic site, the flood tide was in an easterly direction and the

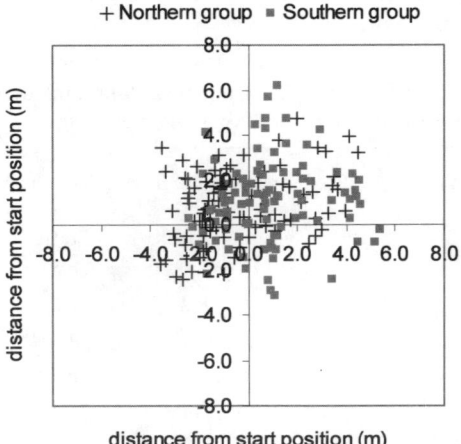

Fig. 3 Cage movement for two cage groups at a Scottish fish farm measured using ten minute averaged DGPS over a period of 16 hours. The position (0,0) is the group starting position and similar movement was measured in each group (northern group – 50% data ±2.2 m, 95% ±5.0 m; southern group – 50% ±2.4 m, 95% ±4.5 m)

ebb tide in a westerly direction. Mean and maximum speeds of the current during the period were 6.1 and 29.1 cm s^{-1} respectively, with a fresh W wind between 7 and 12 m s^{-1}. Cage movement was measured overnight using DGPS drifting buoys previously described attached to a corner of two cage groups. A small amount of cage group movement was measured with generally 95% of data within 5 m of the starting position (Fig. 3). Little correlation was found between current direction and cage movement, indicating the groups were primarily wind rode throughout the period.

Despite the small amount of cage movement observed, it is still useful to determine the effect on model predictions. In particular, flux predictions over short periods are usually compared with sediment trap observations so testing the cage movement effect on this aspect is useful. For the modelling undertaken here, the effect of cage movement on flux predictions close to the cages where traps are normally deployed was minimal (Table 4). A very slight difference was obtained at the 100 m station but it is unlikely this difference would be detected in reality. At these outer stations, the magnitude of natural background sedimentation over farm derived material is much higher and this would tend to disguise this difference. ITI predictions in Table 4 show low sensitivity of the benthic response model to cage movement and differences between model runs could not be distinguished from the natural benthic variability expected (i.e. the Envelope of Acceptable Precision (EAP)). The effect of cage movement in these scenarios is irrelevant when compared with using different record lengths of hydrographic data described earlier.

Although the data presented are a useful test of cage movement effects on model predictions, this short data set can only be taken as a preliminary test. There were some problems with positional data analysis and implementation

Table 4 Predicted vertical flux and ITI using two static and dynamic cage groups at a Scottish fish farm (EAP is Envelope of Acceptable Precision)

Group	Station	Predicted vertical flux (g m^{-2} yr^{-1})		Predicted ITI ($\pm$ EAP)	
		Static	Dynamic	Static	Dynamic
N	0 m	12974	11974	1 $\pm$ 4	1 $\pm$ 4
	25 m	6169	5681	4 $\pm$ 4	4 $\pm$ 4
	50 m	2314	2181	6 $\pm$ 5	7 $\pm$ 5
	100 m	301	311	26 $\pm$ 5	26 $\pm$ 5
S	0 m	12743	12319	1 $\pm$ 4	1 $\pm$ 4
	25 m	5858	5620	4 $\pm$ 4	4 $\pm$ 4
	50 m	2183	2132	7 $\pm$ 5	7 $\pm$ 5
	100 m	193	255	30 $\pm$ 7	27 $\pm$ 7

into the model. The inherent error in the positioning system needs to be taken into account and this was undertaken here by averaging the positional data to obtain an interval equal to current data. Given the movement measured was mostly less than 5 metres, use of GPS technology alone will not provide the required accuracy for an experiment of this type. These data cannot easily be correlated with the force associated with either current or wind and as a result, common use of cage movement data in anything other than intensive studies seems unlikely.

3.4
Feeding Systems and Waste Particle Release Times

Automated feeding systems detect feed wastage in cages and switch off feeding. Hand feeding however, is still in use as a preferred husbandry practice at some sites or where automated systems are impractical. The default settings of DEPOMOD assume a continuous release of food and faeces over a 24-hour period. Long-term predictions of benthic effects use a tidal current record repeated back-to-back to simulate a period much longer than the measurement period. Releasing particles at specific times of the day would not be accurate once the model passes into the second loop of the run. This is avoided by releasing particles continuously over 24 hours, placing less emphasis on the time of day. However, much shorter simulations of flux such as those used for comparison with sediment trap observations can show some sensitivity to particle release times. This sensitivity was tested using a continuous release over 24 hours, for both automated and hand feeding methods. In the model, automated feeding included hourly releases of equal amounts of feed between 09:00 and 16:00 local time with 100% of the daily faecal mass discharged in this period. Hand feeding was simulated with two feeds per day with 70% and 30% of feed and associated faecal mass in the morning (09:00–10:00) and afternoon (15:00–16:00) events respectively.

For 24-hour simulations (Fig. 4a,b), model predictions were sensitive to feeding method implying that validation studies should consider these aspects. In addition, differences between spring and neap simulations are particularly important at the outlying stations (50 and 100 m) where an order of magnitude difference was predicted. Validation studies would therefore be best designed to incorporate both of these periods to test the model across the expected range of deposition. In practice, where only one study is planned, a worse-case scenario would concentrate on the neap period. Modelling a long period of 206 days with measured current data allowed the model to accurately couple discharge events and current data. Given such a long simulation period, the release of waste material is likely to take place over a whole range of hydrographic conditions. This resulted in the feeding method showing minimal sensitivity for the longer simulation periods (Fig. 4c).

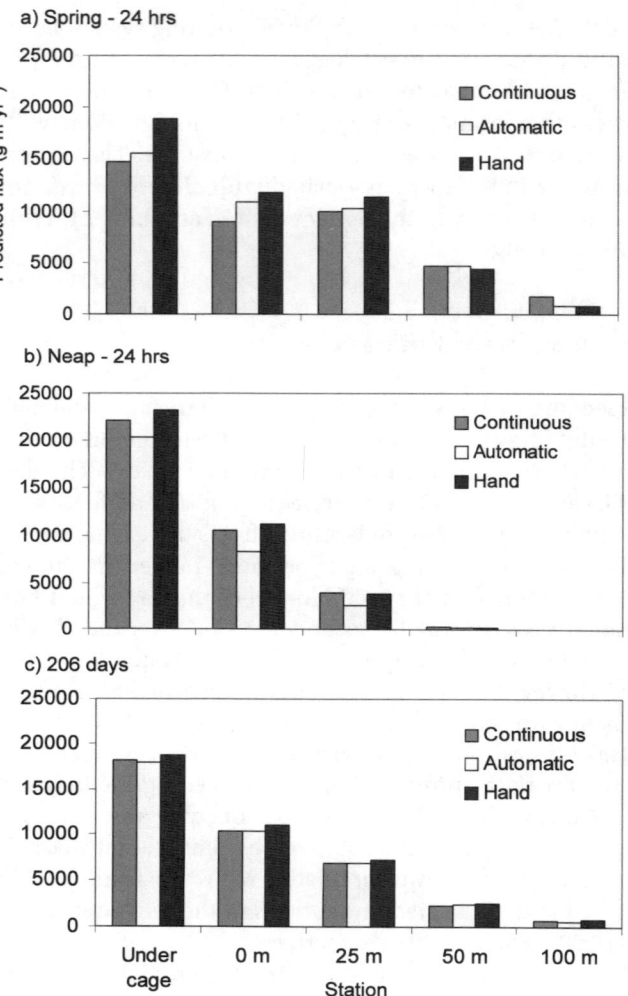

Fig. 4 Difference in predictions of flux over a 24 hour period during (**a**) spring and (**b**) neap periods for different methods of feeding. Predictions over a prolonged period of time (**c**) 206 days resulted in little difference between feeding methods

Additional husbandry observations such as the distribution of feed pellets hitting the surface during feeding are useful for sediment trap studies. In addition, location of feed barrels and feed loading/unloading points can be noted if near the experimental cage. Recording this level of detail can often give useful clues if particular sediment traps contain elevated levels of waste material.

4
Conclusion

The predictive capability and level of validation of aquaculture discharge models has advanced significantly in the last few years. Rather than just modelling particle trajectories from sea surface to sea bed, important biochemical and physical processes such as decay and resuspension are being included with attempts to validate these satisfactorily. Similarly, some benthic effect models have been developed and validated, but are still in relatively early stages of development. The main challenges for this field of modelling lie with linking these aspects, especially as the organic gradients are so steep. Improved accuracy and the lower cost of instrumentation is resulting in better spatial and temporal resolution of data, thereby assisting the advance of models. Benthic landers and underwater digital equipment capable of undertaking real time, highly accurate measurements will influence model development and validation. An increase in the level of modelling used in the regulation of fish farms globally is also expected.

Acknowledgements We would like to thank the European Union for funding MERAMOD (EU project Q5RS-2000-31779), the recent DEPOMOD developmental work supported by the Scottish Environment Protection Agency and discussions with the modelling group there. The thorough research undertaken by Louise Wilson provided insight on fish farm hydrographic data.

References

1. Hevia MG, Rosenthal H, Gowen RJ (1996) J Appl Ichthyol 12:71
2. Silvert W, Cromey CJ (2000) Modelling impacts in environmental impacts of aquaculture. In: Black KD (ed) Environmental impacts of aquaculture. Sheffield Academic Press, UK, p 154
3. Doglioli AM, Magaldi MG, Vezzulli L, Tucci S (2004) Aquaculture 231:215
4. Chen YS, Beveridge MCM, Telfer TC (1999) Aquac Int 7:89
5. Dudley RW, Panchang VG, Newell CR (2000) Aquaculture 187:319
6. Cromey CJ, Nickell TD, Black KD, Provost PG, Griffiths CR (2002) Estuaries 25:916
7. Gacia E, Duarte CM (2001) Est Coast Shelf Sci 52:505
8. Holmer M, Perez M, Duarte CM (2003) Mar Pollut Bull 46:1372
9. Cromey CJ, Nickell TD, Black KD (2002) Aquaculture 214:211
10. SEPA (2003) Marine Fish Farming Manual (annex H). Scottish Environment Protection Agency, Stirling, Scotland, www.sepa.org.uk/aquaculture
11. Silvert W, Sowles JW (1996) J Appl Icthyol 10:258
12. Panchang V, Cheng G, Newell C (1997) Estuaries 20:14
13. Allen CM (1982) Proc Roy Soc London A 381:179
14. Magill SH, Thetmeyer H, Cromey CJ (2005) Aquaculture (In press)
15. Chen YS, Beveridge MCM, Telfer TC (1999) Aquac Res 30:395
16. Chen YS, Beveridge MCM, Telfer TC, Roy WJ (2003) J Appl Ichthyol 19:114
17. Tlusty MF, Snook K, Pepper VA, Anderson MR (2000) Aquac Res 31:745

18. Holmer M, Kristensen E (1994) Microbiol Ecol 14:33
19. Stewart ARJ, Grant J (2002) Aquac Res 33:799
20. Dempster T, Sanchez-Jerez P, Bayle-Sempere JT, Gimenez-Casalduero F, Valle C (2002) Mar Ecol Prog Ser 242:237
21. Holmer M, Kristensen E (1992) Mar Ecol Prog Ser 80:191
22. Black KD, Keimer MBC, Ezzi IA (1996) J Appl Ichthyol 12:15
23. Westrich JT, Berner RA (1984) Limnol Oceanogr 29:236
24. Holmer M, Kristensen E (1994) Microbiol Ecol 14:33
25. Findlay RH, Watling L (1997) Mar Ecol Prog Ser 155:147
26. Morrisey DJ, Gibbs MM, Pickmere SE, Cole RG (2000) Aquaculture 185:257
27. Uncles RJ, Elliott RCA, Weston SA (1985) Estuar Coast Shelf Sci 20:147
28. Partheniades E (1965) Proc Am Soc Civ Eng 91:105
29. Puls W, Sündermann J (1990) Simulation of suspended sediment dispersion in the North Sea. In: Cheng RT (ed) Residual currents and long term transport. Coastal and estuarine studies No. 38. Springer, Berlin Heidelberg New York, p 356
30. Teisson C (1991) J Hyd Res 29:755
31. Clarke S, Elliot AJ (1998) Estuar Coast Shelf Sci 47:235
32. O'Connor BA, Nicholson J (1992) An estuarine and coastal sand transport model. In: Prandle D (ed) Dynamics and exchanges in estuaries and the coastal zone. Coastal and estuarine studies No. 40. Springer, Berlin Heidelberg New York, p 507
33. Velegrakis AF, Gao S, Lafite R, Dupont JP, Huault MF, Nash LA, Collins MB (1997) J Sea Res 38:17
34. Amos CL, Feeney T, Sutherland TF, Luternauer JL (1997) Estuar Coast Shelf Sci 45:507
35. Sanford LP, Halka JP (1993) Mar Geol 114:37
36. Futawatari T, Kusuda T (1993) Modelling of suspended sediment transport in a tidal river. In: Prandle D (ed) Coastal and estuarine studies No. 42. Springer, Berlin Heidelberg New York, p 504
37. Sanford LP (1992) Limnol Oceanogr 37:1164
38. Black KD, Fleming S, Nickell TD, Pereira PMF (1997) ICES J Mar Sci 54:276
39. Nickell TD, McHenery JG, Black KD, Katz T, Coates A, Gillan N, Roberts A, Breuer E (2005) J Appl Ichthyol (In press)
40. Morrisey DJ, Gibbs MM, Pickmere SE, Cole RG (2000) Aquaculture 185:257
41. Brooks KM, Mahnken CVW (2003) Fish Res 62:295
42. Burt TN, Turner KA (1983) Deposition of sewage sludge on a rippled sand bed. Hydraulics Research Report IT248, Wallingford, UK
43. de-Jonge VN, van den Bergs J (1987) Estuar Coast Shelf Sci 24:725
44. Washburn L, Jones BH, Bratkovich A, Dickey TD, Chen MS (1991) J Hydraul Eng 118:38
45. Lund-Hansen LC, Valeur J, Pejrup M, Jensen A (1997) Estuar Coast Shelf Sci 44:521
46. Sanford LP, Panageotou W, Halka JP (1991) Mar Geol 97:87
47. Cromey CJ, Black KD, Edwards A, Jack IA (1998) Estuar Coast Shelf Sci 47:295
48. Tappin AD, Harris JRW, Uncles RJ (2003) Sci Total Environ 314:665
49. Terrados J, Duarte CM (2000) J Exp Mar Biol Ecol 243:45
50. Ye LX, Ritz DA, Fenton GE, Lewis ME (1991) J Exp Mar Biol Ecol 145:161
51. Mearns AJ, Word JQ (1982) Forecasting effects of sewage solids on marine benthic communities. In: Mayer GF (Ed) Ecological Stress And The New York Bight: Science And Management. Columbia S Carolina Estuarine Research Federation, p 495
52. Maurer D, Nguyen H, Robertson G, Gerlinger T (1999) Ecological Applications 9:699
53. Gille ST, Romero L (2003) J Atmos Ocean Technol 20:1633

54. Thompson KR, Sheng JY, Smith PC, Cong LZ (2003) J Geophys Res -Oceans 108:3277
55. McClean JL, Poulain PM, Pelton JW, Maltrud ME (2002) J Phys Oceanogr 32:2472
56. Ozgokmen TM, Piterbarg LI, Mariano AJ, Ryan EH (2001) J Phys Oceanogr 31:2691
57. Yanagi T, Murashita K, Higuchi H (1982) Deep Sea Res 29:217
58. Stocker R, Imberger J (2003) Limnol Oceanogr 48:971
59. Tseng RS (2002) Estuar Coast Shelf Sci 54:89
60. Natunewicz CC, Epifanio CE, Garvine RW (2001) Mar Ecol Prog Ser 222:143
61. Johnson D, Stocker R, Head R, Imberger J, Pattiaratchi C (2003) J Atmos Ocean Tech 20:1880
62. Manda A, Takahashi T, Komori S, Kyozuka Y, Nishimura S (2002) Int J Offshore Polar 12:213
63. Grant J, Bacher C (2001) Can J Fish Aquat Sci 58:1003
64. Goudey CA, Loverich G, Kite-Powell H, Costa-Pierce BA (2001) ICES J Mar Sci 58:497
65. Tsukrov II, Ozbay M, Fredriksson DW, Swift MR, Baldwin K, Celikkol B (2000) Mar Technol Soc J 34:29
66. Colbourne DB, Allen JH (2001) Aquac Eng 24:129

Near-Field Depositional Model for Salmon Aquaculture Waste

Dario Stucchi[1] (✉) · Terri-Ann Sutherland[2] · Colin Levings[2] · Dave Higgs[2]

[1]Institute of Ocean Sciences, Fisheries and Oceans Canada, 9860 West Saanich Road, Sidney, B.C. V8L 4B2, Canada
StucchiD@pac.dfo-mpo.gc.ca

[2]West Vancouver Laboratory, Fisheries & Oceans Canada, 4160 Marine Drive, West Vancouver, B.C. V7V 1N6, Canada
SutherlandT@pac.dfo-mpo.gc.ca, LevingsC@pac.dfo-mpo.gc.ca, HiggsD@pac.dfo-mpo.gc.ca

1	Introduction	158
2	Fish Production/Waste	160
2.1	Particle Dispersion Model	163
3	Hydrodynamics	164
4	Diagnostic	165
4.1	Single Net Pen – Analyses of Depth and Currents	165
4.2	Multiple Net Pens	168
5	Predictive	170
5.1	Methods	171
5.2	Data and Model Simulations	172
6	Discussion	174
6.1	Comparison of Observed and Predicted Sedimentation Rates	174
6.2	Model Limitations and Uncertainties	176
6.3	Future Research	178
	References	178

Abstract An analytical near-field depositional model for solids wastes (organic matter from waste feed and faeces) from open net pen culture of finfish is presented. The model is based on the premise that the statistics of the depth-averaged currents, which are assumed to be normally distributed, determine the distribution of wastes on the ocean bottom. Using a farm configuration consisting of a single net pen, the model is used in a diagnostic mode to quantitatively examine the effects on the waste depositional field or footprint of the farm that result from changing the depth under the net pens and changing the statistics (standard deviations) of the depth-averaged velocity. The model is also used to examine the changes to the farm footprint that result from orientating a two by four linear grouping of net pens perpendicular to and parallel to the principle current direction. The model was tested on an operating Atlantic salmon farm by comparing the predicted organic matter fluxes from the farm with the vertical fluxes of organic matter measured by sediment traps. Based on a rather limited data set, predicted organic matter fluxes were found to be about four to five times higher than observed sedimentation rates. Further, the model predictions were sensitive to the value used for feed waste. The limitations and uncertainties in the model assumptions, parameterizations and in the methodologies used to validate the model are discussed and recommendations for future research are provided.

Keywords Aquaculture · Model · Solid wastes · Sedimentation

1
Introduction

There exists a range of environmental impacts resulting from the culture of salmon in open net cages in the marine environment. The impacts occur over a range of space and time scales and include alterations to ecosystems caused by the enhanced release of nutrients and carbon to the water column, organic wastes to the sediments, discharge of contaminants, theraputants and cross-transmission of pathogens and parasites [1]. One of the most conspicuous impacts of open net cage fish farms is on the sediment geochemistry and benthic fauna underneath and in the immediate vicinity of net cages – the near-field environment [2, 3]. Benthic impact typically results from the enhanced sedimentation of organic-rich wastes (faeces and waste feed) from the farm. The degree of impact to the benthic community and habitat is influenced by a combination of factors such as production levels, feed characteristics (including ingredient composition and digestibility as well as physical characteristics such as pellet length and diameter, etc.), feeding efficiency, bathymetry, circulation, and the assimilative capacity of the benthic environment.

Mathematical or numerical models describing the distribution and impact of solid wastes from marine salmon farms can be useful tools for the management, monitoring and study of the aquaculture industry and its impacts. Firstly, the models can aid in the selection of suitable locations, in the configuration of the net pen structures, or in setting site-specific production limits. Secondly, because the models map both the shape and size of the depositional field of the farm wastes or footprint of the farm at any stage of the production cycle, they can assist in the design of monitoring schedules and selection of monitoring sites. Since high-resolution spatial grid sampling is technically difficult and expensive to accomplish, model footprints can aid in targeting transects of organic enrichment gradients. Finally, the models can be used as research tools by helping to identify and understand the key processes responsible for a wide range of impacts associated with the open pen culture of salmon.

Modelling the changes in sediment chemistry and benthic impact caused by the organic-rich waste solids from marine salmon farms is complex and crosses over many disciplines. To address the effect of solid waste deposition on the benthos, an overall model must adequately represent all the important processes (e.g. resuspension, benthic assimilation) that lead to and cause benthic impact. It is important to note that this type of model would address the effect of solid waste deposition, while additional models would be required to address the effects of other solid wastes (fouling debris) or other nutrient inputs (dissolved components). In this paper, we implicitly assume that the solid wastes are the main cause of the impacts on the benthos, but if other farm-related factors are important, then the model will be of limited usefulness.

The overall modelling approach consists of a sequence or progression of sub-models or modules each addressing the important processes and components of the problem. The starting point for all models is the fish production/waste module (Fig. 1) This initial module uses information about the farm configuration (net pen dimensions and layout), and fish production (species, biomass, size), and then transforms the feed input to a solid waste output from the net pens. The next modelling step in the sequence is the hydrodynamic and waste settling model, or particle tracking model, which disperses the wastes onto the ocean bottom. Using the sinking rate characteristics for the wastes, together with information about the ocean currents and the quantity of wastes (from the fish production/waste module), the particle tracking model then maps the accumulation or sedimentation rate of the wastes onto the bottom. The sedimentation rate is a key determi-

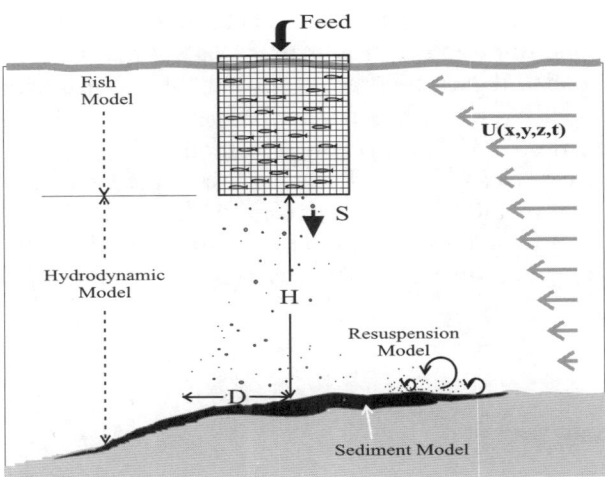

Fig. 1 Schematic diagram showing the components and main processes relating to the dispersion and transport of solids wastes from open net cage finfish farms. The letters, S, H, and U, respectively, refer to the sinking rate, vertical fall distance, and the spatially and temporally varying horizontal current velocity

nant of benthic impact, and is often used as the independent variable against which impact is assessed. Once wastes settle to the bottom, the currents, if sufficiently strong, may transport wastes through resuspension and saltation processes. Cromey et al. [4] and Panchang et al. [5] have modelled the resuspension process.

The final component or step in this progression of models is the benthic module. In this step, measures of benthic impact such as free sulfide in the sediments [6] or indices of benthic diversity [7] are compared to model-derived sedimentation rates in order to establish semi-empirical relationships. If significant relationships can be demonstrated, then the model predictions of sedimentation rates may be used to predict the degree and spatial extent of benthic impact at other locations having similar substrates, oceanographic and hydrodynamic conditions.

Many models of the distribution of organic wastes and their ecological effects have been developed and continue to be developed. Models range from the simple sedimentation model of Gowen et al. [8] to the complex site-specific model developed by Panchang et al. [5]. Most recently, the Scottish authorities have developed and commercialized a comprehensive aquaculture waste modelling suite called DEPOMOD [7]. DEPOMOD is notable in that several of the modules have been validated to some extent with field observations [4, 7].

In Fisheries and Oceans Canada, we have developed a simple analytical near-field depositional model for the waste solids from marine open cage salmon farms. Our model was developed primarily as a diagnostic tool which could be used to obtain a quantitative understanding of the key variables and how their variations altered the predicted depositional field. However, where local conditions and hydrodynamics satisfy the underlying assumptions of the model, it may be used to predict the footprint from a salmon farm. The model is primarily a particle dispersion model, which includes the fish production and hydrodynamic sub-models (Fig. 1) and is analytical and thus relatively simple and efficient. In this paper we present our model and explore its capabilities as a diagnostic tool. We also test the model as a predictive tool by comparing the data obtained from model simulations for an operating salmon farm with observations obtained from sediment traps.

2
Fish Production/Waste

The first sub-model is the fish production/waste module, wherein a given feeding rate for a given biomass of fish is converted into an estimate of the production of solid wastes. The organic matter fraction of the solid wastes is the waste characteristic of interest in this paper. Organic matter, as measured by the ash-free dry weight of the sample, represents the sum of the protein, lipid and carbohydrate constituents in the sample.

In simple terms, of all of the organic matter that is provided to the fish in the feed, OM_{feed} most is ingested ($OM_{ingested}$) by the fish and a small portion is uneaten ($OM_{waste\ feed}$) or wasted.

$$OM_{feed} = OM_{ingested} + OM_{waste\ feed} \tag{1}$$

The fate of the organic matter that is ingested may be further partitioned into three components: fish growth, respiratory/soluble wastes and faecal wastes – the first three terms on the right hand side of Eq. 2:

$$OM_{feed} = OM_{growth} + OM_{resp.\ \&\ sol.\ waste} + OM_{faeces} + OM_{waste\ feed} \tag{2}$$

The components of interest for the model are the organic matter in the solid wastes, namely the organic matter contained in the faecal wastes and the uneaten feed.

The amount of organic matter in the faecal wastes may be calculated (Eq. 3) using the apparent digestibility coefficient (ADC) for organic matter and the amount of organic matter ingested by the fish. The methodology used to determine the apparent digestibility coefficient is described by Hajen et al. [9, 10] and Forster [11].

$$OM_{faeces} = OM_{ingested} \times (1 - ADC) \tag{3}$$

Apparent digestibility coefficients for organic matter in present day high energy extruded diets for Atlantic salmon of 88% are not unreasonable (Higgs, Rowshandeli and Oakes, unpublished data). Thus about 12% of the ingested organic matter is not digested and consequently is excreted in the faeces. From Eq. 1, the amount of organic matter ingested is simply the total amount of organic matter provided in the feed minus the amount of organic matter wasted in the uneaten feed pellets.

The uneaten feed or waste feed is a direct source of solid wastes to the bottom. Improvements in feed monitoring systems through the use of underwater camera technology and other means have reduced feed wastage rates, and present day estimates are believed to be 5% or less [12]. However, direct measurements of the feed wastage rate are seldom made. Pearson and Black [1] have reported that feed wastage rates range from 5% to 15%, while Findlay and Watling [13] calculated wastage rates of 5% and 11% from observations. For the purpose of this modelling exercise, we have adopted a feed wastage rate of 8%, the average of the Findlay and Watling [13] calculated wastage rates.

Thus, waste feed accounts for about 8% of the total organic matter dispensed to the fish (extruded salmon diets contain $\sim$ 86% organic matter on an air-dry basis), while the faeces account for approximately 12% of the organic matter that is actually ingested by the fish. In relation to the total quantity of organic matter dispensed to the fish, the faecal organic matter wastes account for about 11% after correction for the organic matter stemming from waste feed. Hence, in total, about 19% of the total organic matter in the feed used in the salmon aquaculture operation may settle onto the sediments as faeces and uneaten feed pellets.

All particle dispersion models require that the solid wastes (waste feed and faeces) be characterized by their settling rates and mass fraction in order to calculate their horizontal displacement and sedimentation rates. Several investigators have undertaken laboratory experiments to determine the sinking rates of feed pellets. Cromey et al. [7] found a relationship between sinking rate and feed pellet diameter and determined a mean sinking rate of 10.8 cm s^{-1} ± 2.7 cm s^{-1} corresponding to a mean pellet diameter of about 7 mm. Findlay and Watling [13] reported sinking rates in the 5–15 cm s^{-1} range, while Panchang et al. [5] reported a mean of about 10 cm s^{-1}. The sinking rate will depend on the size and composition of the feed pellet as well as the pelleting conditions used in its production. For this general modelling exercise we have used a general sinking rate of 10 cm s^{-1}.

The determination of the sinking rate for faecal material is not as straightforward as for feed pellets. Faecal material is often not in the form of a well-defined pellet or solid mass but rather exists as a gelatinous mass that may disintegrate. Thus, the methodologies used to collect and handle the faecal material are important factors to consider. In the laboratory, Cromey et al. [7] found that the sinking rates of faecal material were normally distributed ($\mu = 3.2$ cm s^{-1}, $\sigma = 1.1$ cm s^{-1}). Further, Panchang et al. [5] observed a similar mean sinking rate and found that 70% of the observations fell within the 2–4 cm s^{-1} range. In addition, Chen et al. [14] determined a mean sinking rate of 5.3 cm s^{-1} with sinking rates ranging from 4 to 6 cm s^{-1}. An interesting observation was made by Wong and Piedrahita [15], when they determined the settling velocity characteristics of the faeces of rainbow trout, in that about 25% of the faecal solids were non-settleable, i.e. this fraction did not settle within the 150 min time frame of their tests. Given the range of sinking rates that have been reported for faecal material by others above, we chose in this paper to partition the faecal solids into three fractions as detailed in Table 1.

Table 1 Sinking rates and mass fraction for faecal solids used in the model

Fraction	Sinking rate (cm s^{-1})	Mass fraction (%)
1	4	15
2	3	70
3	2	15

2.1
Particle Dispersion Model

In our model, the dispersion of the solid wastes from the salmon farm is approached from a statistical perspective. Simply stated, the statistics of the circulation are assumed to determine the distribution of the solid wastes on the bottom. The important hydrodynamic variable is the depth-averaged horizontal current. A flat bottom or constant depth is assumed throughout the model domain.

The calculation of the horizontal displacement D of a particle sinking at a constant rate S through a depth H is straightforward and common to all models used to predict the dispersion of solid wastes [8]:

$$D = \left(\frac{H}{S}\right) \overline{U} \tag{4}$$

The horizontal displacement is simply the product of the time it takes the particle to fall to the bottom (vertical fall distance H divided by the sinking rate S of the particle) and the depth-averaged horizontal current speed $\overline{U}$ as given by:

$$\overline{U} = \frac{1}{H} \int_0^H u(z)\, dz \tag{5}$$

where $u(z)$ is the horizontal velocity.

First, we considered the one-dimensional problem of calculating the distribution of particles on the bottom from a point source of identical particles. The particles were assumed to fall a constant vertical distance to the bottom. Also, we assumed that the depth-averaged velocity had a normal or Gaussian distribution. The probability density function $f(x)$ that describes the distribution of the particles on the bottom is then completely determined by the statistics (the mean μ_u and standard deviation σ_u) of the depth-averaged velocity field $\overline{U}$ as described by:

$$f(x) = \frac{1}{\sigma_x \sqrt{2\pi}} e^{-1/2\left(\frac{x-\mu_x}{\sigma_x}\right)^2} \tag{6}$$

where $\mu_x = \left(\frac{H}{S}\right)\mu_u$ and $\sigma_x = \left(\frac{H}{S}\right)\sigma_u$

For the two-dimensional problem, we assumed that the velocity field was represented by a bivariate normal distribution and that the velocity components u and v could be rotated so that they were uncorrelated. The probability density function for the distribution of the particles on the bottom is accordingly given by the equation (Eq. 7) in which it was assumed that the two velocity components were uncorrelated:

$$f(x,y) = \frac{1}{\sigma_x\sqrt{2\pi}}e^{-1/2\left(\frac{x-\mu_x}{\sigma_x}\right)^2}\frac{1}{\sigma_y\sqrt{2\pi}}e^{-1/2\left(\frac{y-\mu_y}{\sigma_y}\right)^2} \qquad (7)$$

where $\mu_x = \left(\frac{H}{S}\right)\mu_u$ and $\mu_y = \left(\frac{H}{S}\right)\mu_v$

$\sigma_x = \left(\frac{H}{S}\right)\sigma_u$ and $\sigma_y = \left(\frac{H}{S}\right)\sigma_v$

A fish cage is not a point source but rather an area source. Integration of Eq. 7 over the area of the bottom of the net pen transforms the point source into an area source.

Each size fraction of the solid wastes (waste feed, and several faecal fractions) is characterized by its own sinking rate and mass fraction. Each waste fraction falling through the bottom of the net pens is then distributed on the ocean bottom using the statistics (standard deviation and mean of the two horizontal velocity components) of the depth-averaged currents. The computation of the distribution of each particle fraction for each grid point in the model domain was carried out and then summed for all particle fractions to yield the total solids waste flux at all grid points in the model domain.

3
Hydrodynamics

The means and standard deviations of the depth-averaged velocity are the key parameters in the model that represent the effect of the currents. Most of the moored current meter observations around salmon farms in British Columbia have been collected at two discrete depths (i.e., 15 m and near the bottom). Consequently, these measurements only provide a coarse description of the depth-averaged current. For the model simulations, we employed current data that were collected using an acoustic Doppler profiling current meter (ADCP) as part of our near-field impacts study. In this regard, we deployed a 600 kHz ADCP in a water depth of 56 m, close to an operating salmon farm and obtained profiles of currents in 1 m bins from near the bottom to near the surface over a period of 51 days in the fall of 2001. The depth-averaged current (from the bottom of the net pens to near the ocean bottom) was computed for the ADCP current profiles and then the statistics of current velocity were calculated. The principal direction of the flow was in an east/west (EW) direction and most of the variance was in this velocity component (σ_{EW} = 4.4 cm s^{-1} versus σ_{NS} = 1.1 cm s^{-1}). The residual or mean flows were small (< 0.5 cm s^{-1}). The distribution of the each of the velocity components was examined and in a qualitative sense the distributions appeared to be Gaussian or normal.

4
Diagnostic

4.1
Single Net Pen – Analyses of Depth and Currents

It is clear from Eq. 4 that the depth under the net pens and the currents are the fundamental factors that control the dispersion of waste solids from an open cage salmon farm. To explore in a quantitative manner the effects of depth under the net pen, i.e. the distance between the bottom of the net pen and the ocean bottom, and the currents on the characteristics of the footprint, we undertook simulations on a hypothetical single net pen, Atlantic salmon farm and present the results in this section. In this simulation, the net cage dimensions were 15 m wide by 15 m long and 20 m deep, which gave a net pen bottom area of 225 m^2 and volume of 4500 m^3. The production or harvest biomass for the single net pen, with an assumed stocking density of 10 kg m^{-3}, was then estimated to be 45 tonnes. The model simulations were for the average flux of organic matter over an entire production cycle or grow-out period of 18 months in duration.

Using the wastage rates for feed and faecal material that were derived earlier and the farm biomass production of 45 tonnes, we calculated the waste flux coming out of the bottom of the net pen. The total amount of feed supplied to produce the 45 tonnes of fish biomass was simply obtained as the product of the total harvested biomass and the effective FCR (feed conversion or feed to gain ratio) of 1.2 or 54 tonnes of feed (the effective FCR is defined as the ratio of the total weight of feed (tonnes) that is needed to produce a tonne of fish or harvest biomass and takes into account feed loss and mortality). Using an organic matter concentration in the feed of about 86% on an air-dry basis, the total organic matter dispensed over the production cycle was calculated to be 46.44 tonnes. The total amount of organic matter waste solids that was produced over the entire production cycle was then estimated to be about 8.8 tonnes, which was calculated as the product of the total amount of organic matter that was supplied and the total organic mater wastage rate of 19% (8% from uneaten feed + 11% from the faeces), as derived and discussed earlier. Finally, the average organic matter flux contributed over the production cycle from the bottom of the cages was calculated to be 72 g m^{-2} d^{-1}. This value was obtained by dividing the total for the waste organic matter produced by the bottom area of the net pen and production time.

Once the waste flux from the bottom of the net pen was computed, the model then required that the solid wastes (waste feed and faeces) be characterized by their settling rates and mass fractions in order to calculate their horizontal displacement as described in Eq. 4.

The simulations for the single net pen were run at four different depths (5 m, 20 m, 40 m and 80 m) from the bottom of the net pens to the ocean bottom. There were several features of the computed footprints that were noted to be common to all of the simulations. First, the shape of the footprint was determined by the relative sizes of the standard deviations of the two velocity components. For these simulations, the standard deviation in the EW direction was four times larger than in the north/south (NS) direction. Consequently, the footprint was observed to be elongated in the EW direction relative to the NS direction (Fig. 2). Second, in all of the simulations, the maximum depositional flux was found to be directly under the net pen. Although not shown because the current means in the EW and NS direction were set to zero, the effect of a mean flow would have shifted the footprints in the direction of the mean flow. Third, most of the wastes fell within a short distance (< 60 m) from the net pen.

Changing the depth under the net pens while maintaining the same current characteristics and feed inputs was found to alter the footprint in two different ways. First, the maximum depositional flux, which was directly under the centre of the net pen, decreased as the depth under the net in-

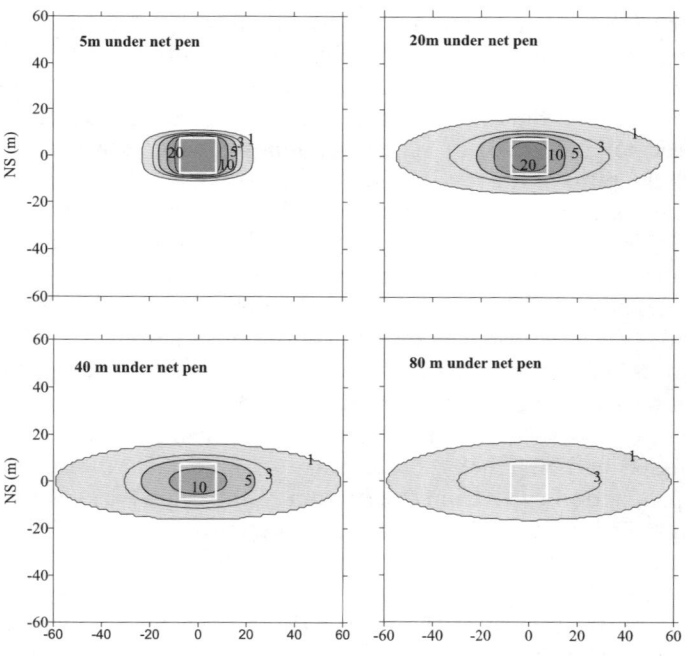

Fig. 2 Simulated spatial contours of organic matter flux (g m^{-2} d^{-1}) produced over an 18 month period by 45 tonnes of Atlantic salmon held in a single net pen (*square*) for different depths under the net pen

creased (Fig. 2). For depths of 20 m and 80 m under the net pens, the maximum depositional fluxes of organic matter were calculated to be 27 and 5 g m^{-2} d^{-1}, respectively. Second, the area of the depositional field increased as the depth under the net pen increased and this is evident from the area of the 1 g m^{-2} d^{-1} contour given in Fig. 2. However, the area within a given contour of organic matter flux did not increase for all depths under the net pen. For example, the area bounded by the 10 g m^{-2} d^{-1} contour was larger for a distance of 20 m under the net pen than for 5 m and 40 m under the net pen and was zero (not present) for the simulation with 80 m under the net pens.

Thereafter, simulations for a range of depths (0 m to 100 m) under the net pen were computed while maintaining the same velocity statistics, and the variations in the peak flux and area of the footprints were computed (Fig. 3a and b). At 0 m under the net pen (the net pen was sitting on the bottom) the peak flux was found to be 72 g m^{-2} d^{-1}, which was the same value as the organic matter flux out of the bottom of the net pen calculated earlier. As the depth under the net pen increased, the peak organic matter flux declined in a non-linear fashion. The decline was rapid at first and then more gradual at increased depths under the net pen Fig. 3a. The inverse relationship between the maximum flux and depth under the net pen may in part be explained by the inverse relationship between the bivariate probability density (flux) and standard deviations of particle distribution on the bottom, namely σ_x and σ_y (Eq. 7). The standard deviations of particle distribution on the bottom may be rewritten in terms of depth under the net pen, H, particle sinking velocity, S, and the standard deviations in the velocity field, σ_u and σ_v, such that the maximum flux (when $x = 0$ and $y = 0$ in Eq. 7) for a point source is proportional to $S^2/(H^2 \sigma_u \sigma_v)$. This relationship shows that not only is the maximum flux of a point source inversely related to H^2, but it is also inversely related to the product of the standard deviations of the velocity components.

The area of the footprint bounded by a given contour of organic matter flux was observed to change in an interesting fashion. When the depth under the net pen was 0 m (net pen was on the bottom), the area of the footprint was the same as that of the net pen (15 m × 15 m = 225 m^2), since we assumed that all wastes exited through the bottom of the net pen in this model. As the depth under the net pen increased, the area for a given contour of organic matter flux (3, 5, 10 and 20 g m^{-2} d^{-1}) initially was noted to increase to maximum and then decrease and eventually disappear Fig. 3b. The depth at which the maximum area was achieved was found to be different for each organic matter flux contour level.

Similar simulations were conducted to examine the effects on the footprint of varying the standard deviations of the current components while maintaining a constant depth under the net pens. The effects on the footprint of varying the currents were found to be equivalent to those for varying the

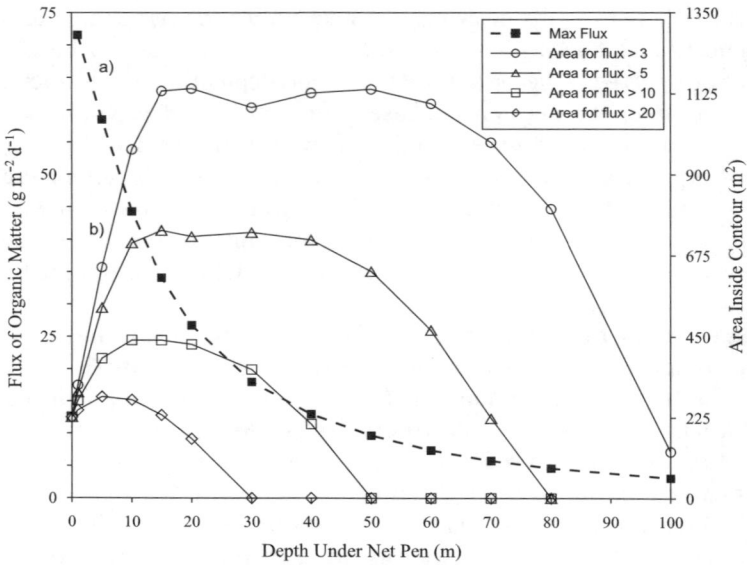

Fig. 3 Plot of (**a**) maximum organic matter flux and (**b**) area of the footprint inside contours of various organic matter fluxes as a function of depth under the net pen

depth under the net pen. With respect to this, an increase in the standard deviations of the currents produced a decrease in the maximum depositional flux of wastes and vice versa. Moreover, doubling the standard deviations of the EW and NS currents had exactly the same effect as doubling the depth under the net pen. As the standard deviations of the current components approached zero, the peak organic matter flux to the bottom approached the value of the organic matter flux from the bottom of the net pen (i.e. $72 \text{ g m}^{-2} \text{ d}^{-1}$), and the area of the footprint approached that of the bottom of the net pen (i.e. 225 m^2). The curves of maximum waste flux and areas within a given contour versus currents were noted to be identical to those presented in Fig. 3a.

4.2
Multiple Net Pens

As a final exercise, we examined the effects of changing the orientation of a linear grouping of net pens relative to the orientation of the currents on the size and geometry of the depositional field, and maximum waste fluxes. In these simulations, we used the same current statistics as before, namely, $\sigma_{EW} = 4.4 \text{ cm s}^{-1}$ and $\sigma_{NS} = 1.1 \text{ cm s}^{-1}$, but instead of a single 15 m × 15 m by 20 m deep net pen, we used eight pens of a similar size configured in a two by four linear grouping. Two simulations were conducted. One was with the net

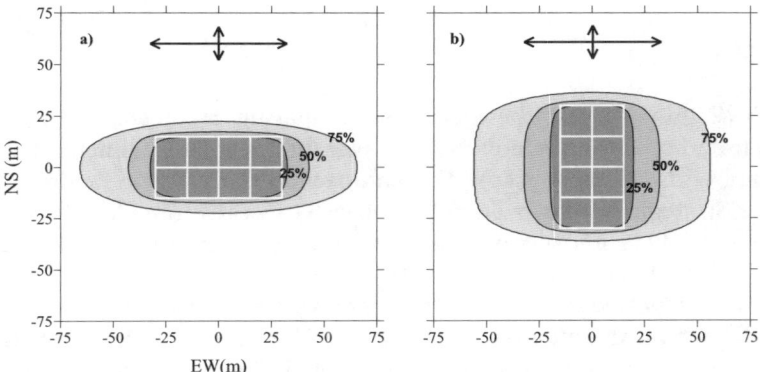

Fig. 4 Contours of organic matter flux containing 25%, 50%, and 75% of total organic matter wastes from a 2 × 4 linear grouping of 8 net pens (shown in *white*) oriented **a** parallel to and **b** perpendicular to the principle current axis. The EW and NS standard deviations in the currents are shown by the current arrows

Table 2 Model prediction of areas containing 25%, 50%, and 75% of the total organic matter (*OM*) for the configuration of net pens aligned with or perpendicular to the main direction of the current flow. Also tabulated are the contour levels of organic matter flux corresponding to the areas

Wastes %	Parallel to flow		Perpendicular to flow	
	Area (m^2)	Contour of *OM* flux (g m^{-2} d^{-1})	Area (m^2)	Contour of *OM* flux (g m^{-2} d^{-1})
25	1825	28.4	2251	20.7
50	2671	13.7	3756	9.4
75	4882	4.4	7185	3.1

pen group aligned in the EW direction or parallel to the main direction of the flow. The second was with the group aligned in a NS direction or perpendicular to the main direction of flow (Fig. 4).

The differences in the depositional field were found to be significant. First, the peak fluxes of organic matter were larger for the farm configuration having net pen groups aligned with, as opposed to perpendicular to, the dominant EW current axis (57 vs. 45 g m^{-2} d^{-1}, respectively). Second, the wastes were dispersed over a larger area when the net pen grouping was oriented perpendicular to the main direction of the flow (Fig. 4). For example, the area of the depositional field containing 50% of the wastes was approximately 40% larger for the orientation perpendicular to the flow than for the parallel orientation (Table 2).

5
Predictive

In this section, we test the model by comparing predicted vertical waste fluxes for an operating salmon farm to measurements of sedimentation rates obtained with sediment traps. The data stemmed from our Environmental Science Strategic Research Fund (Fisheries & Oceans Canada) study of the near and far-field impacts on sensitive habitats at several farm sites on the central coast of British Columbia.

The site is located in the outer reaches of Knight Inlet – a well- studied high run-off fjord on the mainland coast of British Columbia. The site was used because the bathymetry near the farm is relatively flat and the location also satisfied most of the data and information requirements for application and testing of the waste dispersion model. The study site is located in an embayment off the main channel of the fjord and as such is sheltered from strong wind and waves of the main channel.

The current meter measurements from this site were collected in October and November 2001 with an ADCP placed about 75 m offshore from the net pens. The currents, which were presented earlier, were oriented in an EW direction, more or less in line with the orientation of the shore line. The current regime in this embayment is generally weak having a median speed (depth-averaged) of about 3 cm s^{-1} and peak speeds seldom exceeding 15 cm s^{-1}.

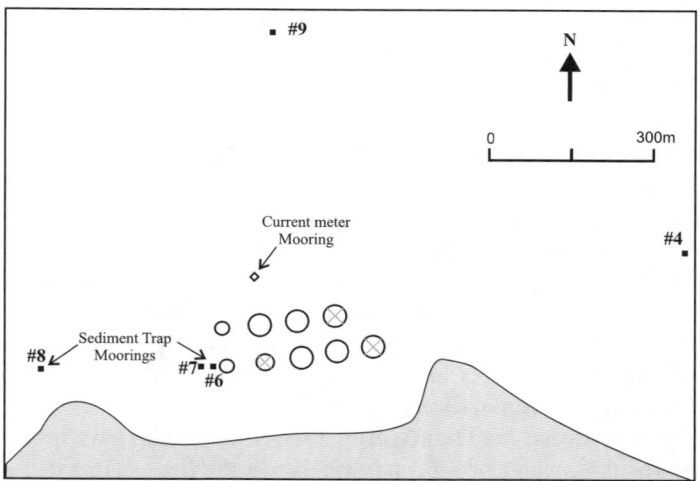

Fig. 5 Sketch of farm layout showing the location of the net pens (*open circles*), sediment trap moorings (*solid squares*), and current meter mooring (*open diamond*). (*Open circles* containing an × are net pens without fish.) *Gray area* is the shoreline

The site is of variable depth ranging from 45 to 70 m along an EW direction within 300 m of the net pens. The bathymetry to the east of the farm site deepens from 50 to 70 m within a few hundred meters and further deepens to more than 100 m in depth about 1 km to the east of the farm site. To the west, the bathymetry is relatively flat, shoaling gradually from 50 m to 35 m within 1 km.

The farm configuration consisted of two parallel rows of cylindrical or polar pens (11 and 19 m radius, and all pens were 20 m deep) oriented approximately in an EW direction and also aligned with the southern shoreline (Fig. 5). The row closest to the shore contained five pens while the outermost row contained only four pens. The distance between pens ranged from 50 m to 75 m, and the distance from the bottom of the pens to the seabed was about 35 m in the offshore row of pens and 28 m for the row closest to the shoreline. Only six of the nine pens contained fish. In February 2001, the farm was in the early stages of its production cycle and contained Atlantic salmon with an average weight of 1 kg. The total daily feeding rate during our experiment was 2850 kg.

5.1
Methods

To measure the vertical particle fluxes, we used cylindrical sediment traps constructed from PVC (polyvinyl chloride) pipe, which had an inner diameter of 12.5 cm and overall height of 48 cm [16]. The traps had baffle grids at their openings and on the bottom of the cylinders in their sample collection cups. We deployed the cylindrical sediment traps in pairs (side A and B) and positioned them 5 m above the bottom. The traps were deployed in February, 2001 for a short duration (4–5 days) experiment. Most of the sediment traps were in the 38– 42 m depth range except for the traps at the eastern far-field station (stn #4). Located in the sloping trough leading out to Knight Inlet, the near bottom sediment trap at stn#4 was at a depth of 66 m. Sediment traps were moored at 5 m, 30 m and 330 m at stations #6, #7 and #8, respectively along a transect downstream (to the west) of the net pens (Fig. 5). We also deployed sediment traps at two far-field locations (#4 and #9), 580 m and 516 m distance from the net pens. Station 8 was also considered a far-field station along the downstream transect.

Various analyses were performed on the contents of the sediment traps but for the purpose of this paper, we present and discuss only the measurement of total solids and organic matter fluxes (Table 3).

Suspended particulate matter (SPM) was determined gravimetrically through filtration. The sediment trap material was then filtered onto preweighed 1 μm glass fibre filters (47 mm) and dried at 55 °C for 48 h. The filters containing the sediment trap material were then re-weighed following a 2-h desiccation period. A salt correction was applied by running filtered seawater

Table 3 Sedimentation data for total solids and organic matter (*OM*) from traps positioned 5 m above bottom during the February 2001 study and model predictions of organic matter flux

Stn #	Distance from net pen (m)	Trap depth (m)	Mean solid flux [% OM] ($g\,m^{-2}\,d^{-1}$)	Organic matter flux ($g\,m^{-2}\,d^{-1}$)	Observed–background ($g\,m^{-2}\,d^{-1}$)	Model results ($g\,m^{-2}\,d^{-1}$)
6	5	41	10.4 [37]	3.8	2.8	15.0
7	30	38	6.8 [25]	1.7	0.7	3.2
8	331	39	4.3 [23]	1.0	0.0	0
4	580	66	14.0 [19]	2.6	1.6	0
9	516	42	3.3 [n/a]	n/a	n/a	0

through ten replicate filters. A vertical flux was obtained for each sediment trap by standardizing the total SPM by trap area and deployment time. Organic content was determined through loss on ignition by placing filters in a muffle furnace for 2 h at 550 °C and then in a dessicator for 2 h.

5.2
Data and Model Simulations

Of the five sedimentation rates that were measured, three were from far-field locations and two were from near-field locations taken close (30 m or less) to the net pens. Determinations of organic matter fluxes were possible from four locations; two far-field and two near-field. The measured solid fluxes ranged from 3.3 to 14.0 $g\,m^{-2}\,d^{-1}$, and fluxes of organic matter ranged from 1.0 to 3.8 $g\,m^{-2}\,d^{-1}$ (Table 3). Generally, the sedimentation rates for organic matter and total solids declined with distance from the farm net pens with the most rapid decline occurring within a short distance from the net pens (Fig. 6). The highest organic matter concentration (37%) was measured at the location closet to the net pen and the lowest (19%) at the station furthest away (Table 3). The sedimentation rates at far-field stations were usually the smallest, except for the large values observed at station #4.

Because the model did not include a background or natural sedimentation, we corrected the measured organic matter sedimentation rates before comparing them to the model estimates of organic matter flux. Two of the three far-field near-bottom sedimentation rates for total solids were reasonably similar, cf. 3.3 $g\,m^{-2}\,d^{-1}$ at station #9 versus 4.3 $g\,m^{-2}\,d^{-1}$ from station #8. The near-bottom sedimentation rates for total solids and organic matter from station #4 were, however, much larger compared to the other far-field station(s) and therefore were problematic. The sediment traps at station #4 were different from the others in that they were placed in deeper water and in a trough sloping downward to the main inlet. We conjecture that a trap placed at this location may have measured the effects of sediment focusing. Sediment focusing occurs when sediments in the shallower regions are eroded, re-suspended and then settle in the deeper regions of an embayment, producing higher sedimentation rates in the deeper area of the embayment than in the shallower regions. We noted that the measurements at station #4 were much higher and did not include them in the analysis. Thus, in order to compare the measured sedimentation rate for organic matter with the model predictions we subtracted the lowest background flux as measured at station #8 ($1.0 \, \mathrm{g \, m^{-2} \, d^{-1}}$) from all of the near-bottom measurements.

We set up the model using the farm configuration described above and set the depth under the net pens to 30 m. The standard deviations for the velocity field used earlier in the diagnostic simulations were used, namely $\sigma_{EW} = 4.4 \, \mathrm{cm \, s^{-1}}$ and $\sigma_{NS} = 1.1 \, \mathrm{cm \, s^{-1}}$. In addition, the means, although small, were also used $\mu_{EW} = 0.5 \, \mathrm{cm \, s^{-1}}$ and $\mu_{NS} = -0.3 \, \mathrm{cm \, s^{-1}}$. The production rate of faecal material was as discussed earlier, and faecal particle settling rates and corresponding mass fraction were as presented in Table 1. A feed wastage rate of 8% was used with a sinking rate of feed pellets set to $12 \, \mathrm{cm \, s^{-1}}$. Direct comparison between the model and measurements was possible for only two locations, stations #6 and #7 (Table 3). The predicted organic matter sedimentation rates along the transect made up of stations #6, #7 and #8 showed that the fluxes declined rapidly and reached zero well before reaching far-field station #8, which was considered to be at background levels (Fig. 6).

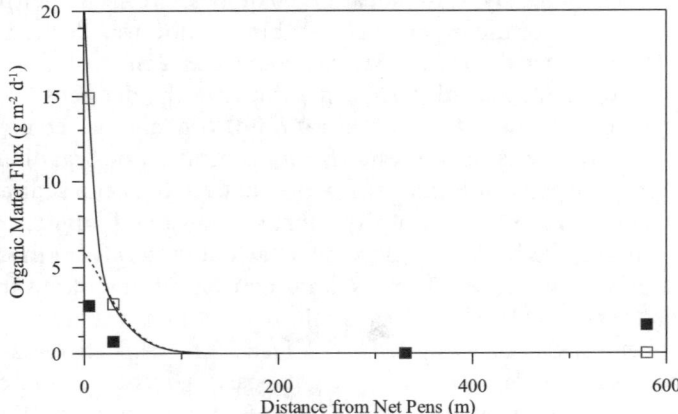

Fig. 6 Observed (*solid square*) and predicted organic matter fluxes for 8% feed wastage rate (*open square* and *solid line*) and for 0% feed wastage rate (*dashed line*) 5 m above bottom as a function of distance from the net pens

6 Discussion

6.1 Comparison of Observed and Predicted Sedimentation Rates

Detailed discussion and conclusive comparisons between model predictions and observations are challenging considering that there were only four data points. Nevertheless, at the two near-field sampling locations, i.e. stations #6 and #7, the model simulations over-predicted the measured organic matter fluxes by 530% to 460%, respectively (Table 3). Conversely, the sedimentation trap estimates may have significantly underestimated the actual organic matter fluxes. It is likely that the true values were somewhere in between the model and sedimentation estimates since there are many factors that could account for differences between the observed and predicted waste fluxes. Some of these are discussed below.

In relation to accuracy of the model estimates, one important factor that can significantly influence the results is the value that is used for the feed wastage rate. To explore the sensitivity of the model predictions to this factor, we re-ran the simulation with the wastage rate set to zero (Fig. 6). At the location closest to the edge of the net pen (station #6, 5 m away), there was a large reduction in predicted flux of organic matter, namely from 15.0 to 5.5 g m^{-2} d^{-1}, while at the next closest sampling location (station #7, 30 m away) the predicted flux change only slightly. The large difference found in the predicted fluxes between the two simulations (8% and 0% feed wastage rate) demonstrates the importance of accurately determining the feed wastage rate, especially near (5 m) the net pens. The observation that the predicted flux changed only slightly at the location 30 m away from the net pens indicates that almost all of the organic flux at this location was derived from the faecal wastes and that the waste feed pellets, because of their faster sinking velocity, were being deposited in close proximity to the net pens.

In support of this view, the concentration of organic matter in the near-field sediment traps was of interest. The concentration of organic matter in the anthropogenic (farm-produced) fraction collected by the sediment traps was estimated by subtracting out the background organic matter and solid fluxes (as represented by the measurements at station #8) from the measurements at stations #6 and #7. The resultant concentrations of anthropogenic organic matter were 46% and 28% at stations #6 and #7, respectively. Organic matter concentrations of waste feed and faeces are respectively about 86% (air-dry basis) and about 55% (dry weight basis). Given that anthropogenic solids are composed of waste feed and faeces, the organic matter concentration should have been 55% of dry matter or higher depending upon the relative proportions of feed and faeces in the wastes.

Interestingly, the concentration of organic matter collected in the sedimented material at station #6 (46%) approached that in faecal wastes, whereas our model indicated that the majority of the organic matter should have originated from waste feed. If this had been the case, the concentration of organic matter would have been substantially larger than 55%. A likely explanation for the low organic matter concentration found in the closest trap (#6) is that very little waste feed was actually being captured in the trap. This would explain why the sedimentation trap data may have significantly underestimated the true organic matter flux. Likewise, sediment trap measurements taken at the perimeter of net pens of several BC farms [17] also show low organic matter concentrations (25–50%). Brooks and Manhken [12] have suggested this indicates that there is little waste feed in the solids collected by the sediment traps.

In addition, the accuracy in the placement of the sediment traps relative to the net pens can introduce significant uncertainty in the comparisons because the decay in the sedimentation rate is very steep in the first 30 m from the perimeter of the net pen (Fig. 6). An inaccuracy of several metres in the relative positions is probable because of the accuracy of the GPS (global positioning system) position of the sediment trap and net pens and because of the movement of the net pens with the currents [18].

Finally, the accuracy of the sediment trap measurements themselves is largely unknown because there is no standard in use for calibration of sedimentation trap data. The trapping efficiency is certainly known to be affected by the current velocity and sinking rate of particles [19]. Furthermore, placement of sediment traps in shallow depths close to shore and the bottom boundary may lead to contamination of the measurements from resuspension events or sediment laden run-off events. Indeed, it appears that sediment traps as presently designed and deployed do not present a true picture of organic matter flux from net pens involving the deposition of large particles, and improved methodology in this regard is therefore warranted.

Other possible sources of error in the model simulation results may be associated with the distribution of fish biomass in the net pens. The exact distribution of biomass amongst the six pens was not known. Hence, in the model runs, the biomass was assumed to be equally distributed. A different distribution of biomass in the pens would have produced significantly different waste flux predictions. Lastly, it should be mentioned that the current meter data used in the model simulations were not collected concurrently with the sediment trap data. The sediment traps were deployed in the winter (February 2001) while the current meter data were collected in the fall (October and November 2001). We chose to use the ADCP data because they provided the only complete observations of the currents from the bottom to the surface. However, we recognize that there will be significant seasonal and event-driven variations in the currents as well as spatial variations that could have influenced the accuracy of the model predictions.

6.2
Model Limitations and Uncertainties

Models have many limitations because they simplify the processes that they attempt to explain and reproduce. Some of the limitations are model-specific while others are specific to our understanding of the important processes and our ability to observe and characterize them. In this section, we discuss both types of model limitations with respect to our statistical model of the deposition of solid wastes from open cage salmon farms.

The statistical model we have presented and discussed herein contains several assumptions that limit its applicability and give rise to uncertainty in its predictions. A principal constraint on the applicability of the model is the assumption of a constant ocean depth. This constant depth assumption is required because the depth-averaged current is a key model parameter, and as the depth changes so will the depth-averaged current. The coastal waterways and fjords of British Columbia are characterized by rapidly changing bathymetry and many farm sites are located over steeply sloped bottoms. Nevertheless, some farm sites are located over relatively flat bottoms where the constant depth assumption is reasonable.

The assumption of a horizontally uniform velocity field also adds to the uncertainty of the model predictions especially for farms located very close to shore or along convoluted shorelines. However, the length scale for the dispersion of solid wastes is quite small (< 100 m). Consequently, it may be argued that currents, though spatially varying, may not change dramatically over such short distances. To address this issue of spatial variability of the currents at short horizontal length scales would require a substantial effort. Specifically, a large-scale circulation model of the region with high resolution grids around the farm sites would be required to represent the spatial variability of the currents. Large-scale circulation models have their own inherent limitations. From a practical perspective, depth-averaged currents are seldom measured at salmon farm sites because of costs and regulations that specify that current measurements should be determined at only two discrete depths. While acoustic Doppler current profilers are commonly used by the oceanographic community, they are not commonly used by the salmon aquaculture industry in British Columbia.

The alteration of the natural flow or shadowing effects caused by the physical structures of the farm net pens is recognized as an important factor [18]. Obstruction of the flow by the net cages generates secondary flows around and under the net pens, which may differ significantly from the primary flow field measured before the placement of the farm structures. With the trend in British Columbia towards larger farms (> 200 m in length), the effect of the farm structures on the flow field is likely to be increasing. None of the aquaculture waste models that we know of take into account the flow alterations caused by the farm structures themselves. Our statistical model is based on the mathematics of the normal or Gaussian distribution. Consequently, the departure of current observations from the normal distribution will undermine confidence in the model predictions. For each site that will be modelled, the statistical distribution of the currents should be analysed to determine how well they fit a normal distribution.

This next set of limitations and uncertainties is applicable to most models that presently exist and is not specific to the statistical model reported herein. Most models assume that the waste solids follow Stokes' Law of settling particles in which the size and density of the particles generally determine their sinking rate. Fish faeces are often ejected as gelatinous or loosely consolidated mucous masses that may break apart into smaller particles as they sink. Conversely, the organic-rich solid wastes may flocculate with inorganic particles or other organic solids and form larger clumps or flocs that sink more quickly to the bottom. In both cases, the view of coherent particles sinking at a constant rate may be oversimplified and not representative of the actual dynamics of the settling organic wastes. Further complicating the parameterization of the sinking rates of waste particles is the methodology used to collect and handle the waste in laboratory experiments designed to measure their sinking rates.

Finally, there is another mechanism that may be important to the dispersion of wastes, namely, the episodic sloughing of wastes and organisms that accumulate on net pens. Net pens often require cleaning when the accumulation restricts flow through the net pen. The wastes that accumulate and aggregate on the net pens may become unstable or become dislodged by high current or wave activity and slough-off the net and fall to the bottom in large clumps. This is a difficult process to model, and one that has not been documented. Therefore its importance remains speculative.

The amount of waste feed is also a source of uncertainty in all models as discussed previously. Waste feed is generally believed to be in the form of uneaten feed pellets, but a small percentage (< 0.5%) may results from the abrasion of the feed pellets as they travel through the automated feeding system [12]. Wastage rates for feed are believed to be < 5%, but the wastage rate is not a parameter that is routinely measured. The setting of feed wastage rate is discretionary and not well known.

In some models the wastes are assumed to exit through the bottom of the net cages with the waste flux being equally distributed through the bottom surface of the net cage. In DEPOMOD [7], the wastes exit through all the surfaces of the net pens. While evidence exists to show that most of the wastes exit through the bottom of the net pen, some do exit through the sides [20]. Also, little is known about how the flux of wastes through the bottom is distributed.

6.3
Future Research

The findings of the present paper clearly indicate that there is a need to improve the values used for the parameters that comprise our model so that total organic matter fluxes from salmon farms can be predicted more accurately and the distribution of organic matter wastes from salmon farms from uneaten feed and faeces can be more accurately characterized in relation to current flows, depths underneath the farm sites and other factors. Present methods that are used to estimate organic matter fluxes from salmon farms using sediment traps appear to be very inaccurate. Thus, there is clearly a need to accurately obtain estimates of organic matter fluxes in the field. This is essential for validating the accuracy of the model predictions. Validation of the model parameters should be conducted comprehensively during the summer and winter, the seasons of greatest and least feed intake in salmon farms, for several BC farm sites. These sites, in turn, should occupy a range of bathymetric and oceanographic conditions.

Earlier in this volume, Strain and Hargrave [21] calculate that even in depositional areas most of the organic matter released from the farm does not actually accumulate on the seabed but appears to be transported away from the site. If we are to progress in our understanding of the impact that organic enrichment will have on the sediment chemistry and benthos we need to investigate the processes that affect the fate of the organic matter once it is deposited on the seabed. Future research should be conducted by a multidisciplinary team of researchers with expertise in modelling, sediment geochemistry, benthic ecology and fish nutrition.

Acknowledgements We would like to acknowledge the support of this study provided through the Environmental Sciences Strategic Research Fund of Fisheries and Oceans Canada. We also acknowledge and thank Stolt Sea Farms Inc. for the cooperation and assistance provided by their staff during the course of our field work and for providing information regarding farm production and configuration. We also thank the officers and crew of the Canadian Coast Guard Vessel VECTOR for their assistance in the field operation. Finally, we thank the dedicated technical support staff that carried out the field program and undertook the analyses of samples.

References

1. Pearson TH, Black KD (2001) The environmental impacts of marine fish cage culture. In: Black KD (ed) Environmental impacts of aquaculture. CRD, Boca Raton, Florida, p 213
2. Brooks KM, Stierns AR, Backman C (2004) Aquaculture 239:81

3. Wildish DJ, Dowd D, Sutherland TF, Levings CD (2004) Near-field organic enrichment from marine finfish aquaculture. In: A scientific review of the potential environmental effects of aquaculture in aquatic ecosystems, vol III. Fisheries & Oceans Canada. Can Tech Rep Fish Aquat Sci 2450:66
4. Cromey CJ, Nickell TD, Black KD, Provost PG, Griffiths CR (2002) Estuaries 25:916
5. Panchang VG, Cheng G, Newell C (1997) Estuaries 20:14
6. Hargrave BT (1994) A benthic enrichment index. In: Hargrave BT (ed) Modelling benthic impacts of organic enrichment from marine aquaculture. Can Tech Rep Fish Aquat Sci 1949:1
7. Cromey CJ, Thomas TD, Black KD (2002) Aquaculture 214:211
8. Gowen RJ, Smyth D, Silvert W (1994) Modelling the spatial distribution and loading of organic fish farm waste to the seabed. In: Hargrave BT (ed) Modelling benthic impacts of organic enrichment from marine aquaculture. Can Tech Rep Fish Aquat Sci 1949:19
9. Hajen WE, Beames RM, Higgs DA, Dosanjh BS (1993) Aquaculture 112:321
10. Hajen WE, Beames RM, Higgs DA, Dosanjh BS (1993) Aquaculture 112:333
11. Forster I (1999) Aquaculture Nut 5:143
12. Brooks KM, Mahnken CVW (2003) Fish Res 62:255
13. Findlay RH, Watling L (1994) Toward a process level model to predict the effects of salmon net-pen aquaculture on the benthos. In: Hargrave BT (ed) Modelling benthic impacts of organic enrichment from marine aquaculture. Can Tech Rep Fish Aquat Sci 1949:47
14. Chen YS, Beveridge MCM, Tefler TC (1999) Aquaculture Res 30:395
15. Wong KB, Piedrahita RH (2000) Aquaculture Eng 21:233
16. Iseki K, Whitney F, Wong CS (1980) Bull Plankton Soc Jpn 27:27
17. Brooks KM (2001) An evaluation of the relationship between salmon farm biomass, organic inputs to sediments, physico-chemical changes associated with those inputs and the infaunal response – with emphasis on total sediment sulfides, total volatile solids, and oxidation-reduction potential as surrogate endpoints for biological monitoring. Final report produced for the Technical Advisory Group. BC Ministry of Environment, Nanaimo p 184
18. Cromey CJ, Black KD (2005) Modelling the impacts of finfish aquaculture (in this volume). Springer, Berlin Heidelberg New York
19. Baker ET, Milburn HB, Tennant DA (1988) J Mar Res 46:573
20. Sutherland TF, Martin AJ, Levings CD (2001) ICES J Mar Sci 58:404
21. Strain PM, Hargrave BT (2005) Salmon aquaculture, nutrient fluxes and ecosystem processes in southwestern New Brunswick (in this volume). Springer, Berlin Heidelberg New York

Organic Enrichment from Marine Finfish Aquaculture and Effects on Sediment Biogeochemical Processes

Marianne Holmer[1] (✉) · Dave Wildish[2] · Barry Hargrave[3]

[1] Institute of Biology, University of Southern Denmark, Campusvej 55, 5230 Odense M, Denmark
holmer@biology.sdu.dk

[2] Fisheries and Oceans Canada, Biological Station, 531 Brandy Cove Road, St. Andrews, New Brunswick, E5B 2L9, Canada
wildishd@dfo-mpo.gc.ca

[3] Fisheries and Oceans Canada, Marine Environmental Sciences, Bedford Institute of Oceanography, Dartmouth, Nova Scotia, B2Y 4A2, Canada
hargraveb@rogers.com

1	Introduction	182
2	Measurements of Organic Enrichments in Sediments	183
3	Vertical Gradients in Metabolic Processes in Sediments	187
4	Using Changes in Sediment Biogeochemistry as Indicators of Organic Enrichment	194
5	Organic Enrichment and Changes in Benthic Macrofauna	197
6	Organic Enrichment Effects in Biogenic Sediments	199
7	Biogeochemical Conditions in Seagrass Beds Enriched by Aquaculture Waste Products	201
8	Conclusions	203
	References	204

Abstract Organic enrichment of sediments underlying fish farms in temperate and tropical coastal zones is reviewed to identify similarities and important biogeochemical differences. Improvements in technology have allowed farms to move from depositional sites to more erosional offshore locations. However, low cost farms are still being located in sheltered areas, in particular in the tropics. Important differences in the response of sediment geochemical variables to organic enrichment are associated with finfish aquaculture located under highly diverse hydrographic and sedimentological conditions in different coastal areas. In temperate latitudes where farms are often located over soft bottom, organic enrichment increases sediment microbial activity and may alter benthic community structure. Enhanced anaerobic activity may lead to accumulation of sulfides with adverse effects on aerobic bacteria, plants and fauna due to progressive oxygen depletion. In warm temperate waters, such as the Mediterranean and tropical latitudes, many farms are located in more advective areas with coarse-grained carbonate-rich sediments. Effects of organic enrichment in these areas are less well described, but studies have also shown sulfide accumulation in sediments indicative of deteriorated benthic habitats.

Keywords Aquaculture · Organic enrichment · Sediment biogeochemistry · Sea grass communities

1
Introduction

Many industrial uses of the coastal zone result in increasing release of nutrients. In an area of restricted exchange with offshore water, this often leads to nutrient and organic matter enrichment (eutrophication) [1, 2]. Meyer-Reil and Köster [3] described critical changes associated with eutrophication in coastal waters. Progressive stages of enrichment include increased inorganic and organic nutrients, microbial biomass and enzymatic decomposition of substrates, nitrification, denitrification and benthic oxygen and nutrient fluxes. Evidence is also accumulating to show that with increasing eutrophication the ratio of autotrophic to heterotrophic microbial processes is reduced as increasing amounts of organic matter are respired in sediments than in the water column [4]. As organic enrichment of aquatic ecosystems increases, the balance between pelagic and benthic metabolism appears to shift to become dominated by benthic processes.

In oligotrophic and mesotrophic coastal marine systems, where high turbidity does not limit light and phytoplankton production, material flow and cycling predominantly occur in the water column. This is illustrated in Chesapeake Bay where almost two-thirds of total annual oxygen consumption occurred in the water column [5]. In eutrophic, nutrient-rich areas however, heterotrophy predominates based largely on stored organic matter in sediments. In coastal areas, this fundamental shift in ecosystem structure may be reflected seasonally. For example, during spring and late summer following input of organic matter sedimented from algal blooms benthic respiration increases [6]. This natural seasonal cycle may be enhanced when organic matter released from aquaculture sites results in peaks in benthic mineralization, aerobic and anaerobic respiration in late summer [7–10]. A shift in the balance between pelagic and benthic respiration could occur on an inlet-wide scale in coastal areas as a result of finfish aquaculture activity if increases in sedimentation of fine-grained particles and associated organic matter are sufficient to cause sulfide accumulation in sediments [9, 10]. When finfish and shellfish aquaculture facilities are located in coastal areas that receive other sources of organic waste, soluble and particle matter products released as a result of aquaculture operations are added to what may be an already high supply of organic matter. The question then becomes – what is the maximum capacity of an inlet or water body to assimilate additional organic matter?

Many sources, both natural and anthropogenic, contribute to organic matter in coastal sediments [2]. Urban effluents such as discharges from sewage, pulp and paper and fish processing plants may supply organic matter in addition to natural sources of input (e.g. sedimentation of phytoplankton blooms, burial of seagrass or macrophyte debris to sediments). Organic matter may also be supplied by rivers and shoreline erosion. In urbanized port and industrialized coastal areas, the water column and sediments are impacted by organic matter present in domestic sewage and industrial effluents. Organic matter from any of these potential sources will contribute to sediment biological oxygen demand (BOD) and nutrient fluxes. Although no single method exists for differentiating sources of organic matter, stable isotopic analysis appears to offer a general methodology that might be useful. Nitrogen stable isotopes have been used to show changes in food web structure as a result of eutrophication in coastal wetlands [11]. Stable carbon isotopes were used to determine that $\sim 40\%$ of sediment organic carbon up to 150 m from salmon farm sites in Tasmania was derived from fish pen wastes [12]. Other studies [13] have also shown that fatty acids, sterols and stable carbon/nitrogen isotopes in sediments under pens changed rapidly over time during fallowing indicating that these compounds may be sensitive indicators of changes in organic matter associated with sedimented aquaculture waste products. Total organic matter or elemental carbon : nitrogen ratios, which might be thought useful for differentiating terrestrial and marine sources, are relatively insensitive indicators [7, 14]. This is largely because substantial amounts of refractory material in sediment add to the organic content but do not affect BOD or decomposition rates.

2
Measurements of Organic Enrichments in Sediments

Different methods have been used to quantify organic matter accumulation and to determine the consequences for underlying benthic community structure arising from excessive organic matter deposition [14–20]. Measures of organic matter in sediments can be used directly to show enrichment since the accumulation of organic compounds reflects the net (residual) product of all processes of addition and loss. Grain size is a critical factor determining sediment organic content. Fine-grained sediments have a high weight-specific surface area and these generally occur in depositional areas where bottom stress and currents are low. However, the proximate composition (relative amounts of carbohydrates, proteins and lipids) of major organic components is variable depending on source material. Particulate matter settled from the water column collected in sediment traps often has an order-of-magnitude higher carbon content (5 to 30%) than underlying surficial sediments.

Total organic matter (often referred to as total volatile solids) in marine sediments is usually measured as percent weight loss on ignition (550 °C; ash weight). Carbon and nitrogen can be measured directly using an elemental analyzer to determine actual organic carbon content after correction for inorganic carbon. Carbonates, the major source of inorganic carbon can be removed by acid treatment. Up to 15% of sediment weight may be present as organic matter in fine-grained coastal sediments with low sand (high silt/clay) content if only natural sources of organic matter enrichment occur. Organic carbon can account for up to 20% of this amount. Thus organic carbon usually represents < 5% of sediment dry weight in coastal sediments even when the fraction of fine silt/clay particles is high [14]. In organically rich sediments total organic matter may reach 20% or more of sediment dry weight and the proportion of organic carbon can increase to 30 to 40% [20, 21].

A general linear relationship may be observed between percent organic matter and percent organic carbon in a given area if the range of concentrations observed is great enough (Fig. 1). As mentioned above, the overall total amount of organic matter and organic carbon in a deposit will be influenced by the grain size distribution and the net effect of differential rates of supply and removal.

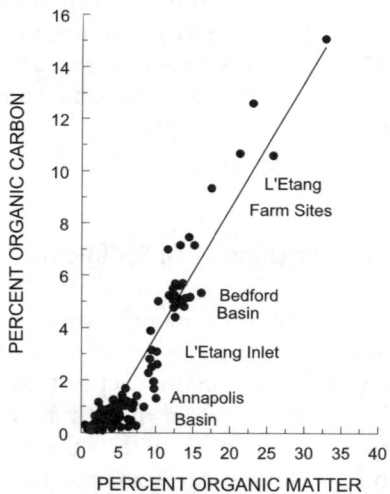

Fig. 1 Sediment organic matter determined as weight loss on ignition (550 °C for 4 h) and organic carbon (elemental analysis) for surface (upper 2 cm) sediment. Data is summarized [6, 7, 20] from farm sites and reference (> 500 m from farm sites) locations in the Bay of Fundy where salmon aquaculture is carried out and an urbanized inlet (Bedford Basin) in Nova Scotia. *Line* indicates the linear regression y = 1.11 + 0.47x (r^2 = 0.88, n = 244)

When temporary anoxic conditions occur, loss of organic carbon through aerobic decomposition is reduced and preservation is increased due to reduced macrofauna consumption. Combined effects of feeding and bioturbation by macrofauna enhance loss of organic matter to sediments under oxic conditions [22]. With prolonged or permanent, anoxia, numbers of sulfate reducing bacteria and rates of sulfate reduction increase [10]. This leads to sulfide (H_2S, HS^-, $S^=$) accumulation with associated low redox potentials and FeS formation resulting in black sediment. White *Beggiatoa* mats can form on surface sediment when anoxic sediments contact oxygenated overlying water as often occurs in high sedimentation areas in the "footprint" of salmon cages [7–9].

High levels of sediment organic matter (> 20%) or organic carbon (> 10%) are seldom observed in marine sediments. Exceptions occur in urbanized coastal areas such as industrialized harbours and upwelling areas on continental margins where sustained high rates of effluent discharges or primary production increase organic matter sedimentation. In general, organic carbon in coastal marine sediment not impacted by excessive organic matter accumulation, depending on grain size, is usually < 5% of dry weight. Values of 1 to 2% are common in many temperate coastal areas where tidal resuspension ensures oxic conditions for aerobic bacterial decomposition. In shallow coastal marine areas there may also be significant seasonal changes within this narrow range reflecting variations in organic matter supply and loss [7–10]. In temperate latitudes, maximum decomposition and consumption of sediment organic matter by microorganisms and benthic fauna usually occurs during summer months when temperatures are at seasonal maxima. Variations in seasonal values may be a useful indicator of the amount of labile (freshly deposited) organic carbon from natural sources.

The amounts of waste feed lost from marine finfish aquaculture farms depend on the feeding efficiency, feeding methods and strategies – factors that vary among cultured species and are subject to change due to the continuous optimization of the food within the industry. The sedimentation of waste products is strongly dependent upon the hydrodynamic conditions around and within a farm site with modifications by fouling organisms and attraction of wild fish and invertebrates. Organic enrichment of sediments surrounding finfish operations is not spatially uniform [14]. The presence of seagrasses may also trap waste products due to reduced water flow within meadows [21].

Changes in the amount and composition of accumulated farm waste products in the sediments can be observed as an alteration in the quantity and biochemical composition of sediment organic matter. These measures provide powerful tools for assessing modification of trophic state at the ecosystem level [23, 24]. For instance, clear differences in biopolymeric carbon concentrations have been observed between fish farm and control sediments [25–27] suggesting that these compounds are sensitive indicators for the presence of farm wastes. Chlorophyll *a* concentrations are also generally higher under fish cages and lipid concentrations are elevated in comparison with control stations [28]. Very high lipid concentrations detected in fish farm sediments appear to be strictly related to the farming activity and can thus be used to indicate direct organic input from aquaculture sources.

Particulate effluents from finfish farms consist of faecal material and uneaten fish feed pellets. The amount of particulate effluents depends on the farmed species, size of the farms (numbers and biomass of fish) and husbandry practices that determine the efficiency of feed utilization. There is a global tendency to increase the size of the farms by increasing the number of cages at a site. Organic loading of the sediments has generally been found in the close vicinity of the farms (< 50 m), and this zone of enrichment may increase as the farms increase in size, unless the farms are moved to more advective locations. Particles are transported by water currents and the resulting dispersion and sedimentation at various distances from the cages reflects site specific hydrodynamic features (Sect. 7). High settling rates lead to increased particle deposition near farms where accumulated organic matter may cause oxygen depletion. When the oxygen demand exceeds the oxygen diffusion rate from overlying waters, sediments become anoxic with profound effects on the benthic flora and fauna [1, 22, 29].

The productivity in the farms also influences the local accumulation of organic matter in sediments. Annual variations in fish growth rates occur under both temperate and tropical conditions. Maximum growth occurs in temperate regions in the summer when temperature is maximum (as discussed in Sect. 2). Minimum growth occurs in the winter when the temperature is low. In some areas fish are harvested and cages are removed during periods of icecover [8, 9, 29]. Higher temperatures in the tropics lead to production cycles that are often much shorter (3–4 months) and each growth period may be separated by a fallowing period [21]. These annual changes in farm productivity result in annual changes in sediment loading, and at highly advective sites organic enrichment may be reduced by resuspension during fallowing [8, 29].

The development of sediment profiling imagery (SPI) and the REMOTS™ technology has made it possible to observe stages of benthic enrichment and the formation of anoxic conditions in upper sediment layers [29]. SPI has been used successfully to investigate environmental changes along organic pollution gradients by quantifying changes in visual appearance of sediments [29–31]. The method has been applied to assess fish farming impacts in the Mediterranean [32]. A large number of visible attributes (depth of dark sediment, signs of out-gassing, bioturbation marks) showed significant correlations with geochemical and biological variables related to effects of organic enrichment. Sediment microbial populations are particularly sensitive to changes in environmental conditions and trophic state [26, 33], especially when subjected to nutrient input related to anthropogenic activity [34, 35].

It is well known that input of labile organic waste products to sediments increases microbial activity and sulfate reduction rates are particularly sensitive to stimulation through enrichment [7–10]. As a result, the determination of sulfide levels in the sediments using electrochemical methods is a widely used monitoring tool for assessing benthic environmental effects of Atlantic salmon fish farming in North America and Canada [7, 14, 20]. In Northern Europe pH and redox are measured along with a visual inspection of the thickness of organic matter accumulated, smell, colour and consistency of the sediments and presence of gas bubbles. In some areas these observations are the basis for environmental monitoring programs (eg. the MOM approach) [36, 37]. SPI can further support these measurements since the equipment is relatively easy to deploy and can be replicated over large areas to provide a detailed assessment of the extent of organic enrichment around a fish farm [32].

3
Vertical Gradients in Metabolic Processes in Sediments

Bacterially-mediated metabolism is responsible for geochemical gradients that exist in the upper surface layer of all marine sediments. Bacteria utilize organic compounds through inter-related metabolic processes (Fig. 2). Aerobic processes such as heterotrophic respiration and aerobic chemosynthesis require oxygen; anaerobic processes such as iron reduction, sulfate reduction and methanogenesis do not. The coupled reactions depend on the downward diffusion of dissolved oxygen from overlying water and oxic surface layer sediments and the upward movement from deeper sediment layers of reduced inorganic and organic products that result as end-products of anaerobic respiration.

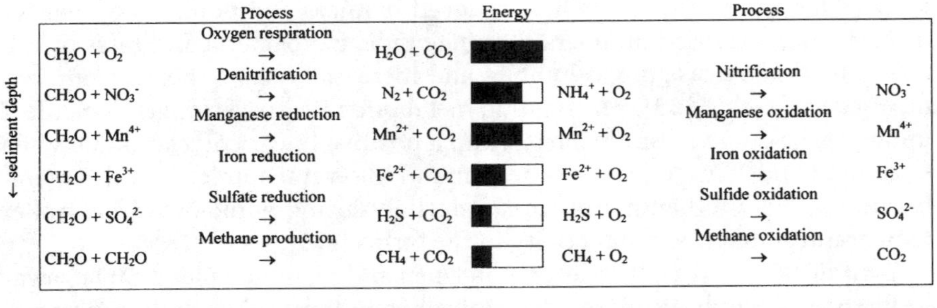

↓ sediment depth	Process			Energy	Process		
		Oxygen respiration					
	$CH_2O + O_2$	→	$H_2O + CO_2$	▬▬▬			
		Denitrification			Nitrification		
	$CH_2O + NO_3^-$	→	$N_2 + CO_2$	▬▬	$NH_4^+ + O_2$	→	NO_3^-
		Manganese reduction			Manganese oxidation		
	$CH_2O + Mn^{4+}$	→	$Mn^{2+} + CO_2$	▬	$Mn^{2+} + O_2$	→	Mn^{4+}
		Iron reduction			Iron oxidation		
	$CH_2O + Fe^{3+}$	→	$Fe^{2+} + CO_2$	▬	$Fe^{2+} + O_2$	→	Fe^{3+}
		Sulfate reduction			Sulfide oxidation		
	$CH_2O + SO_4^{2-}$	→	$H_2S + CO_2$	▬	$H_2S + O_2$	→	SO_4^{2-}
		Methane production			Methane oxidation		
	$CH_2O + CH_2O$	→	$CH_4 + CO_2$	▬	$CH_4 + O_2$	→	CO_2

Fig. 2 Vertical distribution of microbial processes in marine sediments. The relative energy output from organic matter oxidation and the oxidation of reduced mineralization products is indicated

The oxic-anoxic gradients within surface sediment layers (often millimetres in silt/clay sediment) are striking. They are often visible as colour changes from an oxidized (light coloured) surface layer to anoxic (dark brown to black) deeper layers. However, considerable small-scale horizontal variations may exist in the depth of the surface oxic layer due to sediment mixing and bio-irrigation (bioturbation) by macrofauna [22, 31, 38]. In depositional areas, where water currents are low (< 2 cm s^{-1}) and oxygen penetrates sediment primarily by diffusion, the oxic layer is only a few millimetres deep. Respiration by aerobic organisms rapidly consumes oxygen and subsurface anoxic sediments occur close to or at the sediment-water interface. Vertical chemical zonation within the sediment is readily seen in these types of deposits where decomposition and consumption of organic matter by bacteria and benthic fauna is most intense at the sediment-water interface [32, 36, 37].

Benthic metabolic rates can be measured as oxygen consumption and TCO$_2$ release across the sediment-water interface by the use of benthic chambers [7, 19, 20, 39]. Under optimal conditions the chambers are deployed in-situ but cores can also be collected by divers and incubated on shore [40]. However, rates of sediment oxygen uptake (SOU) for determination of organic matter turnover are limited by the consumption of oxygen in reoxidation processes. Results must be interpreted with care particularly in fish farm sediments where high sulfide concentrations and rates of sulfate reduction could lead to increased rates of oxygen uptake due to sulfide oxidation [10]. TCO$_2$ release, on the other hand, is considered to represent total faunal and bacterial respiration and is used as a measure of the terminal mineralization of organic matter [40]. Both types of measurements, however, should be considered with caution in biogenic sediments where carbonate dissolution may occur during incubations. It is possible to correct for this process by measuring calcium release at the same time [40]. SOU and TCO$_2$ release have been found to be enhanced up to several orders of magnitude in the vicinity of fish farms (0–50 m, Table 1) and to decrease with distance from the farms (> 50 m [7–9, 41–43]). The zone with highest rates of oxygen and TCO$_2$ fluxes is often restricted to the immediate vicinity of the farms (< 50 m away). This is also the area where reduced biomass and numbers of benthic macrofauna, increased numbers of sulfate reducing bacteria and rates of sulfate reduction, release of gas bubbles and increased rates of benthic nutrient flux occurs [7–9, 19, 38]. The benthic metabolic rates are strongly dependent upon the productivity in the farms, and a positive linear correlation between feed input and TCO$_2$ release has been found in several studies [8, 42]. In temperate regions the benthic processes decline during winter with low water temperatures and low productivity in the farms [8, 9].

Benthic SOU and TCO$_2$ release at net-pen and reference sites > 500 m away in the Bay of Fundy increased dramatically at a threshold sulfide concentration between 200 and 350 µM S$^=$ (Fig. 3). Highest levels of S$^=$, gas exchange and NH$_4^+$ release occurred at farm sites that had experienced high rates of

Table 1 Sediment oxygen uptake (SOU), total carbon dioxide release (TCO$_2$) and sulfate reduction rates (SRR) measured at fish farms under the cages, close to the cages and at control sites

Species/location	Under cage (mmol m^{-2} d^{-1})	Close to cage (mmol m^{-2} d^{-1})	Control (mmol m^{-2} d^{-1})	Reference
SOU				
Rainbow trout *Denmark*	278	–	46	[60]
Atlantic salmon *Canada*	52–100	26–60 (5 m)*	10–26	[7]
Milkfish *Philippines*	!Frymesh 61	61 (15–25 m)*	78	[42]
	Starter 90	76	92	
	Grower 120	114	72	
	Finisher 261	63	44	
Atlantic salmon *Scotland*	468		9	[43]
TCO$_2$				
Rainbow trout *Denmark*	610	–	74	[8]
Milkfish *Philippines*	!Frymesh 141	44 (15–25m)*	69	[42]
	Starter 88	40	47	
	Grower 141	164	124	
	Finisher 641	101	41	
SRR				
Rainbow trout *Sweden*	31–553			[41]
Rainbow trout *Denmark*	92	–	24	[8]
Milkfish *Philippines*	!Frymesh 14	19 (15–25 m)*	24	Holmer (unpublished)
	Starter 32	26	15	
	Grower 36	14	2	
Seabass/seabream *Cyprus*	64	55 (200 m)*	43	Holmer pers. comm.
Seabass/seabream *Greece*	38	18 (10 m)*	7	Holmer pers. comm.
Seabass/seabream *Italy*	212	39 (10 m)*	–	Holmer pers. comm.
Seabass/seabream *Spain*	43	35 (10 m)*	16	Holmer pers. comm.

* Distance from edge of cages
! The type of food used in the farms

organic loading [20]. While some reference sites had slightly elevated rates of benthic respiration and ammonium flux, showing a response to organic enrichment at distances > 500 m away from cage sites, most far-field samples show lower benthic fluxes characteristic of other coastal areas in the Bay of Fundy. Despite the high metabolic rates generally encountered in fish farm sediments, organic matter tends to accumulate under the net cages, and thus create enriched sediments, which may persist after farming activities stop [36, 37, 44].

Respiratory quotients calculated as the ratio between TCO_2 release and oxygen consumption are often greater than 1 in fish farm sediments suggesting that anaerobic processes are important for the organic matter turnover. This is consistent with parallel measures of high rates of sulfate reduction (Table 1). Fish farm sediments are generally characterized as reduced (excess electron activity and negative redox potentials), and sulfate reduction is an important mineralization process. This has been confirmed by several studies in temperate and tropical latitudes, where sulfate reduction rates were enhanced more than one order of magnitude compared to unimpacted sediments (Table 1). Rates are particularly high under the net cages where sulfate reduction accounts for all the CO_2 released (e.g. measured as TCO_2 flux) indicating that the sediment metabolism is strictly anaerobic. Sulfate reduction

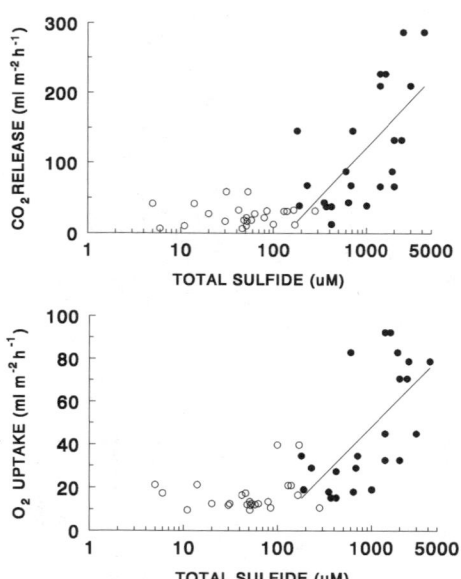

Fig. 3 Comparison of sediment-water fluxes of dissolved CO_2 and O_2 (benthic respiration) measured under salmon cages (*solid points*) and at reference sites (*open circles*) > 50 m away in the Western Isles region of the Bay of Fundy, New Brunswick, Canada. Redrawn from [20]

rates also decrease with distance from the cages (Table 1) and often more rapidly than the total sediment metabolism. The contribution of anaerobic processes to organic matter decomposition thus declines with distance from cages.

One reason for this decrease is the increasing presence of larger benthic macrofauna, which oxidize the sediments and thereby suppress sulfate reduction rates [45]. Sulfate reduction rates in sediments from fish farm sites (Table 1, [8]) are among the highest measured in any marine ecosystem. It appears that organic enrichment from finfish farm wastes stimulates sulfate reducing bacteria more than other sources of organic matter such as phytoplankton detritus. This may reflect combined effects of high sulfate concentrations in overlying water and high rates of organic matter deposition which lead to the formation of anoxic surface sediment layers. Sulfate, as an electron acceptor, is only seldom fully depleted due to high concentrations in seawater and sediment pore water [46, 47]. Fish farm sediments are quite special in this regard, as high rates of organic loading may exhaust sulfate locally in the surface layers [8]. There are also observations of coexistence of sulfate reduction and methanogenesis [48], two processes which do not co-occur in natural sediments. In natural sediments sulfate reducing bacteria out-compete methanogenic bacteria due to a larger energy output during organic matter oxidation [46]. The large pools of organic matter available in fish farm sediments may allow both processes to coexist. If sulfate is exhausted in deeper layers of fish farm sediments, methanogenesis becomes the predominant metabolic pathway.

Recently, bacterial iron reduction has been identified as an important mineralization process in coastal sediments [49], but it has not yet been measured in fish farm sediments. Rates of iron reduction appear to be strongly controlled by the availability of oxidized iron and a positive relationship between the sedimentary pools of oxidized iron (Fe^{3+}) and the rates of iron reduction has been reported in a wide range of coastal sediments [49, 50]. It is likely that iron reduction is an important process in sediments with high iron pools (e.g. where terrestrial material predominates in sedimented material). However, if there is an accumulation of reduced sulfides, as often occurs in fish farm sediments, pools of oxidized iron may be reduced when iron is used for reoxidation of reduced sulfides and/or precipitate with sulfides in iron-sulfur compounds such as FeS and pyrite [49] (Fig. 4). Sulfate reduction is thus favoured over iron reduction where there is a continuous supply of sulfate from the overlying water independent of the oxygen concentration in the water. Sulfate may, however, become limiting in the deeper sediment layers where diffusion is insufficient to maintain the supply due to lack of bioturbation [8].

Few measurements have been undertaken to quantify rates of methane production from sediment at marine finfish farm sites but observation of gas bubbles released from deposits under net-pens suggests that the process

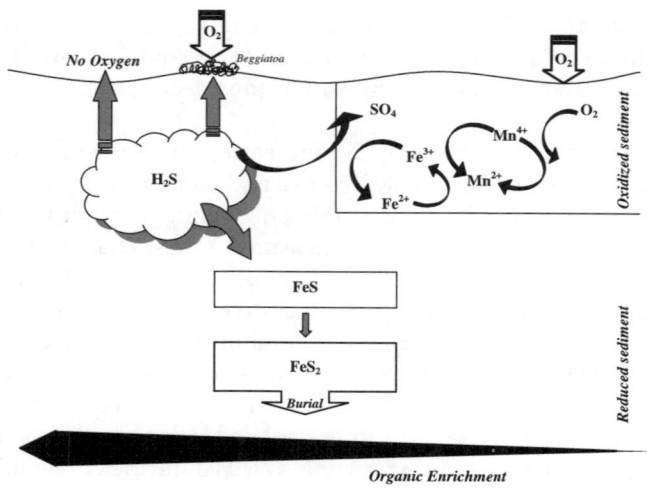

Fig. 4 Major sulfur pathways along an organic enrichment gradient in marine sediments

is important [37]. It has been concluded that dissolved sulfides are transported vertically in sediments by more than molecular diffusion and that in many sulfide-rich deposits gas bubble ebullition driven by methane production below the sulfate reduction zone is an important process [10]. Fish farmers try to avoid this "outgassing" as the methane bubbles often carry high concentrations of sulfides. When the toxic gas is released, as may occur immediately under net-pens with excessively high rates of sedimentation, fish may be stressed or die. Exposure to low (< 30 μM) sublethal concentrations of hydrogen sulfide for brief (20 min) periods results in damage to gills, liver and metabolic enzymes in Atlantic salmon [51]. Hydrogen sulfide concentrations within this range have been measured in water immediately beneath fish cages in Scotland and Ireland [52]. Presence of bubbles is a clear sign of overloading as defined in the MOM monitoring program [36]. The stimulation of sulfate reduction due to increased organic matter loading is one of the major potential occurring problems in fish farm sediments, as it can lead to accumulation of toxic sulfides and to elimination of the benthic macrofauna and even meiofauna [14]. The presence of sulfides also increases oxygen consumption tremendously due to the spontaneous reaction between sulfides and oxygen [1, 39]. Sulfide diffusion from sediments into the overlying water column can also result in the formation of oxygen depleted bottom water: a problem first observed in stratified fjords with sills in Norway and Scotland [53]. The accumulation of sulfides in fish farm sediments is controlled by the potential for re-oxidation and the availability of iron in the sediments (Fig. 4). This in turn is strongly correlated with the activity of the benthic fauna since irrigation, feeding and burrowing activity increase the input of oxygen into the deeper sediment layers during their irrigation and burrowing activity [31, 54–56]. Oxygen is used directly for reoxidation of sulfides back to sulfate or to intermediate oxidized sulfur compounds [40] (Figs. 2 and 4). Oxygen is also utilized preferentially over manganese for reoxidation of reduced iron and oxidized iron is then used for sulfide oxidation [40]. The availability of iron is therefore of importance both for reoxidation of sulfides by oxidized iron and for precipitation of sulfides. Low iron concentrations usually occur in biogenic carbonate sediments and in sandy deposits with low mineral content. These sediments are especially prone to sulfide toxicity when enriched with organic matter from fish farms [42].

The marine nitrogen cycle is very important in the coastal zones, where if light limitation does not occur nitrogen may be the limiting nutrient of primary productivity (Sect. 3). Fish farms are characterized by large releases of dissolved inorganic and organic nitrogen compounds, which are dispersed in the water column. Nitrogen is also an important part of the particulate waste product stream (Sect. 2) [57] and nitrogen accumulates in the underlying sediments [8, 9, 21]. Marine detritus is generally limited in nitrogen compared to carbon and this leads to burial of more carbon than nitrogen and the creation of a large pool of refractory organic matter in the sediment [58]. Since marine bacteria are generally limited by nutrient availability [59], high concentrations of nitrogen and phosphorus in fish farm sediments may be one of the main reasons for the stimulation of microbial activity. As discussed above, organic matter in sediments under net pens derived from fish faeces and waste feed appears to be more labile than that derived from natural detritus [28]. Nitrogen compounds undergo important transformations in marine sediments, for example nitrate is produced during nitrification or taken up directly by sediments from the water column (Fig. 2). However, nitrification has been found to be almost eliminated in fish farm sediments due to the prevailing anoxic conditions [60]. Denitrifying bacteria can utilize nitrate directly from the water column if nitrification is inhibited, but they appear to be sensitive to high concentrations of sulfides and their metabolism may be restricted in fish farm sediments [60].

As an alternative pathway, nitrate has been observed to be reduced to ammonium through dissimilatory nitrate reduction in fish farm sediments [60]. A newly discovered process – anaerobic ammonium oxidation or ANAMMOX where ammonium is oxidized to N_2 through reduction of nitrate has not yet been detected in fish farm sediments (Dalsgaard pers. comm.). This process and denitrification may be important processes for nitrogen removal in anoxic marine environments, but this has yet to be verified in fish farm sediments from different areas. Since there are few transformations of nitrogen in fish farm sediments most mineralized nitrogen is released across the sediment-water interface as ammonium, where it is available for primary production [7, 8, 57]. One major concern of this release in coastal zones is that it occurs during summer months where fish growth and feeding rates are high and primary production is often limited by nitrogen [61]. A small release of nitrogen may thus lead to blooms of phytoplankton in the nutrient limited period. In shallow water where light reaches the bottom and benthic microalgae or seagrasses occur (discussed below), dissolved ammonium may be utilized by the primary producers at the sediment water interface ([42, 62] Dalsgaard pers. comm.). Benthic microalgae may form dense mats at the sediment surface and growth of macroalgae may be stimulated by nutrient enrichment. Algal mats may have negative impacts on the performance of seagrasses and are a clear sign of eutrophication [63].

4
Using Changes in Sediment Biogeochemistry as Indicators of Organic Enrichment

Due to the major changes in sediment biogeochemistry as a result of organic enrichment it is important to identify the impacted area underneath and near the cages, and if unacceptable conditions are found then provide suggestions for improvements of the farming practice. The historical background for the development of the organic enrichment gradient concept is given in [1]. Initially the approach was suggested as a successional model for changes in benthic macrofauna in response to increasing amounts of organic wastes, where the enrichment gradients were applicable both in space and time. The changes include (1) the appearance of organic enrichment-tolerant, opportunistic, indicator species; (2) changes in macrofaunal bioturbation activity which controls the mixed layer depth of the sediment; and (3) a reduction in the mean size of the macrofauna as organic enrichment increases. Most previous studies of macrofauna within the area of sediments affected by wastes deposited from net pens have followed the spatial and temporal gradients identified in the organic enrichment gradient successional model of Pearson & Rosenberg [1, 64].

Rhoads & Germano [65] and Nilsson & Rosenberg [31] used SPI to characterize organic enrichment in sediments. This method uses underwater photographs of the sediment profile to determine characteristic structural features, such as the depth of the redox potential discontinuity, RPD (where a predominantly oxic, becomes an anaerobic sediment), and the presence/absence of macrofaunal artefacts (e.g. faecal pellets, tubes, feeding pits, mounds, burrows, voids). The results can be used to identify the four stages of enrichment described in the Pearson & Rosenberg model [64]. Hargrave et al. [20] conducted a study in the area of the Bay of Fundy to identify which of 20 possible environmental variables best characterized the degree of organic enrichment in an area of intensive salmon aquaculture in the Bay of Fundy. Total sulfide was the most sensitive and practical variable to detect organic enrichment effects at farm versus reference locations. Benthic metabolic rates measured as oxygen uptake and carbon dioxide release from the undisturbed sediment surface were also significantly different under farm cages and in adjacent reference areas unaffected by organic enrichment (Fig. 3). The transition from "background" benthic metabolic rates at reference sites to elevated rates under cage sites occurred at a threshold of approximately 1300 μM S$^=$. Although redox potentials were a less sensitive variable for detecting enrichment effects due to high variability in measurements, this sulfide concentration was associated with an Eh$_{NHE}$ potential range (< 0 mV) characteristic of the transition from aerobic respiration to sulfate reduction in sediments. Wildish et al. [66, 67] also showed that both Eh and S$^=$ could also be used to indicate organic enrichment based on a four level-classification described in the Pearson & Rosenberg gradient model (Fig. 5).

The limits of both variables in the Bay of Fundy and characteristics for the four enrichment categories were determined for the upper (< 2 cm) sediment layer at study sites in the Bay of Fundy as: Normally oxic marine sediments (Eh >+ 100 mV, total S$^=$ < 300 μM) with low accumulation rates. Benthic epifauna feed on suspended material and dissolved oxygen is supplied by relatively high rates of advection. Oxygen penetration occurs to variable depths in sediment, depending on the amount of biogenic re-working. Aerobic respiration predominates over anaerobic metabolism. There is little sulfide accumulation in surface layer sediments that are usually light red/grey

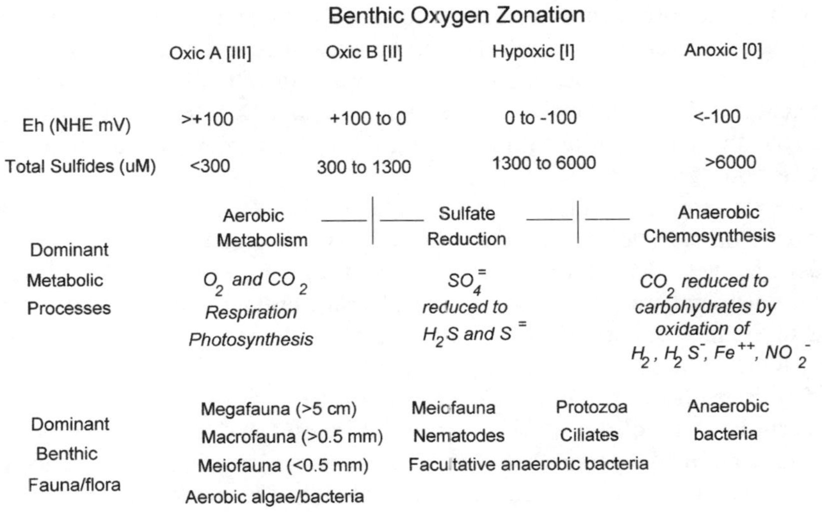

Benthic Oxygen Zonation

	Oxic A [III]	Oxic B [II]	Hypoxic [I]	Anoxic [0]
Eh (NHE mV)	>+100	+100 to 0	0 to -100	<-100
Total Sulfides (uM)	<300	300 to 1300	1300 to 6000	>6000
	Aerobic Metabolism		Sulfate Reduction	Anaerobic Chemosynthesis
Dominant Metabolic Processes	O_2 and CO_2 Respiration Photosynthesis		$SO_4^=$ reduced to H_2S and $S^=$	CO_2 reduced to carbohydrates by oxidation of H_2, H_2S^-, Fe^{++}, NO_2^-
Dominant Benthic Fauna/flora	Megafauna (>5 cm) Macrofauna (>0.5 mm) Meiofauna (<0.5 mm) Aerobic algae/bacteria		Meiofauna Protozoa Nematodes Ciliates Facultative anaerobic bacteria	Anaerobic bacteria

Fig. 5 Benthic organic enrichment zonation based on oxygen gradients determined from relationships between oxidation-reduction (Eh) potentials (mV), total sulfide concentrations (μM), dominant benthic metabolic processes and taxonomic groups of benthic fauna and flora

in colour depending on inorganic source material and the relative abundance of iron and manganese.

Transitory sediments have lower Eh potentials and higher sulfide concentrations (Eh 0 to + 100 mV, total $S^=$ 300 to 1300 µM). Sediments are depositional but accumulation rates are low due to periodic resuspension and horizontal transport. This results in variable proportions of silt and clay due to heterogeneous conditions of deposition and erosion. There is also usually a high diversity of benthic epifauna and infauna utilizing the spatially heterogeneous sediment structure. Due to physical advection, macrofauna bioturbation and irrigation that move both sediment pore water and particles vertically across the sediment-water interface, oxygen penetrates from several millimetres to many centimetres. Aerobic and anaerobic metabolism co-occurs within micro-niches in the heterogeneous sediment matrix. Some sulfide accumulation results from anaerobic sulfate reduction in surface layers and in these cases sediments may be visibly darker in colour (medium to dark brown or grey). Hypoxic sediments (Eh -100 to 0 mV, total $S^=$ 300 to 6000 µM) reflect higher rates of sediment accumulation and $S^=$ accumulation, where resuspension is infrequent. Such sediments usually occur in depositional areas where currents are low (< 2 cm s^{-1}) with a correspondingly higher proportion of silt/clay (> 90%). Due to the fine-grain texture and higher water content (> 70%) in these deposits there may be few epifauna. Communities of infauna are highly diverse and often dominated by polychaetes. Oxygen penetration is limited (millimetres to centimetres). Diffusion and bioturbation are the major processes supplying dissolved oxygen and there is only a thin surface oxidized layer. Limited oxygen availability leads to an increase in the relative importance of anaerobic metabolism with a corresponding increase in sulfide accumulation. Small scale variations in the oxic/anoxic boundary at the sediment surface may lead to a discontinuous distribution of white sulfur bacteria (*Beggiatoa*) mats appearing as white patches on dark brown to black sediment.

Anoxic sediments (Eh <- 100 mV, total $S^=$ > 6000 µM) represent the highest level of organic enrichment where the surface sediments become fully reduced. These deposits occur in areas of very high sediment accumulation where deposits are primarily (> 95%) silt/clay. Epifauna are generally absent due to the lack of hard surfaces for attachment. Anoxic sediments may be totally without fauna (azoic) but some infauna such as some ciliates, nematodes and polychaetes (e.g. *Capitella* sp.) tolerant to sulfides can be present [14]. Anaerobic metabolic processes predominate and end products (H_2S, CH_4, H_2) may be out-gassed across the sediment-water interface. High concentrations of sulfides produce a strong odour of H_2S. The percent cover by sulfur bacterial mats is higher than in hypoxic sediments, or may be absent if the bottom water is hypoxic/anoxic. Underlying sediments are black and usually without colour variation.

5
Organic Enrichment and Changes in Benthic Macrofauna

Benthic macrofauna can have a profound influence on sediment physical-chemical conditions through their feeding, burrowing and irrigation activities. Seasonal changes in benthic faunal biomass and community structure will therefore affect rates of a variety of biogeochemical processes. For example, seasonal or permanent hypoxia may occur when seawater with low oxygen concentrations in benthic boundary layer seawater come into contact with surface sediments. Mortality of macrofauna has been observed when this has occurred resulting in critically low dissolved oxygen in bottom water in stratified inlets [1, 22]. Other changes, however, may be due to natural seasonal cycles of macrofauna recruitment and mortality which alter species abundance and biomass distributions. Both effects are associated with changes in community structure and must be accounted for if organic enrichment effects on benthic fauna are to be adequately demonstrated.

As far as we are aware there have been no seasonal studies of macrofaunal communities at active farm sites that would allow separation of natural variations in species composition and biomass from those due to local organic enrichment. Observations by Pohle et al. [68] in two areas (Back and Lime Kiln Bay) of the Bay of Fundy where the salmon culture industry is centred in the Bay of Fundy suggested that changes in the benthic macrofaunal community occurred over a five year period (1994–1999) after expansion of salmon aquaculture in the area. Since grab stations were 200 m distant from the nearest fish farm the evidence suggested far-field organic enrichment effects linked to salmon culture. Studies of the recovery process after fish farming have shown that the benthic fauna follow a series of successes and catastrophes, where some sites may recover quickly with a diverse community, whereas others show very slow recovery or show oscillations between a diverse and a poor community [69]. Much more research is needed to understand the dynamics of the benthic communities.

A shift in benthic communities in fish farm sediments due to organic enrichment may also influence the turnover of organic matter [38]. Benthic macrofauna are often divided into functional groups based on their feeding mode. The two main groupings are suspension- and deposit-feeders, where the deposit-feeders are further separated into surface- and subsurface-feeders [54, 55]. The most important role of the suspension feeders is to enhance the benthic-pelagic coupling by transferring suspended organic matter to the sediments, primarily as biodeposits. Surface- and subsurface deposit feeders do not increase the organic supply to sediments, but are very important for the mechanical breakdown of the organic matter and an increase in the surface area enhancing the microbial colonization [38, 54, 55]. They also mix freshly deposited organic matter deeper into the sediments during their feeding and burrowing activities. Surface-deposit-feeders typically live in L- or U-shaped burrows, which they ventilate to maintain oxidized conditions. Large subsurface-deposit-feeders, such as the lug-worm *Arenicola marina*, are intense bioturbators and both percolate and mix deep layers of surface sediments during their ventilation and feeding activities [70].

The most powerful effects on biogeochemical conditions in sediments are created by the activity of surface- and subsurface-deposit-feeders. This fauna stimulates the decomposition of organic matter by increasing the oxygen-exposure time through re-introduction of buried organic matter to oxic conditions during their irrigation and burrowing activities [38, 71]. The decomposition of refractory organic matter is particularly enhanced, probably since oxygen is required to breakdown complex chemical bounds [71]. Activities of fauna may also bring other electron-acceptors deeper into the sediment and thereby increase the oxidized surface areas. For example, nitrification may be significantly stimulated, as has been found in burrow walls of a number of polychaetes [38, 54, 55]. Enhanced nitrification in burrow walls also increases coupled nitrification-denitrification processes. The removal of nitrogen is higher from bioturbated sediments compared to sediments without fauna [72]. Iron reduction has also been found to be enhanced by the burrowing activities of several types of benthic fauna [56, 73]. Feeding and irrigation by the fauna increases the reoxidation of reduced iron and thereby also microbial iron reduction. The oxidation of the sediments also suppresses sulfate reduction [56, 70]. Sulfate reduction rates may, however, be stimulated in the deeper layers where the irrigation is reduced but the organic content enhanced by macrofauna burrowing activity [70]. In soft sediments bioturbation increases the depth of the RPD, where diffusion alone will only result in oxygen penetration to a few mms depth.

A shift in the benthic fauna from large, intense bioturbators to small opportunistic species with progressive hypoxic conditions may have significant impacts on the nature of microbial communities and the predominant processes for organic matter decomposition and sediment oxidation [1, 60]. Only a few studies of such a shift have been performed in fish farm sediments and only for fine-grained sediments dominated by polychaetes [74]. As discussed above, organic matter in fish farm sediments is more labile than natural sources, and microbial activity is high. In contrast to natural sediments, the activities of benthic fauna do not appear to stimulate microbial activity further. This could be due to the high lability of the organic matter and also to the fact that the microbial community is saturated with substrates. In contrast, in natural sediments faunal feeding and burrowing activities provide nutrient-starved microbes with fresh organic matter [38]. This hypothesis is supported by observations of accumulation of small organic acids in pore water [8], which does not occur in natural sediments, where the turnover is very rapid [75].

The taxonomic composition and physical and biological activities of benthic fauna have a major influence in determining which microbial processes become dominant in the decomposition of fish farm waste products, as well as on oxidative processes in sediments. One study showed that large polychaetes reduced rates of sulfate reduction by up to 50% [74]. The colour of the surface sediment changed from black to brown, and the white layer of *Beggiatoa*, which was present on defaunated cores, was absent. Experiments with the smaller polychaete *Capitella capitata* did not show similar positive effects on sulfate reduction, as rates were similar to those in defaunated sediment [45]. This suggests that only large species of benthic fauna are able to maintain oxic conditions in surface sediments, while a shift to smaller species enhances reducing conditions.

Benthic fauna may also have a significant impact on other geochemical processes such as the retention of phosphate. Phosphate in oxic marine sediments is bound to oxidized iron (Fe^{3+}). However, as sediments and iron become reduced, phosphate is released from the sediment and can be transported into the water column. Reduction of the sediments may also affect nitrogen loss through denitrification. This may be reduced since nitrification is inhibited by lack of oxygen and the coupled nitrification-denitrification processes [61]. Denitrification may also be inhibited by high concentrations of sulfides [60]. This provides an incentive for management of fish farms to minimize the organic loading of sediments as much as possible to maintain oxic conditions, a diverse benthic faunal community and the associated bioturbation activity.

6
Organic Enrichment Effects in Biogenic Sediments

Most of the biogeochemical investigations of fish farm sediments have been carried out in near-shore temperate coastal regions dominated by terrigenous sediments. However, there is currently rapid growth in fish farms located over biogenic sediments such as in the tropics and in the Mediterranean [76]. Sediments in temperate areas are dominated by mineral particles derived from terrestrial material transported to the estuaries by rivers. Biogenic sediments, on the other hand, are derived from calcareous deposits either from coral reefs or epiphytic debris. These biogenic sediments are characterized by relatively low organic matter (typically < 1%) and low amounts of iron which has important consequences for sediment biogeochemical conditions. In comparison to terrigenous sediments, low iron in biogenic deposits may limit microbial iron reduction since sulfide buffer capacity is reduced through formation of FeS or reoxidation by iron oxides. Knowledge of biogeochemical conditions in biogenic sediments is, however, quite limited [77]. Studies have shown that rates of mineralization in biogenic sediments often are lower than expected given high temperatures that characterize coastal sediments in these sub- and tropical ecosystems, probably reflecting more refractory organic matter [78, 79]. Mangroves and seagrasses are important components of the coastal zones, and detritus from these plants is generally less labile than phytoplankton detritus which usually dominates organic input to sediments in temperate systems. Biogenic sediments are often coarse grained and highly advective with low nutrient pools.

The severity of benthic impacts of fish farming activities in biogenic sediments is controlled by many factors, including bathymetry and water circulation as found for terrigenous sediments. A study of a series of fish farms located at shallow depths (< 3 m) in the Philippines showed very high rates of sedimentation and highly enriched sediments [42, 79]. This had a major impact on the biogeochemical conditions. Surprisingly, the mineralization of organic matter was negatively correlated with sedimentation rates, suggesting that decomposition was not able to keep up with the organic loading and is a clear sign of overloading. Microbial activity in the sediments may have been reduced due to lack of electron acceptors or high concentrations of sulfides [48]. Reduced decomposition leads to an accumulation of organic matter, which may persist also when the farming activities stop and it may thus take years to recover [44]. There was a release of ammonium and phosphate from the sediments, while nitrate was taken up probably by denitrification or dissimilatory reduction of nitrate to ammonium. The nutrients are dispersed in the area by the local currents and tides, and may be transported to the nearby coral reefs. Coral reefs have been found to be very sensitive to nutrient enrichments, which may alter the ecological integrity and lead to coral reef degradation. Sulfate reduction was the overall dominating metabolic process for organic matter decomposition, and due to the very low iron pools, high concentrations of sulfides occurred in the sediments, and the benthic fauna appeared to be very sparse (Holmer pers. obs.).

Other studies in the Philippines have shown negative effects on an important benthic species, the shrimp *Alpheus marcellarius*, which was absent close to the farms (Heilskov and Holmer pers. comm.). Nearby seagrass meadows were also negatively affected since seagrass abundance and seagrass diversity decreased close to the farms. This may have been a result of impoverished sediment conditions and reduced transparency of the water column due to phytoplankton growth (Marbá pers. comm.). The development of the aquaculture industry in this area has been extremely rapid and over a 10-yr period and nutrient concentrations in the water column have increased significantly. In the spring of 2002 an extreme weather condition with low tides and reduced circulation of the water caused a major decline in water column oxygen concentrations probably accelerated by high sediment oxygen consumption. There was at the same time a bloom of a nitrogen fixing and toxic dinoflagellate, which caused a massive die-back of wild and cultured fish [42].

In the Mediterranean most fish farms are located in deeper water and at locations with rapid water exchange. This leads to larger dispersion of waste products and the benthic impacts are less severe [26, 32]. Preliminary results from studies in the Mediterranean however, show that the sediments change from autotrophic to heterotrophic in proximity to the net cages (Dalsgaard pers. comm.). Oxygen consumption is enhanced and there is a release of ammonium and phosphate from the sediments. Denitrification is generally low in these sediments, but stimulation of denitrification occurs from utilizing nitrate from the water column [61] which is higher close to the cages. Sulfate reduction rates are also significantly enhanced and this leads to an accumulation of sulfides in sediments. Such changes may have major impact under the oligotrophic conditions experienced in the Mediterranean and in particular on the benthic vegetation, as discussed below.

7
Biogeochemical Conditions in Seagrass Beds Enriched by Aquaculture Waste Products

Of particular concern with aquaculture operations under oligotrophic conditions is that the natural transparency of the water is high creating a potential for benthic primary production. Locations with seagrasses are often characterized by high rates of advective water exchange and these conditions favour seagrass growth. This may lead to a potential conflict when selecting sites for fish farms and conservation of seagrass meadows, as the same sites are excellent for fish farming. This problem exists, where diverse and productive seagrass meadows are found in the tropical coastal zones, and also in the Mediterranean, where the slow-growing seagrass *Posidonia oceanica* has a wide distribution [21, 62, 80–82].

There is limited information available on the impact of fish farms on seagrass meadows, but reduced light conditions due to enhanced phytoplankton growth and increased sedimentation of organic matter as a result of fish farming, may lead to negative impacts on the meadows (Fig. 6) [82–85]. These areas retain particles because of reduced water flow over the canopies and enhanced rates of sedimentation have been found in P. oceanica meadows near fish farms [62]. Organic matter loading within seagrass meadows increases anaerobic mineralization in the sediments, and in particular high sulfate reduction rates have been measured leading to an accumulation of sulfides in the root zone (Holmer unpubl.). Seagrasses are adapted to growth in reduced sediments by having an internal system of air-channels to transport oxygen down to the below-ground tissues [86]. The capacity for oxygen transport is, however, limited and decreasing oxygen concentrations in the water column may have significant effects on oxygen availability. During oxygen de-

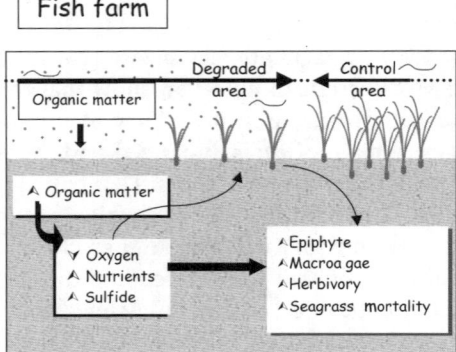

Fig. 6 Possible effects of organic loading of seagrass meadows around fish farms. Redrawn from [62]

pletion events, oxygen transport in the seagrasses is only maintained during the day with active photosynthesis and oxygen production in the leaves [85]. Increased oxygen consumption in the sediment most likely intensifies the demand for oxygen transport, and the presence of sulfides in the pore waters may further constrain oxygen availability.

Sulfide is a potent phytotoxin, as it inhibits essential enzymes, and reduced growth and increased mortality has been observed in seagrasses during oxygen depletion events and during exposure to sulfides [87]. Seagrass sediments are often biogenic with low iron contents [88]. This means that the potential for reoxidation of sulfides and binding of sulfides to iron is limited, and that seagrasses readily become exposed to anoxic conditions and high sulfide concentrations in the pore waters. Chemical binding of the limited amounts of iron in the sediments to sulfide may further constrain the growth of the seagrasses due to iron-limitation. Iron limitation has been found in seagrasses in carbonate sediments [89], and additions of iron to organic-enriched carbonate sediments has been shown to increase the growth of seagrasses ([77], Holmer pers. comm.). One result was a stimulation of the iron-demanding enzymes for nutrient uptake (Holmer pers. comm.), suggesting that the irreversible binding of iron to sulfides was one of the factors controlling reduced seagrass performance in organic enriched sediments.

Enhanced mineralization in seagrass meadows impacted by fish farms leads to increased nutrient availability in the root zone and in the water column just above the sediment. Seagrasses growing under oligotrophic conditions are usually nutrient limited, and utilize regenerated nutrients. If the nutrient content in water and sediment increases near fish cages leading to increased production and biomass, a secondary effect might be that herbivore pressure on the seagrasses increases significantly [62, 80]. Observations of large numbers of sea urchins have been done at several fish farms in the

Mediterranean [80]. It is not yet known if these animals utilize seagrass tissue or the attached epiphytes in their diet, but they reduce the leaf area leading to a decline in the overall photosynthetic capacity of the seagrass meadow.

Negative effects of aquaculture operations on surrounding seagrass meadows appear to be much more severe compared to unvegetated sediments. The seagrasses die in the vicinity of the cages probably due to a synergistic effect of the organic loading and herbivory. The seagrass community suffers from increased mortality and reduced recruitment at considerable distances – up to several hundred metres from farms [62, 80, 84]. This may reflect the fact that seagrass meadows enhance the sedimentation of organic matter and thus also fish farm waste products compared to unvegetated sediments. Further investigations of these conditions are necessary in the coastal zones where aquaculture operations are expanding, as many seagrasses including *Posidonia oceanica* in the Mediterranean are threatened species. Their slow growth rate implies that it will take decades or longer to recover lost meadows [62, 81]. Similar studies in the Philippines showed decreased productivity and a significant decline in the number of seagrass species near the fish cages. Thus net-pen aquaculture may be a major threat to the biodiversity of the coastal zones in the tropics (Marbá pers. comm.).

8
Conclusions

The severity of environmental impacts from aquaculture operations on the marine environment in any one area may be controlled by several external factors such as nutrient limitation of primary productivity, naturally occurring sources of biological oxygen demand and oxygen availability. Nitrogen is generally considered a limiting nutrient for marine primary production at seasonally stratified temperate latitudes and organic enrichment of the underlying sediments leading to a high release of nitrogen compounds during the summer should be avoided. On the other hand, phosphate loading may be more important under tropical conditions where the availability of phosphate controls primary production of the diverse seagrass flora in the coastal zone. Release of phosphate from aquaculture may alter the structure and productivity of macrophyte communities in these areas. Benthic changes resulting from excessively high sedimentation rates associated with finfish aquaculture are most readily observed immediately under and adjacent to net-pens, but in some cases spatial effects may be more wide-spread. If organic matter sedimentation rates are sufficiently high, there may be shifts in dominant metabolic processes as sediments are altered from oxic, to suboxic and anoxic states. Under fully anoxic conditions sulfate reduction and methanogenesis become the dominant benthic metabolic processes and accumulation and out-gassing of toxic sulfides may occur. Such increases in sulfide levels may result in changes in major benthic groups of bacteria, meiofauna and macrofauna along an organic enrichment gradient extending away from farms in the direction of the prevailing current. The challenge for future marine aquaculture development is to minimize organic enrichment of the benthic environment as much as possible. This is particularly the case in warm temperate and tropical areas where carbonate sediments occur with limited iron available for buffering sulfide toxicity.

References

1. Gray JS, Shiu-sun WR, Or YY (2002) Mar Ecol Prog Ser 238:249
2. Cloern JE (2001) Mar Ecol Prog Ser 210:223
3. Meyer-Reil L-A, Köster M (2000) Mar Poll Bull 41:255
4. Rizzo WM, Dailey SK, Lackey GJ, Christian RR, Berry BE, Wetzel RL (1996) Estuaries 19:247
5. Kemp WM, Sampou PA, Garber J, Tuttle J, Boynton WR (1992) Mar Ecol Prog Ser 85:137
6. Hargrave BT (1980) Factors affecting the flux of organic matter to sediments in a marine bay. In: Tenore KR, Coull BC (eds) Marine Benthic Dynamics. Univ. S. Carolina Press, Georgetown SC, p 243
7. Hargrave BT, Duplisea DE, Pfeiffer E, Wildish DJ (1993) Mar Ecol Prog Ser 96:249
8. Holmer M, Kristensen E (1992) Mar Ecol Prog Ser 80:191
9. Holmer M, Kristensen E (1996) Biogeochem 32:15
10. Roden EE, Tuttle JH (1992) Limnol Oceanogr 37:725
11. McClelland JW, Valiela I (1997) Limnol Oceanogr 42:930
12. Li-Xun Y, Rice DA, Fenton GE, Lewis ME (1991) J Exp Mar Biol Ecol 145:161
13. McGhie TK, Crawford CM, Mitchell IM, O'Brien D (2000) Aquaculture 187:351
14. Brooks KM, Mahnken CVW (2003) Fish Res 62:255
15. Warwick RM (1986) Mar Biol 92:557
16. Warwick RM (1987) Mar Biol 95:193
17. Weston D (1990) Mar Ecol Prog Ser 61:233
18. Holmer M (1991) Impacts of aquaculture on surrounding sediments: generation of organic-rich sediments. In: DePauw N, Joyce J (eds) Aquaculture and the environment. European Aquaculture Society Spec. Publ. 16, Ghent, Belgium, p 155
19. Findlay RH, Watling L, Mayer LM (1995) Estuaries 18:14
20. Hargrave BT, Phillips GA, Doucette LI, White MJ, Milligan TG, Wildish DJ, Cranston RE (1997) Water, Air Soil Pollut 99:641
21. Holmer M, Duarte CM, Heilskov A, Olesen B, Terrados J (2003) Mar Poll Bull 46:1470
22. Nilsson HC, Rosenberg R (2000) Mar Ecol Prog Ser 197:139
23. Dell'Anno A, Bompadre S, Danovaro R (2002) Limnol Oceanogr 47:899
24. Pusceddu A, Dell'Anno A, Danovaro R, Manini E, Sara G, Fabiano M (2003) Estuaries 26:641
25. Mirto S, La Rosa T, Danovaro R, Mazzola A (2000) Mar Poll Bull 40:244
26. La Rosa T, Mirto S, Favaloro E, Savona B, Sarà G, Danovaro R, Mazzola A (2002) Water Research 36:713
27. Vezzulli L, Chelossi E, Riccardi G, Fabiano M (2001) Aquacult Int 10:123

28. Danovaro R, Corinaldesi C, La Rosa T, Luna GM, Mazzola A, Mirto S, Vezzulli L, Fabiano M (2003) Chem Ecol 19:59
29. Karakassis I, Tsapakis M, Hatziyanni E (1998) Mar Ecol Prog Ser 162:243
30. O'Connor BDS, Costelloe J, Keegan BF, Rhoads DC (1989) Mar Poll Bull 20:384
31. Nilsson HC, Rosenberg R (1997) J Mar Syst 11:249
32. Karakassis I, Tsapakis M, Smith CJ, Rumohr H (2002) Mar Ecol Prog Ser 227:125
33. Danovaro R (2000) Int J Env Poll 13:380
34. Jensen MJ, Lomstein E, Sørensen J (1990) Mar Ecol Prog Ser 61:87
35. Hansen LS, Blackburn TH (1992) Mar Biol 112:147
36. Ervik A, Hansen PK, Aure J, Stigebrandt A, Johannessen P, Jahnsen T (1997) Aquaculture 158:85
37. Hansen PK, Ervik A, Schaanning M, Johannessen P, Aure J, Jahnsen T, Stigebrandt A (2001) Aquaculture 194:75
38. Welsh DT (2003) Chem Ecol 19:321
39. Findlay RH, Watling L (1997) Mar Ecol Prog Ser 155:147
40. Thamdrup B, Canfield DE (2000) Benthic respiration in aquatic sediments. In: Sala OE, Jackson RB, Mooney HA, Howarth RW (eds) Methods in Ecosystem Science. Springer, Berlin Heidelberg New York , p 86
41. Hall POJ, Anderson LG, Holby O, Kollberg S, Samuelsson O (1990) Mar Ecol Prog Ser 61:61
42. Holmer M, Marbà N, Terrados J, Duarte CM, Fortes MD (2002) Mar Poll Bull 44:685
43. Nickell LA, Black KD, Hughes DJ, Overnell J, Brand T, Nickell TD, Breuer E, Harvey SM (2003) J Exp Mar Biol Ecol 285:221
44. Karakassis I, Hatziyanni E, Tsapakis M, Plaiti W (1999) Mar Ecol Prog Ser 184:205
45. Heilskov AC, Holmer M (2001) ICES J Mar Sci 58:427
46. Canfield DE (1993) In: Wollast R, Chou L, Mackenzie F (eds) Interactions of C, N, P and S Biogeochemical Cycles. Springer, Berlin Heidelberg New York, p 333
47. Cranston R (1994) In: Hargrave BT (ed) Modelling benthic impacts of organic enrichment from marine aquaculture. Tech Rep Fish Aquat Sci 1949:93
48. Holmer M, Kristensen E (1994) Mar Ecol Prog Ser 107:177
49. Thamdrup B (2000) Advances in Microbial Ecology 16:41
50. Jensen MM, Thamdrup B, Rysgaard S, Holmer M, Fossing H (2003) Biogeochem 65:295
51. Kiemer MCB, Balck KD, Lussot D, Bullock AM, Ezzi I (1995) Aquaculture 135:311
52. Black KD, Kiemer MCB, Ezzi IA (1996) J Appl Ichthyol 12:15
53. Aure J, Stigebrandt A (1990) Aquaculture 90:135
54. Kristensen E (2000) Hydrobiologia 426:1
55. Kristensen E (2001) Geochem Trans 12:92
56. Nielsen OI, Kristensen E, Holmer M (2003) Aquat Mar Ecol 33:95
57. Hall POJ, Holby O, Kollberg S, Samuelsson MO (1992) Mar Ecol Prog Ser 89:81
58. Hedges JI, Keil RG (1999) Mar Chem 65:55
59. King GM, Blackburn TH, Fenchel T (1998) Bacterial Biogeochemistry. The Ecophysiology of Mineral Cycling. Academic Press, New York, p 307
60. Christensen PB, Rysgaard S, Sloth NP, Dalsgaard T, Schwærter S (2000) Aquat Microb Ecol 21:73
61. Rysgaard S, Christensen PB, Nielsen LP (1995) Mar Ecol Prog Ser 126:111
62. Holmer M, Pérez M, Duarte CM (2003) Mar Poll Bull 46:1372
63. Valiela I, McClelland J, Hauxwell J, Behr PJ, Hersh D, Foreman K (1997) Limnol Oceanogr 42(5):1105
64. Pearson TH, Rosenberg R (1978) Oceanogr Mar Biol Annu Rev 16:229

65. Rhoads DC, Germano JD (1986) Hydrobiologia 142:291
66. Wildish DJ, Akagi HM, Hamilton N, Hargrave BT (1999) Can Tech Rep Fish Aquat Sci 2286:iii + 31 p
67. Wildish DJ, Hargrave BT, Pohle G (2001) ICES J Mar Sci 58:469
68. Pohle G, Frost B, Findlay R (2001) ICES J Mar Sci 58:417
69. Karakassis I, Hatziyanni E, Tsapakis M, Paliti W (1999) Mar Ecol Prog Ser 184:205
70. Banta GT, Holmer M, Jensen MH, Kristensen E (1999) Aquat Microb Ecol 19:189
71. Aller RC (1994) Chem Geol 114:331
72. Pelegri SP, Blackburn TH (1994) Mar Biol 121:253
73. Gribsholt B, Kristensen E (2002) Mar Ecol Prog Ser 241:71
74. Heilskov AC, Holmer M (2003) Vie et Milieu – Life and Environment 54:153
75. Hee CA, Pease TK, Alperin MJ, Martens CS (2001) Limnol Oceanogr 46:1845
76. FAO (2002) The state of world aquacultures 2002. ISBN 92-5-104842-8
77. Chambers RM, Fourqurean JW, Macko SA, Hoppenot R (2001) Limnol Oceanogr 46:1278
78. Kristensen E, Andersen FØ, Holmboe N, Holmer M, Thongtham N (2000) Aquat Microb Ecol 22:199
79. Holmer M, Andersen FØ, Holmboe N, Kristensen E, Thongtham N (2001) Wetlands Ecology and Management 9:141
80. Ruiz JM, Perez M, Romero J (2001) Mar Poll Bull 42:749
81. Duarte CM (2002) Environ Conserv 29:192
82. Cancemi G, De Falco G, Pergent G (2003) Estuar Coast Shelf Sci 56:961
83. Delgado O, Grau A, Pou S, Riera F, Massuti C, Zabala M, Ballesteros E (1997) Oceanol Acta 20:557
84. Delgado O, Ruiz J, Pérez M, Romero J, Ballesteros E (1999) Oceanol Acta 22:109
85. Dimech M, Borg JA, Schembri PJ (2000) Biol Mar Medit 7:361
86. Greve TM, Borum J, Pedersen O (2003) Limnol Oceanogr 48:210
87. Holmer M, Bondgaard EJ (2001) Aquat Bot 70:29
88. Hemminga MA, Duarte CM (2000) Seagrass Ecology. Cambridge University Press, Cambridge, UK
89. Duarte CM, Merino M, Gallegos M (1995) Limnol Oceanogr 40:1153

# Lithium-Normalized Zinc and Copper Concentrations in Sediments as Measures of Trace Metal Enrichment due to Salmon Aquaculture

P. A. Yeats[1] (✉) · T. G. Milligan[1] · T. F. Sutherland[2] · S. M. C. Robinson[3] · J. A. Smith[4] · P. Lawton[3] · C. D. Levings[2]

[1]Department of Fisheries and Oceans, Bedford Institute of Oceanography, P.O. Box 1006, Dartmouth, N.S. B2Y 4A2, Canada
yeatsp@mar.dfo-mpo.gc.ca, milligant@mar.dfo-mpo.gc.ca

[2]Department of Fisheries and Oceans, DFO-UBC Centre for Aquaculture and Environmental Research, 4160 Marine Drive, West Vancouver, B.C. V7V 1N6, Canada
sutherlandt@pac.dfo-mpo.gc.ca, levingsc@pac.dfo-mpo.gc.ca

[3]Department of Fisheries and Oceans, St. Andrews Biological Station, 531 Brandy Cove Rd., St. Andrews, N.B. E5B 2L9, Canada
robinsonsm@mar.dfo-mpo.gc.ca, lawtonp@mar.dfo-mpo.gc.ca

[4]Amec Earth and Environment Ltd, 25 Wagonners Lane, Fredericton, N.B. E3B 2L2, Canada
jamey.smith@amec.com

1	Introduction	208
2	Lithium Normalization	209
3	Application to Lime Kiln Bay and the Letang Estuary	210
4	Other Applications	212
5	Discussion	214
6	Conclusions	219
	References	220

Abstract The results of metal analyses carried out on surficial sediment samples collected in the coastal waters of southwest New Brunswick and the Broughton Archipelago in British Columbia have been used to investigate the use of heavy metals as tracers of salmon farm wastes. New Brunswick in particular has seen rapid expansion of open cage salmon aquaculture in recent years. While techniques have been developed to identify benthic impacts directly beneath the cages, no method has been developed to determine the fate of dispersed wastes. We show that Zn and Cu, two elements associated with aquaculture operations, can be used to identify farm wastes in sediments at some distance from the cage sites. Geochemical normalization for grain size is needed in order to see the small tracer signals. Excess Zn and Cu levels are found in the sediment at varying distances from the salmon cages in depositional areas in southwest New Brunswick and in the Broughton Archipelago. Evidence that links these observations to salmon aquaculture development is described.

Keywords Sediment · Trace metals · Farm waste tracer · Salmon aquaculture · Far-field effects

1
Introduction

Determining where wastes from finfish aquaculture are deposited is an important consideration for the sustainable development of the aquaculture industry. In some cases these wastes can lead to excessive accumulation of organic matter in sediments, increased sediment anoxia and decreased macrofaunal abundance and diversity [1]. While it is simple to see the accumulation of the waste from the farms immediately below the cages in relatively shallow harbours, identification of these wastes in sediments farther away from the farms is more difficult. It is generally difficult to distinguish between the organic wastes from the fish farms and organic matter from other pollution sources or natural organic matter that exists in all sediments. It is thus very difficult to prove that any observed organic matter enrichment originates with farm wastes. A chemical tracer of finfish farm wastes would be a very useful tool for characterizing the deposition of farm wastes in both near- and far-field sediments, and would support environmentally sustainable development of the aquaculture industry.

The purpose of this paper is to investigate the application of Zn and Cu as tracers of wastes from salmon aquaculture operations. Zn's potential as a tracer originates from the fact that Zn is an essential component of the salmon diet [2] that is added to the feed as a supplement. That Zn is highly enriched in the wastes is evident from observations of as much as 400 mg kg^{-1} of Zn in organically-enriched sediments immediately under cages in the Quoddy area of southwestern New Brunswick [3]. Copper is another potential tracer and elevated Cu concentrations have also been observed beneath the Quoddy farm sites, possibly as a result of the use of Cu-based antifoulants by the industry. Zn and Cu can be readily, precisely and inexpensively measured in sediments by inductively coupled plasma mass spectroscopy (ICPMS), which provides simultaneous analysis of Zn, Cu and a number of other metallic elements, making them good candidates for tracer applications. But Zn and Cu are naturally occurring components of all sediments. This makes their use as tracers more difficult because the tracer signal must be distinguished from the natural background. In this paper we will first describe the method we have used to separate natural and tracer Zn and Cu concentrations, and then several of their applications as tracers of farm wastes. Heavy metals are not the only potential tracers of farm wastes; other chemicals added to the environment, such as pigments, pesticides or antibiotics could be potential markers, as could characteristics of the organic matter, such as lipid content or isotope ratios. Metals, however, are appealing as tracers because they are chemically stable, inexpensive to measure, and background concentrations can be adequately assessed.

2
Lithium Normalization

The natural variability of Zn and other heavy metal concentrations in sediments is largely related to grain size and mineralogy of the sediments. Grain size is generally the most important factor and it is very difficult to make any conclusions about sediment metal concentrations without compensating for the effects of grain size. Heavy metal concentrations invariably increase with decreasing grain size. Geochemical normalization of sediment concentrations for grain size has been found to be superior to granulometric normalization, because the geochemical normalization can also compensate for mineralogical changes [4]. Various elements have been used for the geochemical normalization of heavy metal variability, but for high latitude areas dominated by glaciation or other physical weathering, Li normalization is preferred [4].

For the assessment of Zn as a fish farm waste tracer in southwestern New Brunswick (SWNB), we have established the 'background' Zn vs. Li relationship for surficial sediments from throughout the region (Fig. 1a) using all available data for SWNB coastal water sites remote from obvious pollution sources.

The data were extracted from Bedford Institute data archives that were generated in a variety of studies. Samples were either grab samples or surface samples from sediment cores. Analyses are all based on total digestions of the metals in $HNO_3/HCl/HF$, but a mixture of inductively coupled plasma mass spectrometry (ICPMS) and atomic absorption spectrophotometry was used for the metal determinations. A similar approach could be used for any other region of interest. Some care, however, has to be taken to ensure that the metal vs. Li relationships are appropriate for the study area. The relationships are not the same in all areas. We have calculated the Zn vs. Li linear regression for these data and estimated a 95% confidence band about the regression line using the standard error of the regression multiplied by the t factor for $p = 0.05$. The number of samples used to generate this regression is large ($n = 120$) and the correlation coefficient is high ($r = 0.952$). As a result, the Zn concentrations are well constrained by the relationship with Li. Similar heavy metal vs. Li correlations can be made for Cu, Ni, Co, Cr and V, all of them having very highly correlated linear relationships. If sediments associated with the aquaculture operations are enriched with Zn, they should fall on the high side of the regression line. Similarly, results for any other element, such as Cu, that is also enriched in farm waste, should exceed the 95% confidence limit of its metal vs. Li relationship.

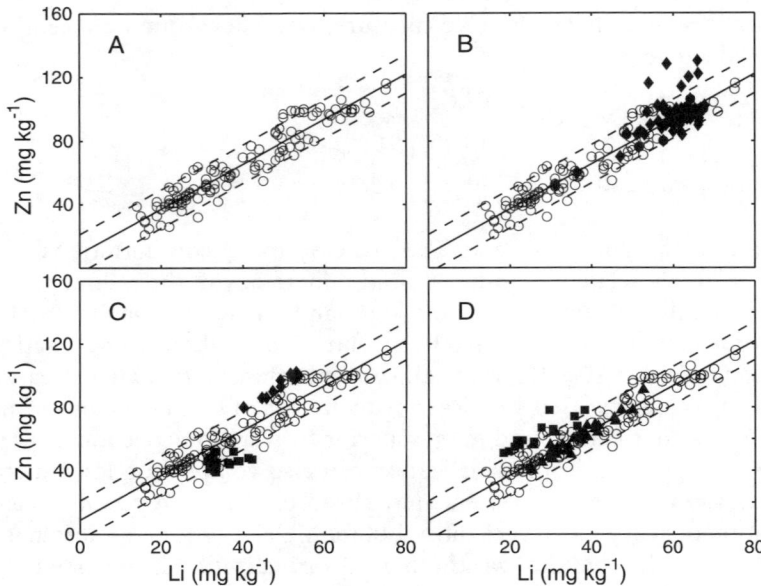

Fig. 1 Zn vs. Li plots for SWNB aquaculture areas. (**a**) Background data for unimpacted areas from the western Bay of Fundy, showing the Zn vs. Li regression line and a 95% confidence interval for the regression. (**b**) Background data with samples from Passamaquoddy Bay and the Letang Estuary added (*diamonds*). (**c**) Data for intertidal samples from Hinds Bay (*diamonds*) and Clam Cove (*squares*) superimposed on the background relationship. (**d**) Similar plot for the Maces Bay aquaculture area showing farm site (*squares*) and control station (*triangles*) samples

3
Application to Lime Kiln Bay and the Letang Estuary

In the initial test of this tracer application, we collected and analysed surficial sediment samples in Lime Kiln Bay and several other areas of the Letang Estuary with salmon farming operations. Samples were dried at 60 °C and then sieved at 180 μm prior to digestion in $HNO_3/HCl/HF$ and analysis by ICPMS. When the data from these samples are added to the background Zn vs. Li plot (Fig. 1b), we find a small number of samples that lie on the upper side of the regression outside the 95% confidence band and none that are below the confidence band. When we plot the locations of these anomalously elevated concentration samples on a map of the area (Fig. 2, all samples with excess Zn > 12 mg kg^{-1}), we find that most of the samples are clustered in Lime Kiln Bay and the inner part of Back Bay, two areas of the Letang Estuary with salmon farms.

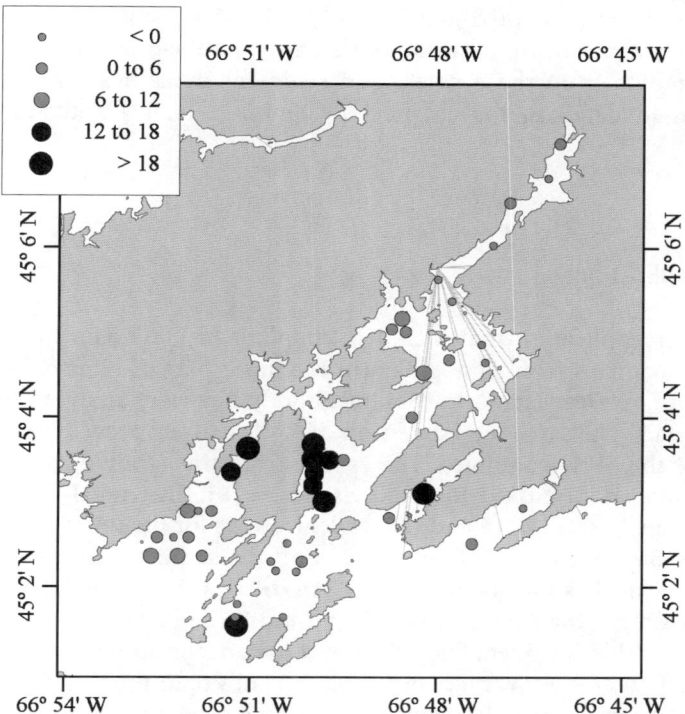

Fig. 2 Map of the Letang area showing zinc concentrations in excess of background (observed concentrations minus those predicted from the Zn vs. Li relationship). Concentrations > 12 mg kg^{-1} are significantly ($P < 0.05$) greater than background

Bliss Harbour, another area with a similar density of farms but much better flushing, shows one isolated anomaly. If we conduct the same exercise with Cu, we find a very similar pattern. In fact all the samples with elevated concentrations of both zinc and copper are from either Lime Kiln or Back Bays with the exception of one sample in Black's Harbour. Black's Harbour has potential inputs from both a sewage treatment plant (sewage is a known source of Cu) and a fish processing plant (potentially a Zn source), but no aquaculture activities.

Elevated Zn and Cu in the surficial sediments are found in depositional areas that have shown increased deposition of material in a flocculated form from 1990 to 1999 [5]. The deposition of surface-active trace metals has been shown to be highly correlated with sediments derived from flocculated suspensions [6]. Four dated sediment cores from Lime Kiln Bay show low (background) levels of Zn and Cu in locations remote from farms and elevated levels close to the farms, exactly the pattern seen in the surface grab samples. The enrichment in the cores near aquaculture operations are restricted to the top parts of the cores, with the initial enrichment occurring at approximately 1980, coincident with the beginning of the salmon aquaculture industry in this area. Radioactive dating and metal measurements from these cores, and sediment dynamics for the area are discussed in detail in the chapters by Smith et al. [7] and Milligan and Law [5].

4
Other Applications

In another application that included sampling in Lime Kiln Bay, intertidal sediment samples from Hinds Bay (in the inner part of Lime Kiln Bay) and Clam Cove on Deer Island in Passamaquoddy Bay were analysed for Zn and other metals. The same total digestion, ICPMS analysis procedures were used for all of the SWNB samples. The results from this study of intertidal sediment biota are discussed in detail elsewhere [8]. The metal measurements show elevated Zn (Fig. 1c) and Cu concentrations for the intertidal sediments of Hinds Bay, but not for Clam Cove. Both these bays have salmon farms, with the farm sites much closer to the intertidal sampling locations for Clam Cove ($\sim$ 100 m) than for Hinds Bay (> 1 km). The sediments in Clam Cove are coarser (this is evident from the lower Li concentrations), suggesting that these are higher energy, more erosional beaches than those in Hinds Bay. In both of these bays, samples were collected in a geometric pattern, with samples collected from east to west across the beaches and from the low water to high water levels. The Zn : Li ratios for the samples from these different parts of the beaches (Table 1) suggest increasing accumulation of farm wastes across the Clam Cove beach from the more exposed to more sheltered side, with highest ratios midway up the beach. Gradients for Hinds Bay were much smaller (none are significant at $P < 0.1$), but maximum concentrations at mid beach were again observed.

Table 1 Zn : Li ratios for intertidal sediment samples from Clam Cove and Hinds Bay

Location	Clam Cove			Hinds Bay		
	n	x	s.d.	n	x	s.d.
East	6	1.34	0.09	6	1.87	0.03
Central	6	1.24	0.15			
West	6	1.16	0.07	6	1.88	0.04
High	6	1.19	0.13	4	1.88	0.05
Middle	6	1.32	0.13	4	1.89	0.04
Low	6	1.23	0.11	4	1.85	0.04

Sediment samples from a part of Maces Bay with salmon farms have also been analysed for heavy metals. In this case, the metal vs. Li relationships show that samples collected in the vicinity of salmon farms have significantly elevated concentrations of Zn (Fig. 1d), but not Cu. Samples from control stations farther from the farm sites are indistinguishable from those for the general background relationships. The interesting aspect of this sampling is that the affected samples were collected close to the cage sites (within 150 m) but only six months after the initial stocking of the cages. Zn accumulation apparently occurs rapidly. Here, as well as at our other sampling locations in SWNB, the increases in Zn (and Cu) concentrations that we observe are very small. They do indicate that farm wastes are deposited in these locations, but do not indicate any significant degradation of sediment quality, in other words no concentrations that exceed the probable effects levels of the Canadian Council of Ministers of the Environment sediment quality guidelines (www.ccme.ca/assets/pdf/e1_062.pdf).

So far, all of the examples are for sediments in salmon farming areas in southwestern New Brunswick in inlets off the Bay of Fundy. They are all rather shallow (water depths generally < 20 m) environments and subject to the high tidal energies of the Bay of Fundy (modelled current velocities from < 5 to > 30 cm s^{-1}). Another set of samples from a salmon aquaculture site in an area known as the Broughton Archipelago (BA), on the south central coast of British Columbia, represents a very different oceanographic environment where the farm sites are located within complex waterways with steeply-sloping bathymetry.

Despite these regional differences, we still see, in the example illustrated in Fig. 3, a very similar picture of zinc enrichment in the sediments associated with a single fish farm operation located over a soft sediment shelf at the entrance of a small bay.

In a reference area of the BA (Fig. 3a), we find a linear Zn vs. Li relationship. Samples from the nearby lease site collected after the farm system had ceased operation showed varying levels of enrichment (Fig. 3b). Differences in the excess Zn concentrations in the BA compared to those found in SWNB may be attributed to the high-resolution sampling grid focused directly under and around the former BA net pen system. Zn concentrations at the cage site were similar to those found at farm sites in SWNB [3]. Cu (Figs. 3c and 3d) also shows elevated concentrations at the BA farm site but the correlation between Zn and Cu was not as strong as that found in the Letang. Analytical techniques employed in the east and west coast studies were different: no sieving of the west coast samples was performed, and they were digested using a nitric acid dissolution (strong acid leachable metals—SALM) procedure that will not result in a complete digestion of mineral matrices. Thus, quantitative comparisons of the east and west coast results are not possible. Nevertheless, the east and west coast results are qualitatively very similar—linear, highly correlated Zn (and Cu) vs. Li regressions are observed for the

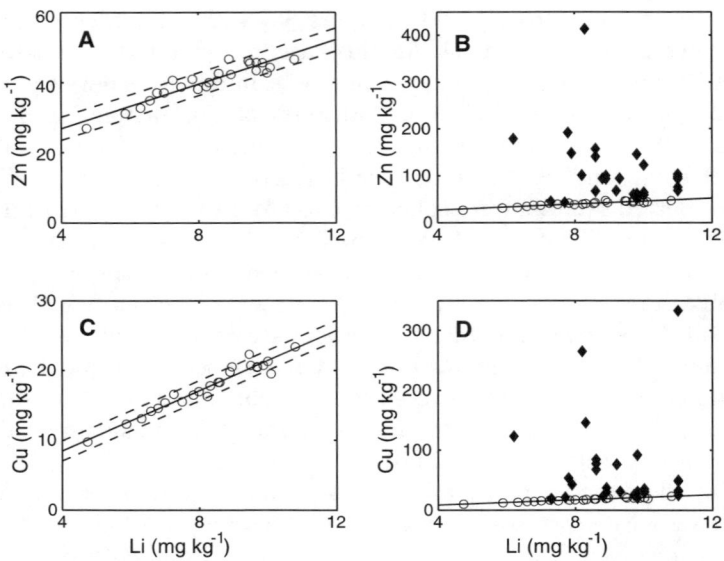

Fig. 3 Metal vs. Li plot(s) for the Broughton Archipelago. (**a**) Background Zn data showing the Zn vs. Li regression line and a 95% confidence interval for the regression. (**b**) Background Zn data with samples from vicinity of farm site added (note change in scale for the Y-axis). (**c**) Background Cu data and (**d**) background Cu data with farm site samples added

reference sites, and some but not all of the samples from the vicinity of the cage sites show elevated concentrations.

5
Discussion

It is difficult to assess the quantities of organic matter from fish farm wastes that are deposited in sediments away from the immediate vicinity of the farms from the organic carbon content of the sediments because of contributions of organic matter from other sources (sewage, fish processing plants, primary productivity, and so on) to the overall accumulation of organic matter in the sediments. The organic matter in sediments is also chemically and biologically reactive, resulting in losses of carbon from the sediments over rather short time-frames, which will further complicate the picture. Our observations have shown that Zn and Cu are useful markers that can distinguish farm wastes from those from other sources.

There is no definitive way of proving that the elevated Zn concentrations we see in Lime Kiln and Back Bays, in the Maces Bay, in Hinds Bay, or at the

BA site, originate from salmon aquaculture. There are, however, a number of pieces of evidence that indicate that this is the source. First, data collected from sediments directly under some cages have high concentrations of Zn (> 400 mg kg^{-1} at sites in both SWNB and the Broughton Archipelago [3, 9]). Chou et al. [3] found similarly elevated Zn concentrations in some types of salmon feed pellets used in SWNB, and high concentrations of Zn have been measured in feed pellets elsewhere [10, 11]. Second, concentrations of Zn in sediments have increased over the period of development of the salmon aquaculture industry. This is most evident from the dated SWNB sediment core results (see Chapter 11 in this volume [7] for a discussion of sediment core geochronologies), but also from the temporal changes in fine-grained sediment accumulation and the correlation between Zn and floc deposition (see Chapter 12 [5] for discussion of effects of aquaculture on sediment characteristics). Finally the observation of high Cu concentrations in the Letang area samples with elevated Zn is further evidence, because Chou et al. [3] also found an association of elevated Cu and Zn in sediments from under cages, as did Smith et al. [7] in dated sediment cores. Similarly, in the Broughton Archipelago, elevated levels of Cu are found immediately under cages [12] and in some of the far-field samples. Although elevated Cu concentrations do not appear to be associated with salmon feed, they may be an indirect effect of the farm operation, because of the use of Cu as an antifoulant.

The simplest way to identify enrichment of an element such as Zn would be to compare observed concentrations to general background concentrations. Because of the natural variability in background concentrations, this method will only identify very elevated concentrations. For example, Loring et al. [13] have used a background concentration of 150 mg kg^{-1} Zn to define concentrations in fine-grained sediments in Atlantic Canadian coastal areas that exceed natural concentrations. Using this generic cut-off, none of the SWNB samples exceed background. Alternatively, average concentrations for various nearby areas could be compared (study sites vs. control sites, study sites vs. background areas, and so on), but this can also lead to poor discrimination because of the inherent natural variability, or to misleading results because of changes in grain size. If the average concentrations in Lime Kiln Bay, Back Bay, Bliss Harbour, Passamaquoddy Bay, or Letang Harbour are simply compared to the Bay of Fundy background concentration (Table 2), all five inlets show elevated concentrations, but this comparison does not take account of the obvious differences in grain size (as shown by the lithium concentrations) from the inlets to the open Bay of Fundy.

On the other hand, a comparison of the Maces Bay farm site samples with those from the control sites does not show a significant (t test for small n) difference unless the grain size differences are considered. In all of these cases, Li normalization or other means of accounting for grain size differences are necessary in order to properly characterize differences in concentration from one location to another. Only the Clam Cove–Hinds Bay study shows qualita-

Table 2 Average (x) and standard deviation (s.d.) of Li, Zn and Cu concentrations (mg kg^{-1}) for various areas in southwestern New Brunswick

	n	Li	Li	Zn	Zn	Cu	Cu
		x	s.d.	x	s.d.	x	s.d.
Bay of Fundy	121	39.7	16.0	65.5	13.4	14.4	4.2
Passamaquoddy Bay	13	61.3	6.6	95.1	8.3	16.6	1.9
Back Bay	11	60.5	2.7	98.6	8.2	21.7	3.9
Lime Kiln Bay	7	54.3	8.3	102.6	21.6	26.7	9.9
Bliss Harbour	9	64.2	2.3	97.1	13.0	19.8	4.2
Letang estuary	14	56.1	9.0	99.8	8.1	16.4	3.8
Maces Bay, farm site	13	26.9	6.7	63.2	9.4	13.5	4.7
Maces Bay, control	38	36.2	6.7	59.7	10.9	13.8	2.6
Clam Bay	25	35.6	4.6	48.8	12.4	9.1	1.6
Hinds Bay	13	49.6	4.1	94.7	7.0	25.3	2.2

tively similar results for a comparison of simple average concentrations and normalized concentrations.

The sensitivity of the technique and its ability to detect small increases in concentration is dependent upon the precision of the background Zn vs. Li relationship. For SWNB, we have generated a background relationship based on all available samples from places that are not obviously subject to contamination for all coastal waters of SWNB. This gives us a large dataset ($n = 120$) that covers the whole range of environmental conditions that might be encountered in the area, but it inevitably increases the scatter in the relationship (standard error of the regression = 7.45). In the BA, on the other hand, we have used a small ($n = 24$) set of samples from a small area adjacent to the study site where the environmental conditions of both the farm and reference sites were relatively uniform. This approach reduces the scatter in the relationship (standard error = 1.98), but runs the risk of not truly reflecting the environmental conditions of a very heterogeneous study location. Other choices for establishing the background relationship—such as samples from some depth in sediment cores that were deposited before significant anthropogenic inputs—may also give tighter regressions. Another difference between the SWNB and BA results is the method used for the metal analysis. In SWNB, a total digestion of the sample was used, while in BA, a strong acid leaching technique was used. The latter method will result in partial digestion of the mineral matrix, and to the extent that scatter in the Zn vs. Li relationship is caused by variability in the metal content of the matrix, may also contribute to the observed tighter regression.

Suspended particulate matter (SPM) in the water column should also reflect the Zn (and Cu) enrichment if particulate wastes from salmon farms are being transported to sediments at considerable distances from the farms, such as to the intertidal zone in Hinds Bay. Samples collected in 2000 and 2001 from Lime Kiln Bay shows that this is indeed the case. The Zn and Cu vs. Li relationships for the SPM (Fig. 4) show elevated concentrations compared to a background relationship based on SPM samples from the Bay of Fundy.

The elevated concentrations are restricted in the 2000 survey to samples collected from Lime Kiln Bay and the immediately adjacent waters of the Letang Inlet. Elevated concentrations are not found in Bliss Harbour or waters outside the Letang. In 2001, all samples were collected in Lime Kiln Bay or its immediate vicinity. Concentrations of Co, Cr, Ni, V, and so on are not enriched in Lime Kiln Bay SPM compared to background metal vs. Li relationships for Bay of Fundy SPM samples, but Pb is. In the Broughton Archipelago, a sediment trap deployed adjacent to a net pen was used to study material in transit from the pens to the sediments [11]. Measurements of metals in the sediment trap material show elevated levels of Zn that are likely due to feed material and Cu that could come from some other farm-related source.

For SPM, Li may not be simply a grain size normalizer. All the inorganic mineral grains in the SPM samples should be fairly uniformly small, so the Li concentration will reflect the total mineral content of the SPM. With the high tidal energy of the Bay of Fundy, resuspension of bottom sediment particles will be high and the inorganic mineral content of the SPM higher than normally seen for coastal water SPM. This is reflected in the Li content, where a sample with Li of approximately 80 mg kg^{-1} would indicate SPM that is almost 100% inorganic mineral material. On this basis, most of the particles in our data set are > 50% inorganic, and organic material (farm wastes and/or plankton) would make up less than half of the total. In less turbid waters we would expect the organic fraction to be much greater.

It is difficult to estimate the distance from sediment sample locations to the nearest farm cages for an area such as the Letang Estuary, with very

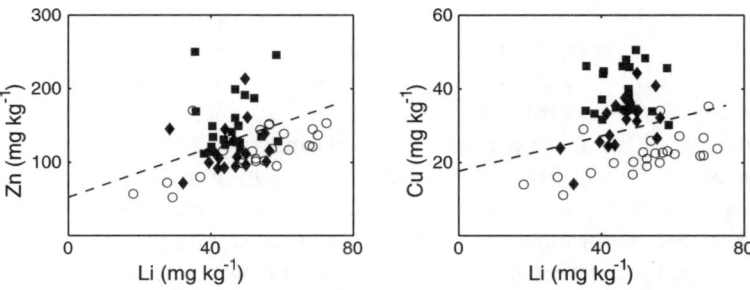

Fig. 4 Zn and Cu vs. Li plots for SPM samples showing background data for the Bay of Fundy (*circles*) and Letang data from 2000 (*diamonds*) and 2001 (*squares*). Upper bounds of the 95% confidence interval for the background relationships are shown by the *dashed lines*

complicated geography, hydrodynamics, and aquaculture leases dispersed throughout the area. None of the sites in Lime Kiln or Back Bays with elevated Zn concentrations, however, are more than a few hundred metres from the nearest cages. Similarly, the elevated Maces Bay samples are within approximately 150 m of cages, but the Hinds Bay intertidal samples are more than a kilometre from the nearest cages. Tidal energy and current velocities are obviously critical in determining where the wastes accumulate. Letang and Bliss Harbours have a similar density of farm sites to that in Lime Kiln Bay and a similar proximity of the sampling sites to cages, but little evidence of waste accumulation in the sediments. The main difference is the restricted nature of the circulation in Lime Kiln Bay. Estimated current velocities from a numerical model [14] show mean velocities of < 5 cm s^{-1} for Lime Kiln vs. $10-30$ cm s^{-1} for Bliss and Letang Harbours. This is similar to the previously-mentioned observations for beach sediment from Clam Cove and Hinds Bay, where accumulation occurred on the beach that was farther from the cage sites but in a lower energy environment. At the Broughton Archipelago site, most of the samples out to ~ 180 m from the cage site (more than half were within 30 m) show elevated Zn concentrations. Average current speeds recorded by the farm operators were 7.6 cm s^{-1}.

Trace metals have been shown to associate with flocs in suspension, and are found at highest concentrations in the sediment in areas dominated by deposition of material as flocs [5, 6]. The accumulation of flocs and their associated metals in the sediment requires low bottom stress. Few regions in SWNB exist where either tidal or wave stress does not exceed that required for floc break-up, therefore limiting regions of high Zn or Cu concentration to the most sheltered areas such as Lime Kiln Bay.

None of the samples we collected in SWNB have Zn or Cu concentrations that are greater than the Canadian Council of Ministers of the Environment's probable effects level (PEL) for harm to aquatic biota. We did not, however, collect any samples directly under the cages. Samples collected beneath farm cages in SWNB by others [3, 15] do occasionally show concentrations that exceed the PELs for both zinc (271 mg kg^{-1}) and copper (108 mg kg^{-1}). At the Broughton Archipelago farm we found a few samples from the cage site that exceed the PEL for Zn. These observations are in agreement with the findings of Brooks et al. [16], who also collected samples in 1999 following the termination of the fish farm operation at this site in 1998. Away from the cage location, many of the concentrations were higher than background but none were above the PEL. Observations of Cu concentrations in excess of the PEL were more common and less closely associated with the cage site. Similar observations for both Zn and Cu in sediments around British Columbia aquaculture operations have been made elsewhere [9, 12]. These authors attribute observations of elevated Cu away from the cages to the washing of treated nets at these off-site locations.

6
Conclusions

The results presented in this paper describe the potential for Zn, and perhaps other heavy metals such as Cu, as markers for the particulate wastes from fish farms. Heavy metals are very appealing as chemical tracers because of the ease of collection of samples and low cost of analyses. With recent improvements in ICPMS techniques, reliably accurate and precise measurements of approximately 35 different metals in marine sediment samples are available at a cost of approximately $50–$75 per sample. Ag has been shown to be a very useful tracer of sewage discharges [17] and now we have shown that Zn can be used as a tracer of salmon farm wastes.

Qualitatively, we have been able to show for the Letang area that farm wastes can be detected in sediments from Lime Kiln Bay but not Bliss Harbour and that the wastes are transported to the intertidal sediments of Hinds Bay, a depositional area. From the Maces Bay data set, we can show that Zn from wastes accumulates in the sediments to detectable concentrations over very short time-frames. But can we be more quantitative? The minimum detectable signal is limited by the strength of the input signal and the uncertainty in the grain size and mineralogical normalization. Based on the measured precision of the Zn vs. Li relationship for SWNB, and assuming the waste material has a Zn content of $200\,\text{mg}\,\text{kg}^{-1}$, we should be able to detect sedimenting material that contains 10% farm waste. Using the same assumptions, we can calculate that the material accumulating at the two stations with the highest Zn concentrations contained 30 and 35% farm waste if diluted with background concentration material. But the Zn content of the waste is very poorly constrained. The content of the feed presumably varies substantially with time and location as different feed mixtures are used. Also, we know little about the ability of the fish to retain Zn or about the relative release rates of Zn and organic matter as the wastes decompose. Once in the sediments, Zn should be less mobile than organic matter, especially in anoxic sediments where immobile Zn sulfides would be produced. These uncertainties will limit the ability of the technique to make quantitative predictions of the amount of farm waste in a sediment sample, but not the ability to predict qualitatively where farm wastes have been deposited in the sediments, and the relative deposition at different sites in the same area. In fact, the observed amplification of the Zn signal in the waste compared to feed [10, 11] should increase the signal, improving the usefulness of Zn as a qualitative tracer of fish farm wastes. Thus, measurements of excess Zn concentrations in sediments from a detailed sampling of bottom sediments should provide a very good picture of where organic wastes from salmon farms are deposited in sediments.

Acknowledgements We would like to thank the crew of the CSS Vector and Sooke Post (BA) and the CSS Hart, CSS Pandalus and MV Scott (SWNB). This project was funded through the Environmental Strategic Science Research Fund.

References

1. Pohle G, Frost B, Findlay R (2001) ICES J Mar Sci 58:417
2. Richardson NL, Higgs DA, Beams RM, McBride JR (1985) J Nutr 115:553
3. Chou CL, Haya K, Paon LA, Burridge L, Moffatt JD (2002) Mar Pollut Bull 44:1259
4. Loring DH (1991) ICES J Mar Sci 48:101
5. Milligan TG, Law BA (2005) The effect of marine aquaculture on fine sediment dynamics in coastal inlets (in this volume). Springer, Berlin Heidelberg New York
6. Milligan TG, Loring DH (1997) Water Air Soil Poll 99:33
7. Smith JN, Yeats PA, Milligan TG (2005) Geochronologies for fish farm contaminants in sediments from Limekiln Bay in the Bay of Fundy (in this volume). Springer, Berlin Heidelberg New York
8. Robinson SMC, Auffrey LM, Barbeau MA (2005) Far-field impacts of eutrophication on the intertidal zone in the Bay of Fundy with emphasis on the soft-shell clam, *Mya arenaria* (in this volume). Springer, Berlin Heidelberg New York
9. Brooks KM, Stierns AR, Mahnken CVW, Blackburn DB (2003) Aquacult 219:355
10. Naylor SJ, Moccia RD, Durant GM (1999) N Amer J Aquacul 61:21
11. Sutherland TF, Martin AJ, Levings CD (2001) ICES J Mar Sci 58:404
12. Brooks KM, Mahnken, CVW (2003) Fish Res 62:295
13. Loring DH, Rantala RTT, Milligan TG (1996) Can Tech Rep Fish Aquat Sci 2111
14. Trites RW, Petrie L (1995) Can Tech Rep Hydrog Ocean Sci 163
15. Parker WR, Aube JG (2002) Environment Canada Technical Report EPS-5-AR-02-01
16. Brooks KM, Stierns AR, Backman C (2004) Aquaculture 239:81
17. Ravizza GE, Bothner MH (1996) Geochim Cosmochim Acta 60:2753

… Hdb Env Chem Vol. 5, Part M (2005): 221–238
DOI 10.1007/b136012
© Springer-Verlag Berlin Heidelberg 2005
Published online: 6 July 2005

Sediment Geochronologies for Fish Farm Contaminants in Lime Kiln Bay, Bay of Fundy

John N. Smith (✉) · Philip A. Yeats · Timothy G. Milligan

Fisheries and Oceans, Canada Marine Environmental Sciences Division, Bedford Institute of Oceanography, Dartmouth, B2Y 4A2, Canada
smithjn@mar.dfo-mpo.gc.ca, yeatsp@mar.dfo-mpo.gc.ca, milligant@mar.dfo-mpo.gc.ca

1	Introduction	222
2	Methods	223
3	Fish Farm History	224
4	**Results and Discussion**	224
4.1	Sedimentation Rates and Geochronologies	225
4.2	Contaminant Geochronologies	228
4.3	Historical Record of Fish Farm Impacts	229
4.4	Effects Levels	232
4.5	Additional Aquaculture Tracers	233
5	Conclusions	237
	References	237

Abstract Sedimentation rates were measured on gravity cores collected near finfish cages in Lime Kiln Bay, N.B. in the Western Bay of Fundy using the radionuclide tracers ^{210}Pb and ^{137}Cs. Sediment cores collected close to aquaculture sites that have undergone extensive salmon fish farm activity over the past 20 years exhibit elevated levels of Zn and Cu in the upper 50 cm of the cores. Sediment geochronologies indicate that the threshold horizons for elevated Zn and Cu levels conform to the initial introduction of fish farms into Lime Kiln Bay in 1981. The source of the Zn is the fish feed while Cu is associated with chemical agents used to reduce fouling of the cages by marine biota. The highest contaminant levels for Zn ($> 250\ \mu g\,g^{-1}$) and Cu ($> 70\ \mu g\,g^{-1}$) were measured in sediment cores collected within the "footprint" of previously abandoned sites. The contaminant signals decrease with increasing distance away from the cages to values approaching background levels at distances greater than 200 m from the original cage locations. Zn and Cu concentrations have remained elevated in sediments for the five-year period between the removal of the cages and the date of core collection, suggesting that remobilization of these metals from sediments following the termination of aquaculture operations may be minimal. P is present at elevated levels in sediments close to aquaculture sites because it is an important constituent of marine organic material associated with fish feed pellets and fish farm wastes. Elevated Cd, Mo and U levels were also observed in sediments deposited under salmon cages during past periods of aquaculture operations. These elements are soluble in seawater, but can be authigenically precipitated in sediments under reducing conditions. In the present study they have been used as indirect tracers of fish farm activities owing to their transfer from seawater to sediments under the anoxic sediment conditions generated by the high sediment flux of reactive organic material from aquaculture operations.

Keywords Aquaculture wastes · Copper · Sedimentation rates · Radionuclide and trace metals · Zinc

1
Introduction

Commercial finfish aquaculture operations can impose significant stresses on environments in which large quantities of fish are confined to relatively small volumes of seawater. Much of the environmental impact of fish farms is local, and may include oxygen depletion of the water column and enhanced organic and nutrient loadings in underlying sediments that may, in turn, result in anoxia and depauperation of the benthos. However, fish farms are also capable of generating long-lived, persistent contaminants that may have far-field environmental effects. Although the discharge of these types of contaminants from aquaculture sites is an environmental concern, it may also represent an opportunity to trace the far-field influence of aquaculture using these contaminants as tracers, and may thereby provide a new tool for quantifying industry impacts.

One method for studying discharges of particle-reactive contaminants from industrial sources and their dissemination in the environment involves the analysis of contaminant uptake in sediments. For example, Yeats et al. [1] have observed high levels of Zn and Cu in surface sediments in the Letang region of the Western Bay of Fundy, which they ascribe to releases from proximal salmon aquaculture sites. Their conclusions are based partly on the fact that equally high levels of Cu and Zn are absent from similar sediment regimes that are more remote from aquaculture sites. Further, sediments under salmon aquaculture cages in New Brunswick have been shown by Chou et al. [2] to be enriched in Zn and Cu. Similar results have been reported both in sediments under salmon aquaculture cages in British Columbia [3] and in sediment traps adjacent to salmon net pens [4]. Recent measurements of elevated levels of Zn in salmon food pellets [2, 4] support the widely held belief that the source of the Zn in the sediments is the salmon feed. Although Cu is added to salmon feed as a micronutrient at quite low levels [5], the most likely source of the Cu in sediments underlying aquaculture sites is the antifouling agents used to reduce fouling of the nets by marine biota [6].

In an effort to further characterise the discharge of Zn and Cu from finfish aquaculture sites, a series of sediment cores was collected from the Letang region and analysed for metals and radionuclides. ^{210}Pb and ^{137}Cs distributions were used to determine sedimentation rates for each core and these were then used to determine contaminant geochronologies. However, in addition to known fish farm contaminants, other inorganic contaminants were also observed that can provide additional tracers for aquaculture operations. This study documents a method for comparing the history of fish farm activities with the record of contaminant inputs recorded in the sediments and attempts to extend the understanding of both near-field and far-field effects of aquaculture operations.

2
Methods

Lehigh gravity cores (10 cm I.D., 1.5 m long) were collected proximal to aquaculture fish farms in Lime Kiln Bay in the Letang region at the locations indicated in Fig. 1 and Table 1 in 2000 and 2002. The sediment cores were extruded and subsampled at 1 cm intervals and analysed for metals, organic constituents and radionuclides. Metals were analysed following acid digestion using multi-element ICP-MS techniques, while Hg was analysed separately by aqua regia digestion and cold vapour atomic absorption (AA) analysis. Percent organic matter was measured by weight loss by ignition at 520 °C. ^{210}Pb was determined on freeze-dried sediment samples by alpha counting of ^{210}Po electrodeposited onto nickel discs, while the ^{226}Ra supported ^{210}Pb levels were determined either on acid-digested sediment samples using a radon gas emanation method or on dried sediments using Ge gamma ray detection methods [7, 8]. ^{137}Cs was measured on dried sediment samples using

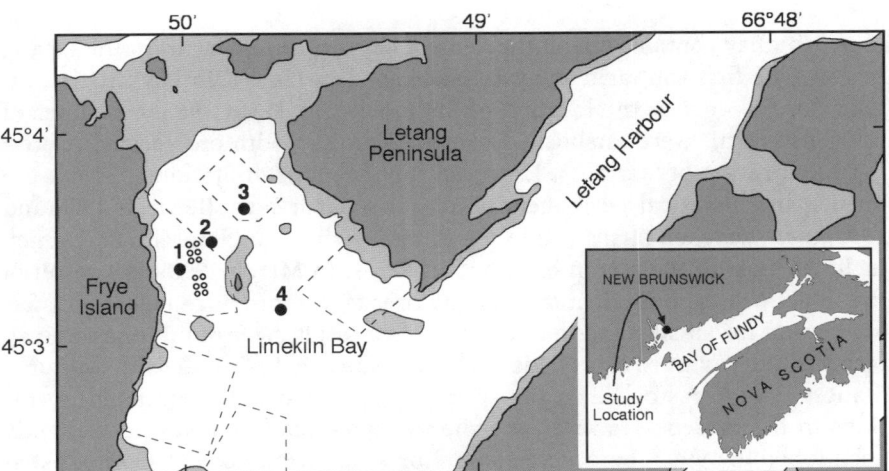

Fig. 1 Gravity cores were collected at stations indicated in Fig. 1. The *small circles* between Stations 1 and 2 represent the positions of 12 fish farm cages at the time of core collection. *Dashed lines* represent boundaries for fish farm leases registered in Lime Kiln Bay since 1981

Table 1 Sedimentation rates and ^{210}Pb inventories for Lehigh gravity cores collected in Lime Kiln Bay. Sediment accumulation rates (g cm^{-2} y^{-1}) can be approximated by the product of the sedimentation rate and an average in situ density of 0.6 g cm^{-3}

Station	Year of core collection	Water depth (m)	Latitude (degrees)	Longitude (degrees)	^{210}Pb Inventory (dpm cm^{-2})	Sedimentation Rate[1] (cm y^{-1})
1	2002	12.9	45.0612	−66.8334	68.1	0.239[2]
2	2002	12.8	45.0621	−66.8319	55.3	0.205[2]
3	2002	13.3	45.0637	−66.8297	65.2	0.298[2]
4	2000	14.8	45.0595	−66.8278	84.8	0.553

[1] Mean value below upper 10 cm of core.
[2] Sedimentation rate in upper 10 cm determined from CF Model varies with depth and is approximately twice the given rate.

a hyper-pure Ge gamma ray detector with a 1 cm diameter well. Counting uncertainties in the ^{210}Pb measurements were generally about 5%, while those in the ^{137}Cs and ^{226}Ra measurements were about 10%. Detailed porosity profiles were determined for each core and used to determine the integrated mass of dry sediment (mass-depth; g cm^{-2}) above each depth (cm) horizon.

3
Fish Farm History

Lime Kiln Bay contains one of the densest clusters of fish farms in the Letang region. The first fish farm site was established in Lime Kiln Bay in 1981, either close to or over the location of Station 1 (Fig. 1). In 1985, a number of other fish farms were established in Lime Kiln Bay, with one located overlying the location of Station 3. However, the locations of Stations 2 and 4 are outside the historical perimeters of all of the farm sites. Between 1986 and 1998, there were ten farms within the general vicinity of Lime Kiln Bay, which includes most of the region indicated in Fig. 1. In March 1998 (as a result of the infectious salmon anemia [ISA] outbreak), the province enforced a fallow period for all sites in Lime Kiln Bay. As a result, by around June 1998, all farms in the bay were empty. Eight of these sites, including the site overlying Station 1, were restocked in the spring of 1999 and have continued operations to the present. However, the site overlying Station 3 was permanently abandoned in 1998. There were some shorter fallow periods (1–3 months) at various locations in the spring of 2001 (B. Danner, pers. comm.).

4
Results and Discussion

4.1
Sedimentation Rates and Geochronologies

Sedimentation rates and geochronologies for the four cores were determined from ^{210}Pb and ^{137}Cs sediment-depth distributions (Fig. 2) using previously-reported methods [7–9]. ^{210}Pb is a natural radionuclide that is produced in the atmosphere from the decay of radon gas, and it enters the sediments at a constant rate via attachment to particles. ^{137}Cs is an artificial radionuclide first introduced to the environment as the result of atmospheric nuclear weapons tests beginning in the early 1950s. Geochronologies for each core were determined following the protocols outlined in Smith [10].

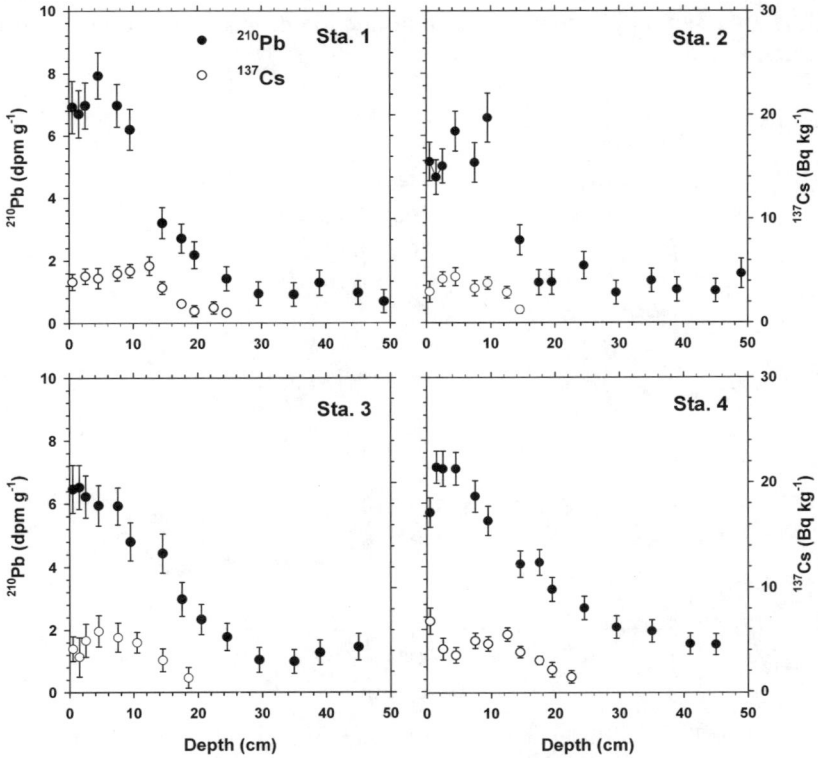

Fig. 2 ^{210}Pb and ^{137}Cs sediment-depth distributions for four cores collected at locations in Lime Kiln Bay indicated in Fig. 1. Cores 1–3 were dated using a Constant Flux ^{210}Pb method, while core 4 was dated using the CSF method

The sediment cores were dated using both a constant sedimentation rate-constant flux (CSF) model and a constant flux (CF) model [8]. In each case, the ^{226}Ra background supported ^{210}Pb was subtracted from the total ^{210}Pb to give excess ^{210}Pb (^{210}Pb$_{ex}$). For the CSF model, ^{210}Pb$_{ex}$ was then plotted as a function of depth (cm) and mass-depth (g cm^{-2}) on a semi log scale and the sedimentation rate (cm y^{-1}) and mass accumulation rate (g cm^{-2} y^{-1}) were determined from the slopes of least squares exponential fits to the experimental data. The fit to a ^{210}Pb sediment-depth distribution for the Station 4 core is illustrated in Fig. 3, where the ^{226}Ra background is 1.0 dpm g^{-1} and the sedimentation rate is estimated to be 0.55 cm y^{-1}. The ^{137}Cs geochronology for the core was then determined from the ^{210}Pb sedimentation rate (Fig. 3). In order to validate the ^{210}Pb geochronology, the ^{137}Cs threshold must be in reasonable agreement with a ^{210}Pb time-stratigraphic horizon of 1954, which for the Station 4 core occurs at a depth of about 25 cm (Fig. 3). If this is not the case, then the core must be assumed to have undergone mixing by various physical processes or bioturbation or be otherwise undatable. Confirmation of the agreement of the ^{137}Cs threshold with a ^{210}Pb date of approximately 1954 is an essential element of the radionuclide dating of recently-deposited sediments [10].

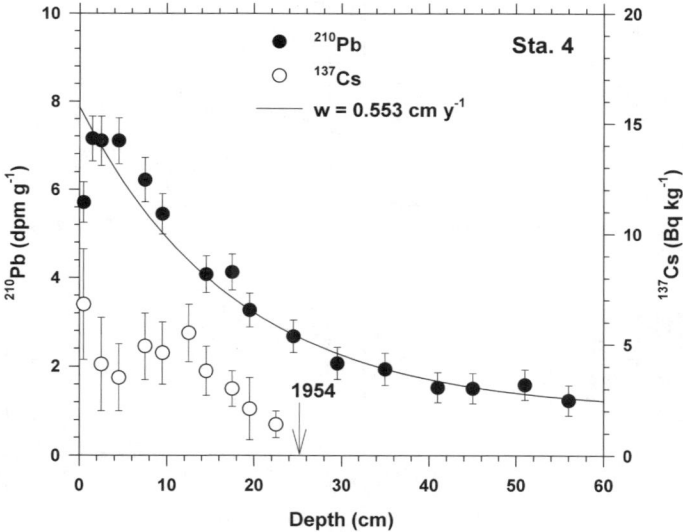

Fig. 3 ^{210}Pb and ^{137}Cs sediment-depth distributions for cores from Station 4 in Lime Kiln Bay. The exponential fit to the ^{210}Pb data represents a sedimentation rate of 0.55 cm y^{-1} (0.42 g cm^{-2} y^{-1}). ^{137}Cs threshold conforms to a date of about 1954, determined from ^{210}Pb dating, indicating that sediment mixing is minimal and that the ^{210}Pb geochronology is applicable to contaminant distributions

The results from the Station 4 core were consistent with a CSF model, but the ^{210}Pb results (Fig. 2) from the other three cores exhibited irregularities in the upper 10 cm that clearly represented a departure from exponential profiles. Although these features resemble those generally associated with bioturbation, sediment mixing is not believed to be important in the present case. Sediments in the four cores collected in the sheltered regions of Lime Kiln Bay are enriched in organic matter and are anoxic below the upper few cm, conditions that discriminate against most types of benthic infauna that are responsible for sediment mixing. Sediments in the upper 15 cm are distinguished by textural unconformities and abrupt changes in porosity that probably represent recent, intermittent inputs of waste material from fish

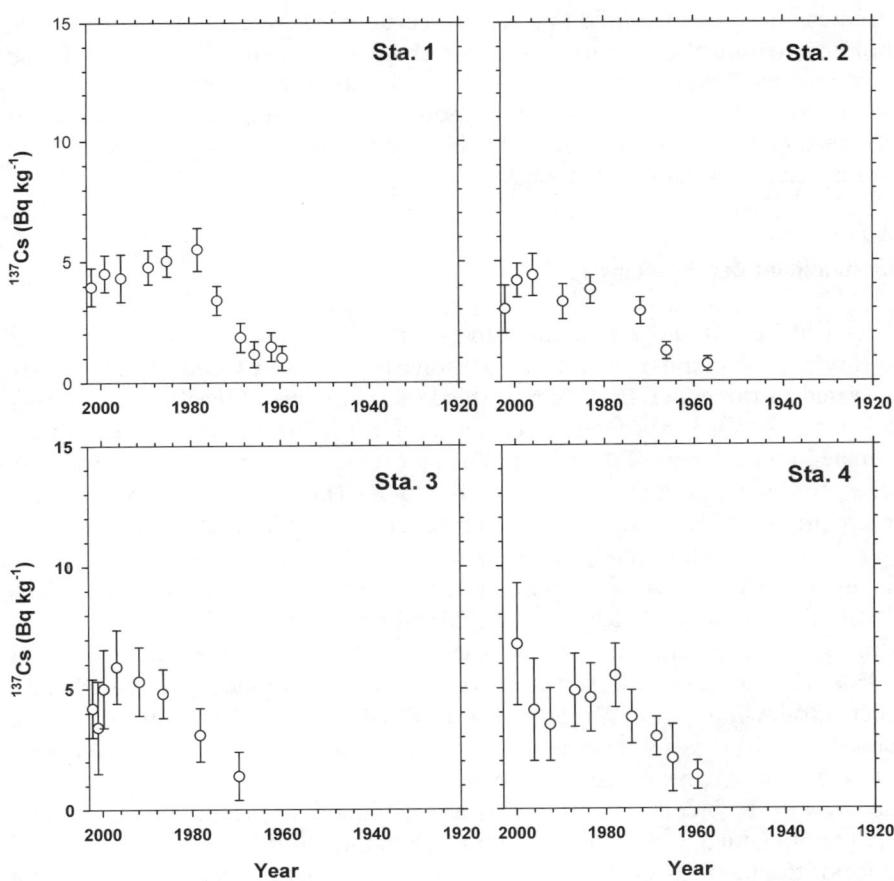

Fig. 4 ^{137}Cs activities are plotted on the geochronological scale determined from ^{210}Pb modelling for each core. ^{137}Cs distributions exhibit thresholds in the 1950s, indicating that the geochronologies are reasonably accurate and that sediment mixing is minimal

farms. These irregular depositional events were interpreted using a "constant flux" (CF) model [8] under the assumption that the recent inputs of material resulted in small modulations in the sedimentation rate, but did not significantly alter the net flux of ^{210}Pb. The CF model (technically, a "mapping" operation [11]) involves the integration of the excess ^{210}Pb distribution over the entire length of the core. The CF geochronology is extremely sensitive to the value measured for the total ^{210}Pb inventory in the core, and in the present case the CF dates are considered accurate only by a factor of $\pm 20\%$. The CF model indicated that sedimentation rates in the upper 10 cm of cores from Stations 1–3 were usually a factor of 2 greater than the mean values (listed in Table 1) for the deeper sections of the cores. In Fig. 4, the ^{137}Cs data for each core are re-plotted as a function of sediment deposition date calculated from the ^{210}Pb CF time-stratigraphy. For cores from Stations 1–3, the CF Model predicted a ^{137}Cs threshold that was in reasonable agreement with a 1954 age horizon, within the $\pm 20\%$ uncertainty estimated for the ^{210}Pb dates. These results provide validation of a CF model for the three Lime Kiln Bay cores (Stations 1–3) and support the assumption that the irregular ^{210}Pb structures in the upper layers of the cores are the result of depositional anomalies that are unrelated to sediment mixing.

4.2
Contaminant Geochronologies

Levels of Zn, Cu and Pb for the Lime Kiln Bay cores are plotted in Fig. 5 as a function of sediment depth. At Stations 1–3, Zn and Cu levels are clearly elevated in the upper 10–20 cm of the cores while metal levels increase only slightly in surface sediments at Station 4. Metal geochronologies have been estimated for each core (Figs. 6 and 7) using the same ^{210}Pb time-stratigraphies used to determine the ^{137}Cs geochronologies (Fig. 4). Pb distributions for each core provide a useful test of the general interpretation of contaminant geochronologies for Lime Kiln Bay cores, because there is no evidence that Pb is introduced as a fish farm contaminant. The Pb geochronologies are similar in all four cores. Pb levels exhibit near-background levels of 20–25 µg g^{-1} during the 1920 s and increase to levels of 30–35 µg g^{-1} during the 1940 s and 1950s. This commonly observed increase in Pb contamination in dated sediments following World War II [12] is ascribed to inputs from anthropogenic sources such as gasoline emissions to the atmosphere or industrial inputs. However, in marine sediments proximal to shipping and fishing activity, the source of the Pb contamination may also be linked to marine paints. Pb was frequently added as an additive to marine paints in order to enhance adherence of the paint to ships hulls [13], a practice that was discontinued in the 1970s. The slight decrease in Pb concentrations in post-1980, upper sediment layers reflects the recent declines in Pb usage as an additive to fuel, marine paints and other commercial products. The similar Pb geochronologies for

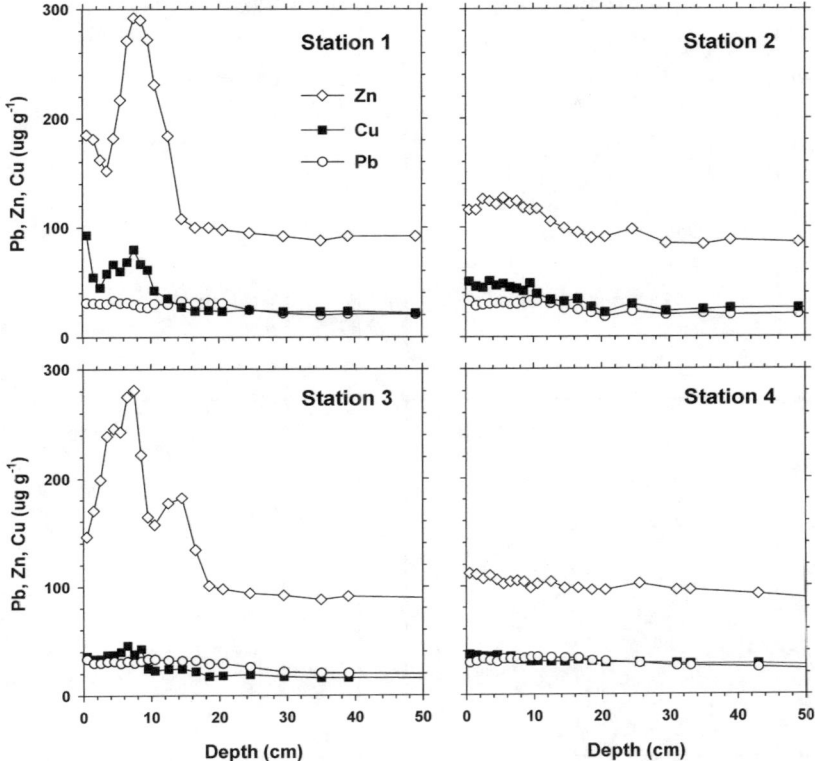

Fig. 5 Zn and Cu concentrations in the upper parts of cores from Stations 1–4 in Lime Kiln Bay are greatly enhanced compared to Pb concentrations, indicating that these contaminants may have an aquaculture source

each of the four cores from Lime Kiln Bay provide additional confidence in the accuracy of the ^{210}Pb time-stratigraphies.

4.3
Historical Record of Fish Farm Impacts

Zn and Cu concentrations in the core from Station 1 have relatively low values of approximately 90 µg g^{-1} and 20 µg g^{-1}, respectively, in deeper sediments (> 50 cm depth) near the base of the core. These concentrations are close to background levels for these metals measured in pre-1920 strata from cores collected from other fine particle depositional sites in the Bay of Fundy. Zn and Cu concentrations increase sharply in sediments deposited in the late 1970s and early 1980s and attain their maximum values in sediments deposited in the late 1980s. A minimum occurs in the metal geochronologies in the late 1990s, followed by metal concentrations that increase towards the

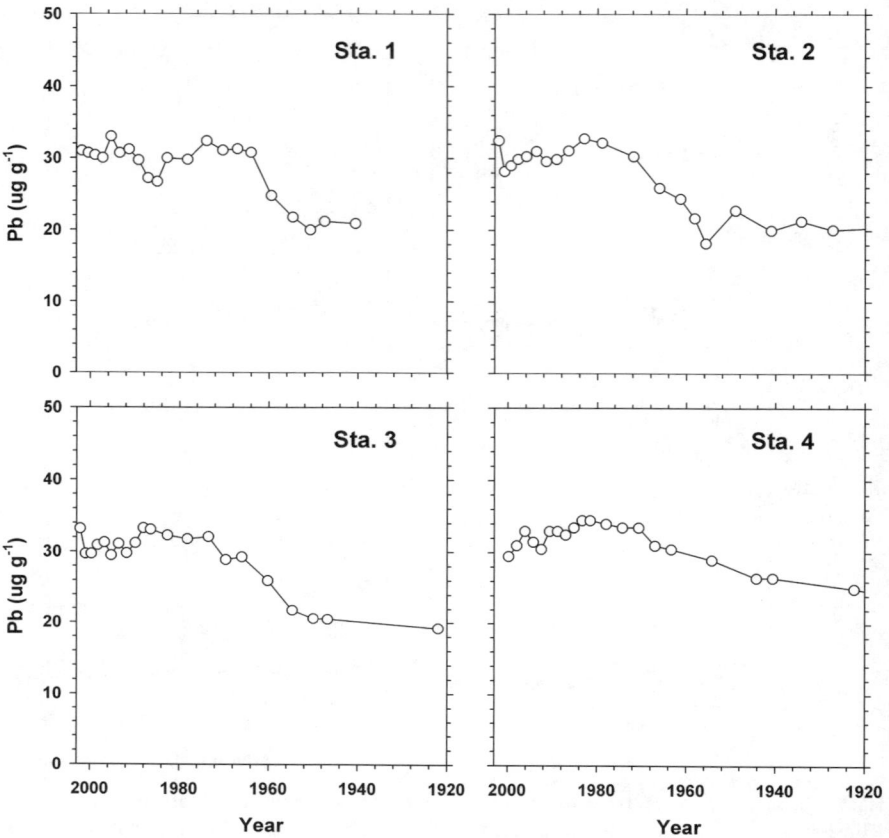

Fig. 6 Pb concentrations plotted as a function of ^{210}Pb deposition date for Lime Kiln Bay cores exhibit elevated levels as a result of anthropogenic inputs beginning in the 1950s

surface. These results are consistent with the history of fish farm activity proximal to the site of Station 1. Aerial photos from 1996 (K. Haya, pers. comm.) show that fish farm cages were positioned over the location of the Station 1 core at that time. It is presumed that these or similar cages that were introduced in 1981 are responsible for the Zn and Cu maxima and threshold horizons (Fig. 7). Removal of the cages during the ISA outbreak in 1998 led to the steep decline in Zn and Cu concentrations observed in the underlying sediments at that time. Cages were re-introduced to this site in 1999, but were positioned to the east of Station 1 by a distance of about four metres. The positions of the 12 circular aquaculture cages at the time of core collection in 2002 are outlined in Fig. 1 by multi-beam images generated by textural signals in the sediments. The re-introduction of the cages has resulted in a recently increasing Zn and Cu signal in surface sediments and

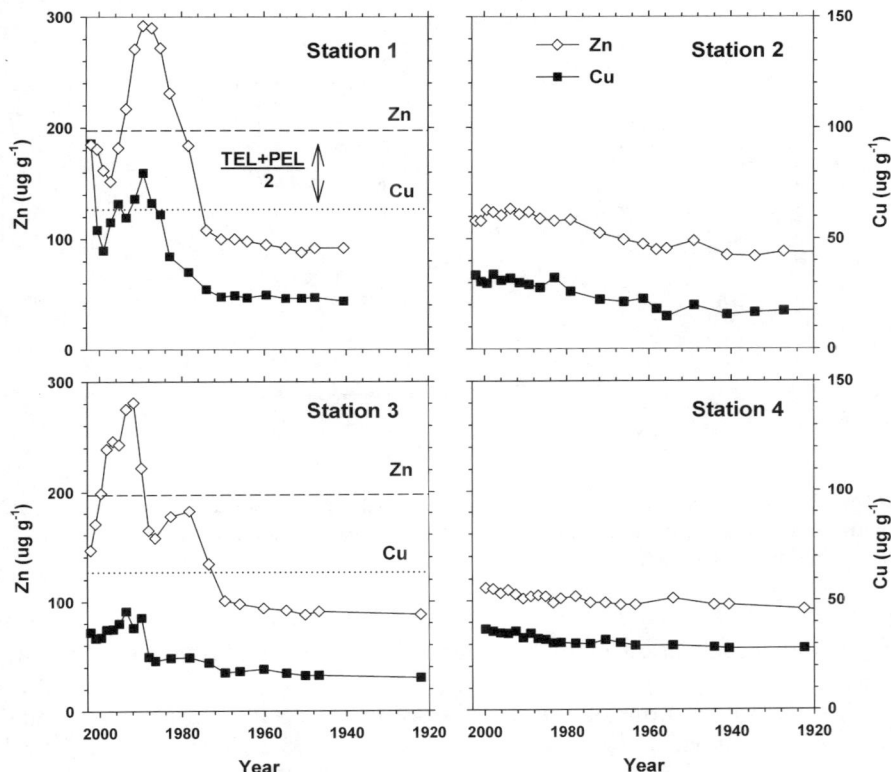

Fig. 7 Cu and Zn concentrations plotted as a function of ^{210}Pb deposition date for Lime Kiln Bay cores 1 and 3 exhibit elevated levels in the 1980s and 1990s owing to releases from aquaculture sites. Post-1980 increases in metal concentrations are much less pronounced at Stations 2 and 4. *Dotted lines* represent sediment effects level benchmarks for Zn and Cu

a minimum corresponding to the 1998–99 hiatus in fish farm activity in Lime Kiln Bay.

Station 2 was located approximately 50 m from the set of cages immediately adjacent to Station 1 (Fig. 1). It is known that the Station 2 location has never been a site of active fish farm activity, although it is near the boundaries of existing sites. This is reflected by the absence of a pronounced maximum in the Zn and Cu concentrations during the 1980s, similar in magnitude to that observed in the core from Station 1. There are smaller broad maxima in Zn and Cu in the recently deposited sediments at Station 2 that may reflect contaminant transport from nearby aquaculture cages. However, these Zn and Cu distributions are similar to those observed for Pb, which is clearly not an aquaculture-derived contaminant. Additional tracers are required to

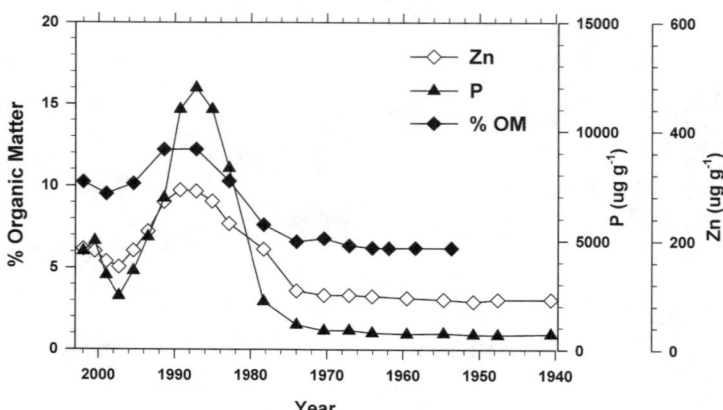

Fig. 8 Elevated P levels in the Station 1 core are synchronous with the Zn and % organic matter maxima and reflect the transport of marine organic matter to the sediments from fish farm operations

determine if the Station 2 core has been impacted by discharges from the fish cages.

Fish farm operations at Station 3 began in 1985 and continued until the permanent removal of the cages at this location in 1998. This is generally consistent with the geochronology for the peaks in the Zn and Cu geochronologies for core 3, (Fig. 7) which exhibit maxima in the 1990s, although the Cu and Zn thresholds are about ten years earlier (mid-1970s) than the predicted date of 1985. This latter observation may be a result of dating uncertainties that reflect inadequacies in the application of a ^{210}Pb CF model to the most recently deposited sediments in the core. In contrast to the core from Station 1, Zn and Cu levels do not increase near the sediment surface. This observation reflects the fact that fish cages were not re-introduced to this aquaculture site following the ISA outbreak in 1998. The Station 4 core is from a location that has never been a site of active fish farm activity. Zn and Cu concentrations exhibit only small increases in post-1980 sediments at this location, indicating only minor far-field contamination from fish farm sites in Lime Kiln Bay.

4.4
Effects Levels

The above results are consistent with those of Uotila [14], who observed elevated Zn ($> 400\,\mu g\,g^{-1}$) and Cu ($> 130\,\mu g\,g^{-1}$) in sediments underlying fish farm cages in Southwestern Finland. Similar to this study, Zn and Cu concentrations decreased to levels of 150 µg g^{-1} and 20–30 µg g^{-1}, respectively, at distances of about 20 m from the cages. Chou et al. [2] have also observed elevated Zn and Cu levels in sediments below fish cages in the Western Bay of Fundy. Brooks and Mahnken [3] reviewed studies of Zn and Cu measured in sediments near fish pens. Zn concentrations were elevated to levels of 200 µg g^{-1} in sediments underlying cages at salmon farms in British Columbia, decreasing exponentially with distance to background levels (25 µg g^{-1}) between 30 and 75 m down-current from the pen. This concentration gradient is similar to that observed in comparing the results from cores at Stations 1 and 2 in the present study. However, their results also suggest that Zn is remobilized from sediments following the ending of fish farm operations. Brooks and Mahnken [3] report levels decreasing from 200 µg g^{-1} to background levels within one year of the abandonment of the site. They suggest that this natural chemical remediation occurs as a result of a change from anoxic to oxic conditions in the surficial layers of the sediments following the reduction in deposition of organic material. They reason that Zn is initially bound as a sulfide in anoxic sediments and subsequently undergoes oxidation to the more soluble sulfate form which is then released into solution. However, there is no evidence of this effect in the cores collected in the present study. High levels of Zn associated with sediment deposition during historical periods of fish farm activity in the 1980s and 1990s have been retained within the sediments until the time of core collection.

Numerous criteria have been proposed to characterize the metal levels in sediments at which biological effects can occur. For example, Jones et al. [15] propose a marine quality sediment standard based on the mean of a threshold effects level (TEL) and a probable effects level (PEL) for each contaminant, which corresponds to values of 198 µg g^{-1} and 63 µg g^{-1} for Zn and Cu, respectively (Fig. 7). These levels are exceeded for both metals during the period of active fish farming in the 1990s in core 1. They are exceeded only for Zn in core 3 and are not exceeded at any time for cores from stations 2 and 4. These results suggest that biological effects may be observed below or immediately adjacent to the cages, but will probably not be observed at locations greater than 50–100 m from the cage locations.

4.5
Additional Aquaculture Tracers

The association of different contaminants with the aquaculture industry can be evaluated from their distributions in sediments from Station 1 where the highest contaminant signals are observed. The percent organic matter signal (measured by combustion) increases from a background level of 5% to a maximum of 12%, which is synchronous with the maxima in Zn (Fig. 8). Most of this additional organic matter is presumed to be undigested fish food or faecal material, but may also include material that has accumulated or accreted on the fish pen structures themselves. Phosphorous concentrations also increase from background values of about 1000 µg g^{-1} to a maximum of 12 000 µg g^{-1}, which is synchronous with the maximum in the percent organic matter. The increase in P is associated with the introduction of a large quantity of marine organic matter (fish feed and wastes) into a near-shore depositional environment that is normally dominated by inputs of terrigenous organic matter. Commercial fish feeds may contain surplus P, because not all P in the feed is available (digestible and absorbable) to the fish. Dietary P that is consumed by fish, but not digested and absorbed, will appear in the waste as faecal P. This may result in higher P levels in fish wastes compared to fish feed, as has been observed in rainbow trout aquaculture [16]. The increase under the Atlantic salmon cages of an order of magnitude in the P concentration, compared to pre-aquaculture, background levels is much greater than the increase (by factors of 1.5–2) observed for Zn, Cu and organic matter concentrations in the same sediment core. This result suggests that P may be a useful tracer of organic material from finfish aquaculture operations, that provides a higher signal-to-noise ratio for recording aquaculture impacts than that provided by total organic matter levels alone.

The three metals, Mo, Cd and U also co-vary with % organic matter in sediments from Station 1 (Fig. 9), but for reasons entirely different from those noted above. All three metals are soluble in seawater, but are much less soluble in the pore waters of reducing sediments. The different solubilities in oxic seawater and reducing pore waters can result in a net flux of these trace elem-

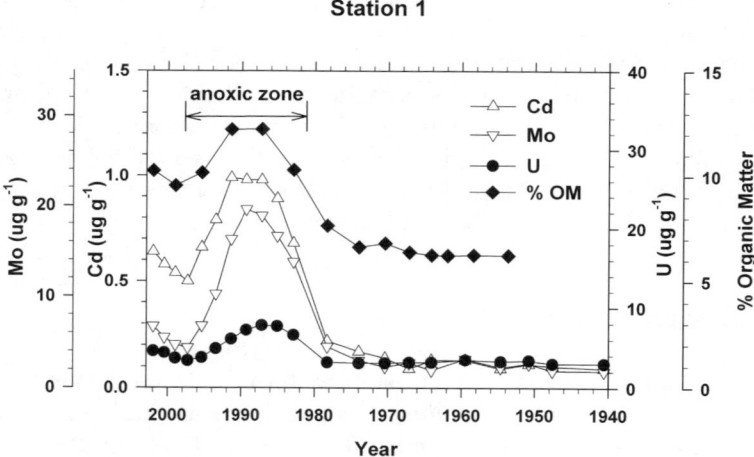

Fig. 9 Maxima in Mo, Cd and U levels conform to the maximum in % organic matter and reflect the fact that these elements are transferred from seawater to sediments where they are authigenically precipitated under anoxic conditions caused by the high flux of marine organic matter

ents into anoxic sediments, where precipitation of authigenic phases occurs. Owing to the low concentrations of these elements in background detrital phases, their concentrations in authigenic phases can dominate in reducing sediments. For this reason they have been used as tracers of paleo-redox phenomena in oceanic sediments [17–21].

The actual mechanism governing the authigenic formation of these metals in sediments is a subject of active investigation [22]. Cd can be deposited in the sediments on particles, remobilized into pore waters following burial, and subsequently precipitated from solution in even weakly reducing sediments as CdS [23, 24]. Cd is also a bioactive element that is concentrated in marine organisms such as phytoplankton and algae and in the liver (at levels of 114 μg kg^{-1} wet wt.) of cod [25]. Its principal input to Lime Kiln Bay sediments may be with marine organic material from fish farm wastes, as was the case for P. In contrast to Cd, U does not precipitate as a sulfide, but the reduction of the soluble U(+5) and U(+6) states to the less soluble U(+3) and U(+4) states can be enhanced by metal-reducing bacteria [26, 27]. These processes are presumably accelerated in Lime Kiln Bay sediments by the high rates of bacterial activity that accompany the high flux of reactive organic material from fish farm wastes. Mo follows the Mn cycle by co-precipitation from seawater with Mn and is then subject to pore water remobilization when manganese oxides are redissolved [28]. It has been hypothesized that Mo undergoes molecular diffusion within sediment pore waters as a stable molybdate ion. In sulfidic sediments, this can be converted to a more particle-reactive thiomolybdate (MoO$_4^-$ $\rightarrow$ MoS$_x$O$^-_{(4-x)}$) that is subsequently scavenged onto particle surfaces within the sediment reducing zone [22, 29, 30].

The overall rates of precipitation of these elements are kinetically controlled and are governed by factors such as rates of metal complexation and transport in pore waters, mixing and sedimentation rates, and the reactivity of the deposited organic matter. However, in general, elevated Mo, Cd and U concentrations can be considered to be residual tracers of sediment anoxia that can remain as sediment artifacts following the removal of the source of reducing material. In this sense, these redox-sensitive trace elements differ from Zn, Cu and P, which are direct tracers of fish farm wastes or material accreted to cages, but which do not require anoxic conditions to persist in sediments.

Mo, Cd and U concentrations increase by factors of 10, 5 and 3, respectively in sediments at Station 1 that accumulated during the approximately 20-year period of aquaculture operations at this location (Fig. 9). The mean rate of authigenic trace element formation can be estimated by subtracting the background, detrital component of each trace element [31] from the total inventory under the peak and dividing this authigenic inventory by the period of deposition ($\cong$ 20 y). The authigenic formation rates of 4×10^{-2} μmol cm^{-2} y^{-1}, 2×10^{-3} μmol cm^{-2} y^{-1} and 7×10^{-3} μmol cm^{-2} y^{-1} for Mo, Cd and U are an order of magnitude greater than similar estimates for their formation rates in the sediments of the Laurentian Trough in the Gulf of St. Lawrence [22]. This enhancement of trace element precipitation in Station 1 sediments compared to other anoxic marine environments is an indirect anthropogenic impact associated with the high flux of reactive organic material from fish farm wastes.

The P and Mo geochronologies for each core that are illustrated in Fig. 10 exhibit features similar to those of the Zn and Cu geochronologies (Fig. 7). The multiple peaks in P and Mo concentrations in sediments at Station 3 reflect similar peaks in Zn and Cu, confirming this historical evidence for fish farm sites at this location. The smaller peaks in Zn and Cu levels measured in Station 2 sediments are accompanied by a maximum in P, but not in Mo concentration. These results indicate that biogenic debris from aquaculture cages was indeed deposited at Station 2, but that the flux of marine organic matter was insufficiently high to produce the level of sediment anoxia required

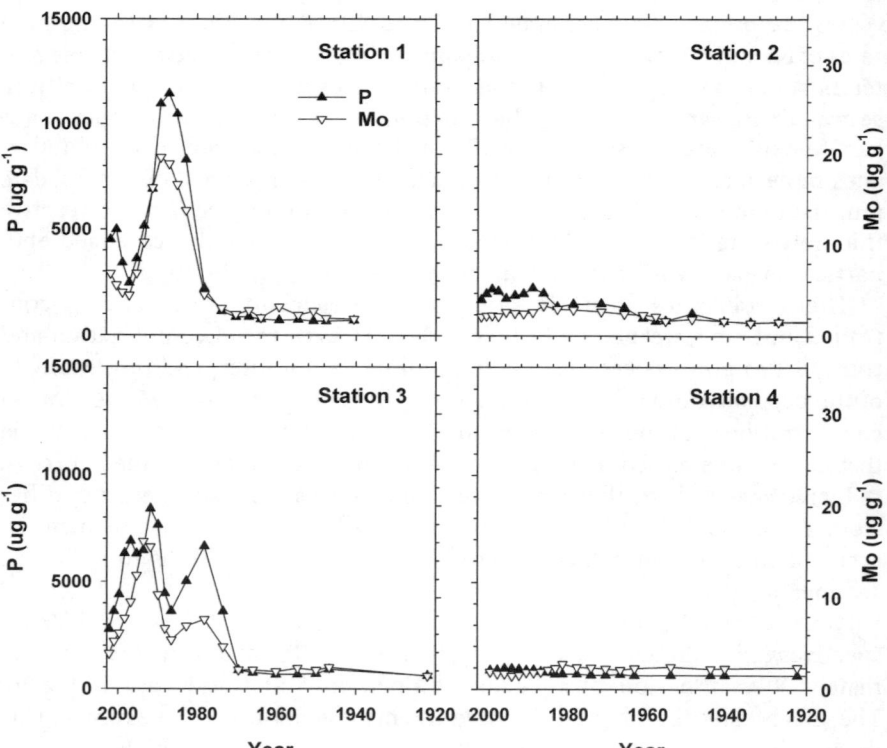

Fig. 10 Maxima in P and Mo levels in cores from Stations 1 and 3 are similar, relative to background levels, to those for Zn and Cu (Fig. 7). These two tracers support Zn and Cu evidence indicating that minor fish farm contamination occurs at Station 2 and negligible contamination occurs at Station 4

to precipitate authigenic Mo. Finally, the small increase in Zn and Cu concentrations in surface sediments at Station 4 is unaccompanied by a significant concentration increase in either P or Mo. This result suggests that the input of fish farm debris was negligible at this location in the past and that the recent small increase in Zn and Cu concentrations is associated with other anthropogenic sources.

5
Conclusions

Sedimentation rates measured using ^{210}Pb and ^{137}Cs on four sediment cores collected proximal to Atlantic salmon fish farm cages in Lime Kiln Bay in the Western Bay of Fundy had values in the range of 0.3 cm y^{-1} to 0.6 cm y^{-1}. Cores collected in the imprint of former cage sites exhibited increases in organic matter and sedimentation rates (determined using the ^{210}Pb constant flux model) in the upper 15 cm, in chronological agreement with the onset of fish farm activities in the 1980s. Four different classes of inorganic tracers of fish farm activities were detected in sediments underlying fish farm cages; (1) Zn represents contaminants introduced into fish feed and deposited in the sediments with unused food pellets and fish wastes; (2) Cu is used in antifouling paints applied to cage surfaces; (3) P is an important constituent of marine biological material that is the basis for fish feed and is transported to the sediments with biogenic debris; (4) Cd, Mo and U are redox-sensitive elements that are soluble in seawater but are authigenically precipitated in the highly anoxic sediment regimes generated by the high flux of reactive organic matter from fish farm operations. This suite of tracers was used to show that sediments approximately 50 m from a set of cages had received fish farm wastes in the past, but not to the extent that the sediment column had been made sufficiently anoxic to precipitate redox-sensitive tracers like Mo. In general, only Zn exhibited levels under fish farm cages that were high enough to cause occasional biological effects, while levels of Cu only infrequently exceeded biological effects horizons.

References

1. Yeats PA, Milligan TG, Sutherland TF, Robinson SMC, Smith JA, Lawton P, Levings CD (2005) Lithium-normalized zinc and copper concentrations in sediments as measures of trace metal enrichment due to salmon aquaculture (in this volume). Springer, Berlin Heidelberg New York
2. Chou CL, Haya K, Paon LA, Burridge L, Moffatt JD (2002) Mar Poll Bull 44:1259
3. Brooks KM, Mahnken CVW (2003) Fish Res 62:295
4. Sutherland TF, Martin AJ, Levings CD (2001) ICES J Mar Sci 58:404

5. Chow KW, Schell WR (1978) In: Fish feed technology. Aquat Develop Coord Prog Rep No. ADCP/REP/80/11. FAO, Rome
6. Lewis AG, Metaxas A (1991) Aquacult 99:269
7. Smith JN, Schafer CT (1999) Limnol Oceanogr 44:207
8. Smith JN, Walton A (1980) Geochim Cosmochim Acta 44:225
9. Smith JN, Ellis KM, Nelson DM (1987) Chem Geol 63:157
10. Smith JN (2001) J Environ Rad 55:121
11. Robbins JA, Holmes C, Reddy K, Newman S (2000) In: Proc South Pacific Environmental Radioactivity Conf, 19–23 June 2000, Noumea, New Caledonia
12. Gearing J, Buckley DE, Smith JN (1991) Can J Fish Aquat Sci 48:2344
13. Buckley DE, Smith JN, Winters GV (1994) App Geochem 10:175
14. Uotila U (1991) In: Makinen (ed) Marine aquaculture and environment. Nordic Council of Ministers, Copenhagen, p 121
15. Jones DS, Suter II GW, Hull RN (1997) Rep ES/ER/TM-95/R4. USDOE, Oak Ridge, TN, p 31
16. Naylor SJ, Moccia RD, Durant GM (1999) N Amer J Aquacult 61:21
17. Turekian KK, Bertine KK (1971) Nature 229:250
18. Bertine KK, Turekian KK (1973) Geochim Cosmochim Acta 37:1415
19. Calvert SE, Pedersen TF (1993) Mar Geol 113:67
20. Zheng Y, Anderson RF, van Green A, Kuwabara J (2000) Geochim Cosmochim Acta 64:4165
21. Nameroff TJ, Balistrieri LS, Murray JW (2002) Geochim Cosmochim Acta 66:1139
22. Sundby B, Martinez P, Gobeil C (2004) Geochim Cosmochim Acta 68:2485
23. Gobeil C, Silverberg N, Sundby B, Cossa D (1987) Geochim Cosmochim Acta 51:589
24. Rosenthal Y, Lam P, Boyle EA, Thomson J (1995) Earth Planet Sci Lett 132:99
25. Environmental and Food Agency of Iceland (1999) AMSUM Report, Environmental and Food Agency of Iceland. Office of Marine Environmental Protection, Ármúli, Iceland
26. Fredrickson JK, Zachara JM, Kennedy DW, Duff MC, Gorby YA, Li S-MW, Krupka KM (2000) Geochim Cosmochim Acta 64:3085
27. Zheng Y, Anderson RF, van Green A, Fleisher MQ (2002) Geochim Cosmochim Acta 66:1759
28. Shimmield GB, Price NB (1986) Mar Chem 19:261
29. Adelson JM, Helz GR, Miller CV (2001) Geochim Cosmochim Acta 65:237
30. Chaillou G, Anschutz P, Lavaux G, Schafer J, Blanc G (2002) Mar Chem 80:41
31. Loring DH (1982) Can J Earth Sci 19:930

The Effect of Marine Aquaculture on Fine Sediment Dynamics in Coastal Inlets

T. G. Milligan (✉) · B. A. Law

Department of Fisheries and Oceans, Bedford Institute of Oceanography, P.O. Box 1006, Dartmouth, N.S. B2Y 4A2, Canada
milligant@mar.dfo-mpo.gc.ca, lawb@mar.dfo-mpo.gc.ca

1	Introduction	239
2	Methods	243
3	Results	245
4	Discussion	247
5	Conclusion	249
	References	250

Abstract The formation and deposition of large, fast-sinking aggregates by flocculation governs the distribution of fine particulate material within the coastal zone. Three major factors control the development of a flocculated suspension: (1) particle number or concentration, (2) particle adhesion efficiency or stickiness, and (3) particle break-up, most often due to an applied shear. The steady state equilibrium size distribution of a flocculated suspension reflects a dynamic balance between particle aggregation and dis-aggregation; changes to concentration, composition, or turbulence can hence affect the distribution of fine particulate material, both inorganic and organic. Owing to the close association of many surface-active contaminants with flocs, the aggregation dynamics of the particulate material will strongly influence their fate. The introduction of waste feed, faecal material, and their resulting degradation products from open cage aquaculture operations in the coastal zone will potentially increase both particle concentration and particle stickiness. As a result, the natural flocculation and depositional equilibrium of an inlet can shift towards increased deposition of fine-grained particulate material within flocs and the sequestering of contaminants within the sediment. Evidence for such a shift in fine-sediment dynamics and contaminant transport has been found in the Western Isles region of New Brunswick.

Keywords Aquaculture · Contaminants · Far-field effects · Flocculation · Sediment

1
Introduction

The seabed beneath salmon aquaculture pens is altered by the increased flux of material from above. Using various parameters such as organic carbon, redox potential (Eh), and sulphide, the area influenced by increased sediment accumulation has been identified and found to be dependent on the sedimentary conditions at the site [1]. Areas with high bottom stress naturally have much lower accumulation rates due to both initial dispersal of falling material and subsequent reworking of the accumulated waste [2, 3]. The rate of accumulation of wastes beneath salmon cages forms the basis of the environmental effects monitoring carried out in Canada. Accumulation rates sufficient to induce anoxia in the sediments require alterations to husbandry practices so that degradation of the region below the cages is reduced. Low accumulation rates are equated with low impact. What is not considered, however, is the fate of the material that is removed from the cage site and presumably added to the normal population of particulate material in a bay or inlet. This material would be subject to the same processes controlling the fate of other sediment in the region. Determining its fate requires a mechanistic understanding of how this additional material interacts with the naturally occurring sediment, and how it would impact the fine sediment dynamics of the receiving basin.

Bottom sediments preserve within their disaggregated inorganic grain size (DIGS) distributions a record of the physical transport processes responsible for their formation. In simple terms, the accumulation of sediment at any point in an inlet can be considered a balance between deposition from suspension and removal by currents and waves. Land and shore erosion produces inorganic sediment grain size distributions in which the amount of material in each logarithmically increasing size class follows a power-law relationship, from sub-micron-sized clays to coarse sands and gravels [4–6]. The exponent in these relationships depends on the composition of the source rock and expresses the relative abundance of fine versus coarse grains [6]. During transport, repeated settling and suspension modifies these size distributions. In particular the coarsest material is lost to deposition, so that the maximum grain size of a sediment defines the maximum turbulent energy to which the sediment is exposed [7, 8]. The degree of sorting indicates how frequently the material is resuspended, the extreme case of which is beach sands which have very well sorted, narrow size distributions, the modal diameters of which reflect wave energy [8].

The third process that controls the size distribution of bottom sediment is the mode of deposition. Fine-grained sediment is deposited as either single grains or in agglomerations of many particles, called flocs, and the relative proportions of these depositional modes are representative of the environmental conditions. Conceptually, a simple box model consisting of two boxes representing single grains and flocs illustrates this effect (Fig. 1). Removal of single grains from suspension is dependent on their individual settling velocity (Stokes settling) whereas removal of aggregated grains is controlled by the settling velocity of the flocs, on the order of $1\,\mathrm{mm\,s^{-1}}$. The most important controls on the extent of aggregation of sediment include: (1) particle

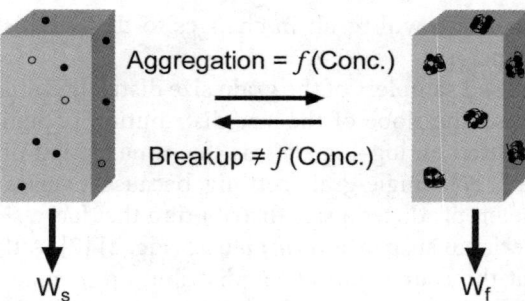

Fig. 1 Simple box model showing relationship between single grain and floc settling from suspension. w_s is the settling velocity of single grains and w_f is the floc settling velocity. Note aggregation is a second order function of concentration whereas floc break-up is not

number or concentration, (2) particle adhesion efficiency or stickiness, and (3) particle break-up [9, 10]. Aggregation of fine sediment and other detrital material increases the effective settling velocity of, for example, a 1 μm particle by over 3 orders of magnitude, resulting in the deposition of mud under low stress conditions. Alterations to the balance between single grains

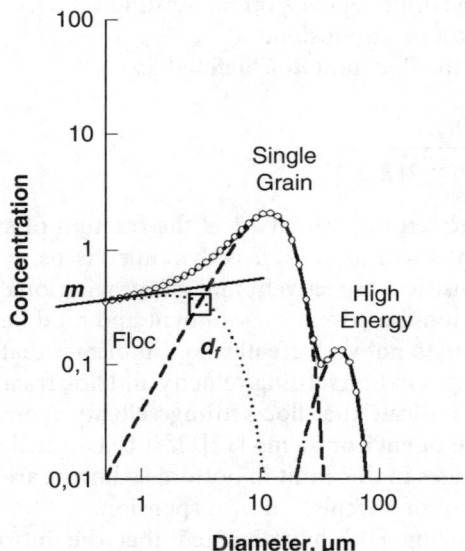

Fig. 2 Idealized disaggregated inorganic grain size distribution curve for a bottom sediment (*circles*) showing floc (*dotted*), single grain (*dashed*) and high-energy bedload (*dot-dash*) components. The *solid straight line* shows the source slope (m) of the parent suspension. Floc limit (d_f), indicated by the *box*, is the diameter at which the flux of floc and single grain settled material is equal. Changes in d_f indicate a change in the fraction of material in suspension that is found in flocs

and flocs in suspension will result in changes to the size distribution of the deposited sediment.

Flocs are unbiased samplers of the grain size distribution in the parent suspension [11, 12], so the slope of the size distribution of sediment deposited as flocs, when plotted on log–log axes, is the same as that of the material in suspension (Fig. 2) [7]. Single-grain settling, because it varies with the square of particle diameter, produces a size distribution that has a slope equal to the slope of the material in suspension (m) plus 2 (Fig. 2) [7]. A third component, which appears at the coarse end of the distribution, is made up of material that has been sorted as a result of high bottom stress and is characterized by a steep slope. The result is a bottom sediment size distribution consisting of three distinct populations of inorganic grains, the relative amounts of which record the depositional conditions of its formation. An "inverse floc model" [13] can be applied to the disaggregated inorganic grain size of the bottom sediment to infer the degree of flocculation in the overlying water. The floc-deposited portion of a suspension intersects the single-grain-deposited portion at a diameter known as the "floc limit", the diameter for which flux to the seabed in flocs equals the single-grain flux. Particles with smaller diameters are deposited predominantly within flocs, and particles with diameters larger than the floc limit deposit primarily as single grains. The value of the floc limit depends on the settling velocity of flocs and on the extent of flocculation in suspension.

Mathematically the floc limit d_f is defined as

$$d_f = \sqrt{\frac{w_f f 18 \mu}{(1-f)(\rho_s - \rho) g}} \qquad (1)$$

where w_f is the floc settling velocity, f is the fraction of suspended mass in suspension contained within flocs, μ is dynamic viscosity, ρ_s and ρ are sediment and fluid densities respectively, and g is gravitational acceleration [13]. Under the assumption that viscosity, sediment and fluid densities, and gravitational acceleration do not vary greatly, Eq. 1 indicates that floc limit changes in response to changes in floc settling velocity and floc fraction in suspension. Numerous studies indicate that floc settling velocity typically has a value of 1 mm s^{-1} in a range of environments [12]. If it is assumed settling velocity is constant, then changes in floc limit in bottom sediment are directly related to changes in the extent of flocculation in suspension.

Milligan and Loring [14] hypothesized that the introduction of waste feed and faecal material from aquaculture operations in Back Bay (New Brunswick) could increase both particle concentration and particle stickiness, resulting in an increase in extent of flocculation in suspension. Bacterial degradation has been shown to increase floc formation through the formation of extracellular polymeric substances [15, 16]. These long chain polysaccharides and other molecules form bridges that enhance flocculation. As a result, the natural flocculation and depositional equilibrium may shift towards increased deposition of fine particulate material within flocs. Missing from this earlier work in Back Bay was a way to infer the extent of flocculation in suspension as a function of time, so a link to the onset of intense aquaculture was impossible to establish. Here, we examine changes to the floc limit of sediment in the Western Isles region of Southwest New Brunswick during a period of rapid expansion of open cage aquaculture, extending the work of Milligan and Loring [14], which looked at the effect of flocculation on trace metal transport and deposition and the impact of aquaculture development on fine sediment dynamics.

2
Methods

The Western isles region of New Brunswick is one the most productive salmon aquaculture areas in Canada, consisting of several interconnecting bays and inlets (Fig. 3). The region is part of the Bay of Fundy macro-tidal environment, with an M2 tidal range of 6–8 m. Development of the salmon aquaculture industry in the area started in 1979 with the first lease in Lime Kiln Bay being occupied in 1981. Production in the Letang area grew rapidly and peaked in the late 1990s when an outbreak of infectious salmon anaemia caused the closure of some sites and the relocation of some leases. Subsequently, new management practices were introduced to reduce the probability of disease outbreak, and production has continued to grow in the region, although the number of cages in Lime Kiln Bay has been reduced.

Surficial sediment samples from the Western Isles region of New Brunswick were collected in 1990 [17] and 1999 (Fig. 3). Bottom samples were collected using a 15×15 cm Eckman grab, and the top 1 cm of sediment was sub-sampled for grain size and metals analysis. From 1990, 57 stations were analysed, and 33 from 1999. The 33 stations in 1999 were located in depositional areas identified in 1990. Two cores were collected in Lime Kiln Bay using 20- and 10-cm diameter Benthos gravity corers in 2000 (C1) and 2002 (C2) respectively (Fig. 3). Figure 3 also shows the location of the core from Back Bay (C3) reported in Milligan and Loring [14] collected by divers in 1994 as part of a much larger study of near-field aquaculture impacts conducted by Hargrave et al. [18]. Core sub-samples for grain size, trace metal, organic material and radiochemical analyses were obtained every 1 cm by extrusion for all cores.

DIGS analysis was carried out using previously described methods [19]. Sediment samples were digested in an excess of 35% H_2O_2 to remove organic material and then resuspended in 1% NaCl. Samples were disaggregated using a sapphire-tipped ultrasonic probe before counting on a Coulter Multi-

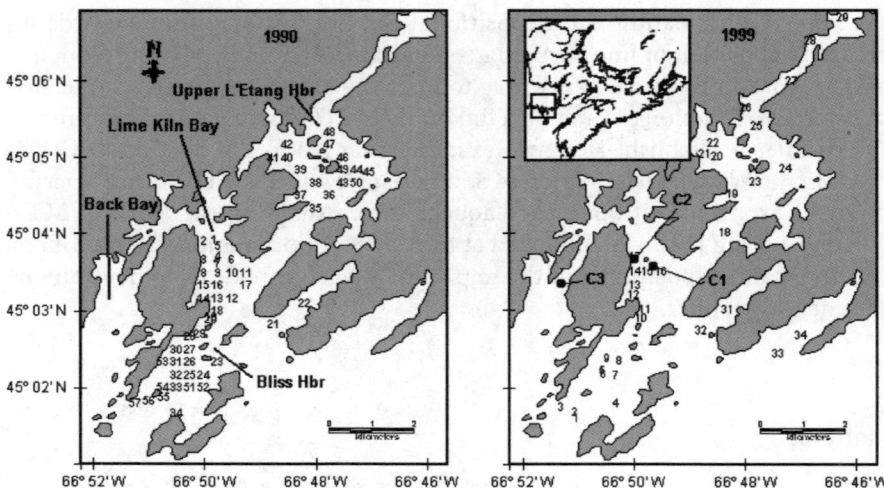

Fig. 3 Station locations for 1990 and 1999 surficial sediment surveys in Southwest New Brunswick. Also shown are locations of cores *C1*, *C2*, and *C3* collected in 2000, 2002, and 1994 respectively

sizer IIE. Up to four aperture tubes (30, 200, 400 and 1000 μm) were used in the analysis, yielding a size range of 0.8–500 μm. Results were expressed as equivalent weight percent calculated from the volume analysed using a specific gravity of 2650 kg m^{-3}. The DIGS spectra were plotted as log equivalent weight (%) versus log diameter to preserve the shapes of the distributions over a wide range of concentrations [20]. An inverse floc model [13] was applied to the DIGS. Using a non-linear least squares fit of the DIGS spectra, the floc limit (d_f) was calculated for each sample [13]. Sensitivity analysis of this method using similar size distributions from fine-grained turbidites showed no significant error when up to 10% noise was introduced into the DIGS data [13], which is equal to the expected precision of the Coulter Counter [21]. A nearest neighbour algorithm was used in MapInfo's Vertical Mapper V3.0 to contour values of floc limit in the Bay.

Organic concentration and trace metal concentrations were determined for the 1999 surficial sediment samples, and cores C1 and C2 from Lime Kiln Bay. Organic material values were calculated from the loss on ignition at 280 °C and 520 °C. Total organic material is the sum of the values from the two ignitions. Trace metal analysis was carried out on the surficial sediment subsamples collected in 1999, and the core samples from 2000 and 2002. Samples were dried and sieved at 180 μm, then digested with HNO_3/HCl/HF and analysed by inductively coupled plasma mass spectroscopy (ICPMS). Results of the trace metal analysis are discussed in other chapters of this volume [22, 23]. Lastly, ^{210}Pb and ^{137}Cs dating was carried out on the two cores collected in Lime Kiln Bay and results are reported in Smith et al. [23].

3 Results

The wide variation in bottom stress in the macro-tidal environment of the Western Isles region is reflected in the DIGS. Bottom type varies from high energy, well sorted sands in regions where tidal current and wave stress are at a maximum, such as in the entrance to Letang Harbour, to areas dominated by floc settling such as Lime Kiln Bay [17]. Contour plots of floc limit from 1990 identify depositional areas, with higher values representing areas dominated by floc settling (Fig. 4). In regions of high bottom stress the grain size distributions are dominated by coarse single grains and floc limit is low. Post-depositional reworking of the sediment in high-stress areas can decrease floc limit.

In 1990, floc limits in areas dominated by fine-grained sediment were similar. With the exception of western side of Lime Kiln Bay, the DIGS was characterized by a well-defined floc tail and easily identifiable change in slope from the floc tail to the single grain peak [17]. The DIGS for samples from the western side of Lime Kiln Bay did not exhibit a clear change in slope between the floc and single grain portions of the distribution, indicating that the area was already receiving a high flux of flocculated sediment. Samples from this part of Lime Kiln Bay all showed an increase in floc limit between 1990 and 1999. Floc limit at these stations and at two others in Bliss Harbour had increased by a factor of up to 2, with a maximum d_f of 18 μm (Fig. 5).

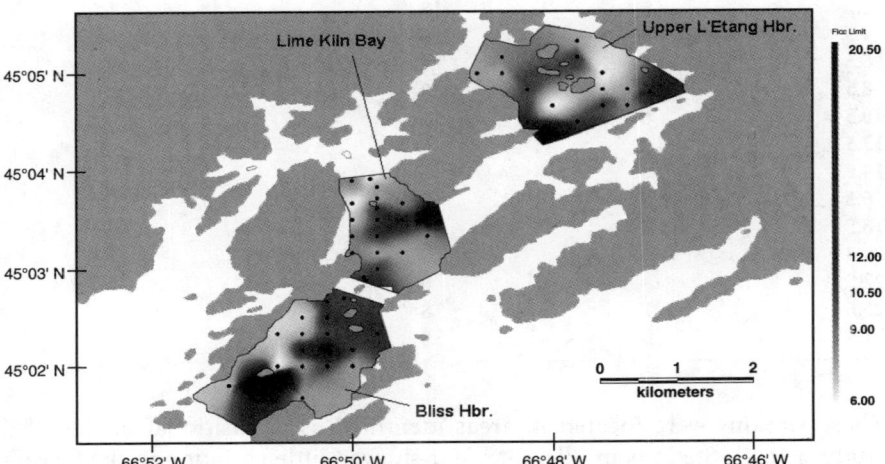

Fig. 4 Contour map of floc limit (d_f) calculated from bottom sediment disaggregated inorganic grain size (DIGS) collected in the Letang area of Southwest New Brunswick in 1990. High floc limit indicates floc-dominated depositional zones

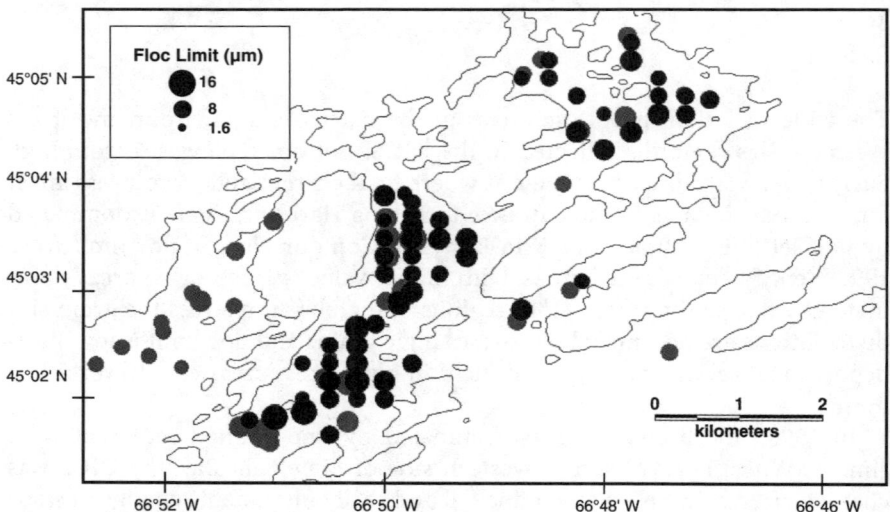

Fig. 5 Floc limit (d_f) for surficial sediments in Letang Inlet area for 1990 (*black*) and 1999 (*grey*). Floc limit has doubled in the depositional areas in Lime Kiln Bay

Table 1 Values of floc limit (d_f) calculated for sediment subsamples from cores collected in Lime Kiln Bay in 2002 (*C1*, *C2*) and Back Bay in 1994 (*C3*)

Core Depth (cm)	C1 d_f (μm)	C2 d_f (μm)	C3 d_f (μm)
0.5	7	16	16
2.5	7	18	14
4.5	9	16	9
6.5	8	21	9
8.5	6	18	11
10.5	6	11	8
12.5	5	6	6
14.5	6	6	9
16.5	6	5	7
18.5	7	7	
20.5	8	5	
22.5	14	6	
25.6	7	8	
31.5	6	5	

These stations were located in areas identified as depositional in the 1990 study (Fig. 4). Stations in Bliss Harbour showed little variation in the 9 years between sampling. The increase in floc limit found in Lime Kiln Bay was of the same order as that observed in the core from Back Bay [14].

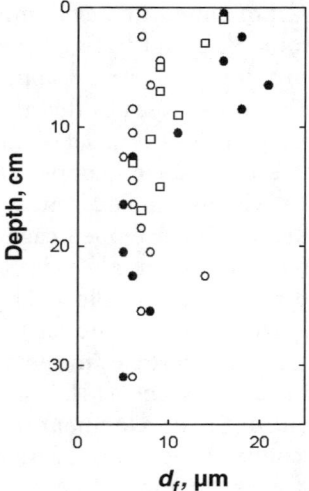

Fig. 6 Plot of floc limit (d_f) for cores collected in Lime Kiln Bay (*solid circle* C1, *open circle* C2,) and Back Bay (*open square* C3)

DIGS from the cores collected at station C2 in Lime Kiln Bay showed an increase in floc limit from a background value of 7 μm at 16.5 cm depth to a maximum 29 μm at 2.5 cm depth (Table 1, Fig. 6). The core collected at C1 showed no systematic change in floc limit with depth. Increasing floc limit up-core in core C2 was associated with an increase in organic material, zinc, and copper in the sediment. The change in floc limit at C2 was of the same order as that found in Back Bay in 1994 (C3) where floc limit increased up-core from a background value of 6 μm to a maximum of 16 μm (Table 1, Fig. 6). The difference in the depth in the core at which floc limit starts to increase likely reflects the 6-year interval between sampling dates, as sediment accumulation would be temporally offset. This increase in floc limit in the fine-grained sediments in Letang Inlet indicates that the fraction of material in suspension that is found in flocs has increased.

4
Discussion

The change in floc limit in the surficial sediment samples collected in Letang Inlet demonstrates that fine-grained sediment deposition in the region changed within the 9-year interval between surveys. Analysis of the cores collected in Lime Kiln Bay shows that the onset of changes in floc limit occurred at the same time as changes in the accumulation rate of organic material in the core collected in the part of Lime Kiln Bay dominated by floc settling. Deposition rates obtained from ^{210}Pb and ^{137}Cs at site C2 increase from on the order of 0.2 cm year^{-1} to around 0.4 cm year^{-1} approximately 20 years before sampling [23]. Dating of the cores indicates that the change in accumulation rate occurred about the same time as the onset of intense cage culture in Lime Kiln Bay [23]. Organic material, zinc and copper all show a similar increase over time [23]. The observed changes in floc limit and sediment accumulation rate support the hypothesis that intense open cage culture of salmonids can change the aggregation dynamics in an inlet with the result being increased deposition of fine-grained sediment [14]. Flocculation is a controlling factor in the transport of metals and organic material to the seabed, so increased floc deposition is associated with increases in the deposition of both organic material and surface active contaminants [14].

Three major factors control the development of a flocculated suspension: (1) particle number or concentration, (2) particle adhesion efficiency or stickiness and (3) particle break-up, most often due to an applied shear. Open cage aquaculture releases large amounts of small organic particles to the environment. The bacterial degradation of this organic material both in the water column and on the bottom provides a pool of small sticky particles that can accelerate floc formation. The production of polysaccharides and other long chain molecules and fibrils during bacterial breakdown of organic material have been shown to enhance flocculation and increase the uptake of several different metals onto particle surfaces and hence into the matrix of aggregates [24–26].

Increased aggregation as a result of higher particle concentration and particle stickiness therefore moves more particles into flocs, which increases the depositional flux of fine-grained inorganic sediment, organic material, and surface active contaminants. The result of increased flocculation in suspension, as observed in bottom sediment size distribution, is an increase in floc limit. Levels of trace metals in the surficial sediment in the Letang area were, for the most part, within expected values based on lithium normalization. Elevated levels of cadmium, copper, molybdenum, uranium and zinc were found in the 1999 surficial sediments [22]. Zinc and copper have been shown to be associated with open cage salmon culture and are thought to be introduced to the water through the feed (zinc) and anti-fouling treatment of the nets and other structures (copper) [22]. Intermittent high concentrations of zinc have also been reported from the surficial sediments on tidal flats in the vicinity of aquaculture operations in this area [27]. Core C2, collected in Lime Kiln Bay, showed elevated levels for the same metals as those found in the surficial sediment samples [23]. Increased organic matter deposition and the elevated levels of cadmium, copper, molybdenum, uranium and zinc were correlated with changes in the accumulation rates of ^{210}Pb and ^{137}Cs [23]. The corresponding increases in metal, organic material and accumulation rate all suggest that the deposition of flocculated material has increased within the same timeframe as the development of aquaculture in Lime Kiln Bay.

Less obvious, but perhaps more significant, is the possible impact on the benthos due to increased fine particle deposition. Higher concentrations of material in suspension have been shown to accelerate the deposition of sediment [11, 28]. The result of this accelerated clearance will be a greater flux of sediment and organic material to the seabed. In an area such as the Western Isles region of New Brunswick, both tidal and storm energy can lead to the intermittent removal of fine-grained sediment from the bed. In high energy areas, floc limit can be ambiguous because post-depositional reworking can destroy the link between floc packaging in suspension at the time of deposition and grain size in the seabed. However, high concentrations of flocs could exist on the bottom for biologically significant periods of time in these regions.

Studies of benthic macrofauna in the Letang area [29, 30] have shown that the benthic macro-faunal species composition in the Letang area has changed from one dominated by suspension feeding organisms to deposit feeders. Suspension feeding species, such as the amphipod *Leptocheirus pingus*, have been extirpated from Lime Kiln Bay and Scotch Bay in the period 1975–2001. The loss of suspension-feeding organisms could be the result of interference with filtration or burrow infill. Anecdotal evidence from the area suggests that the macro-flora community has also changed. The increase in floc limit found in the sediments points to greater flocculation in the water column, which in turn has led to greater vertical flux of fine-grained sediment and organic material. The resulting increase in sediment in the bottom boundary layer would have an impact on the species composition. The loss of suspension feeders from a wide area suggests that muddier conditions can exist throughout the inlet. While this increased flux of sediment from the water column will not necessarily be reflected in an increase in floc limit in high-energy areas, the presence of higher concentrations of flocculated sediment could impact these regions as well. The net result of adding fine-grained, organic-rich particles to the water column is increased flocculation which can lead to a greater flux of mud into low energy areas.

5
Conclusion

The DIGS of surficial and core sediment samples in the Letang Region of New Brunswick show an increase in the floc limit over a 10–20 year period. This change suggests that the aggregation dynamics in the area have been altered in favour of floc formation as the result of open cage salmon aquaculture.

Presumably, increased particle concentration and the creation of bacterial degradation by-products that enhance particle stickiness through the formation of long chain polysaccharides and other macromolecules result in more rapid floc formation, and hence deposition. Increased floc limit is correlated with increased organic loading and excess zinc and copper, two metals associated with salmon aquaculture. A succession of the benthic macro-fauna to species tolerant of high sediment loading has also occurred over a similar time scale [29, 30]. While the exact mechanism for the change in flocculation dynamics can not be determined at this time, it is apparent that open cage salmon aquaculture is having an impact over an area greater than that previously reported.

Despite the absence of change in floc limit in some areas between 1990 and 1999, high concentrations of flocculated sediment may promote the accumulation of floc deposited material over seasonal and/or tidal time scales. The result is a potential impact on benthic habitat due to short term, localized sediment loading that could not be identified in this study. The change to sediment deposition reported here is a far-field effect and is not restricted to the cage footprint. Aquaculture sites located in erosional areas, while having low impact on the benthos directly below the cages, can contribute significant amounts of fine particulate material to the environment. In enclosed bays this could lead to a net flux of both inorganic sediment and organic material into the inlet, increasing sediment concentration in the benthic boundary layer and leading to the accelerated deposition of mud.

Acknowledgements Over the years there have been several different vessels without which this work would have been impossible. Thanks are extended to all their crews. Andrew Stewart helped with the collection and analysis of the samples in 1999. The insightful comments of Paul Hill and two anonymous reviewers are very much appreciated.

References

1. Findlay RH, Watling L (1997) Mar Ecol Prog Ser 155:147
2. Gowen RJ, Smyth D, Silvert W (1994) Can Tech Rep Fish Aquat Sci 1949:19
3. Hevia MG, Rosenthal H, Gowen RJ (1996) J Appl Ichthyol 12:71
4. Bagnold RA, Barndorff-Nielsen OE (1980) Sedimentology 27:199
5. Bennett JG (1936) J Ind Fuel 19:22
6. Kranck K, Milligan TG (1985) Geomar Lett 5:61
7. Kranck K, Smith PC, Milligan TG (1996) Sedimentology 43:589
8. Kranck K, Smith PC, Milligan TG (1996) Sedimentology 43:597
9. Jackson GA (1995) Deep Sea Res II 42:159
10. Hill PS (1996) Deep Sea Res I 43:679
11. Kranck K (1980) Can J Earth Sci 17:1517
12. Hill PS, Milligan TG, Geyer WR (2000) Cont Shelf Res 20:2095
13. Curran KJ, Hill PS, Schell TM, Milligan TG, Piper DJW (2004) Sedimentology 51:1
14. Milligan TG, Loring DH (1997) Water Air Soil Pollut 99:33

15. Muschenheim DK, Kepkay PE, Kranck K (1989) Neth J Sea Res 23:283
16. Santschi PH, Balnois E, Wilkinson KJ, Zhang J, Buffle J (1998) Limnol Ocean 43:896
17. Milligan TG (1994) Can Tech Rep Hydrogr Ocean Sci 156
18. Hargrave BT, Phillips GA, Doucette LI, White MJ, Milligan TG, Wildish DJ, Cranston RE (1995) Can Tech Rep Fish Aquat Sci 2062
19. Milligan TG, Kranck K (1991) In: Theory, Methods and Applications of Particle Size Analysis;Syvitski JP (ed). Cambridge University Press, NY p 109
20. Kranck K, Milligan TG (1991) In: Theory, Methods and Applications of Particle Size Analysis, Syvitski JP (ed). Cambridge University Press, NY p 332
21. McCave IN, Jarvis J (1973) Sedimentology 20:305
22. Yeats PA, Milligan TG, Sutherland TF, Robinson SMC, Smith J, Lawton P, Levings CD (2005) Lithium normalized zinc and copper concentrations in sediments as measures of trace metal enrichment due to salmon aquaculture (in this volume). Springer, Berlin Heidelberg New York
23. Smith JN, Yeats PA, Milligan TG (2005) Geochronologies for fish farm contaminants in sediments from Limekiln Bay in the Bay of Fundy (in this volume). Springer, Berlin Heidelberg New York
24. Cowen JP, Bruland KW (1985) Deep Sea Res 32:253
25. Leppard GG (1997) Sci Total Environ 165:103
26. Quigley MS, Santschi PH, Hung CC, Guo L, Homeyman BD (2002) Limnol Ocean 47:367
27. Robinson SMC, Auffrey LM, Barbeau MA (2005) Far-field impacts of eutrophication on the intertidal zone in the Bay of Fundy with emphasis on the soft-shell clam, Mya arenaria (in this volume). Springer, Berlin Heidelberg New York
28. Milligan TG, Hill PS (1998) J Sea Res 39:227
29. Pohle G, Frost B, Findlay R (2001) ICES J Mar Sci 58:417
30. Wildish DJ, Pohle GB (2005) Benthic macrofaunal changes resulting from mariculture (in this volume). Springer, Berlin Heidelberg New York

Far-Field Impacts of Eutrophication on the Intertidal Zone in the Bay of Fundy, Canada with Emphasis on the Soft-Shell Clam, *Mya arenaria*

S. M. C. Robinson[1] (✉) · L. M. Auffrey[2] · M. A. Barbeau[3]

[1]Department of Fisheries & Oceans, Biological Station, 531 Brandy Cove Road, St. Andrews, New Brunswick, E5B 2L9, Canada
robinsonsm@mar.dfo-mpo.gc.ca

[2]210 Herve Street, Dieppe, New Brunswick, E1A 6Y7, Canada
lauffrey-arsenault@rogers.com

[3]Department of Biology, University of New Brunswick, Bag Service #45111, Fredericton, New Brunswick, E3B 6E1, Canada
mbarbeau@unb.ca

1	Introduction	254
1.1	Defining the Problem	254
1.2	Project Development	256
2	Objectives	256
3	Species Descriptions	257
4	Study Site	258
5	Results and Discussion	258
5.1	Could Nutrients from a Salmon Farm Reach the Intertidal Zone?	259
5.2	Is There Evidence that Salmon Farms Could Be Implicated in the Formation of Intertidal Algal Mats?	260
5.3	How Extensive Are the Algal Mats?	262
5.4	Do the Algal Mats Have an Impact on Clams?	263
5.4.1	Recruitment	263
5.4.2	Production	264
5.4.3	Reproduction	266
5.4.4	Behaviour	267
6	Future Implications	268
6.1	Is There a Cost to Society from the Algae?	268
6.2	Is There Something We Can Do with the Algae?	270
6.3	What Have We Learned About Designing a Field Study on the Effects of Algal Mats?	271
7	Conclusions	272
	References	272

Abstract A study was conducted in the Bay of Fundy, Canada in 2000–2003 to examine the effects of *Ulva*-dominated algal mats on the population dynamics of the soft-shell clam (*Mya arenaria*) and the possible role of Atlantic salmon farming in this interaction. A far-field linkage of the salmon farms to the intertidal zone was determined by the use of zinc : lithium tracers. A combination of laboratory and field-based experiments were conducted to assess the effects of algal mats on the recruitment, production (growth and survival) and behaviour of the clams. The results indicated that increased zinc concentrations were found in intertidal sediments located > 1 km from the nearest salmon site. This implies that if the fine particulates from the salmon site were being transported that distance, then the dissolved fraction could also travel that far, providing it was not absorbed by other nutrient sinks along the way. The algal mats were found to negatively affect clam recruitment and behaviour. The experimental results were not as clear for negative effects on biological production, but growth was found to be lower than expected and survival on the beach was low. The economic cost of the eutrophication to the beaches might be substantial (estimated > 100 000 $CAD per clam beach). One solution would be an integrated culture philosophy where additional crops are intentionally grown and harvested so that a potential liability could become an asset to the marine coastal economy.

Keywords Mya arenaria · Ulva-Enteromorpha · Algal mats · Salmon aquaculture · Bay of Fundy

1 Introduction

1.1 Defining the Problem

There is a growing recognition, on a global basis, that the fresh waters and near-shore marine coastal zone of most industrialized countries have been undergoing a period of increased enrichment of inorganic and organic nutrients over the last several decades, generally known as eutrophication [1, 2]. Eutrophication, in its most basic form, can be defined as "an increase in the rate of supply of organic matter to an ecosystem" [3]. The typical increases in anthropogenic activities near coastal zones (e.g. housing developments, agriculture, golf courses, food processing plants, fossil-fuel combustion) are directly correlated with increased nutrient levels, such as nitrogen and phosphorus, in coastal waters [2, 4]. This addition of nitrogen and phosphorus from a large suite of anthropogenic activities has resulted in changes to the marine coastal ecosystems to an extent that we are just beginning to understand. Fresh-water models of eutrophication do not seem to apply directly to marine systems and we are discovering that there are a suite of biological and physical filters that can control and modify the effects of the inorganic nutrients [1]. While many communities are now controlling much of the organic loading of municipal waste, far fewer municipalities are also treating the inorganic fraction through bacterial denitrification processes. As a result, the marine loading of nitrogen and phosphorus is increasing in direct proportion to the coastal human population growth. A significant source of terrestrial nutrients comes from the agricultural industry's reliance on chemical-based fertilizers that continue to increase in use as well as drop in price [3]. The impact of nutrient enrichment depends on the capacity of the environment to receive and assimilate these wastes [5–7].

The addition of nutrients to the coastal zone is not always from terrestrial sources. As farming in the marine environment has evolved in the form of fish and shellfish cultivation, there have been additional inputs of carbon, nitrogen and phosphorus from aquaculture. While arguably, there has been a decrease in wild stocks of fish and therefore a drop in their contribution to the local nitrogen pool, the sheer concentration of some caged fish in aquaculture operations is capable of increasing the local concentrations of nitrogen several times higher than ambient levels [8]. The cultured Atlantic salmon (*Salmo salar*) industry in the Bay of Fundy was estimated to produce 1225 tonnes of nitrogen per annum in 2001 [6, 9].

A common biological response to additional nutrient loading in many marine locations is an increase in the ephemeral green algal populations that are capable of exploiting the inorganic nutrient surplus. In the intertidal and near-subtidal areas, this increase in algal biomass is often reflected by the formation of dense green macro-algal mats composed of mainly *Ulva* sp. formerly *Enteromorpha* sp. [10] and *Cladophora* sp. [11, 12]. The growth of these seaweed mats has been observed to increase over time, particularly in the North Sea, as a result of escalating nutrients through anthropogenic inputs [13]. Formation of large mats of *Ulva* species develop within a few weeks as the plant grows from the sediment in the spring. The algae develop through the production of spores, which results in germlings that grow into the adult plants. The germlings have been shown in laboratory studies to be able to survive for at least 10 months at both 5 and 15 °C and can continue growing after 8 months in the dark [13].

The degree to which intertidal algae present a problem to the intertidal ecosystem likely depends on the overall scale and biomass involved. A study looking at moderate levels of drifting algal mats, found that they had relatively little effect on benthic diatom and microbial populations [14]. Although some light limitation was found under the algae, the diatom-bacterial colonies seemed to compensate and remain productive allowing for coexistence with the drift algae. There was no oxygen depletion problem, even at night, which prevented redox-related nutrient outflux from the sediment.

However, higher levels of *Ulva*-dominated algal mats have been shown to create serious problems in the culture of the carpet shell clam, *Tapes decussatus*, in Venice Lagoon, Italy [15]. The appearance of large amounts of *Ulva rigida* and *Gracilaria* sp. in the summer caused the clams to suffocate. The algae had to be physically raked off, at least weekly and at the expense of the farming operation, in order to maintain the culture of clams. If the algae were not removed, decomposition of the algae created frequent anoxic events [16]. To combat this, decision support systems with complex models were developed by the industry to provide advice for harvesting strategies for the algae. The model suggested that harvesting should begin as soon as the algae reached certain threshold values. In another study, seasonal blooms of *Ulva expansa* in California were shown to adversely affect the growth rates of *Macoma nasuta* [17]. This experimental study showed that clams in plots that were cleared of algae had significantly higher growth rates in comparison to uncleared plots. The author concluded that seasonal algal blooms were capable of producing discrete patterns of infaunal distribution in intertidal sedimentary habitats. Finally, in a study in Maine on the interactions of soft-shell clams (*Mya arenaria*) and algal mats, the results indicated that the clams were more likely to die under the algal mats presumably because their thick siphons could not penetrate the algal cover [18].

1.2
Project Development

Salmon farming in the southwestern portion of the Bay of Fundy developed in the mid-1980s. Over the last two decades, the industry has grown continuously and, in 2002, produced 45 000 t of Atlantic salmon with an estimated worth of $ 270 million. [19]. The economic value to the local area is substantial, and it is estimated that 25% of the jobs in the local area (Charlotte County) are associated with this industry. As the production has expanded, the interaction of this industry with other species has increasingly been of concern to different groups (i.e. environmental, fishery and tourism). It is often thought that there are negative environmental consequences due to culture of salmon in high densities. From the literature on nutrients and macro-algal responses, it appears that a potential exists for a far-field link between nutrients generated from salmon farms in the Bay of Fundy and algal-mediated impacts on local intertidal fauna. One of the specific questions is how the stocks of the soft-shell clam (*Mya arenaria*) will respond to increased eutrophication. This species is commercially harvested and is an important income source for many local inhabitants.

2
Objectives

The goal of this study was to examine the potential far-field effect of salmon culture on the biological interactions between the green algal mats and the commercially harvested soft-shell clam. Specifically, our objectives were to test for negative interactions on recruitment, growth and behaviour of the algal mats on soft-shell clams. Secondly, if negative interactions were found, an analysis of possible solutions would be done.

3
Species Descriptions

Soft-shell clams (*Mya arenaria*) (Fig. 1), also known as "steamers", are circumboreal. They are distributed along the North Atlantic coast of North America from Labrador to South Carolina, the Pacific coast from Alaska to California, and along the European Atlantic coastline [20–22]. The bivalve inhabits muddy to gravel substrates from the intertidal zone down to 200 m in depth, but are most abundant in the intertidal and shallow subtidal areas [21, 22]. They can live up to 12 years and attain shell lengths of approximately 170 mm [22]. The dioecious adult soft-shell clams spawn in early summer, and the resulting larvae remain in the water column from 2 to 4 weeks, depending on the water temperature, feeding on phytoplankton [21]. Following the planktonic larval stage, the pediveligers settle on the substrate, burrow in the sediment and permanently inhabit that area unless conditions deteriorate or they are washed away by bedload transport [23]. As soft-shell clams grow, their larger siphon allows them to burrow deeper into the sediment (up to 300 mm) and avoid predators such as green crabs (*Carcinus maenas*) and

Fig. 1 Soft-shell clams (*Mya arenaria*). Shell lengths of specimens range from 20–60 mm. The dark colour is due to staining by hydrogen sulfide

moonsnails (*Euspira* sp.). Soft-shell clams are suspension-feeders, and their filter-feeding activity has been associated with the control of phytoplankton blooms linked to eutrophication [24]. Their large abundance in the intertidal zone has made them a target species for fishing operations in many countries [22].

The green Chlorophycophyte algae known as "sea lettuce" and "link confetti" (*Ulva* species) are composed of over a dozen species on the Atlantic coast of North America. Recently, the genus *Enteromorpha* was shown to be synonymous with the genus *Ulva* through analysis of nuclear ribosomal, internally transcribed spacer DNA and the genus was reduced to *Ulva* as it was the older name [10]. The alga is bright green to yellowish green in colour and is tubular or flat in appearance. The body of the plant can grow to 1 m in length and it may occur in clumps or singly. It ranges from Ellesmere Island to Florida and is often seen in tide pools, growing on wharfs and pilings, ships hulls and in mats on intertidal areas [25–27]. The genus is worldwide in distribution and is found in brackish water to high-salinity seawater. It is reportedly eaten as a vegetable in Hawaii, China and the Philippines [25]. In some cases, it can have protein levels up to 20% by weight [28]. In some intertidal areas, it has been observed to form dense mats several centimetres thick over sections of a beach and these can later roll up into algal ropes due to tidal and wind action [27] at the end of the growing season in October.

4
Study Site

Three study sites were used to follow the interactions of salmon sites, algal mats and clam populations: Clam Cove on Deer Island, Blockhouse Beach in St. Andrews and Hinds Bay at the head of Limekiln Bay (Fig. 2). Clam Cove has a salmon site within 150 m from the intertidal zone while at Hinds Bay, the nearest salmon site is approximately 1200 m away. The Blockhouse beach is not located near any salmon farms, but does receive nutrient enrichment from the urban runoff of the town of St. Andrews (population ca. 3000) including an adjacent golf course (Fig. 2).

5
Results and Discussion

This chapter presents a series of questions that examines new and historic data from the study area and reviews the scientific literature for evidence of interactions between clams, *Ulva* mats and salmon farms. The limited discussion on nutrient transport focuses primarily on nitrogen and its related

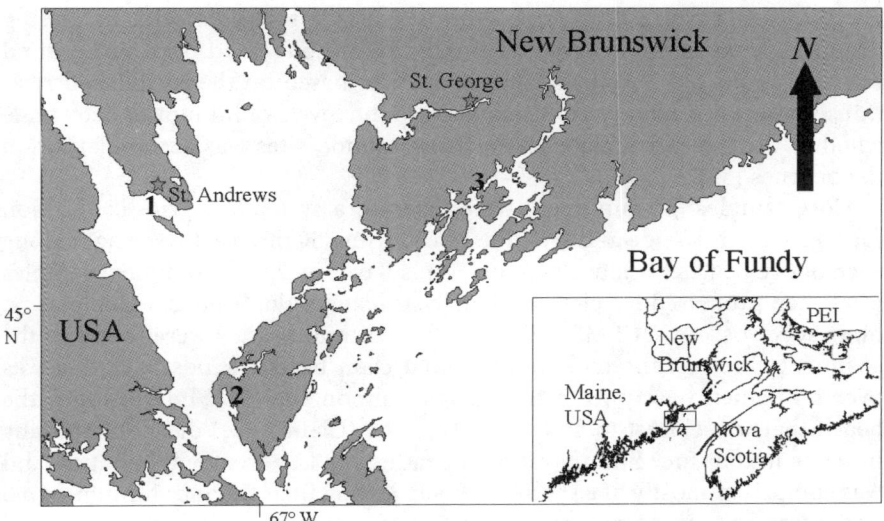

Fig. 2 Map of southwestern New Brunswick, Canada, showing the location of the study sites. 1 Blockhouse Beach, 2 Clam Cove on Deer Island, 3 Hinds Bay

forms (ammonia, nitrite, nitrate), but it should be emphasized that phosphorus and carbon are also released. We focus specifically on nitrogen because it is thought to be the limiting factor in marine algal blooms [29].

5.1
Could Nutrients from a Salmon Farm Reach the Intertidal Zone?

The task of directly determining how much nitrogen reaches the intertidal zone is difficult due to a number of confounding issues. As inorganic nitrogen is released from a salmon farm and is transported by currents to intertidal areas, there are a number of potential biological sinks that could absorb it such as benthic diatoms living on the superstructure of the salmon site, phytoplankton, some marine bacteria and various macrophytes in the subtidal and intertidal areas. Some nitrogen may also be lost to geochemical processes as well. Without a label, such as a radioactive tag or identifiable isotope, it is almost impossible to determine whether or not a particular nitrogen molecule originated from a salmon site or some other source.

An alternative approach would be to use a passive and conservative tracer to try and establish a path from the salmon site to the beach. Since the salmon industry in the area and the local regulators were not supportive of the addition of dyes, etc., to the water column to track water flow around cultured animals, we used ingredients in the salmon feed as a tracer. Zinc is added to the fish diets as a trace metal supplement in order to promote healthier growth in the animals. Observations in the study area have shown that there are decreasing concentrations of zinc in surface sediment with increasing distances from salmon cages and that levels are increased over background concentrations [30]. We used zinc as a tracer based on the established relationship between dietary zinc and background levels of lithium to determine whether the fine particulate matter from salmon sites was accumulating on the beaches [31].

Core samples (50 mm deep) were taken in a systematic grid-like fashion (at high-tide, mid-tide and low-tide levels) using 50 mL plastic syringes along each of two beaches, Clam Cove and Hinds Bay (Fig. 2). The sediment samples were then processed for elemental analysis using inductively coupled plasma mass spectroscopy (ICPMS). Elevated zinc : lithium ratios were found in the sediment cores on the most depositional clam beach (Hinds Bay) that was over a kilometre away from the nearest salmon site [31]. Interestingly, the beach that was closest to an aquaculture site (Clam Cove) did not show any increase in the zinc : lithium ratio. The latter site was erosional in nature and was comprised mostly of a sandy-mud suggesting that the zinc : lithium signal was found in the fine particulate fraction (i.e. the wave action had removed the fine particulates from the beach). Although sediment grain size was not measured on the core samples used for ICPMS at the time, previous subtidal sediment samples at the Clam Cove site indicated the modal grain size was 170 μm [32]. In contrast, the depositional beach (Hinds Bay) was comprised of a very soft mud which made walking difficult. Previous subtidal samples indicated the modal grain size was 15 μm [32].

Increased zinc : lithium ratios in the fine particulate fraction, which are indicative of particulate food material being transported to and deposited on beaches over 1 km away, support the assertion that the dissolved fraction could also be transported at least this far provided it was not absorbed by biological sinks along the way (i.e. phytoplankton, macro-algae, etc.). Also, if the dissolved fraction could travel thousands of metres to Hinds Bay, then we can also assume that it would also be transported 100–200 m to the erosional beach of Clam Cove. While this is not definitive proof that excess ammonia, nitrite or nitrate from the salmon site would reach a distant beach, the observation supports the argument that the transport is quite likely.

5.2
Is There Evidence that Salmon Farms Could Be Implicated in the Formation of Intertidal Algal Mats?

The beach at Clam Cove is a traditional harvesting area for the clam diggers in the area and has a spatial area of approximately 20 ha. It is an enclosed cove with relatively steep slopes leading down to the beach. The growth of the *Ulva* sp. mats are a relatively new phenomenon, based on personal observation and historical photographs from the coastal aerial photography program of the province of New Brunswick. Aerial photographs for the cove were available for the years 1984, 1994, 1999, 2000 and 2001. Examination of these photographs and observations since 1988 (Robinson, personal observation) indicated virtually no algal mats present until 1994 and a substantial increase to over 30% algal coverage by June 1999. Subsequent photographs in July 2000 showed an increase in coverage to over 45% with a decrease in June 2001 to approximately 35%. The timing of the appearance of the algal mats was concurrent with the development of the salmon farm that was located at the mouth of this cove. To test the hypothesis that the bloom was a result of an increase in human habitation, we obtained the census records from Statistics Canada to compare the increase in algal blooms with possible increases in human population on Deer Island during this time. The results showed a slowly decreasing trend in human habitation during the late 1970s to 2001, suggesting that increasing housing development and associated domestic sewage and runoff was not the cause of the algal blooms (Fig. 3). The fact that the beach remained open to commercial clam harvesting during this time means that the faecal counts were low and that any septic systems were not compromised and leaking nutrients into the cove.

The Clam Cove situation is unique as the salmon farm is in close proximity to a very enclosed beach. With a semi-diurnal tidal range of approximately 8 m, there can be no doubt that nutrient-rich water from the salmon farm would reach the beach. The lack of human habitation and a small watershed for the inlet also makes this a special situation. The nearest other point source for additional nutrient input that could cause an increase of the macro-algal blooms would be a fish processing plant slightly further south (1 km). This has been in operation for over 50 years and it is unlikely that this would be making much of a contribution to the recent algal blooms over the past decade.

On more open beaches, such as the Hinds Bay site, other non-point sources of nitrogen input can be present in addition to single-point sources. Most clam beaches have streams that drain the foreshore. The streams can be seasonal, where their contents comprise snowmelt in excess of rainfall from storm drain systems, or they can be a permanent feature of the beach where they contain the runoff water from upland marshes and other wetlands. For the permanent streams, the importance of the nitrogen input to the beach is a reflection of the loading they receive prior to entering the intertidal zone and the volume of water involved. Nutrient samples taken from the streams entering the beaches in our study sites indicated that concentrations of nitrate in the streams were often 4–5 times higher at peak concentrations (ca. 20 μM) than that found in seawater at low tide (ca. 4 μM) [33]. Flow volumes were not measured during the study so it is impossible to determine the absolute load created by this nutrient source to the beach.

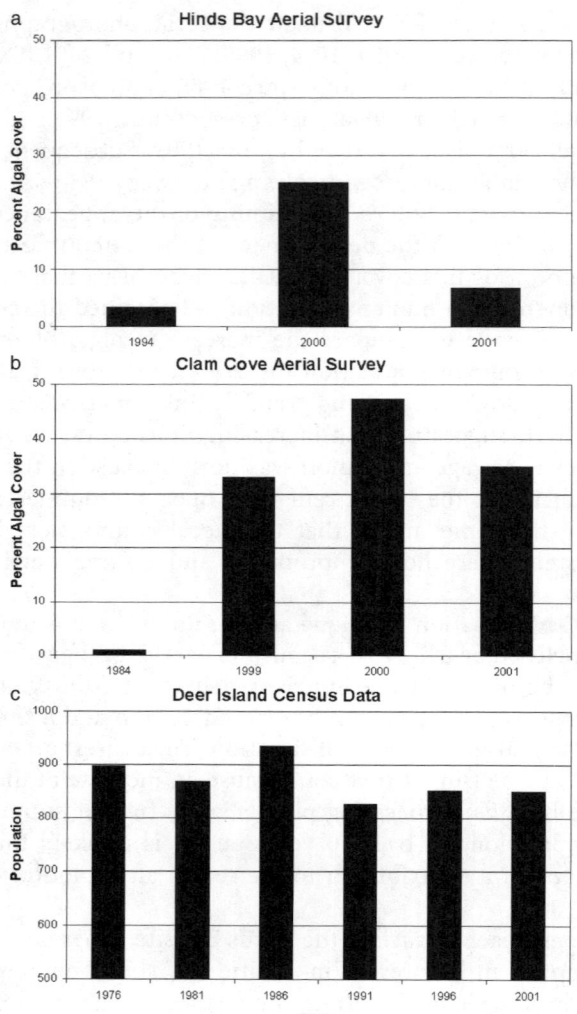

Fig. 3 Spatial coverage of two beaches (**a, b**) by macro-algal mats of *Ulva* based on analysis of aerial photographs. **c** Population census of Deer Island, New Brunswick, from 1976–2001 based on data from Statistics Canada

5.3
How Extensive Are the Algal Mats?

The significance of algal mats in the intertidal zone of the beaches is dependent on their spatial coverage and the biomass present. At the Clam Cove site, up to 40% of the 19.7 ha of the beach could be covered with *Ulva* sp.-dominated algal mats based on the data from aerial photographs. The biomass samples taken during this study in the intertidal zone showed that there could be up to $35\,g\,m^{-2}$ dry weight or 6.9 t dry weight of algae on the beach. Assuming an 8.5 : 1 wet weight to dry weight conversion (Robinson, unpublished data), this would equate to 58.6 t fresh weight of algal mats on this beach. In Hinds Bay, the area of the beach was approximately 3.7 ha of which a maximum of 25% was covered with algal mats in July 2000. With a mean algal dry mass of $30\,g\,m^{-2}$, 0.28 t dry weight of algae would occur on the beach. Using the same wet weight to dry weight conversion, this equates to a fresh weight algal biomass of 2.4 t. Analysis of the nutrient content of the algae revealed an average carbon concentration of approximately 20% on a dry weight basis. This implies that the standing stocks of carbon on the beaches represented by the opportunistic algal mats were 1.38 t and 0.06 t for Clam Cove and Hinds Bay beaches, respectively.

5.4
Do the Algal Mats Have an Impact on Clams?

5.4.1
Recruitment

Algal mats have been suggested to have negative impacts on the recruiting invertebrates. They can act as a direct barrier preventing metamorphosing larvae from settling or preventing drifting, early juveniles from reaching the sediment. The mats can also change the environmental sedimentary conditions underneath and cause anoxic conditions leading to death of the recruits [34]. In a study in July 1997 (Robinson, unpublished), *Ulva*-dominated alga mats were sampled on a sandy-mud beach (Blockhouse in St. Andrews, Fig. 1) for recently settled/drifting juveniles of the soft-shell clam, *Mya arenaria*. Using scissors, samples of the algal mat (1 m^2) were taken, placed in plastic bags and transported to the laboratory. The number of trapped soft-shell clam juveniles was counted and measured in each sample using a dissecting microscope. Mean densities of juveniles in the *Ulva* mat were 6044 ± 3044 individuals m^{-2} (mean $\pm$ 1 SD, $n = 5$) and the clams had a modal shell length of 2 mm. These results confirm that the *Ulva* sp. algal mats in this area were capable of capturing juvenile clams, although we cannot say with certainty what the fate of the entrapped juveniles were as we did not follow their progress. Small blue mussels (*Mytilus edulis*) were also observed in the mat at the time.

Our observations also corroborate speculations in an earlier study on the ability of algal mats to entwine juvenile *M. arenaria* [27] and to possibly remove them from the population. Juvenile *M. arenaria* have been demonstrated to move substantially over hundreds of metres due to bedload transport [23, 35]. During this movement, they would be quite likely to encounter algal mats. Movement of juvenile clams has been demonstrated in other species as well. Other studies have shown that the coquina clam (*Donax variabilis* and *D. parvula*) recruits into the subtidal zone and then subsequently migrates into the upper intertidal zone as they get larger in size [36]. Many early juvenile bivalves and some gastropods show periodicity in their drifting ability [37]. Some species are tied closely to lunar cycles (i.e. *Ensis directus, Macoma balthica, Cerasoderma edule*).

Research on intertidal algal mats in the Baltic Sea reported that, in the presence of algal mats, recruitment to sandy beaches of the dominant bivalve, *Macoma balthica*, was significantly reduced as was the survival of the polychaetes *Pygospio elegans* and *Manayunkia aestuarina* [34]. With the change in the sediment and habitat type, populations of the amphipod *Corphium volutator* tended to increase. Several species of gastropods have also been shown to recruit into algal mats. In Japan, the trochid snail *Conotalopia mustelina* has been reported to abundantly recruit into algal mats in the late summer [38].

In addition, drifting red algae have been shown to limit the recruitment of bivalves to the bottom. A field experiment [39] demonstrated that algal cover prevented any bivalve larvae (*Macoma balthica, Cardium glaucum* and *Mya arenaria*) from settling under algal cover compared to densities ranging from 1500 to 5500 individuals m^{-2} in control areas without algal mats. They suggested that the algal mats could be efficient inhibitors of larval settlement and were acting as larval filters. However, some studies have found that the protoplasmic detritus formed by the bacterial breakdown of *Ulva* can be used as a food source by some larvae of some infaunal animals (i.e. *Ruditapes phillipinarum*) [40], so the implications are not always negative.

Overall, it appears that algal mats tend to capture and retain juvenile bivalves. In Clam Cove on Deer Island, assuming that the same potential exists for capture (in terms of encounter rate and juvenile supply) as that found at the Blockhouse, 476×10^6 juveniles could potentially be captured in the *Ulva* sp. mats.

The second effect of algal mats on recruitment to clams is their ability to change the environmental conditions underneath, often to an anoxic state. This anoxia was usually found under the algal mats in our study sites and the production of hydrogen sulfide could be detected by odour.

5.4.2
Production

The production of a cohort of clams is a function of the cumulative growth and survival rates of the animals. Although many studies have examined the growth rate of soft-shell clams from widely varying geographic locations (for example [41, 42]), very few have considered it in relation to the occurrence of intertidal algal mats. The only exception was a study in New England [18] where shell lengths in natural populations of soft-shell clams were measured before and after an experiment on the effect of removal of algal mats. They found that clams under algal mats appeared to grow slower than areas clear of algae.

We conducted a mark-recapture experiment to assess the effect of *Ulva* algal mats on growth and survival of soft-shell clams. Eight to ten labelled clams (25.86 ± 3.04 mm shell length (mean $\pm$ SD)) were placed in plastic mesh hydroponic plant pots (300 mm $\times$ 300 mm $\times$ 300 mm) which were deployed flush with the sediment surface in Clam Cove, Blockhouse and Hinds Bay study areas. Pots were placed in both areas with and without algal cover ($n = 4$ pots). The clams were monitored for growth and survival at three periods from May to October 2001. Results showed that clam shell growth in pots under algal mats was not significantly different ($p > 0.05$, ANOVA) than clam growth in pots in areas clear of algae [33]. The growth of all the clams was relatively low (i.e. 2–4 mm in 6 months or about 7–15% in shell length). In comparison, the growth of clams of similar sizes on beaches throughout Nova Scotia and New Brunswick were reported to increase their shell length an average of 29$\pm$9% ($\pm$ SD) over 6 months [41, 43–46]. Survival in the pots from both the algal-covered and clear areas dropped continuously during the study to about 40% after 6 months. Clam survival was also not significantly different between treatments.

The implications of these results are that clam survival and growth on these beaches does not appear to be particularly high and that food does not appear to differ between treatments for those animals that survived. A study in Maine [47] reported that predation is more important in regulating clam populations than is food. However, another study in southern Maine inferred that soft-shell clam growth under algal mats was inhibited [18]. In Massachusetts, a positive correlation was observed between the levels of nitrogen released in the coastal waters and the growth of the soft-shell clam [48]. These authors concluded that the increased nitrogen levels were responsible for increasing the biomass of phytoplankton, which resulted in increased growth rates of the clams. The quality of the food, defined by C : N ratios was consistent among sites.

There are a number of possible explanations for the above observations. The first is that our experiment may have been confounded by factors that we did not anticipate. It is possible that small sample sizes, pot confinement, varying coverage time and biomass of algal mats, which were not consistently monitored, affected our findings. It is also possible that the change in the habitat by the algal mats in the mid-low intertidal zone extended beyond the actual borders of the mat itself. Predators have been reported to use the edge of the algal mat as cover, resulting in higher densities [49, 50]. Therefore, predation pressure on both groups of clams may have been similar. We noticed more green crabs during low tide in algal mats. The potential of more phytoplankton from increased nitrogen loading in the water from the salmon sites may not be realized in the intertidal zone if the nitrogen is being incorporated into the algal mats and other macroalgae. More research is required on the phytoplankton/seston dynamics in the intertidal zone in the presence and absence of algal mats to answer these questions.

5.4.3
Reproduction

Although the ultimate effect of algal mats on soft-shell clams may be the mortality of the affected individuals, there may also be other indirect effects as environmental conditions for the clams degrade. Several studies have shown that bivalves and other marine invertebrates are capable of changing their reproductive patterns depending on the relative predictability of their food supply [51, 52]. Studies have demonstrated that under nutritional stress, mussels produce fewer and smaller eggs with lower protein and lipid content [53]. Similar findings have also been reported in prawns where higher diet quality (i.e. essential fatty acids) increased fecundity [54]. Other studies have found that mussels decreased their fecundity in relation to declining water quality [55, 56]. Chemical pollutants have been shown to cause fecundity decreases in mussels [57] and in barnacles [58]. Disease agents have also been shown to decrease fecundity in soft-shell clams [59].

Therefore, because of the potential sub-lethal effects on the soft-shell clam populations, we investigated any reproductive impairment that might be associated with the *Ulva*-dominated algal mats. Samples of clams were taken in June during the spawning season at the Blockhouse Beach and Clam Cove (Fig. 2) and transported back to the laboratory where tissues were separated into reproductive (gonad and digestive complex) and somatic tissue (mantle, gills, foot, crystalline style). Tissues were weighed, dried at 80 °C for 48 h and then reweighed. The results indicated that there were no differences in the size-specific somatic tissues in clams that were from areas that were clear or covered with algae. However, at one site (Blockhouse Beach), the clams from the clear areas had significantly higher ($p < 0.05$) size-specific reproductive tissue weights (Fig. 4).

The trend of lower gonad mass for clams in the presence of algal mats suggests two possible scenarios: clams under algae (1) spawned earlier or (2) received less food and invested relatively more energy in survival than reproduction. Either could be detrimental to population sustainability of soft-shell clams. If clams under the algal mats were spawning out of synchrony with the local population, then problems with fertilization success and using other external cues to time the release of gametes would affect the recruitment success of the animals. In addition, if the impacted clams are producing proportionally less eggs than their non-impacted conspecifics on the beach, then the total larval supply produced by that population would be diminished, again possibly resulting in lowered recruitment rates. However,

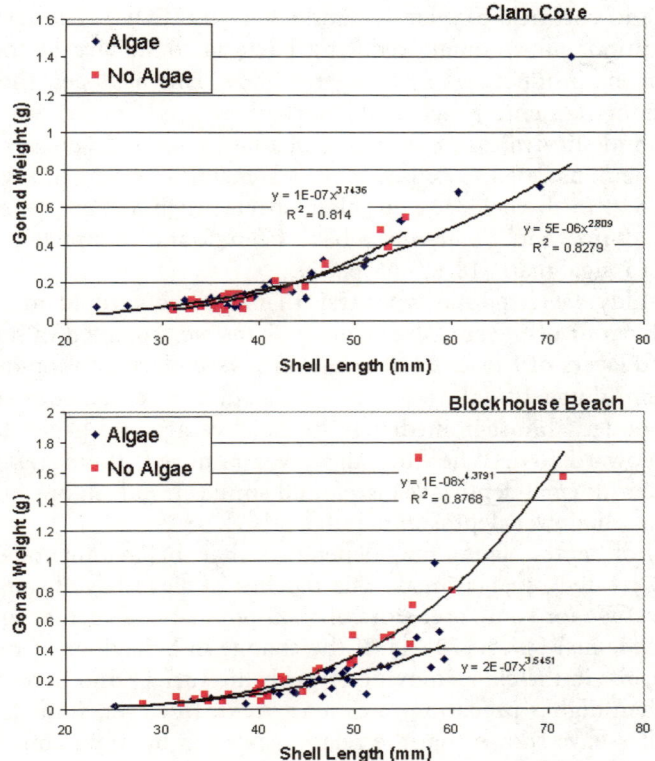

Fig. 4 Gonad weight of different sized clams at the Blockhouse Beach and Clam Cove in June 2001

only one of the two study sites showed this effect of a reduction in relative gonad weight and, therefore, more work needs to be done to clarify this observation.

Finally, in this study done in the mid-intertidal zone, the density of large soft-shell clams was generally found to be more abundant in areas without algal cover than in areas with algal mats [33]. The low abundance of adult clams combined with a potential lower gamete production suggests reduced gamete output for the population on the beach during spawning, as large individuals have more gametes than small clams [60].

5.4.4
Behaviour

One of the usual physical responses of the sediment in the intertidal and subtidal zones to accumulation of algal mats is the depletion of oxygen through biological and chemical oxygen demand (BOD and COD). The resulting hypoxic conditions allow sulfate-reducing bacteria to survive at the surface, resulting in the production of hydrogen sulfide. This has been shown to affect behaviour (i.e. emergence at the surface by the infauna), physiological condition, and survival of the benthic infauna [61–65]. Hypoxic conditions have been demonstrated to cause siphon extension in clams in several studies [61, 66, 67]. In the field, various clam species such as *Cerastoderma* spp., *Macoma balthica* and *M. arenaria* have been found completely unburied under macro-algal mats [18, 63–65, 68, 69].

In this study, we tested the hypothesis that algal mats would affect the burrowing behaviour of the soft-shell clams. Clams were planted in sediment in aquaria and layers of *Ulva* sp. (20 or 60 mm) were placed on top of the sediment. A semi-diurnal tidal cycle was simulated. The results showed that the clams responded almost immediately to the coverage and began to migrate vertically upward [69]. When the algae was removed, the clams burrowed down to their normal depth. We also found similar trends in burial depths in the field (i.e. shallower depths under algae) [69].

Survival of benthic fauna may depend on their behaviour under hypoxic conditions created by algal mats. The patches of algae have been shown to provide a refuge for some predators of clam populations such as green crabs (*Carcinus maenas*) [49, 50, 70, 71]. If the change in behaviour in response to hypoxia results in the clams moving towards the surface and becoming more exposed, then higher predation rates may occur. However, some laboratory-based studies have shown that the relationship can be more complex as the hypoxia can affect the predator's behaviour as well [67, 72, 73]. More research is needed to understand the relationship between hypoxia, algal mats, predation and the consequences for intertidal clam populations.

6
Future Implications

6.1
Is There a Cost to Society from the Algae?

The algae that covers a beach not only imparts an ecological cost to the intertidal infauna and epifauna, but also a cost to the local economy of a rural society. By displacing the soft-shell clams, one of the most dominant intertidal organisms in terms of biomass, the loss of a harvestable food/income source trickles down to the clam fisherman, creating a financial loss. The loss of the primary income to a clam digger generated from harvesting the beach then impacts on the local businesses where the money would have ultimately been spent.

To put this in perspective, we have used the data from Clam Cove as it is a beach with a long history of clam harvesting. The area of the intertidal zone in Clam Cove is approximately 19.7 ha. We will assume that a commercial density of legal-sized clams (44 mm) is 75 m^{-2}, that diggers are 75% efficient at harvesting and that each clam weighs an average of 15 g wet weight. The total weight of clams on the beach will be 222 t. All values are calculated in Canadian dollars. If we assume that a commercial value of these clams are $ 3.36 kg^{-1}, then the value of the crop of clams on the beach available to the digger (i.e. 75% of $ 746 365) would be $ 559 773. Aerial observations on the Clam Cove beach indicated that the percentage coverage of *Ulva* ranged from 30–40% in recent years. If we assume that 30% of the beach is covered with algae and that the clams are lost to the fishery (i.e. dead or unavailable for harvest), this implies that the direct economic cost of algae to the clam digger (30% of $ 559 773) would be $ 167 932. This is a significant loss of income and employment from this one beach for the region and represents the loss of 5–6 full-time jobs from the community. When this is extrapolated to other beaches with similar problems, the cost of algae can multiply quickly. If there were ten similar beaches in the area, this would represent an annual loss to the economy of over $ 1 million. Considering the often high unemployment in some rural communities, the effect on the economy would be significant.

There is, of course, some inherent variability around these estimates. The clam density within a beach is often quite patchy and therefore, numbers of clams may sometimes be less than 75 m^{-2} depending on recruitment, predation, suitability of the habitat or previous harvesting practices. However, densities of 50–75 clams m^{-2} have been observed on commercial clam beaches in the past ([74]; Robinson, personal observation). In addition, historical landing records gathered by the Department of Fisheries and Oceans Canada indicate that catches in the 1950s were almost ten times higher (4500 t vs. 500 t) than those recorded recently [75]. Therefore, it is likely that clam densities in the past were even higher than 75 m^{-2} on many beaches and likely more evenly distributed.

There is also the possibility that the algae may drift away before it severely impacts the infauna or it may not have the same coverage from year to year. While this may be true for low density blooms, heavy algal blooms seem to persist spatially in the same areas from year to year, presumably due to the physical circulation patterns of the water. The algal mats may also be self-fertilizing because when they die, the nutrients can remain in that area and fertilize the following year's crop [63, 76]. In addition, if a heavy bloom was present in the previous year, the chances are that the sediment would already be depopulated of much of the large infauna due to the adverse physical conditions caused by the algae.

6.2
Is There Something We Can Do with the Algae?

The feasibility of dealing with an algal bloom resulting from the nitrification of the coastal zone will centre on whether the algae are perceived as a potential asset or liability. In this study, the focus was on salmon farms and the potential impacts of their nutrient output. However, in most cases, the source of anthropogenically produced nitrogen stems from several sources, salmon farms being just one in certain coastal areas. Usually, it is difficult to partition the degree of responsibility among the different contributors and even more so to get them to acknowledge their role. Often it becomes a trade-off of economic values. How much does the clam industry generate versus how much do aquatic farming activities add to the economy? Or how much will it cost to fix a municipal sewage system?

Therefore, perhaps the solution to this dilemma is to look at the algae as an added crop that we are already paying to produce. In terms of salmon farming, the nitrogen being delivered into the water is an excretory product of the fish, which is being supported by the large amounts of salmon feed delivered to the farm site. Research shows that fish are only capable of retaining 25–40% of the nitrogen in the feed [77] within their tissues, depending on the diet formulation. The remainder is released in dissolved or particulate form. We are left then with a typical biological "leaky" system that must be made more efficient if we are to recapture this energy (i.e. nitrogen, phosphorus, carbon). The challenge is to do this in a way that could provide an additional source of revenue for either the originating source of the nutrients or perhaps some affiliated industry.

We will again use the Clam Cove study site (19.7 ha) as a possible example of this type of recycling scenario. From this study, we found that the average density of *Ulva* on the beach in the algal mats could be 0.04 kg m^{-2} dry weight. Assuming an 85% water content of the algae, this equates to 0.27 kg m^{-2} wet weight. If we assume the same coverage of the beach that we used for clams of 30% (see above), the total wet weight of *Ulva* on the beach would be 53 190 kg. If we assume that 75% of this can be harvested and sold for $ 0.10 kg^{-1} as a soil additive, the value of the algae on the beach to a harvester would be $ 3989. While this is not a great deal of money in itself for a stand-alone industry, when it is viewed in light of a more integrated system, the economic benefits with the algae, clams and salmon become more apparent. The roughly $ 4000 of algae would translate into 25 days employment of 8 h at $ 20 h^{-1}. If efficient harvesting systems are devised for the algae that are relatively benign to the other intertidal biota, then the economic return (i.e. number of days or a higher wage) from the algal harvesting may very well be taken on by the clam industry itself. Furthermore, if an efficient method could be found to separate the entrapped juvenile clams from the algae and then replanted, it would provide even more economic dividends. In southwestern

New Brunswick, as in many other areas of the world, this integrated system of marine food production is still in its infancy as technology and legislation are just beginning to deal with a system approach to management rather than single-species/issues models.

Harvesting of algae from the intertidal zone so that clam production can occur is already taking place in some parts of the world as an advanced approach to deal with the problems of eutrophication. In Venice Lagoon, Italy, the algae *Ulva rigida* and *Gracilaria* sp. are raked off weekly in order to sustain harvesting of the carpet shell clam, *Tapes decussates* [15, 16]. The feasibility of the operation seems apparent as the harvesting operation has invested in developing decision support systems with complex models to advise them when to harvest the algae.

6.3
What Have We Learned About Designing a Field Study on the Effects of Algal Mats?

From our experience, we recommend that sampling programs or experiments be conducted in two types of sites: impacted sites where algal mats occur and reference sites where algal mats are not present. Within each type of site (impacted versus reference), there should be replicate sites—a nested design [78]. Furthermore, within the impacted sites, areas covered with algae as well as areas clear of algae should be compared. Such a design with appropriate comparisons, i.e. (1) between algal-covered areas and clear areas at the impacted sites and (2) between clear areas at the impacted sites and reference sites, enables one to tease out algal mat effects from site effects [69]. For example, if significant differences in the response variable are found between algal-covered areas and clear areas at impacted sites, but not between clear areas at impacted sites and reference sites, then one has very strong evidence of an effect of algal mats. If significant differences are observed both between algal-covered areas and clear areas at impacted sites, and between clear areas at impacted sites and reference sites, then the design allows one to assess how much of the variability in the response variable is due to algal mats and to the type of site. If algal-covered and clear areas at impacted sites are not different, but clear areas at impacted sites are different than reference sites, then one only has evidence of an impact at the impacted sites that is not necessarily related to algal mats. One drawback of this design is that the sites are the experimental units to test for an effect of algal mat and type of site, and it may be difficult to get multiple replicate sites. However, one should strive for this design, since it would provide the strongest evidence for an effect of algal mats.

Many other considerations need to be taken into account that are relevant to most sampling and experiment studies. These include conducting a pilot study to assess the variability in the response variable [79]. With this information, one can do a-priori power analyses to determine the number of replicates needed to detect a certain effect size [80–82]. As well, a pilot study provides information on the spatial distribution of the study organism to determine the best sampling design (e.g. random, stratified random) [83].

7
Conclusions

The results of this study suggest that dissolved wastes such as nutrients released from point sources of enrichment, such as salmon farms, have the potential to reach the intertidal zone and affect the distribution and growth of the *Ulva*-dominated algal mats. The degree to which the salmon farms cause the development of the algal mats appears to be determined by local conditions such as proximity to the beach, the water circulation characteristics and the other nutrient sinks available (i.e. other macrophytes). The resulting algal mats have the ability to affect the population dynamics of the soft-shell clam (*Mya arenaria*) through negative effects on recruitment, growth and survival, and behaviour. In most cases, the effect will be negative with regard to production. The economic and social costs of this impact can be relatively large if it is left untreated and could be in the hundreds of thousands of dollars for a moderately size beach with a significant nutrient input. The loss of the water filtering capacity from the clams in the beach is also an important ecological consideration. One solution to the issue is better control of the nutrient flow from the source, not only salmon farms, but other producers as well. Better design of integrated systems may be one solution where recycling principles are used to grow additional crops to harvest the energy lost from the primary system.

Acknowledgements We would like to thank a number of people who assisted in this project: Phil Yeats, Barry Hargrave, Nicola Johnson, Melisa Wong, Thierry Chopin, Kevin LeBlanc and David Arsenault. The work was supported by the Canada Department of Fisheries and Oceans–Program for Environmental Studies for Sustainable Aquaculture (ESSA) and facilities at the St. Andrews Biological Station. Brian Beal and Chris Pearce provided valuable reviews of earlier drafts of the manuscript.

References

1. Cloern JE (2001) Mar Ecol Prog Ser 210:223
2. Nixon SW (1998) Sci Am 9:48
3. Nixon SW (1995) Ophelia 41:199
4. Rosenberg R (1985) Mar Pollut Bull 16:227
5. Strain PM, Yeats PA (1999) Mar Pollut Bull 38:1163
6. Strain PM, Hargrave BT (2005) (in this volume). Springer, Berlin Heidelberg New York

7. Sowles JW (2005) (in this volume). Springer, Berlin Heidelberg New York
8. Bugden JBC, Hargrave BT, Strain PM, Stewart AJ (2001) Can Tech Rep Fish Aquat Sci 2356:96
9. Chopin T, Bastarache S (2002) Bull Aquacult Assoc Canada 102:119
10. Hayden HS, Blomster J, Maggs CA, Silva PC, Stanhope MJ, Waaland JR (2003). Eur J Phycol 38:277
11. Reise K, Herre E, Sturm M (1989) Helgol Meeresunters 43:417
12. Lavery PS, Lukatelich RJ, McComb AJ (1991) Estuar Coast Shelf Sci 33:1
13. Schories D (1995) Population ecology and mass development of Enteromorpha spp. (Chlorophyta) in the Wadden Sea intertidal at the island of Sylt (North Sea). Dissertation, Hamburg Univ, Hamburg (FRG), Fachber Biologie
14. Sundbaeck K, Carlson L, Nilsson C, Joensson B, Wulff A, Odmark S (1996) Aquat Microbial Ecol 10:195
15. Breber P (1985) Aquaculture 44:51
16. Cellina F, De Leo GA, Rizzoli AE, Viaroli P, Bartoli M (2003) Oceanol Acta 26:139
17. Everett RA (1991) J Exp Mar Biol Ecol 150:223
18. Thiel M, Stearns LM, Watling L (1998) Helgol Meeresunters 52:15
19. Anonymous (2004) New Brunswick salmon growers environmental policy and codes of practice, version 1.0. New Brunswick Salmon Growers Association. Letang, New Brunswick, p 218
20. Miner RW (1950) GP Putnams, New York, p 888
21. Strasser M (1999) Helgol Meeresunters 52:3
22. Abbott RT (1974) American seashells, the marine mollusca of the Atlantic and Pacific coasts of North America. Van Nostrand Reinhold, New York
23. Emerson CW, Grant J (1991) Limnol Oceanogr 36:1288
24. Loo LO, Rosenberg R (1989) J Exp Mar Biol Ecol 130:253
25. Hillson CJ (1977) Seaweeds. Pennsylvania State University Press, University Park
26. Taylor WR (1937) Marine algae of the northeastern coast of North America. University of Michigan Press, Ann Arbor
27. Vadas RL, Beal B (1987) Estuaries 10:171
28. Chapman VJ (1970) Seaweeds and their uses. Methuen, London
29. Bjoernsaeter BR, Wheeler PA (1990) J Phycol 26:603
30. Chou CL, Haya K, Paon LA, Burridge L, Moffatt JD (2002) Mar Pollut Bull 44:1259
31. Yeats PA, Milligan TG, Sutherland TF, Robinson SMC, Smith JA, Lawton P, Levings CD (2005) (in this volume). Springer, Berlin Heidelberg New York
32. Hargrave BT, Doucette LI, Phillips GA, Milligan TG, Wildish DJ (1998) Can Data Rep Fish Aquat Sci 1031:iv
33. Auffrey LM (2003) Effects of green algal mats on soft-shelled clams (Mya arenaria) in southwestern New Brunswick, Canada. MSc, University of New Brunswick
34. Bonsdorff E (1992) Neth J Sea Res 30:57
35. Hunt HL, Mullineaux LS (2002) Limnol Oceanogr 47:151
36. Bonsdorff E, Nelson WG (1992) Veliger 35:358
37. Armonies W (1992) Mar Ecol Prog Ser 83:197
38. Iwasaki K (1996) Venus. Japanese J Malacol 55:223
39. Olafsson EB (1988) Mar Biol 97:571
40. Uchida M, Numaguchi K (1996) J Mar Biotechnol 4:200
41. Appeldoorn RS (1982) J Shellfish Res 2:87
42. Brousseau DJ (1979) Mar Biol 51:221
43. Appeldoorn RS (1983) Fish Bull 81:75
44. Angus RB, Woo P (1984) Can MS Rep Fish Aquat Sci 1842:1

45. Mullen B, Woo P (1985) Can MS Rep Fish Aquat Sci 1877:1
46. Robert G (1979) MS Rep Fish Mar Serv Can :19
47. Beal BF, Parker MR, Vencile KW (2001) J Exp Mar Biol Ecol 264:133
48. Weiss ET, Carmichael RH, Valiela I (2002) Aquaculture 211:275
49. Nicholls DJ, Tubbs CR, Haynes FN (1981) Kiel Meeresforsch 5:511
50. Soulsby PG, Lowthion D, Houston M (1982) Mar Poll Bull 13:162
51. Thompson RJ (1979) J Fish Res Board Can 36:955
52. Eversole AG (1988) J Shellfish Res 7:117
53. Bayne BL, Holland DL, Moore MN, Lowe DM, Widdows J (1978) J Mar Biol Assoc UK 58:825
54. Xu XL, Ji WJ, Castell JD, O'Dor RK (1994) Aquaculture 119:359
55. Chase ME, Bailey RC (1999) J Great Lakes Res 25:122
56. Hillman RE, Boehm PD, Freitas SY (1988) J Shellfish Res 7:216
57. Lowe DM, Pipe RK (1985) Mar Env Res 17:234
58. Wu RSS, Levings CD (1980) Mar Pollut Bull 11:11
59. Barber BJ (1996) J Invert Pathol 67:161
60. Brousseau DJ (1978) Mar Biol 50:63
61. Jorgensen BB (1980) Oikos 34:68
62. Diaz RJ, Rosenberg R (1995) Oceanogr Mar Biol Annu Rev 33:245
63. Norkko A, Bonsdorff E (1996) Mar Ecol Prog Ser 131:143
64. Norkko A, Bonsdorff E (1996) Mar Ecol Prog Ser 140:141
65. Norkko A, Bonsdorff E (1996) PSZNI. Mar Ecol 17:355
66. Rosenberg R, Hellman B, Johansson B (1991) Mar Ecol Prog Ser 79:127
67. Taylor DL, Eggleston DB (2000) Mar Ecol Prog Ser 196:221
68. Norkko J, Bonsdorff E, Norkko A (2000) J Exp Mar Biol Ecol 248:79
69. Auffrey LM, Robinson SMC, Barbeau MA (2004) Mar Ecol Prog Ser 278:193–203
70. Glude JB (1955) Trans Am Fish Soc 84:13
71. Ropes JW (1968) US Fish Bull 67:183
72. Blundon JA, Kennedy VS (1982) J Exp Mar Biol Ecol 65:67
73. Sandberg E, Tallquist M, Bonsdorff E (1996) Mar Ecol 17:411
74. Belding DL (1930) The soft-shelled clam fishery of Massachusetts commission on administration and finance, Boston
75. Chandler RA, Robinson SMC, Martin JD (2001) Can Tech Rep Fish Aquat Sci 2390:11 pp
76. Lavery PS, McComb AJ (1991) Estuar Coast Shelf Sci 32:281
77. Gowan RJ, Bradbury NB (1987) Oceanogr Mar Biol Ann Rev 25:563
78. Underwood AJ (1997) Experiments in ecology. Cambridge University Press, Cambridge, UK
79. Green RH (1979) Sampling design and statistical methods for environmental biologists. Wiley, NY
80. Andrew NL, Mapstone BD (1987) Oceanogr Mar Biol Ann Rev 25:39
81. Zar JH (1999) Biostatistical analysis. Prentice Hall, London
82. Peterman R (1990) Ecology 71:2024
83. Elliott JM (1983) Freshwater Biological Association Science Publication 25:73

Benthic Macrofaunal Changes Resulting from Finfish Mariculture

D. J. Wildish[1] (✉) · G. W. Pohle[2]

[1]Fisheries and Oceans Canada, Biological Station, 531 Brandy Cove Road, St. Andrews, NB, E5B 2L9, Canada
WildishD@mar.dfo-mpo.gc.ca

[2]Huntsman Marine Science Centre, 1 Lower Campus Road, St. Andrews, NB, E5B 2L9, Canada
gpohle@huntsmanmarine.ca

1	Introduction	276
2	Effects of Mariculture on Benthic Macrofauna: a Review	277
2.1	Organic Enrichment is a Temporal Event	277
2.2	Organic Enrichment Indices	279
2.3	Spatial Determination of Near-Field Organic Enrichment	281
2.4	Functional Changes in Sediments Enriched by Mariculture Wastes	282
2.5	Far-Field Organic Enrichment	284
2.6	Conclusions and Research Recommendations	284
3	Temporal Benthic Macrofaunal Change in a Bay of Fundy Tidal Inlet Linked to Mariculture Industrialization	286
3.1	History of L'Etang Inlet Resource Use	286
3.2	Sampling Strategy	289
3.3	Analytical Methods	291
4	Results	292
5	Discussion	298
	References	302

Abstract We present a review of the benthic macrofaunal changes that are circumstantially linked to intensive marine finfish aquaculture, or mariculture. The community structural and functional changes of macrofauna identified are mostly near-field effects, limited to the farm cage footprint. In common with other organic enrichment events in sediments, the mechanism in mariculture-related macrofaunal change is primarily caused by death due to hypoxia in sediments, followed by re-colonization with specialized organic enrichment tolerant macrofauna. Despite recent attention to the field of mariculture ecology, much still remains to be done to fully understand and manage the ecosystem effects of this activity.

In the second part of this chapter we present a case history study from a marine tidal inlet (L'Etang) in the Bay of Fundy, Canada during a period of rapid industrialization, dominated first by pulp mill effluents in the 1970s, and then by salmon mariculture development beginning in the 1980s and continuing today. Circumstantial evidence links the temporal benthic macrofaunal changes found in L'Etang Inlet to far field organic enrichment effects, primarily resulting from pulp mill pollution in the most landward area and salmon mariculture in the seaward end. It is shown that the temporal macrofaunal changes at the seaward end are not due to hypoxia in sediments or to natural seasonal and interannual changes. A new alternative hypothesis in aquaculture ecology is proposed: that the macrofaunal changes are far field effects resulting from the increased sedimentation (quality and amount) associated with intensive mariculture.

Keywords Benthic macrofauna · Finfish mariculture · Near- and far-field effects · Organic enrichment · Temporal change

1
Introduction

The intensive grow-out of fish, such as salmon, in floating marine net pens or cages (mariculture) has, within the last 25 years, become an industrial activity that is common worldwide [1]. Because of the potential for the production of large amounts of organic wastes from mariculture operations [2], which contribute to organic enrichment in the receiving environment, there is already considerable literature dealing with the effects of organic enrichment from mariculture [3, 4].

Ecologically, organic enrichment is a complex process, involving pelagic-benthic coupling and responses in both the water column and sediments. The effects are due to both natural and anthropogenic sources of organic matter. Thus, it is hardly surprising that some species of benthic macrofauna have become specialized to live where high inputs of organic matter occur. Natural examples of organic enrichment include a whale carcass decomposing on the seabed [5], or the annual spring die-off of phytoplankton blooms decomposing in sediments [6]. Common anthropogenic sources of organic enrichment include: municipal sewage, pulp mill effluents, other industrial sources and, more recently, from intensive aquaculture [7].

At present, intensive salmon mariculture is the pre-eminent finfish farming activity in temperate seas [1]. Here the organic matter inputs are from uneaten food and fecal particles which reach the seabed at rates $> 1\,g\,C\,m^{-2}\,d^{-1}$ [2] within the farm footprint. The latter is defined as a generally elliptical area of the seabed, with the major tidal flow being on the long axis of the ellipse where water movements deliver most of the food and fecal particles [8]. The nature of the sedimentary response will depend on the realized amount of sedimentation that is proportional to the farm footprint area, and hence depth and water movements. Sediment particle transport, deposition, and erosion are determined by sediment particle sizes and the ambient velocity patterns of a particular site [9], although the relationship is a complex one. For example, medium particle size sands are the most susceptible to erosion at low velocities, whereas finer silt/clay particles require a higher velocity to erode from a sediment deposit. Deposition, on the other hand, is inversely proportional to velocity and thus depositional environments indicate low velocities. Tidal velocities, too, are variable according to lunar/tidal patterns and may influence both depositional and erosional activities. A method to model the deposition of particulate wastes from fish farming in the footprint area has recently been described [10]. As far as we are aware, a universally applicable guide based on sediment particle characteristics of deposits is still not available to predict flow patterns and hence footprint size at a given site. Consequently, in considering soft sediments we compare net depositional sediments, with a high percentage of silt/clay and low currents, with net erosional sediments, with a low silt/clay content, a high percentage of sand and higher currents. Locations with net depositional sediments are most at risk from fish farm organic enrichment, because at these locations the footprint is much smaller. Hence the sedimentation rate is much higher than at net erosional locations, where the footprint is spread over a larger area.

This chapter is organized into two parts. The first section is a review summarizing the known ecological effects of particulate organic enrichment from aquaculture on benthic macrofauna. By benthic macrofauna we mean those invertebrate animals living on, or in, sediments that do not pass a 0.5 to 1.0 mm^2 mesh sieve. In the second section we compare the macrofaunal community characteristics of L'Etang Inlet, Bay of Fundy, based on before (1975) and after mariculture (1997, 2000) samples. Salmon mariculture was initiated in L'Etang Inlet in 1981 and has been carried out there ever since. Our case history suggests a new form of far field effect in Lime Kiln Bay, which has not previously been linked with salmon culture.

2
Effects of Mariculture on Benthic Macrofauna: a Review

2.1
Organic Enrichment is a Temporal Event

Organic enrichment from intensive mariculture begins as a sudden and massive increase in sedimentation to the footprint area. Sedimentation rates at net depositional locations occupied by salmon farms have been reviewed and may reach high levels, from 1 to 181 g m^{-2} d^{-1} [2, 8]. The influx of feces and waste food, normally orders of magnitude greater than natural background sedimentation, causes a rapid rise in aerobic microbial activity within the surface layer of the sediments. This leads to rapid utilization of dissolved oxygen of pore water, followed by replacement of aerobic by anaerobic microorganisms. The occurrence of hypoxia in pore water at low oxygen levels (< 0.2 mg L^{-1}) is considered to be the primary cause of death of resident organic enrichment intolerant macrofauna, although HS⁻ may also be a contributing factor [11]. The anaerobic microorganisms present include sulfate-reducing bacteria and methanogens, which produce sulfides (H_2S, $S^=$) and methane.

Some species of macrofauna, such as the sludge worm *Capitella capitata*, have become specialized to find organically enriched sediments during larval life. Larval settlement in *Capitella* sp.1 is mediated by chemical attraction to decaying organic matter [12, 13]. Such species are termed organic enrichment tolerant opportunistic species, because they can tolerate hypoxia and high levels of HS⁻. They replace the resident macrofauna that succumb to low dissolved oxygen conditions and/or increasing levels of HS⁻. Typical opportunists found in the footprint of intensive salmon farms are shown in Table 1.

An objective method of selecting species of organic enrichment tolerant opportunists has been described by Pearson et al. [14]. Macrofaunal species diversity within the heavily impacted farm footprint may remain at zero (azoic) or at levels with only the organic enrichment tolerant opportunists present. Dominance levels, by which is meant the extent to which one or a few species of macrofauna are represented by a high proportion of the total abundance, are high within the footprint area.

Further evidence that organic enrichment under fish farms is a temporal event is derived from fallowing studies. Fallowing is the practice of leaving farm sites without fish for various time periods. For the Bay of Fundy salmon culture industry, the purpose of fallowing is to interrupt disease life cycles, e.g. of infectious salmon anaemia. Studies in different geographical locations have determined the geochemical and benthic macrofaunal recovery rates involved in this process. Fallowing recovery studies have been completed in Ireland [15], Norway [16], Hong Kong [17], the Mediterranean [18], the Bay

Table 1 Organic enrichment tolerant opportunistic macrofauna characteristic of marine sediments under/near intensive fish farms

Opportunistic species	Locality	Reference
Capitella capitata, Scolelepis furliginasa	Scotland, UK	Brown et al. [22]
C. capitata, Nematoda	Pacific coast, USA	Weston [23]
C. capitata, Maloceros fuliginosa, Ophryotrocha sp., Nematoda	Scotland, UK	Henderson & Ross [24]
C. capitata	Cephalonia Bay, Greece	Karakassis & Hatziyanni [25]
C. capitata, Ophryotrocha vivipara, Schistomeringos sp., Sigambra tentaculata, Nebalia puggetensis Aorides sp., Pseudotanais oculata	Pacific coast of USA and Canada	Brooks et al. [20] Brooks & Mahnken [26]
C. capitata	Canada	Pohle et al. [7]

of Fundy [19], and the north-west Pacific Ocean [20]. The recovery process follows a temporal version of the organic enrichment succession model of Pearson and Rosenberg [21]. The one exception among the cited studies was that of Karakassis et al. [18] in the Mediterranean, further discussed below. For all reviewed studies, the timing of benthic macrofaunal recovery was found to be variable, from ~ 6 months to > 2 years. Factors that need to be considered in determining the cause of the timing differences include: lack of an endpoint to signify when recovery was complete and that the displaced benthic macrofaunal communities may have been initially different. Brooks et al. [20] studied fallowing at a number of salmon farm sites in the northwest Pacific and found that net depositional locations took longer to recover than net erosional ones.

2.2
Organic Enrichment Indices

The classical organic enrichment gradient succession model of Pearson and Rosenberg [21] includes two components:

- A spatial view along an organic enrichment gradient in a net depositional sediment. The model predicts the appearance of tolerant opportunists as first colonizers, as well as changes in the redox potential discontinuity (RPD) or mixed sediment layer depths, changes in the bioturbation potential and mean size of the macrofauna linked to position along the gradient. Along the gradient four groups of responses by the organism-sediment complex could be recognized (Table 2).
- A quantitative prediction about the relative numbers of species (S), abundance (A), and biomass (B) along the gradient (SAB).

Although most authors describing the macrofauna [7, 18, 22–26] along either a spatial or temporal organic enrichment gradient considered their mariculture-related SAB data to be consistent with this model, a few [18, 27] found inconsistencies. Thus, a salmon farm located in a cove in Maine, USA, with a sandy-mud sediment and low tidal velocities did not show the ordered sequence of four organic enrichment groupings in Pearson and Rosenberg [21]. The authors [27] thought this to be caused by the lack of physical stability in the cove, due to the dominance of stochastic wave events that dominated both the sediment and macrofaunal community. In the fallowing study of Karakassis et al. [18] at a sea bream and sea bass farm in the Mediterranean, the temporal pattern of macrofaunal recovery also did not follow the successional model of Pearson and Rosenberg. The attributed cause was seasonal releases of nutrients from the enriched sediments that resulted in microalgal blooms, which then decayed in the sediments causing further asphyxiation of macrofauna, and hence alternating periods of macrofaunal loss followed by macrofaunal enhancement.

Table 2 The organic enrichment gradient based on different measures described in previous publications

Measure	Successional stages				Ref.
	III	II	I	0	
Microbial	Normal	Oxic	Hypoxic	Anoxic	[28]
Macrofaunal	Normal	Transitory	Polluted	Grossly polluted	[21]
Sediment profile imaging					
Organism-sediment index (OSI)	III	II	I	Azoic	[29]
Benthic habitat quality (BHQ) index	> 10	5–10	2–4	< 2	[30]
Geochemical	Normal	Oxic	Hypoxic	Anoxic	[31]
Eh, mV_{NHE}	>+ 100	0–100	– 100–0	<– 100	
$S^=$, micromoles	< 300	300–1300	1300–6000	> 6000	

The predictive power of the organic enrichment gradient succession model is compromised by inadequacies of benthic biological theory. Thus the following question usually cannot be answered: Is the benthic community (= recurrent group or habitat) and its particular suite of environmental characteristics similar to the benthic community and its response described in Pearson and Rosenberg [21]? The specific problems encountered include:

- A different response is expected if the initial benthic community is different (e.g. from communities in net depositional versus net erosional sediments).
- Although the opportunists which recolonize an organically enriched sediment can be expected to be similar if they occur in the same region, this will not be the case if they are from different geographical regions.
- Because of known seasonal changes in macrofaunal SAB [32] the SAB comparisons must be seasonally equivalent to be valid.
- The physically unstable sediments described by Findlay et al. [27] cannot be temporally accommodated by the Pearson and Rosenberg [21] model.

Despite these problems, the organic enrichment gradient model is still widely used, notably for describing the degree of impact, often with surrogate variables (other than macrofauna) that are more cost effective than classical macrofaunal sampling and analysis (Table 2).

Other organic enrichment indices have been proposed and include:

- Those related to the groupings of the Poole et al. (28) and Pearson and Rosenberg [21] models as shown in Table 2. They include the OSI, BHQ

(both based on sediment profile photographs) and redox potential/sulfide concentration in interfacial sediments approaches.
- The benthic enrichment index (BEI) of Hargrave [33] that is based on the product of organic carbon (as mmol C m^{-2}) and Eh$_{NHE}$ (mV) in interfacial sediments.
- Net carbon burial rates of Cranston [34] that requires the down-core determination of sulfate and ammonium, to determine the mineralization rate.

A range of additional indices have been proposed with respect to other industrial sources and municipal sewage that also produce large amounts of organic wastes. Presently, there is no consensus on the best index available. That depends on the purpose of the ecological investigation, such as research versus compliance monitoring. Choices between indices should be based on the following operational criteria [35]: scientific defensibility, ability to provide a statistical weight of evidence, providing relevant resource management decision points and to be cost effective in time and resources to obtain the ecological information.

2.3
Spatial Determination of Near-Field Organic Enrichment

Many of the earlier benthic ecological studies of fish farms employed transect sampling with sampling locations at increasing distances from the fish farm edge. Thus Brown et al. [22] showed by macrofaunal sampling that the sediment was grossly enriched (stage 0 or anoxic) to the edge of the cage footprint and highly enriched (stage I or hypoxic) up to ∼ 15 m away. Similar results have been obtained by many of the authors listed in Table 1. Based on a footprint ellipse, transect sampling in the main current direction and another transect at right angles, it is possible to estimate the areas affected by the various stages of organic enrichment. Such calculations suggest that salmon farms producing up to 2000 t of salmon per growth cycle, cause anoxic (stage 0) conditions and hence significant changes in benthic macrofaunal communities, in relatively small areas (< 0.05 km^2) where the sediments are net depositional.

The high cost of processing each replicate macrofaunal grab sample for SAB [31] limits the numbers of stations that can be sampled. For this reason, classical macrofaunal grab sampling as a means of determining the precise spatial effects of organic enrichment remains problematic. Recent results [36] suggest that acoustic methods could possibly be adapted to determine the spatial extent of at least highly enriched conditions (stage 0 or anoxic) under fish farms. Some results for a net depositional sediment in Lime Kiln Bay, L'Etang Inlet are shown in Fig. 1. High backscatter, which appears white, indicates that the footprints are approximately under each individual cage. Low

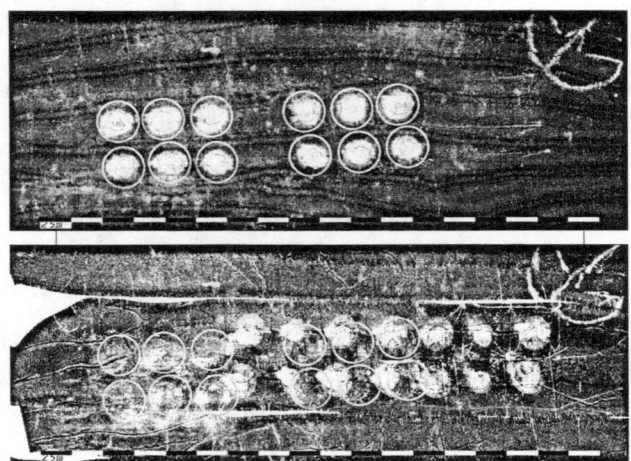

Fig. 1 Sedimentary acoustic backscatter images at a salmon farm in Lime Kiln Bay, L'Etang Inlet. Upper panel November 2000 (EM 3000 multibeam), lower panel July 2001 (Knudsen sidescan). *Circles* indicate the position of individual cages in 1999–2000. In 2001 all of the cages were moved closer to the derelict weir (*upper right hand corner*), and two more cages were added (scale bars = 25 m). From Wildish et al. [36]

backscatter, which appears as dark areas, indicates that a silt/clay sediment is present. The structures in the top right hand corner of each of the panels are the footings of an abandoned herring weir.

2.4
Functional Changes in Sediments Enriched by Mariculture Wastes

A prediction of the organic enrichment gradient successional model is that along the gradient of increasing enrichment, there is an increasing loss of macrofaunal SAB. As shown in Sect. 2.1 a larger-bodied, equilibrium community, is replaced by a smaller, opportunist, r-strategist one. A concomitant narrowing of the oxidized, sediment mixed layer also occurs. Our observations in the footprint of Bay of Fundy salmon farms suggests that at newly established farms the sedimentary responses begin as a narrow surface layer of anoxia in which black, sulfide-rich deposits overly a still oxic deeper layer. If the high inputs of wastes from the farm continue for more than a few years, the black sulfide layer deepens and eventually occupies the whole of the profile (at depths to < 40 cm).

Benthic macrofauna in undisturbed sediments play an important role in the mineralization of organic matter [37]. How this is accomplished depends on the trophic group of the species. Deposit feeders influence the rate of breakdown of organic matter in the following ways:

- By cropping the microalgal flora found on the surface of sedimentary and organic matter particles.
- By production of a gut flora capable of degrading complex, recalcitrant plant compounds.
- By production of feces and pseudofeces that contain partially degraded organic matter, which can be further processed by microbes and after coprophagy.
- By bioturbation activities, which assist in mixing the sediment and in geochemical cycling of oxygen, carbon and sulfate throughout the sediment interfacial layers. Pearson [38] presents a modern detailed review of soft sediment macrofauna based on their functional groups and the role each taxon plays in bioturbation.

Heilskov and Holmer [39] describe experimental manipulations of footprint sediments obtained from a rainbow trout farm in Horsens Fjord, Denmark. The burrowing, surface deposit feeder, *Nereis diversicolor*, and the head down deposit feeder, *Capitella* sp I, were added to defaunated sediments from beneath the cages, in a separate series of experiments. Over two months, the two species stimulated mineralization by 135% (*Nereis*) and 87% (*Capitella*) over azoic controls. Although these results are consistent with the model [21], they are difficult to place in context, because the pre-farm benthic community was not characterized, nor was the degree of enrichment at this farm determined. Similar problems were met in an observational study at a salmon farm in Loch Creran, Scotland, by Nickell et al. [40]. A series of four stations, beginning at the cage edge and extending to 2060 m away were selected. The macrofaunal community structure and SAB followed the predictions of the successional model [21], decreasing with distance from the farm. The mixing intensity in surface sediments, due to bioturbation, was least at the farm edge and greatest at the two intermediate stations. Mixing depth was determined by fitting a diffusion model to sediment chlorophyll a data with depth in core samples. At the most distant location from the farm an altogether different benthic community was found to be present, characterized by deep burrowers such as *Maxmuellaria lankestreri*. This station was an inappropriate reference station for the fish farm.

The functional importance of the sediment near-field effects circumstantially linked to mariculture has not been examined. Thus the extent to which it changes marine food webs and enhances, or reduces, secondary production remains unknown. The sediment under a farm shown in Fig. 1 is net depositional and as the acoustic footprint is approximately the same size as the floating cage structure, the area affected is: πr^2, where $r = 11$ m $= 380$ m^2; and for a total of 14 cages (in 2001)$= 0.5$ km^2.

2.5
Far-Field Organic Enrichment

As far as we aware, the only ecological study that suggested far-field effects linked to mariculture is that of Pohle et al. [41]. This study, carried out in two embayments of the L'Etang Inlet and a nearby reference site, described significant temporal changes of the benthic macrofaunal community at the experimental locations during the period 1994–99. The benthos of Lime Kiln Bay, in particular, changed to a biologically stressed community. The changes were correlated with increased levels of organic matter, attributable to effects of multiple salmon farms 200–300 m away. These changes were not attributable to other factors, such as seasonal differences or differences in sediment characteristics, nor did the reference area show these trends.

2.6
Conclusions and Research Recommendations

Most of the effects found in the literature consulted for this review are near-field: that is they are local, or footprint-limited. For the group of cages which make up an individual farm (producing < 2000 t) the area impacted by anoxia (stage 0) is, $0.05 \sim 0.5 \, km^2$. The ecological effects on macrofauna along an organic enrichment gradient are the same as those occurring from natural or other anthropogenic sources of organic wastes, e.g. pulp and paper mill wastes [21, 28]. The fallowing studies available suggest that sediments impacted by intensive fish farming can recover, if the sedimentation from the fish farm is stopped (Sect. 2.1).

Further research is needed in the following four main areas: benthic biological theory, choice and standardization of environmental monitoring methods, far field and near-field effects. Each of these is presented in further detail below.

1. Benthic biological theory. A universally applicable theory concerning spatial distribution of macrofauna will need to consider a range of limiting factors (42). Excluding biological limiting factors would leave: salinity, temperature, wave activity, tidal velocity, sediment deposits, food fluxes in the water column and sedimentation rates, as primary controls of benthic macrofaunal community structure and function. The last 4–5 of these variables are under the direct control of water movement and hence involve trophic group mutual exclusion [43]. It may be possible to predict the type of functional trophic groups, using an updated version of functional groups [38], that can be expected from water movement characterization (or the corresponding sediment deposits) at a given location. The value of this for organic enrichment studies is that given a set of key physical variables, a predictive classification of benthic macrofaunal com-

munities would be feasible. It would then be possible to relate organic enrichment responses to similar macrofaunal communities anywhere in the temperate parts of the world.

The availability of multi-beam acoustic methods in describing benthic "habitats" and thus macrofaunal communities or recurrent groups, is seen to be a crucially important technology. This is because it is required by the theory outlined above and is the only known method which can map macrofaunal communities at a sufficient speed to make it practically usable. Examples of the successful application of multibeam acoustics are: the mapping of horse mussel reefs in the Bay of Fundy [44] and the mapping of macrofaunal communities on the Scotian Shelf [45]. The work required in the future will center on groundtruthing the acoustic maps with a wide variety of benthic ecological methods, including classical benthic grab sampling. Grab sampling must be done with precise acoustic positioning methods (McKeowon, D, personal communication) to ensure accurate acoustic cross-referencing. Another demand for the use of acoustics is from coastal zone scientists and resource managers attempting to implement ecosystem management strategies, who need them to determine the spatial distribution of rare and common benthic habitats [46].

2. Choice and standardization of organic enrichment monitoring methods. Presented in Sect. 2.3 are the operational criteria for choosing suitable methods. The purposes intended for the method influences its choice and may include: detecting, comparing treatment/reference locations, comparing the before/after status of the same sites, and delimiting the spatial extent of an organic enrichment event. Because of the large investment in time required for identifying and counting macrofauna, it will often be the case that alternative proxy methods will be superior in cost effectiveness. Nevertheless, we expect classical macrofaunal sampling to remain as the method of choice, against which all others will need to be compared. It is the only method that can be used to groundtruth multibeam acoustic data to produce benthic "habitat" maps.

Methods of choice need to be standardized as far as possible among users. Thus, for macrofaunal sampling we suggest sampling with a $0.1 \, m^2$ grab and sieving on a $1.0 \, mm^2$ mesh. We also recommend that more attention be paid to inter-laboratory calibration experiments. For this, the participating groups would collect, sieve and taxonomically analyze from the same locations. We also recommend a system of auditing for benthic environmental methods. This would be particularly applicable where the monitoring objectives were directed towards compliance or regulatory purposes.

3. Far field effects. As mentioned in the Introduction, organic enrichment may involve pelagic-benthic coupling and in this case the major effect may be re-directed by water column transport, and therefore extend beyond the immediate vicinity of enrichment sources. Such an event, in-

volving transport of nutrients to a nearby beach where enhanced growth of macroalgae occurred (= eutrophication), is reported in Chap. 10 of this volume. Another example distant from the input source and reviewed here [41] involved changes in benthic macrofaunal community structure in the L'Etang Inlet. Additional independent before/after impact benthic macrofaunal studies in L'Etang Inlet showed that the species composition had changed over time (Sect. 3 of this chapter). As the sampling locations were well away from farm footprints, hypoxia could not be the cause. We hypothesize that the species composition changes were due to the increased sedimentation rates associated with intensive salmon mariculture as discussed in Chaps. 10 and 11. We recommend that experimental attempts be made to test the null hypothesis that higher sedimentation rates and increased organic enrichment linked to mariculture are not the cause of benthic macrofaunal changes.

4. Near field effects. It is suggested that further studies of benthic macrofaunal recovery involving fish farm fallowing be made because some differences in recovery rates between locations differing in hydrodynamic conditions were noted. There is also evidence that enriched net erosional sediments recover faster than depositional ones. This could be addressed in concert with answering the open question of how to define benthic macrofaunal community recovery. We recommend that a standardized endpoint for the latter be established. There is presently little ecological information on how the macrofaunal changes attributable to organic enrichment in sediments affect ecosystem functioning. For example, what are the effects on the higher levels of the food web and does it affect the productive capacity of commercially valued species, such as fish, in it?

3
Temporal Benthic Macrofaunal Change in a Bay of Fundy Tidal Inlet Linked to Mariculture Industrialization

3.1
History of L'Etang Inlet Resource Use

The L'Etang Inlet is located near the mouth of the Bay of Fundy in the NW Atlantic Ocean. The northern part of Maine and southern part of New Brunswick on either side of the Canadian–US border are at the center of the current Bay of Fundy salmon culture industry. Maine had 44 and New Brunswick 95 farm lease sites in 2002 (with up to 5% non-operational farms). Salmon culture involves growing salmon eggs to smolt size in freshwater hatcheries, followed by seawater acclimation, then transfer to marine netpens (= cages) for grow-out. Within the Bay of Fundy industry the period of grow-out is 18–24 months. The Canadian mariculture industry in the Bay of Fundy produced 8900 tonnes in 2002. Thus, assuming that half of the farms were carrying first year (smolt) fish, excluded from consideration, the mean 2002 production per farm was $8900/45 = 198$ tonnes.

The L'Etang Inlet (Fig. 2a) is a marine tidal water body extending ~ 14 km inland from the Bay of Fundy. Because of a small drainage area of 62 km^2 [47], with few small streams delivering freshwater to it, salinity distribution is dominated by tidal action.

For this reason, salinity remains high, even in the most landward parts of the Inlet. Geologically (Table 3), the L'Etang arose as the pre-glacial estuary of the present day Magaguadavic River, which was blocked by a glacial moraine at the southern end of lake Utopia. Today, the river drains via a gorge in the town of St. George into the present day Magaguadavic estuary shown in Fig. 2a. The position of the L'Etang Inlet at $\sim 45°$N ensures that the climate is cold temperate.

Historically, it is known that paleo-Indian people were present in the general area soon after the end of the last Ice Age (Table 3). The first Europeans did not arrive until 1604. The industrialization of the L'Etang Inlet began soon after the development of the herring weir fishery, starting in 1797. Processing plants to can herring followed, including the establishment of a plant at Black's Harbour in 1893. A more modern plant was built there in 1996 with improved waste disposal facilities. In 1967 a causeway was built across the upper part of L'Etang Inlet for a new highway. This separated the upper

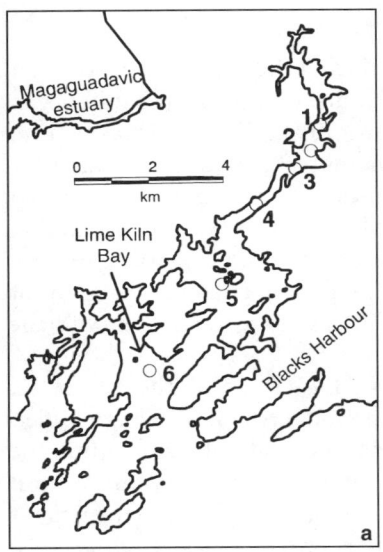
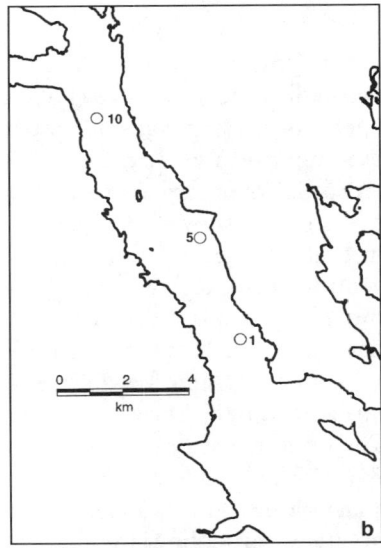

Fig. 2 Locations of sampling sites in L'Etang Inlet (**a**) and St. Croix estuary (**b**)

Table 3 Geological/historical events and industrialization of the L'Etang Inlet (45°05'N, 66°48'W), Bay of Fundy, Canada

Year	Event	Reference or Web site
B.C.		
18 000	Pre-glacial estuary of Magaguadavic River	Wildish et al. [47]
?	Glacial moraine blocks Magaguadavic flow	Wildish et al. [47]
9500	Paleo–Indians present in Maine	Bourque [48]
A.D.		
1604	French landing on island in St. Croix estuary	http://collections.ic.gc.ca/SaintCroixIsland
1797	Herring weir fishery initiated	Doucet & Wilbur [49]
1893	Fish processing plant established at Black's Harbour	www.connors.ca
1996	Modernized processing plant at Black's Harbour	Connors Bros. Ltd. Pers. comm.
1967	Causeway built for Route 1 across upper L'Etang	Wildish et al. [47]
1971	Pulp mill effluent enters upper L'Etang Inlet	Wildish et al. [47]
1988	Pulp mill effluent treated in anaerobic digester Second causeway built, 0.5 km below the first	Brent Kibbe, Pers. Comm.
1978	Salmon farming initiated on Deer Island	Anderson [50]
1981	Salmon farming in L'Etang (Lime Kiln Bay)	George Wolfe, Pers. Comm.
1986	Six salmon farms in Lime Kiln Bay	Wildish et al. [51]
1996	Six salmon farms in Lime Kiln Bay. Epidemic of ISA leads to fallowing from 1997–1999	Wildish et al. [19]
1999	Four salmon farms in Lime Kiln Bay	Wildish et al. [52]

part (2 km) from the rest of L'Etang, much reducing the exchange of seawater above the causeway. In 1971 a semi-chemical sulfite digesting pulp mill began operations, using water for processing pulp from Lake Utopia. Effluent from the pulp mill was discharged directly into the upper part of the L'Etang Inlet, initially with excessive amounts of pulp fibres and BOD (biological oxygen demand), because of inefficient treatment facilities. Following the installation in 1988 of a second causeway seaward of the first (with a one-way gate preventing seawater entry), and an anerobic digester for effluent treatment at the mill, the waste discharge to the upper L'Etang finally complied with the national guidelines for pulp mill effluents.

Salmon farming began in Lime Kiln Bay in 1981. By 1985 there were six farms within the bay with many more throughout the rest of the L'Etang Inlet. The L'Etang, with 1.07 farms km^{-2}, has the highest density of farms within the Bay of Fundy industry. The appearance of a viral disease, infectious salmon anaemia (ISA), in 1996 led to drastic changes in salmon culture management practices throughout the Bay of Fundy industry. Two changes of importance for Lime Kiln Bay were the initiation of a fallowing period, when fish and all wooden cages were removed. Fallowing started after the removal of market fish in the fall of 1997 and lasted until the re-introduction of smolts in spring 1999. The second change involved reducing the number of farms from six to four and all operations converting to single year class stocking (odd-year) to minimize disease spread. Wildish et al. [52] calculated that 2354 tonnes of salmon are produced every market year by the four farms in Lime Kiln Bay, which still exceeds the holding capacity limits for this small bay of 2.5–3 km^2. The holding capacity limits calculated independently by the method of Silvert [53] and Cranston [34] for the whole of L'Etang Inlet are 7000 to 10 000 tonnes, respectively, excluding four farms in the most landward part of the Inlet. Proportionate to its area, the Lime Kiln Bay holding capacity limits calculated for both models are 732 to 1046 tonnes.

3.2
Sampling Strategy

All of the benthic macrofaunal data presented here were originally collected for other purposes. Because comparable samples were taken at the same locations both before and after the initiation of salmon culture in the L'Etang Inlet we designed a post-facto, before/after, study to test whether salmon mariculture had any far-field effect on benthic macrofauna. The null hypothesis tested was that benthic macrofaunal communities in L'Etang Inlet distant from farm locations were not significantly different between 1975 and 1997 or 2000 and hence there were no far-field effects attributable to mariculture. As shown in the previous section, there is good evidence of footprint-limited or near-field effects in Lime Kiln Bay [36]. Because of possible regional long-term changes of macrofaunal community structure reference locations in the St. Croix estuary (Fig. 2b) were also included in the analysis. The nearest reference station in the estuary was at least 10 km distant from fish farms. The three stations worked in the St. Croix estuary should be free of any salmon mariculture influence and at the same time should indicate the extent of any long-term changes or differences due to taxonomic biases.

Two different grabs were used during this study. Earlier sampling (1974, 1975) was completed with a 0.1 m^2 Smith–McIntyre grab [54], later sampling (1997, 2000) with a 0.1 m^2 Hunter grab [55]. Both grabs sample similar areas and volumes of sediment [56]. Samples were sieved with running seawater, the lowest mesh size being 1 mm^2 for all sampling dates. Macrofauna were preserved first in formalin in seawater, then in alcohol and identified to the lowest taxon possible, with the abundance, and in some cases biomass of each, determined. Results are based on at least three replicate grabs per station. 1975 L'Etang Inlet species-abundance data are reported in Wildish et al. [57, 58] with identification by the Identification Centre, St. Andrews and the Canadian Oceanographic Centre, Ottawa. 1974 St. Croix SAB data are available in Wildish et al. [59] with identifications completed by the

Canadian Oceanographic Centre, Ottawa. In the latter case, there were ten replicates available at each of three stations. The contemporary 1997 and 2000 macrofaunal species-abundance (not biomass) matrices were prepared by the Atlantic Reference Centre, Huntsman Marine Science Centre, St. Andrews, and have not previously been published. Six matching before/after stations (see Fig. 2a) were available in L'Etang Inlet, each with at least three replicates (one exception with two). The contemporary L'Etang location 1, unlike all other stations, needed to be moved ~ 500 m downstream to just below the second causeway built at Pull-and-be-Damned Narrows after the 1975 sampling. The species-abundance data for all stations and years were compiled into two location-based digital format matrices [60] used for the data analysis presented herein.

A Simrad EM3000 acoustic survey of L'Etang Inlet from 3–7 November 2000 undertaken by DFO, and multibeam backscatter maps, prepared from these data by the Ocean Mapping Group at the University of New Brunswick (not shown), were used to designate L'Etang stations as net depositional (stations 1–4) or net erosional (stations 5 and 6). This interpretation is in agreement with the limited sediment particle data available (Table 4), the median particle diameter (MD phi) indicating that station 5 in L'Etang Inlet is medium sand, whereas all other stations in both locations are silt or clay. The sediment sorting coefficient (QD phi) indicates that sediments are poorly, or very poorly sorted. For the quartile deviation (Skq phi), a plus indicates that the mean of the quartile lies to the right of the median, a minus indicates it lies to the left. Thus the skewness of the samples is different between L'Etang

Table 4 Mean depth at LW, sediment sorting characteristics and total organic matter (% volatile solids) in St. Croix estuary and L' Etang Inlet. The station numbers are shown in Fig. 2. Data from the following sources: Wildish et al. [56], Wildish et al. [59] and Wildish and Kristmanson [61]

Station/ sites	Depth at LW m	MDø	QDø	Skqø	% Organic carbon*	% Volatile solids
		St. Croix estuary				
1	18	8.40	2.09	−0.26	2.03	5.92
2	28	6.75	3.02	−1.12	2.55	6.44
3	12	5.80	3.16	−0.46	2.76	7.30
		L'Etang Inlet				
1	1					10.10
2	1					6.20
3	2					4.80
4	5	7.82	2.51	+1.65	2.35	1.80
5	13	1.87	3.46	+2.24	0.88	0.91

*Walkley-Black method [62]

and St. Croix estuary sediments, suggesting that re-suspension is commoner at the two lower L'Etang locations (5, 6).

3.3
Analytical Methods

Species/abundance matrices were assessed using PRIMER (Version 4, Plymouth Routines in Multivariate Ecological Research) with its various subroutines. Multivariate methods were employed to classify individual grab samples from each location (three or ten replicate grabs per location) into similar groupings by hierarchical clustering and ordination, using non-metric multidimensional scaling (nMDS), for gradient visualization. The nMDS data were log transformed to minimize data distortion, and the Bray-Curtis similarity coefficient was used to relate overall similarity between samples by taking all taxa into consideration in producing the similarity matrix. Map-like plots of nMDS indicate the relative distance between samples visually and represent the relative similarity of species composition, without defining plot configuration or scale, allowing a gradation of locations to be represented graphically. The extent to which these relations can be adequately represented in a two-dimensional map is expressed as the "stress" coefficient statistic. Pohle et al. [41] provide further details on methodology and theoretical approach.

Changes in community structure were also analyzed with more traditional species independent univariate measures (species richness, H'), more advanced biodiversity relatedness indices [63] recently developed, including average taxonomic distinctness (AvTD, Δ^*) and variation in taxonomic distinctness (VarTD, Λ^{+}), and by trophic type analysis. Changes in community structure were linked to environmental changes by statistical analysis of % organic matter data. The direct linkage between carbon flux induced increases in benthic oxygen demand to benthic impacts, associated with mariculture, was established by Findlay and Watling [64]. STATGRAPHICS PLUS (Version 5.0, Manugistics) was used for plotting and standard tests of significance at the 5% level, based on ANOVA, and the ANOSIM equivalent in PRIMER for multivariate data. Species analysis, based on the species contributions to similarity percentages available from the SIMPER subroutine of PRIMER, was used to identify dominant taxa, defined as those that contribute > 5% to the total abundance at each location.

We also calculated the trophic ratio for each location based on the known trophic type of each species [42, 43]:

TRS% = D/D + S100

Where TRS is the trophic ratio based on species presence/absence, D is the sum of deposit feeding species at each location and date, and S is the sum of suspension feeding species at each location and date.

The trophic group designations used in our analysis are shown in Wildish [60]. We were able to designate trophic groups for ~ 80% of the species. Other trophic groups recognized (e.g. carnivores, omnivores and algal scrapers) were excluded from the calculation.

4
Results

L'Etang Inlet

The MDS plot shown (Fig. 3a) indicates that replicates for each sampling site cluster together and hence are similar in species composition and therefore representative of sampling areas.

A comparison of early (1975) and later (1997 or 2000) samplings at the same locations show a clear temporal change in community structure for all stations, as indicated in Fig. 3a. The 1975 data indicate that different sam-

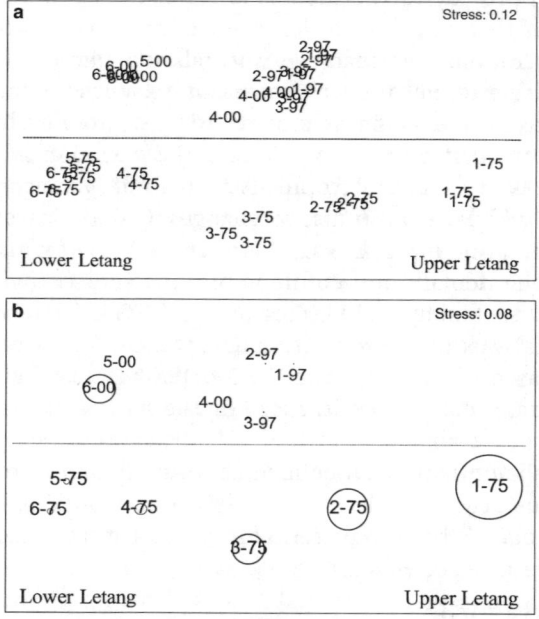

Fig. 3 a Two-dimensional nMDS ordination plot of benthic macrofaunal associations in L'Etang Inlet. Sampling site numbers are indicated by the first digit with the last two numbers indicating the year of sampling. **b** Mean nMDS plot with superimposed % organic matter. Bubble size is proportional to percent volatile solids values

pling locations have a distinct community structure, with only the two lower L'etang stations 5 and 6 not being significantly different. More importantly, a clear gradient in changing community structure is apparent progressing seaward, with sampling locations becoming increasingly different with increasing distance from the upper L'Etang site 1. The 1997 or 2000 data show the same separate clustering of the two lower L'Etang locations seen in 1975, but there is less distinction between areas of central and upper L'Etang.

Figure 3b is the mean nMDS plot with superimposed organic matter data from Table 4 and Pohle et al. [41]. The size of the circles is proportional to % volatile solids. The 1975 data show a clear correlation between increasingly different community structure and decreasing organic loading in the seaward

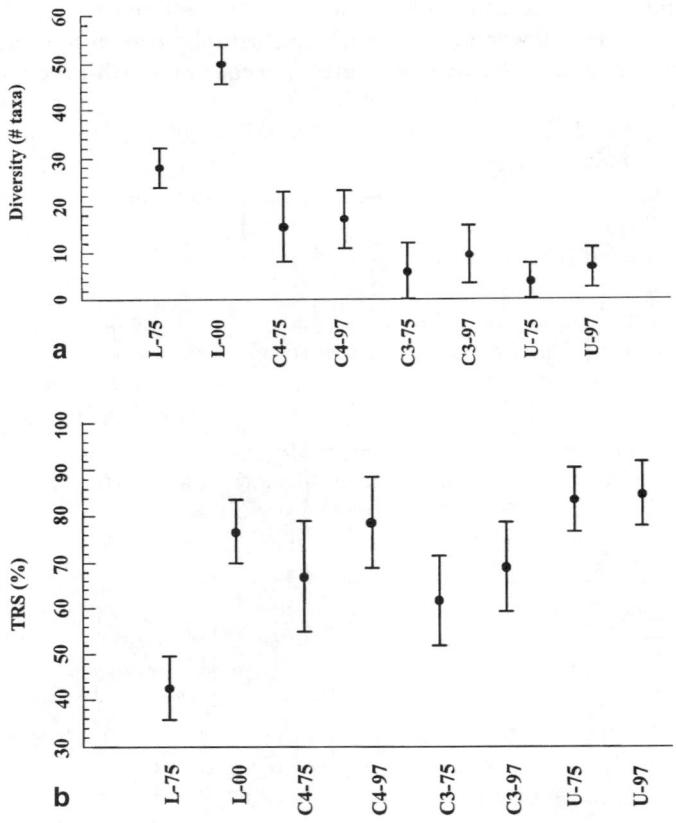

Fig. 4 L'Etang Inlet macrofauna. **a** Number of taxa at sampling locations of upper (U, 1 + 2) L'Etang, central locations C3 and C4, and lower (L, 5 + 6) L'Etang locations. **b** Trophic ratio of deposit/suspension feeders (TRS%) for the same L'Etang locations. The last two digits of sampling locations refer to year of sampling. *Data points* and *vertical lines* represent mean and 95% confidence intervals

direction in L'Etang, starting with untreated pulp mill effluent entering near location 1, where organic loading was highest, then gradually diminishing with increasing distance to lowest levels at site 5. By contrast, a temporal comparison shows high organic levels at sister site 6 in 2000, after aquaculture start-up in 1981 [41]. While we have no 1975 volatile solids for Lime Kiln Bay (location 6), the indistinguishable community structure of locations 5 and 6 suggests low levels, as in site 5 at that time. We also have no organic matter data for central and upper L'Etang areas in 1997/2000. However, organic loading was decreased to acceptable levels with pulp mill effluent being treated in the intervening time. This concurs with the community structure of upper L'Etang locations 1 and 2 in recent times, developing more like that of the central L'Etang (Fig. 3b).

Cluster analysis (not shown) was used to group sample locations for further species composition investigation. This resulted in the two upper (U, 1 + 2) and two lower (L, 5 + 6) L'Etang sampling locations being combined into one data set (with 6 replicates) for each area, with the transitional

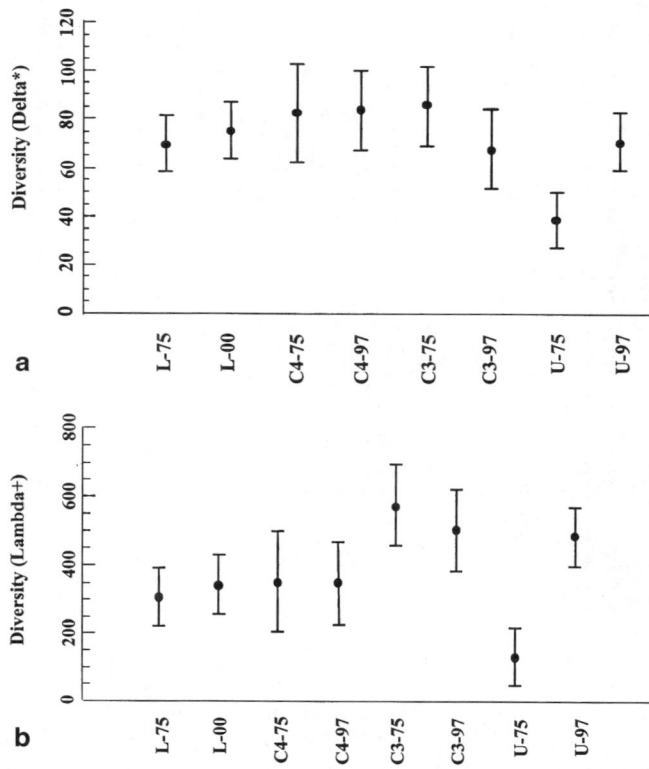

Fig. 5 L'Etang Inlet macrofauna. **a** Taxonomic distinctness (Delta *) for L'Etang sampling locations. **b** Variation in taxonomic distinctness for the same locations and areas as in Fig. 4

central (C) L'Etang locations (3 and 4) being analyzed separately as individual locations (each with three replicates).

A comparison of the numbers of species at these grouped locations (Fig. 4a) shows a rise with distance seaward for both sampling times.

However, the only significant difference between years is for the lower L'Etang locations 5 and 6 (L). Here the change is to a more diverse macrofauna in 1997 or 2000. A plot based on the Shannon diversity index (not shown), incorporating abundance data, confirms both these results. Corresponding trophic ratio analysis of the dominant species (Fig. 4b) showed that the only groupings that were significantly different between the two sampling eras were also for lower L'Etang. This reflects a change from a predominantly suspension feeding mixed community to a predominantly deposit feeding mixed one. The change in diversity is thus correlated with a shift toward deposit feeders. This is consistent with a scenario of increased sedimentation as a result of aquaculture discussed in Chap. 11.

Table 5 Dominant taxa in the L'Etang Inlet as determined in 1975 (D = deposit feeder, S = suspension feeder, C = carnivore)

Dominant taxa	Trophic type	1975 # m^{-2}	% Contribution to total density	1997 or 2000 # m^{-2}
		Upper L'Etang locations (U, #1 + 2)		
Capitella capitata	D	555	68	3
Oligochaeta	D	115	18	2
Polydora cornuta	S	60	7	0
Eteone longa	D	43	7	2
TOTAL			100	
		Central L'Etang location (C, #3)		
Ampelisca abdita	S/D	93	43	80
Hediste diversicolor	D/C	97	40	0
Unciola irrorata	D	37	8	0
Eteone longa	D	20	8	7
TOTAL			100	
		Central L'Etang location (C, #4)		
Unciola irrorata	D	175	46	0
Leptocheirus pinguis	S	305	27	0
Thyasira flexuosa	S	65	8	20
Harpinia propinqua	?	35	5	27
Nephtys incisa	D/C	30	5	10
TOTAL			92	
		Lower L'Etang locations (L, #5 + 6)		
Leptocheirus pinguis	S	2829	75	0
Unciola irrorata	D	459	9	0
TOTAL			84	

Using relatedness-based biodiversity indices, rather than the simpler measures above, revealed another temporal change. Both taxonomic distinctness measures (Fig. 5) showed significantly depressed levels in the upper L'Etang in 1975, compared to all other areas of the Inlet.

Subsequently, the upper L'Etang recovered to comparable levels elsewhere. The depressed 1975 levels correlate with known untreated pulp mill effluent entering the upper L'Etang at that time, as reflected by the highest levels of organic loading (Fig. 3b). The impact and subsequent recovery at this site is confirmed by the enrichment indicator sludge worm, *Capitella capitata*. It changed from dominant in 1975 to near-absent by 1997/2000 (Table 5), with effective effluent treatment at the pulp mill beginning in 1988.

Discriminant Species

Dominance classification will capture a different number of top ranked species according to the species evenness at each location. In Table 5 the 1975 lower L'Etang dominants consist of only two species, the amphipods *Leptocheirus pinguis* and *Unciola irrorata*. Both species were extirpated by 1997 or 2000 in this area, where aquaculture became established in the intervening time. At the adjacent location 4 the same two amphipods had also been extirpated, indicating that impacts may have extended into that area. At location 3 further up the L'Etang, *U. irrorata* and the polychaete *Hediste diversicolor* disappeared, whereas the amphipod *Ampelisca abdita* persisted. For the up-

Table 6 Dominant taxa in L'Etang Inlet in 1997–2000 (trophic types as described in Table 3)

Dominant taxa	Trophic type	1997 or 2000 number (m^{-2})	% Contribution to total density	1975 number (m^{-2})
Upper L'Etang locations (U, 1 and 2)				
Nepthys neotena	D/C	1033	95	0
Central L'Etang location (C, 3)				
Nepthys neotena	D/C	4480	99	0
Central L'Etang location (C, 4)				
Nepthys neotena	D/C	1290	86	0
Lower L'Etang locations (L, 5 and 6)				
Anobothrus gracilis	D	382	11	0
Euchone incolor	?	333	8	0
Ampharete lindstroemi	D	232	7	0
Terebllides stroemi	D	232	7	12
Prionospio steenstrupi	D	343	6	0
Polycirrus medusa	D	233	5	0
Thyasira flexuosa	S	177	5	3
TOTAL			51	

per L'Etang locations there is evidence that the 1975 dominants, including the sludge worm *Capitella capitata*, are much reduced in density and replaced in 1997 (Table 6) by the polychaete *Nepthys neotena*. The latter is also the case at central L'Etang locations 3 and 4. The dominant species in 1997/2000 in the lower L'Etang Inlet (Table 6) were more evenly distributed among the species and the seven taxa meeting the 5% criterion made up only 51% of the total density, indicating that other less dominant species were also significant contributors to density. Two of the seven dominants only were present in the earlier sampling, suggesting that the other five taxa had colonized this area and became dominant in the period between the before/after samplings.

St. Croix Estuary Reference Site

The total number of taxa identified at three locations in the St. Croix site was 129, based on ten grabs per location in 1974, and three grabs per location in 2000. This is comparable to the 156 species identified in the L'Etang Inlet,

Table 7 Dominant taxa in 1974 in the St. Croix estuary (trophic types as described in Table 3)

Taxa	Trophic type	1974 number m^{-2}	% Contribution to total density	2000 number m^{-2}
		Location 1		
Cyclocardia borealis	S	66	73	10
Yoldia sapotilla	D/S	19	15	7
Nepthys sp (?*incisa*)	D/C	12	9	23
TOTAL			97	
		Location 2		
Yoldia sapotilla	D/S	13	19	160
Rhodine gracilor	D	11	17	120
Astarte elliptica	S	12	11	0*
Astarte undata	S	7	10	13
Scoletoma fragilis	D	8	8	0**
Nepthys sp (?*incisa*)	D/C	6	6	40
Unciola irrorata	D	11	5	160
TOTAL			76	
		Location 3		
Cyclocardia borealis	S	86	61	10
Thyasira flexuosa	S	37	15	83
Yoldia sapotilla	D/S	18	9	117
Crenella glandula	S	17	7	30
TOTAL			92	

*Could be mistaken for *A. undata*
**Could be *Scoletoma tetraura*

based on six locations and three grabs per location in 1975 and 1997/2000. However, it was noticeable that the densities for most 1974 St. Croix taxa were low (except for *Cyclocardia borealis*, a large easily recognizable bivalve). The total number of species identified in 1974 (60) was also less than in 2000 (114). We believe that the small, but evident, separation between 1974 and 2000 data in the MDS plot (not shown) is due to taxonomic insufficiency of early sampling. Dominant species in 1974 are shown in Table 7, indicating the generally low density in comparison to 2000 data.

All of the species recognized as dominants in 1974 were also found in 2000, with the exception of *Astarte elliptica*, which is difficult to separate from another *Astarte* species present in 2000 (*A. undata*). The only other exception was *Scoletoma fragilis*, although *Scoletoma tetraura*, was present in 2000 at a density of 55 individuals m^{-2}, with which it could have been confused.

5
Discussion

Are there before/after Macrofaunal Changes?

The benthic macrofaunal community data we have analyzed for the L'Etang Inlet are consistent with rejecting the null hypothesis of no before/after differences. Based on the most sensitive multivariate measures, before/after differences were recognized at all of the L'Etang locations we examined. However, in addition to organic enrichment effects, there are other possible mechanisms that could cause the observed differences. These include (1) seasonal changes in community structure, (2) artificial changes introduced by variable sampling and taxonomic skills employed during this study, as well as (3) long-term changes in community structure which might result from changes in a range of limiting factors. They are examined in more detail below:

1. Seasonal changes in macrofaunal communities can result from annual occurrences of low dissolved oxygen in bottom water in enclosed ecosystems. These events kill resident benthic animals [65]. Another cause of seasonal change is the periodic recruitment of new individuals to the population [66]. However, because of the energetic tidal mixing in the Bay of Fundy the areas considered are not subject to annual anoxia and because we sampled in the same seasonal period (August to October) possible changes due to recruitment should be small.
2. It is clear that differences in sampling and taxonomic skills were employed during this work, although we do not know exactly how it would affect the results. Data from both periods were carefully edited and reviewed to minimize possible incongruencies that could affect the analysis.

3. Long-term changes in community structure of the macrofauna should be minimal if the benthic communities we studied were stable [67], suggesting that no region-wide physical or chemical changes had occurred in the period 1974 to 2000. We have tested this with measures based on macrofaunal species presence/absence (6 and 7). Such measures assess the sum of inequality in taxonomic skill as well as the long-term changes in benthic macrofaunal structure. Another conservative measure of change is TRS, and in the reference locations of St. Croix estuary a mean of 7.7% increase in TRS had occurred between 1974 and 2000 (Table 8). Most of this change we consider to be due to better taxonomic expertise in 2000, favoring the small, less conspicuous, deposit feeding individuals, more likely omitted in earlier sampling.

An independent way of determining long-term change and differences due to taxonomic analytical skills is to determine the persistence of dominants found in early samplings. Dominants should be present at later samplings if no natural physical limiting factors had changed in the interim. This does appear to be the case for all St. Croix estuary locations (Table 7).

We conclude that the changes in before/after macrofaunal communities we recognized in the lower L'Etang Inlet allow us to reject the null hypothesis of no change. The temporal changes found are consistent with far-field organic enrichment effects as discussed further below.

Organic Enrichment Effects on the L'Etang Inlet

Major sources of organic wastes within the L'Etang, sensu stricto, are known to be from a pulp mill and the salmon culture industry [68], and, sensu lato, from municipal and fish canning sources (Blacks Harbour). The before/after locations available to us allowed the examination of the first two of these sources.

Untreated pulp mill effluent first entered upper L'Etang Inlet in April 1971 and this continued until 1988 (Table 3) when the effluent was properly treated both for dissolved and particulate wastes. The purpose of building a second causeway across Pull-and-be-Damned Narrows was to limit the penetration of seawater, containing sulfate, that enabled anaerobic bacteria to proliferate using sulfate as the electron acceptor. The product of this anaerobic metabolism includes foul-smelling H_2S.

Other 1971 results from the L'Etang based on 3 mm^2 mesh data [57, 58] and unpublished 2000 data showed that predominantly suspension-feeding communities were replaced with deposit-feeding ones. Our samples from 1975, based on a 1 mm^2 mesh, confirm this and that the upper L'Etang Inlet was dominated by organic enrichment tolerant species (Table 5). After 1988, the benthic macrofauna began a recovery phase in that area which by 1997 was dominated by the deposit/carnivorous feeding polychaete *Nepthys neotena* (Table 6), absent in 1975. Relatedness indices of the upper L'Etang also show depressed biodiversity levels in 1975, increasing significantly over time (Fig. 5a and b) to levels of other sampling areas. Furthermore, the upper L'Etang recovery correlates with known effluent treatment starting in 1988. Similar changes were also found at the two more seaward locations (3 and 4). They too are the result of organic enrichment, but whether from the landward (pulp mill) or seaward (salmon mariculture) sources in the L'Etang is unclear. This uncertainty is due to the limited temporal samplings available over the 26-year period considered.

In 1975, prior to enrichment of the two lower L'Etang locations (stations L, 5 and 6), the net erosional sediments were occupied by two dominant amphipods (Table 5). Of these, *Leptocheirus pinguis* builds U-shaped burrows in the top 10 cm of the sediment profile. Here they brood their young for extended periods [69], the pleopods of the females being used to irrigate the burrows and bring particulate seston to the filtering apparatus on the anterior limbs. According to Bousfield [70] the co-dominant, *Unciola irrorata*, "is living in tubes of other organisms," such as *L. pinguis,* and therefore would share the same fate as that species. It is also known [71] that *L. pinguis* is susceptible to wave washout in storms because of their near surface position in the sediment. Our results show that these two species were extirpated from these locations by the time of the 1997 or 2000 samplings. These species were also extirpated from the more landward locations (3 and 4), as were other amphipods, including the tube-building *Corophium* sp. Tyler [72] followed the seasonal feeding preferences of 25–40 cm long haddock finding that a principal prey was *Leptocheirus pinguis*. The absence of amphipod species in recent times could have negative implications for commercially fished stocks such as haddock.

We have considered two possibilities to explain amphipod extirpation: wave washout or organic enrichment effects. We believe that the available circumstantial evidence favors the latter. This is because of the fundamental change in the benthic macrofaunal community at the lower L'Etang locations by 2000 to a mixed one dominated by deposit feeders from one dominated by suspension feeders in 1975 (Table 8). If the extirpation of amphipods had been due to wave washout, we would not expect to see these concomitant changes in benthic structure. Changes in the nature of the sediment particles are thought to have occurred in Lime Kiln Bay as a result of salmon mariculture (Chap. 11). Since the onset of aquaculture activity in 1981 suspended particles have become dominated by flocs and this has caused an increase in far-field sedimentation of available carbon sources. This is consistent with our data indicating increased organic loading after 1975 in the lower L'Etang Inlet. Thus amphipod extirpation in that area can be the result of sedimentation changes involving quality (increased floc content) as well as increased sedimentation amounts from far-field effects of salmon mariculture, supported by data on organic loading. Secchi disc depths are presented in Table 9, which

Table 8 Before/after trophic ratio estimates (TRS%) for St. Croix estuary and L'Etang Inlet locations. Species not in the estimate include those of an inappropriate trophic group or species of uncertain trophic type

Location	Year	Total taxa identified	Taxa used in the TRS	TRS%
St. Croix 1	1974	24	19	53
	2000	78	46	61
St. Croix 2	1974	33	27	52
	2000	84	53	58
St. Croix 3	1974	27	21	48
	2000	65	43	58
L'Etang 1	1975	1	1	100
	1997	12	6	89
L'Etang 2	1975	9	7	57
	1997	13	8	63
L'Etang 3	1975	10	7	57
	1997	20	18	67
L'Etang 4	1975	23	17	65
	2000	32	27	74
L'Etang 5	1975	49	31	48
	2000	67	47	72
L'Etang 6	1975	49	25	48
	2000	76	49	69

Table 9 Secchi disc depths (m) in L'Etang Inlet at different locations and sampling years. Early data from Wildish et al. [57] and later from J.L. Martin (personal communication). SE = standard error

Year	Location	Mean	2(SE)	n
1971/72	1	0.30	0.06	11
1971/72	2	0.77	0.17	12
1971/72	4	1.95	0.23	16
1971/72	5	3.24	0.38	16
1971/72	6	4.30	0.47	9
1991	5	2.41	0.02	32
1991	6	3.18	0.24	32
2000	5	–	–	–
2000	6	3.775	0.33	29

suggest that seawater at locations 5 and 6 has a lower opacity in recent years, perhaps due to an increased particulate load. Mean Secchi disc depth values between 1971/72 and later years for locations 5 and 6 are significantly different ($p > 95\%$). However, other factors, such as increased scallop dragging in L'Etang, may have contributed to this effect.

In this section we have described before/after, circumstantial evidence based on macrofaunal density which relates temporal change to organic enrichment incidents in the L'Etang Inlet. In the period 1971–1988 this was caused by pulp mill pollution in the most landward part of L'Etang. At the seaward end of the L'Etang we describe far-field effects circumstantially linked to salmon mariculture wastes in the period after 1981. The macrofaunal changes described herein are not related to known causes of seasonality, as reviewed in the first section, inclusive of those associated with recurrent hypoxia in bottom seawater or phases of recruitment to benthic populations. We propose here a new hypothesis that the temporal changes of macrofauna described in the lower L'Etang Inlet, results from increased amount and quality of sedimentation associated with intensive finfish mariculture.

Acknowledgements We thank Blythe Chang for providing maps and statistical data of the salmon culture industry and Hugh Akagi for help in preparing Table 3 and the figures.

References

1. Tilseth S, Hansen T, Moller D (1991) Aquaculture 98:1
2. Hargrave BT (1994) Can Tech Rep Fish Aquat Sci 1949:79
3. Gowen RJ, Bradbury NB (1987) Oceanogr Mar Biol Annu Rev 25:563
4. Black KD (2001) Environmental impacts of aquaculture. CRC Press, Boca Raton
5. Bennett BA, Smith CR, Glaser B, Maybaum H (1994) Mar Ecol Prog Ser 108:205
6. van de Bund WJ, Olafsson E, Modig H, Elmgen R (2001) Mar Ecol Prog Ser 212:107
7. Pohle GW, Lim SSL, Frost BR (1994) In: MacKinnon BM, Burt MDB (eds) Proceedings of the workshop on ecological monitoring and research in the coastal environment of the Atlantic maritime ecozone. Env Can Atl Reg Occ Rep No 4:92
8. Fisheries and Oceans Canada(2004) A scientific review of the potential environmental effects of aquaculture in aquatic ecosystems, Vol III, Near-field organic enrichment from marine finfish aquaculture (Wildish DJ, Dowd D, Sutherland TF, Levings CD). Can Tech Rep Fish Aquat Sci 2450:1–66
9. Perkins EJ (1974) The biology of estuaries and coastal waters. Academic, London
10. Cromey CJ, Nickell TD, Black KD (2002) Aquaculture 214:211
11. Gray JS, Wu RS-S, Or YY (2002) Mar Ecol Prog Ser 238:249
12. Butman CA, Grassle JP (1992) J Mar Res 50:669
13. Grassle JP, Butman CA, Mills SW (1992) J Mar Res 50:717
14. Pearson TH, Gray JS, Johannsen PJ (1983) Mar Ecol 12:237
15. O'Connor B, Costelloe J, Dineen P, Faull J (1993) ICES CM 1993/F:19
16. Johannsen PJ, Boten HB, Tvedten OF (1994) Aquacult Fish Manage 25:55
17. Lu L, Wu RSS (1998) Env Pollut 101:241
18. Karakassis I, Hatziyanni E, Tsapakis M Plaiti W (1999) Mar Ecol Prog Ser 184:205

19. Wildish DJ, Akagi HM, Hamilton N (2001) Bull Aquacult Assoc Can 101:49
20. Brooks KM, Stierns AR, Mahnken CVW (2003) Aquaculture 219:355
21. Pearson TH, Rosenberg R (1978) Oceanogr Mar Biol Annu Rev 16:229
22. Brown JR, Gowen RJ, McLusky DS (1987) J Exp Mar Biol Ecol 109:39
23. Weston DP (1990) Mar Ecol 61:233
24. Henderson AR, Ross DJ (1995) Aquacult Res 26:659
25. Karakassis I, Hatziyanni E (2000) Mar Ecol Prog Ser 203:247
26. Brooks KM, Mahnken VW (2003) Fish Res 62:255
27. Findlay RH, Watling L, Mayer LM (1995) Estuaries 18:145
28. Poole NJ, Wildish DJ, Kristmanson DD (1978) CRC Crit Rev Environ Control 8:153
29. Rhoads DC, Germano JD (1986) Hydrobiologia 142:291
30. Nilsson HC, Rosenberg R (1997) J Mar Systems 11:249
31. Wildish DJ, Hargrave BT, Pohle G (2001) ICES J Mar Sci 58:469
32. Wildish DJ, Akagi HM, Hamilton N, Hargrave BT (1999) Can Tech Rep Fish Aquat Sci 2286:31
33. Hargrave BT (1994) Can Tech Rep Fish Aquat Sci 1949:79
34. Cranston R (1994) Can Tech Rep Fish Aquat Sci 1949:93
35. Holmer M, Lassus P, Stewart JE, Wildish DJ (2001) ICES J Mar Sci 58:363
36. Wildish DJ, Hughes-Clarke JE, Pohle GW, Hargrave BT, Mayer LM (2004) Mar Ecol Prog Ser 267:99–105
37. Poole NJ, Wildish DJ (1979) In: Berkley RCW, Ellwood DC, Gooday GW (eds) Microbial polysaccharides and their degradation. Academic, London, p 399
38. Pearson TH (2001) Oceanogr Mar Biol Annu Rev 39:233
39. Heilskov AC, Holmer M (2001) ICES J Mar Sci 58:427
40. Nickell LA, Black KD, Hughes DJ, Overnell J, Brand T, Nickell TD, Breur E, Harvey SM (2003) J Exp Mar Biol Ecol 221:285–286
41. Pohle G, Frost B, Findlay R (2001) ICES J Mar Sci 58:417
42. Wildish DJ, Kristmanson DD (1997) Benthic suspension feeders and flow. Cambridge University Press, New York
43. Wildish DJ (1977) Helg Wiss Meeresunters 30:445
44. Wildish DJ, Fader GBJ, Lawton P, MacDonald AJ (1998) Con Shelf Res 18:105
45. Kostylev VE, Todd BJ, Fader GBJ, Courtney RC, Cameron GDM, Pickerill RA (2001) Mar Ecol Prog Ser 219:121
46. Wildish DJ, Stewart PL (2004) Can Sci Advisory Secretariat Res Doc 2004/012:31
47. Wildish DJ, Carson WG, Carson WV, Hull JH (1972) Fish Res Board Can MS Rep No 1177
48. Bourque BJ (2001) Twelve thousand years American Indians in Maine. University of Nebraska Press, Lincoln
49. Doucet R, Wilbur R (2000) Herring weirs: the only sustainable fishery. Image Express, St. George
50. Anderson JM (1984) Atlantic Salmon J 2:26
51. Wildish DJ, Martin JD, Wilson AJ, De Coste AM (1986) Can Tech Rep Fish Aquat Sci 1473
52. Wildish DJ, Akagi HM, Martin A (2003) Can Tech Rep Fish Aquat Sci 2447
53. Silvert W (1994) Can Tech Rep Fish Aquat Sci 1949:1
54. Holme NA, McIntyre AD (1971) (Eds) Methods for the study of marine benthos, IBP Handbook No 16. Blackwell, Oxford
55. Hunter B, Simpson AE (1976) J Mar Biol Assoc UK 56:951
56. Wildish DJ (1983) In: Thomas MLH (ed) Marine and coastal systems of the Quoddy region, New Brunswick. Can Spec Pub Fish Aquat Sci Ottawa 64:140

57. Wildish DJ, Carson WV, Wilson AJ, Hull JH (1974) Fish Res Board Can MS Rep No 1295
58. Wildish DJ, Poole NJ, Kristmanson DD (1977) Fish Mar Serv Tech Rep No 718
59. Wildish DJ, Wilson AJ, Akagi HM (1977) Fish Mar Serv MS Rep No 1462
60. Wildish DJ (1985) J Mar Biol-Assoc Ur 65:335
61. Wildish DJ, Kristmanson DD (1979) J Fish Res Board Can 36:1197
62. Akagi H, Wildish DJ (1975) Fish Res Board Can MS Rep No 1370
63. Warwick RM, Clarke KR (2001) Oceanogr Mar Biol Ann Rev 39:207
64. Findlay RH, Watling G (1997) Mar Ecol Prog Ser 155:147
65. Diaz RJ, Rosenberg R (1995) Oceanogr Mar Biol Annu Rev 33:245
66. Kitching JA, Ebbling FJ, Gamble R, Hoate AA, Mcleod QR, Norton TA (1976) J Anim Ecol 45:731
67. Gray JS, Valderhaug V, Ugland KI (1985) In: Gibbs PE (ed) Proceedings European Marine Biology Symposium. Cambridge University Press, Cambridge, p 245
68. Strain PM, Wildish DJ, Yeats PA (1995) Mar Poll Bull 30:253
69. Thiel M, Sampson S, Watling L (1997) J Nat Hist 31:713
70. Bousfield EL (1973) Shallow-water gammaridean amphipoda of New England. Cornell University Press, Ithaca
71. Wildish DJ (1980) Int J Invert Reprod 2:311
72. Tyler AV (1972) J Fish Res Board Can 29:997

A Review and Assessment of Environmental Risk of Chemicals Used for the Treatment of Sea Lice Infestations of Cultured Salmon

K. Haya[1] (✉) · L. E. Burridge[1] · I. M. Davies[2] · A. Ervik[3]

[1]Department of Fisheries and Oceans, St. Andrews Biological Station, 531 Brandy Cove Road, St. Andrews E5B 2L9, UK
hayak@mar.dfo-mpo.gc.ca, burridgel@mar.dfo-mpo.gc.ca

[2]Fisheries Research Services Marine Laboratory, 375 Victoria Road, Torry, Aberdeen AB11 9DB, UK
i.m.davies@marlab.ac.uk

[3]Institute of Marine Research, 5817 Bergen, Norway
arnee@imr.no

1	Introduction	306
2	Sea Lice Biology	307
3	Therapeutants in Use	309
4	Bath Treatments	310
4.1	Organophosphates	310
4.1.1	Efficacy and Mechanism of Action of Organophosphates	310
4.1.2	Distribution and Fate of Organophosphates	311
4.1.3	Biological Effects of Organophosphates	312
4.2	Pyrethroids and Pyrethrins	314
4.2.1	Efficacy and Mechanism of Action of Pyrethroids and Pyrethrins	314
4.2.2	Distribution and Fate of Pyrethroids and Pyrethrins	315
4.2.3	Biological Effects of Pyrethroids and Pyrethrins	316
4.3	Hydrogen Peroxide	318
4.3.1	Efficacy and Mechanism of Action of Hydrogen Peroxide	318
4.3.2	Distribution and Fate of Hydrogen Peroxide	319
4.3.3	Biological Effects of Hydrogen Peroxide	319
5	In-Feed Treatments	320
5.1	Avermectins	320
5.1.1	Efficacy and Mechanism of Action of Avermectins	320
5.1.2	Distribution and Fate of Avermectins	321
5.1.3	Biological Effects of Avermectins	322
5.2	Chitin Synthesis Inhibitors	324
5.2.1	Efficacy and Mechanism of Action Chitin Synthesis Inhibitors	324
5.2.2	Distribution and Fate of Chitin Synthesis Inhibitors	325
5.2.3	Biological Effects Chitin Synthesis Inhibitors	326

6	Risk Assessment of Sea Lice Therapeutants	328
6.1	Bath Treatments	331
6.2	In-Feed Treatments	333
7	Risk Management	334
8	Conclusion	336
	References	336

Abstract Chemicals (sea lice therapeutants) currently authorized in North America and Europe for the treatment of sea lice infestations in cultured salmon may be classified into two major groups. The classification is based on their routes of administration, and includes bath techniques (organophosphates, pyrethroids and hydrogen peroxide) and additives in feed (avermectins, chitin synthesis inhibitors). The ecological risk posed by the use of the chemicals is reviewed and assessed in this chapter. While the biological effects of sea lice therapeutants on aquatic animals that may live near salmon culture sites have been studied under laboratory conditions, field studies on the efficacy, fate and distribution, and biological effects are limited. In general, the in-feed treatments are more convenient to administer and posed less ecological risk than the bath treatments. As an example, the approach adopted by the UK was used to assess the environmental safety of the sea lice therapeutants. It was concluded that there are considerable differences between the environmental characteristics of fish farm sites and their ability to accept discharges of sea lice treatments without giving rise to unacceptable environmental impacts. Such site-specific risks can be managed through the application of appropriate environmental quality standards for the chemicals concerned, and site-specific assessment of the maximum acceptable rate of use of the treatments.

Keywords Aquaculture therapeutants · Antiparasidics · Ecotoxicolgy · Efficacy · Risk assessment

1
Introduction

Fin-fish marine aquaculture is a relatively new but important food industry with a worldwide production of 3.79 M metric tonnes in 2002 of which 47.5% was salmonid species valued at $4.9 B (US) [1]. Cultured Atlantic salmon, *Salmo salar*, comprised 60% of the salmonid species production, and of which 91% was produced in Canada, Chile, Norway and the United Kingdom. The demand for cultured fisheries products is increasing with the continued decline in catchable wild fisheries and an increase in demand by consumers [2]. The challenges for the aquaculture industry are to expand, remain environmentally sustainable and be socially acceptable [3].

Cultured salmon in the crowded and stressful conditions of aquaculture are susceptible to epidemics of infectious bacterial, viral and parasitic diseases. Sea lice are ectoparasites of many species of fish and are a serious problem for salmon aquaculture industries [4, 5]. The species that infest cultured Atlantic salmon are *Lepeophtheirus salmonis* and *Caligus elongatus*. Infestations result in skin erosion and sub-epidermal haemorrhage, which, if left untreated, would result in significant fish losses, probably as a result of osmotic stress and other secondary infections [6,7]. Sea lice are natural parasites of wild Atlantic salmon and infestations have occurred routinely in European aquaculture since the 1970s [4]. The first severe epidemic in Atlantic Canada occurred in 1994 [8,9]. Sea lice reproduce year round and the aim of a successful lice-control strategy must be to pre-empt an internal infestation cycle from becoming established on a farm by exerting a reliable control on juvenile and preadult stages, thus preventing the development to gravid females [10]. Effective control of sea lice infestations requires good husbandry, linked to the use of natural predators such as wrasse and effective anti-parasitic chemicals [11–13].

Aquaculture, like all forms of intensive food production, will potentially generate environmental costs. Chemicals used in the treatment of sea lice infestations are subsequently released to the aquatic environment and may have an impact on other aquatic organisms and their habitat. The present paper will review the chemical therapeutants available to control sea lice and assess their risks to the aquatic ecosystem. The review will be limited to those chemicals that are currently authorized for use by the salmon aquaculture industry in Europe and North America.

2
Sea Lice Biology

Adult females of *Caligus elongatus* are 6 to 8 mm in length, while males are about half of this size (Fig. 1). The sea lice on cultured fish tend to be a bit smaller than those on wild fish. Likely sources of new infections of fish farms are planktonic stages, which may originate from sea trout, wild salmon or rainbow trout or salmon that have escaped from captivity, although of low prevalence, *L. salmonis* has also been found on other marine species.

The eggs of the sea lice hatch directly into the water from egg strings fastened to the genital segment of the female lice. The larvae are free-swimming nauplii through one moult and then become infective copepodids. These are about 0.7 mm long and 0.3 mm wide, and it is this stage that can recognize and become attached to a host fish. It is, however, observed that adult sea lice can transfer from fish to fish. The dispersion of the nauplii is primarily passive as the larvae drift in the water, but the vertical movements of the larvae (copepodids are positively phototaxic) will also influence their position in a water column. In total, the sea lice pass through 10 stages, with one moult between each stage [14].

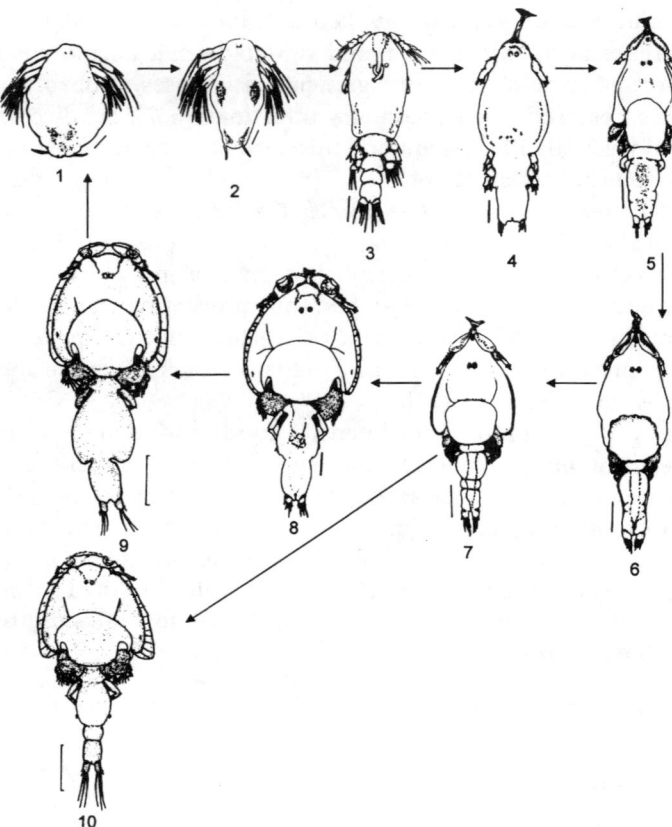

Fig. 1 Life cycle of sea lice, *Caligus elongatus*. Legend: 1 Nauplius I; 2 Nauplius II; 3 Copepodid; 4 Chalimus I; 5 Chalimus II; 6 Chalimus III; 7 Chalimus IV; 8 Pre-adult; 9 Adult (female); 10 Adult (male). *Scale bars*: 1–4 = 0.1 mm; 5–6 = 0.2 mm; 7–8 = 0.5 mm; 9–10 = 1 mm. Drawings courtesy of Bill Hogans

The rate of development of sea lice is dependent on the sea temperature. It takes a male 42 days, and a female 50 days, to develop from egg to adult at 10 °C, a temperature that is normal in salmon farming areas. The sea lice can, however, tolerate relatively large range of temperatures and can hatch and develop at as low as 2 °C [15].

The period during which a copepodid can infect a fish is called the "infective window" and is crucial in the control of sea lice. Larvae can infect fish from the first day after molting, but they appear to be more infective after a few days. Longer than this, and the copepodid exhausts its energy reserves and becomes less successful in infecting susceptible host fish. Calculations on empirical data indicate that the latest day that a larva can infect fish is 32.5 days after hatching at 6 °C and 17 days at 12 °C. Such long infective pelagic stages suggest that *L. salmonis* has a great potential for dispersion and that it can infect fish over a wide area away from the source. Thus, massive infection problems may be encountered by the salmon farming industry. This emphasizes the need for efficient husbandry strategies and chemical agents to control infections on fish farms and to reduce the potential for transfer of lice between farms.

3
Therapeutants in Use

Chemicals currently authorized for the treatment of sea lice infestation may be classified into two groups based on their route of administration (Table 1). Organophosphates (azamethiphos), pyrethroids (cypermethrin and deltamethrin) and hydrogen peroxide are administered by bath techniques,

Table 1 Chemical therapeutants currently authorized in Europe and North America for the treatment of sea lice infestations in salmon mariculture

Therapeutant	Dosage	Quantity used			
		Canada 2002	Norway 2003	Ireland 2003	Scotland 2003
Bath					
Azamethiphos	0.1 mg L^{-1} for 60 min	15.0 kg			24.2 kg
Cypermethrin	0.005 mg L^{-1} for 60 min		59 kg	107.6 L	76.0 kg
Deltamethrin	0.002–0.003 mg L^{-1} for 40 min		16 kg		
Hydrogen Peroxide	500 mg L^{-1} for 20 min				19.3 L
Medicated Feed					
Emamectin Benzoate	0.05 mg kg^{-1} d^{-1} for 7 d	25.0 kg	23 kg	4.97 kg	24.8 kg
Teflubenzuron	10.0 mg kg^{-1} d^{-1} for 7 d 2.0–3.0 mg kg^{-1} d^{-1} for 14 d				36.0 kg

Canada: Veterinary Drugs Directorate, Health Canada, Ottawa
Norway: Norwegian Medical Depot
Ireland: Marine Institute, Galway
Scotland: Scottish Environmental Protection Agency, Dingwall

while avermectins (emamectin benzoate) and chitin synthesis inhibitors (teflubenzuron and diflubenzuron) are administered as additives in medicated feed. The number of chemicals authorized for use is limited because of the high cost of development and licensing for a small market relative to other markets for pesticides and medicinals.

Bath treatments are conducted by reducing the depth of the net in the salmon cage, thus reducing the volume of water. The net-pen and enclosed salmon is surrounded by an impervious tarpaulin and the chemical is added to the recommended treatment concentration. The salmon are maintained in the bath for a period of time (usually 30–60 min) and aeration/oxygenation may be provided. After treatment, the tarpaulin is removed and the treatment chemical is allowed to disperse into the surrounding water. Bath treatments are considered a topical application as the therapeutant is absorbed by the sea lice from the water.

Medicated feed is prepared by adding concentrated pre-mix containing the active ingredient to feed during the milling and pelletization processes. The chemical is administered by calculating the dosage based on the feed consumption rate of the salmon. Generally, the medicated feed is given on the first feeding of the day, as this can counteract any reduction in appetite of infected fish, or any tendency of the fish to discriminate against medicated feed. The therapeutant is absorbed though the gut into the bloodstream of the salmon and is then transferred to the sea lice as they feed on the skin of the salmon. Advantages of in-feed preparations compared to bath treatments are that releases to environment are much slower and less direct. Treatment is less stressful to the fish, the dosage can be more accurately controlled, the oral preparations are not toxic to farmers, and it requires less labor. One disadvantage, however, is that stressed or diseased fish often feed less than healthy fish and therefore may not receive a fully effective dose.

4
Bath Treatments

4.1
Organophosphates

4.1.1
Efficacy and Mechanism of Action of Organophosphates

Four organophosphate compounds have been used in the treatment of infestations of sea lice: malathion, trichlorfon, dichlorvos (DDVP) and azamethiphos [4]. Malathion was tested experimentally but the concentration required to effectively remove lice was so high that the treated fish became lethargic. Trichlorfon also had a narrow margin of safety for salmon [16]. Trichlorfon degrades into the more toxic and effective DDVP, but the rate of transformation is dependent on water temperature and pH [4]. The inconsistency of this transformation, the acute toxic risk to salmon and the increase in use of DDVP resulted in the gradual cessation of use of trichlorfon. For a number of years, DDVP was the treatment of choice against infestations of sea lice. However, frequent use led to the resistance to DDVP in sea lice in some areas [17]. This coupled with a small therapeutic index (dose toxic to salmon/dose used to treat sea lice), about 4, resulted in the product being phased out as an anti-louse therapeutant.

Azamethiphos is an organophosphate insecticide and the active ingredient in the formulation Salmosan. It is used as a bath treatment at 0.1 mg L^{-1} for up to 1 h and has a therapeutic index near 10. Azamethiphos is registered for use in Chile, Ireland, Norway and Scotland. Novartis, the producer of azamethiphos applied to discontinue the use of their product in Canada from April 1, 2002. Aquaculturists may continue to use azamethiphos until April 1, 2005 (Cathy Morris, Health Canada, personal communication). Similar initiatives have occurred in other countries.

Atlantic salmon can tolerate 1-h exposures to 0.5 mg L^{-1} of azamethiphos, and three, 1-h doses repeatedly weekly to 0.3 mg L^{-1} of azamethiphos. Deaths were observed after 1 h exposure to 1.0 mg L^{-1} [4]. Azamethiphos has neurotoxic action, acting as an acetylcholinesterase (AChE) inhibitor. The AChE levels in the brain were reduced by 74% but the depression of AChE was not cumulative in the Atlantic salmon that died during the treatment with azamethiphos,. Exposure at 1 mg L^{-1} for 24 h resulted in 15% mortality of Atlantic salmon after 24 h [18]. Azamethiphos has been shown to be mutagenic in several in vitro tests [19]. The high alkylating potency of azamethiphos explains this mutagenicity and it was recommended that biological effects studies on non-target biota should include tests for delayed effects [20].

The sensitivity of lice to azamethiphos is variable, and some populations of lice are more sensitive to this compound than others [21]. Development of resistance to organophosphates is common and has been shown for azamethiphos [22]. In sensitive populations of lice, azamethiphos is effective in removing > 85% of adult and pre-adult lice but is not effective against the earlier life stages of the parasite [21].

4.1.2
Distribution and Fate of Organophosphates

Azamethiphos is soluble in water (1.1 g L^{-1}) and has a low octanol-water partition coefficient (log K_{ow} = 1.05) [23]. Consequently, azamethiphos is likely to remain in the aqueous phase upon entering the environment. It is unlikely to accumulate in tissue or in sediment. Azamethiphos decomposes by hydrolysis in natural water with a half-life of 8.9 days. Dispersion studies indicated that after release of an experimental treatment (200 µg L^{-1}), the concentration of azamethiphos was below detection (0.1 µg L^{-1}) in a short period of time. It was not detected below 10 m depth and it was suggested that it is unlikely that azamethiphos would accumulate in sediment.

The bioaccumulation of azamethiphos by salmon is low and depletion of total azamethiphos in salmon is rapid and the withdrawal time is 24 h [19] prior to harvesting.

4.1.3
Biological Effects of Organophosphates

4.1.3.1
Laboratory Studies with Organophosphates

Lobster and shrimp were the most susceptible species to azamethiphos in laboratory acute toxicity tests, while bivalves such as scallops and clams were unaffected [24]. Adult American lobsters, *Homarus americanus*, held within the tarpaulin during an operational treatment did not survive [25]. The 48 h LC50's estimated for the first four larval stages and adults of the American lobster are: Stage I 3.57 µg L^{-1}, Stage II 1.03 µg L^{-1}, Stage III 2.29 µg L^{-1}, Stage IV 2.12 µg L^{-1}, and Adults 1.39 µg L^{-1} [26]. There was no statistically significant difference between these values. There is a seasonal aspect to susceptibility of American lobsters to azamethiphos. Female lobsters are significantly more sensitive to azamethiphos in the summer than at any other time of year [27]. For Adult and Stage IV lobsters exposed repeatedly for varying length of time to four concentrations of azamethiphos [28], the no observed effect level (NOEL) was nine exposures of 2 h each over three days to 1 µg L^{-1} of azamethiphos. In addition to observed lethality, several surviving lobsters showed significant behavioral responses after repeated exposure to concentrations greater than 10 µg L^{-1}.

Research commissioned by Ciba Geigy shows that azamethiphos is only lethal to several groups of invertebrates (bivalve mollusks and gastropods), amphipods, and echinoderms) at concentrations greater than the prescribed treatment concentration of 100 µg L^{-1} [23]. Exceptions to this include copepod, decapod and mysid crustaceans. The 24 h LC50 of azamethiphos to the copepod, *Temora longicornis*, is reported to be > 10 µg L^{-1}. The 96 h LC50 for European lobster larvae, *Homarus gammarus*, is 0.5 µg L^{-1} and is in general agreement with the 48 h LC50 for the American lobster [26]. Finally, the 96 h LC50 for the mysid shrimp, *Mysidopsis bahia*, is reported as 0.52 µg L^{-1} [23].

In laboratory studies, American lobsters exposed to azamethiphos (5.0–10.0 µg L^{-1}) became quite agitated, often 'flopping' erratically around the exposure tank [29]. They were also aggressive to other lobsters and reacted very quickly to any movement. They seemed to lose control of their claws and eventually flipped onto their backs and died within hours. Some affected lobsters remained moribund for periods of time ranging from hours to days. The consequences of behavioral responses such as these on organisms and populations in the natural environment are unknown.

Laboratory studies were conducted to investigate possible sublethal effects of azamethiphos exposure on American lobster. Preovigerous females were exposed for 1 h biweekly to 10 $\mu g\,L^{-1}$ azamethiphos and monitored for spawning success and survival [29, 30]. Surprisingly, even with such infrequent exposures, up to 100% of the animals exposed to this concentration died during the experiment: some expired after only three treatments. At a lower concentration, a significant number of the surviving lobsters failed to spawn. A laboratory study indicated that shelter use behavior could be affected by azamethiphos [31]. However, exposure to concentrations of azamethiphos in water was greater than five times the recommended treatment concentration for periods of several hours.

The response of mussels to stimuli was unaffected by exposures to 10.0 $\mu g\,L^{-1}$ for up to 24 h [23]. The inhibition of AChE by azamethiphos is not cumulative in fish [4]. However, cumulative inhibition of AChE may have occurred in lobster in the studies above [29]. Mussel closure rate was affected at concentrations above 100 $\mu g\,L^{-1}$ and exposure to 46.0 $\mu g\,L^{-1}$ resulted in 50% inhibition of AChE activity. AChE activity in herring yolk sac larvae and post-yolk sac larvae was inhibited by 96 h exposure to azamethiphos at 33.4 and 26.6 $\mu g\,L^{-1}$, respectively. Herring larvae were reported to tolerate azamethiphos better than DDVP [4].

4.1.3.2
Field Studies with Organophosphates

During 1995, a study was conducted to determine the effects of single operational azamethiphos treatments on juvenile and adult American lobsters, shrimp, *Pandalus montagui*, clams, *Mya arenaria*, and scallops, *Placopecten magellanicus*, suspended at two depths and varying distances from the treated cage. During two of the treatments, all lobsters held within the treatment tarpaulin died [32]. No other treatment-related mortalities were observed. In addition, no mortalities were observed with lobsters that were suspended at three depths at 20 sites surrounding a salmon cage site that was conducting operational treatments with azamethiphos. Mussels deployed during field trials in Scotland were unaffected [23]. Mortality among lobster larvae was 27% but was not correlated to distance from the treatment cage.

Finally, survival of American lobsters suspended at mid-depth and near bottom at four sites in the salmon farming area of Lime Kiln Bay, New Brunswick, Canada, plus a control site, was monitored for 9 weeks during August–October 1996. There were no apparent differences in lobster survival between the experimental and control sites [25]. No residues of azamethiphos were detected in water samples collected weekly from the five sites (Detection Limit = 50 pg L^{-1}). Diving surveys at a lobster nursery area located near a salmon farm in early August, September and late October of 1996 found no apparent changes in lobster populations over time, and the area was found to have a considerable population of juvenile lobsters.

Measurements of primary productivity and dissolved oxygen were made before, during and after chemical treatments at salmon farms in southwest New Brunswick in August–September 1996. There were no evident effects on dissolved oxygen and chlorophyll *a* levels, indicating no impact on primary production (D. Wildish, St. Andrews Biological Station, St. Andrews, NB, unpublished data).

4.2
Pyrethroids and Pyrethrins

4.2.1
Efficacy and Mechanism of Action of Pyrethroids and Pyrethrins

Pyrethrins are the active constituents of an extract from flower heads of *Chrysanthemum cinerariaefolium*. This mixture of chemically related compounds has been used for their insecticidal activity since the late 19th century [33]. The pyrethrins decompose readily as they are susceptible to catabolic enzymes and sunlight. In the early 1960s, synthetic analogues that were more persistent than the natural pyrethrins were developed and referred to as pyrethroids [34]. It was their high degradability, low toxicity to mammals and high toxicity to crustaceans that led to the initial interest in pyrethins as treatments for sea lice infestations.

The mechanism of action of the pyrethrins involves interference with nerve membrane function, primarily by their interaction with Na channels [35] which results in depolarization of the nerve ending. This interaction results in repetitive firing of the nerve ending in the case of the pyrethroids, cypermethrin and deltamethrin.

A method used for delousing salmon with pyrethrins is to put an oil based 10.0 mg L^{-1} solution of pyrethrins in a 5-m tube, and then pass salmon through the solution in the tube [36]. The time for the salmon to pass through the solution was approximately 5–30 s and overall effectiveness has been reported as 96%. An advantage of this tube method is that the treatment solution can be recovered from the tube and not released to the marine environment.

In the autumn of 1989, a modified version of the bath method for delousing was tested [37]. The technique was based on using pyrethrins mixed in oil instead of adding a synthetic emulsifier to make a water-soluble solution. The oil-based solution was allowed to float on top of the water in a cage and the sea lice are exposed as the infested salmon jump out of the water. The water solubility of salmon mucus was expected to protect the fish but the salmon louse with a lipid layer in the cuticle should selectively absorb the pyrethrin mixture. Tests suggested that three jumps would give acceptable delousing (85%). However, this method was considered too susceptible to changes in fish behavior and to the amount of decomposition of the pyrethrins with variations in sunlight.

The pyrethrins act only on adult and pre-adult life stages [38], and aquaculturalists have preferred several pyrethroids in conventional bath treatment techniques. Deltamethrin and cypermethin are approved for use in Norway, and cypermethrin is approved in Ireland and the United Kingdom. Cypermethrin had temporary registration in the United States but it was recently withdrawn. The application for use in Canada for the treatment of sea lice was not approved by the Canadian Pest Management Authority.

The recommended treatment of salmon against sea lice is a 1-h bath with cypermethrin at a concentration of $5.0\,\mu g\,L^{-1}$, and for deltamethrin it is $2.0-3.0\,\mu g\,L^{-1}$ for 40 min [39] Cypermethrin is effective against all attached stages including adults, and therefore less frequent treatments should be required than with organophosphates; 5–6 week intervals rather then 2–3 week intervals, respectively.

In one of five Norwegian salmon sites that used deltamethrin for the treatment of sea lice, there was a significant decrease in the effectiveness of the treatment with an increase in the number of treatments [40]. Bioassays using preadult stage II sea lice under laboratory conditions verified that resistance contributed to treatment failure, and that the EC50 was 25 times higher than at an area previously unexposed.

4.2.2
Distribution and Fate of Pyrethroids and Pyrethrins

Synthetic pyrethroids are unlikely to be accumulated to a significant degree in fish and aquatic food chains since they are rapidly metabolized [41]. However, pyrethroids such as cypermethrin can persist in sediments for weeks and may be desorbed and affect benthic invertebrates. While there is a large amount of knowledge regarding the ecotoxicology of cypermethrin in freshwater environments [41–43], knowledge is more limited for marine species.

The pyrethrins are unstable with greater than 30% loss in sea water after 1 h [44] and a half life of 5 h [45]. A $10\,mg\,L^{-1}$ solution will lose effectiveness for the treatment of sea lice after 1 h, but will remain effective if an antioxidant (piperonylbutoxide) is added [46]. The concentration of cypremethrin decreases rapidly on release from a cage site after treatment. Data collected in Loch Eil Scotland showed that the highest concentration found was $187\,ng\,L^{-1}$ 25 min after release 25 m from the site in the direction of the current flow [39]. Cypermethrin remained above $0.031\,ng\,L^{-1}$ up to 50 min after release and above $0.074\,ng\,L^{-1}$ up to 30 min [47]. Mussels exposed inside a treated cage ($5.0\,ug\,L^{-1}$ cypermethrin) accumulated $133\,\mu g\,g^{-1}$.

Mussels 2 m from cages accumulated 9.2 ng g^{-1} after seven treatments and cypermethrin was only occasionally detectable 100 m from the cage. There were no effects on shrimp, *Crangon crangon*, used as sentinel species near the cage site. Organisms in the vicinity of the cages would be exposed to concentrations which fall to 50 ng L^{-1} within 1 h of release [39].

In aerobic sediments, cypermethrin biodegrades with half lives of 35 and 80 d, in high and low organic sediment, respectively. It degrades much more slowly in anaerobic sediments [39]. The rapid disappearance of deltamethrin from water (60% in 5 min), its high adsorption on sediment and its low bioconcentration capacities (for example, in daphnia, *Chlorella asellus*, indicate that this molecule will not accumulate through food chains. Nevertheless, its high toxicity and rapidity of action could cause significant harm to limnic ecosystems after direct treatment [48]. The adsorption of pyrethroids onto suspended solids can produce dramatic reductions in the apparent toxicity of the compound. The 96 h LC50 value for rainbow trout is 1.0–0.5 µg L^{-1} [49]. When trout were caged in a pond containing 14–22 mg L^{-1} suspended solids, the 96 h LC50 was 2.5 µg L^{-1}. In a pond sprayed with deltamethrin containing 11 and 23 mg L^{-1} suspended solids, deltamethrin partitioned rapidly to suspended solids, plants, sediment and air with a half life of 2–4 h in water [50]. Because pyrethroids tend to adsorb onto particulate matter, chronic exposures may not occur other than in laboratory studies. Cypermethrin adsorbed by sediment was not acutely toxic to grass shrimp, *Palaemonetes pugio*, until concentrations in sediment were increased to the point where partitioning into the overlying water resulted in acutely lethal concentrations [51]. For example, the 96 h LC50 for cypermethrin to grass shrimp is 0.016 µg L^{-1}, but grass shrimp could tolerate cypermethrin concentrations in sediment of 10.0 µg kg^{-1} for 10 days.

4.2.3
Biological Effects of Pyrethroids and Pyrethrins

The impact of pyrethroids and natural pyrethrins on non-target aquatic animals, especially invertebrates has been reviewed [52]. In general pyrethroids are more toxic to non-target insects and crustaceans than to other phylogenetically distant invertebrates. Among arthropods, however, crustaceans are phylogenetically closer to insects than mollusks and showed noticeable sensitivity. The isopod, *Asellus aquaticus* and the mysid shrimp, have shown even higher sensitivities than other crustaceans to pyrethroids, including cypermethrin and permethrin. Spray operations on ponds have resulted in 95% reduction of arthropod fauna such as crustaceans, insects and arachnids. The residue profile of cypermethrin in water immediately after application, coupled with rapid decay (4–24 h), explained the limited effect of pyrethroids on populations of non-target aquatic invertebrates in some case studies. On the other hand, invertebrates in habitats subjected to frequent treatments are likely to be more affected, especially those species that show greater sensitivity. However, populations of affected organisms generally recovered to pretreatment levels within weeks to month of the exposure.

In freshwater studies, cypermethrin had a significant sublethal impact on the pheromone-mediated endocrine system in mature Atlantic salmon parr [53]. It was suggested that cypermethrin acts directly on the Na channels and inhibits nervous transmission within the olfactory system and thus the male salmon is unable to detect and respond to the priming pheromone. In the marine environment, cypermethrin may reduce the homing abilities of returning adult salmon and increase straying rates between river systems.

Shrimp (*C. crangon*) were deployed in cages at various distances and depth from the cages at two salmon aquaculture sites in Scotland during treatment with cypermethrin. The only mortalities were to shrimp held in treated cages [39]. Shrimp in drogues released with the treated water were temporarily affected but recovered. In an American field study, cypermethrin was lethal to 90% of the lobsters in the treatment cage but no effect was observed in those located 100–150 m away. There was no effect on mussels placed outside or inside the cages. Similar field studies indicated that cypermethrin was lethal to lobsters and some planktonic crustaceans in the treatment tarpaulin but not to mussels, sea urchins or planktonic copepods.

The toxicity (96 h LC50) of cypermethrin to American lobster and sand shrimp (*Crangon septemspinosa*), was $0.04\,\mu g\,L^{-1}$ and $0.01\,\mu g\,L^{-1}$, respectively [54]. The 24 h LC50 was $0.14\,\mu g\,L^{-1}$ for adult lobster. For other marine invertebrates, 96 h LC50 values range from 0.005 to $0.056\,\mu g\,L^{-1}$ for mysid shrimp [43, 46]. The 96 h LC50 for five other marine crustaceans ranged from $0.016\,\mu g\,L^{-1}$ for grass shrimp to $0.20\,\mu g\,L^{-1}$ for fiddler crab. Oysters were relatively insensitive, with a 48 h EC50 of $2.3\,mg\,L^{-1}$ based on larval development. For marine fish, the 96 h LC50 of cypermethrin to Atlantic salmon was $2.0\,\mu g\,L^{-1}$ [54] and for sheephead minnow was $1.0\,\mu g\,L^{-1}$ [43]. Exposure of Atlantic salmon to a $10\,mg\,L^{-1}$ solution of mixed pyrethrins for 6 min was 100% lethal, and some death occurred if the period of exposure was greater than 2 minutes [36].

Larvae are often considered the most susceptible life stage to environmental or chemical stress. The 12 h LC50 of cypermethrin for stage II American lobster larvae at 10 and 12 °C was 0.365 and $0.058\,\mu g\,L^{-1}$, respectively [47]. At sublethal concentrations effects on swimming ability and responsiveness of the lobster larvae were observed. The 48 h LC50 of cypermethrin to the three larval stages (I, II, and III) of the American lobster and to the first post-larval stage (IV) was 0.18, 0.12, 0.06, $0.12\,\mu g\,L^{-1}$, respectively [55]. Thus, cypermethrin was lethal to larval lobsters over 48 h at approximately 3% of the recommended treatment concentration. In a study with larval lobsters and a formulation of pyrethrins and piperonyl butoxide there were significant differences in sensitivity between larval stages [45]. Stage I larvae were more tolerant of the pyrethrins formulation than Stage II, and both were more tolerant than Stages III and IV (48 h LC50 = 4.42, 2.72, 1.39, 1.02 μg L^{-1}, respectively). On the other, hand soft shell clam larvae, green sea urchin larvae and rotifers were tolerant of cypermethrin and 12 h LC50 values were greater than 10 mg L^{-1} [47].

Cypermethrin induced glutathione S-transferase (GST) activity in shore crab, *Carcinus maenas*, exposed to a solution of 50 and 500 ng L^{-1} of cypermethrin or injected intra-cephalothoracically with 10 ng [56]. However, activity of the enzyme returned to base levels after 36 h and there was no clear dose response and so GST activity may not be a useful biomarker of exposure to cypermethrin.

4.3
Hydrogen Peroxide

4.3.1
Efficacy and Mechanism of Action of Hydrogen Peroxide

Hydrogen peroxide is a strong oxidizing agent that was first considered for the treatment of ecto-parasites of aquarium fish [57]. It is widely used for the treatment of fungal infections of fish and their eggs in hatcheries [58]. With the development of resistance of sea lice to dichlorvos [59] it was preferable to use of hydrogen peroxide to treat infestations of both *L. salmonis* and *C. elongatus*. Hydrogen peroxide was used in salmon farms in Faroe Islands, Norway, Scotland and Canada in the 1990's [10]. Hydrogen peroxide (Paramove, Salartect) is still authorized for use in all countries but it is not normally the treatment of choice and there is no record of usage in 2003. There may be renewed interest in the use of hydrogen peroxide, in conjunction with the use of wrasse, as part of a strategy to allow sites to maintain "organic salmon aquaculture" accreditation status.

The suggested mechanisms of action of hydrogen peroxide are mechanical paralysis, peroxidation by hydroxyl radicals of lipid and cellular organelle membranes, and inactivation of enzymes and DNA replication [60]. Most evidence supports the induction of mechanical paralysis when bubbles form in the gut and hemolymph and cause the sea lice to release and float to the surface [61].

The recommended dosage for bath treatments is 0.5 g L^{-1} for 20 min but the effectiveness is temperature dependent and the compound is not effective below 10 °C. Treatments are rarely fully effective, but 85–100% of mobile stages may be removed [62]. The first farm treatments in Scotland in October 1992 removed 83% of the mobile stages of sea lice. The recommended course is to repeat the treatment at 3–4 week intervals. This usually results in low numbers of sea lice for 8 weeks following the third treatment [10]. Hydrogen peroxide has little efficacy against larval sea lice and its effectiveness against preadult and adult stages has been inconsistent [57]. Effectiveness can be difficult to determine on farms as the treatment concentration varies due to highly variable volumes of water enclosed in the tarpaulin. Temperature and duration also influence the efficacy. Ovigerous females are less sensitive that other mobile stages [62]. It is possible that a proportion of the eggs on gravid female lice may not be viable after exposure to hydrogen peroxide [63]. Hydrogen peroxide was less efficacious when treating sea lice infestation on salmon in a cage that had been treated regularly for 6 years than in cages where the sea lice were treated for the first time. This suggested that *L. salmonis* had developed some resistance to hydrogen peroxide [62].

In a laboratory experiment, all adult and pre-adult sea lice exposed to $2.0\,\text{g}\,\text{L}^{-1}$ hydrogen peroxide for 20 min became immobilized, but half had recovered 2 h post-treatment [61]. The recovered sea lice swam normally and may have been able to reattach to the host salmon [64]. Therefore it was recommended that floating lice should be removed. However, re-infection has not been noticed in practice [62] as the removed sea lice generally show little swimming activity. Re-infection in the field is less likely because the free sea lice will be washed away with the tidal flow or eaten by predators. After treatment of a cage with approximately $1.5\,\text{g}\,\text{L}^{-1}$ hydrogen peroxide at $6.5\,°\text{C}$, all the sea lice that were collected from surface water of treated cages were inactive but recovery commenced within 30 min and 90–97% of the sea lice were active 12 h post-treatment [10]. In this study, a higher proportion of pre-adult sea lice was removed than of adult sea lice.

4.3.2
Distribution and Fate of Hydrogen Peroxide

Hydrogen peroxide is generally considered to be the treatment method of lowest environmental risk because it decomposes into oxygen and water. At $4\,°\text{C}$ and $15\,°\text{C}$, 21% and 54%, respectively, of the hydrogen peroxide has decomposed after 7 days in sea water. If the sea water is aerated the amount decomposed after 7 days is 45% and 67%, respectively [61]. Field observations suggest that decomposition in the field is more rapid, possibly due to reaction with organic matter in the water column, or decomposition catalyzed by other substances in the water, such as metals. In most countries, hydrogen peroxide is considered a low environmental risk and therefore of low regulatory priority.

4.3.3
Biological Effects of Hydrogen Peroxide

There is little information of the toxicity of hydrogen peroxide to marine organisms. Most toxicity data are related to the potential effects on salmonids during treatment of sea lice infestations. Experimental exposure of Atlantic salmon to hydrogen peroxide at varying temperatures shows that there is a very narrow margin between treatment concentration (0.5 g L^{-1}) and that which causes gill damage and mortality (2.38 g L^{-1}) [65].

Toxicity to fish varies with temperature; for example, the 1 h LC50 to rainbow trout at 7 °C was 2.38 g L^{-1}, at 22 °C was 0.218 g L^{-1} [57] and for Atlantic salmon increased five fold when the temperature was raised from 6 °C to 14 °C [4]. There was 35% mortality in Atlantic salmon exposed to hydrogen peroxide at 13.5 °C for 20 min. There was a rapid increase in respiration and loss of balance, but if the exposure was at 10 °C there was no effect [61]. Hydrogen peroxide is not recommended as a treatment for sea lice infestations at water temperatures above 14 °C. Whole bay treatments in the winter should reduce the need for treatments in the summer [66].

The method of application of hydrogen peroxide is not standardized but is a balance between achieving consistently effective treatments and toxicity to fish. For example, high concentrations were used (2.5 g L^{-1} for 23 min) to treat a farm for 6 years, which achieved 63% removal of sea lice. Longer exposure periods were used in an attempt to increase removal, but caused 9% mortality in the salmon [62]. There is evidence that the concentrations of hydrogen peroxide used in sea lice treatments can cause gill damage and reduced growth rates for 2 weeks post treatment [67].

5
In-Feed Treatments

5.1
Avermectins

5.1.1
Efficacy and Mechanism of Action of Avermectins

Two avermectin compounds have been used to treat sea lice infestations. ivermectin and emamectin benzoate synthetic derivatives of a chemical produced by the bacterium, *Streptomyces avermitilis*. Ivermectin is manufactured by Merck, Sharp and Dohme (MSD) and the company has made it clear they do not wish to have the product licensed for use as an anti-louse treatment [68]. In Canada, ivermectin has been used under veterinarian prescription to treat sea lice as an 'off-label' drug treatment under veterinary prescription [69]. This means the drug can be prescribed but is not specifically registered for the treatment of sea lice. In the UK and Europe, a similar regulation exists (the Cascade Principle) by which veterinarians can prescribe ivermectin if no other effective licensed product is available. The subsequent availability of emamectin benzoate as a treatment against sea lice infestations should eliminate the need for the use of such 'off-label' prescriptions.

Emamectin benzoate has been available in Canada as an Emergency Drug Release (EDR) from Health Canada since 1999 and is used to treat salmon against sea lice in Canada (DI Alexander, Health Canada. Veterinary Drugs Directorate, personal communication). Emamectin benzoate is registered for use in the UK, Norway, Ireland, Iceland, Chile and the Faroes.

The avermectins are effective in the control of internal and external parasites in a wide range of host species, particularly mammals [70]. The avermectins open glutamate-gated chloride channels at invertebrate inhibitory synapses. The result is an increase in chloride concentrations, hyperpolarization of muscle and nerve tissue, and inhibition of neural transmission [71, 72].

Ivermectin is effective against chalimus as well as adult stages of the parasite giving it a wider efficacy than the organophosphates and hydrogen peroxide [68, 73]. The 'standard treatment' is 0.1 mg kg^{-1} divided into two treatments of 0.05 mg kg^{-1} separated by 3 or 4 days [74]. This treatment regimen reduced the numbers of sea lice by up to 93% [75]. When fish were treated weekly at a dose of 0.02 mg kg^{-1} for 3 month, ivermectin was shown to be effective in preventing re-infection for about 4 weeks after the termination of treatment [73].

The optimum therapeutic dose for emamectin benzoate is 0.05 mg kg^{-1} fish d^{-1} for seven consecutive days [76]. This dose has been shown to reduce the number of motile and chalimus stages of *L. salmonis* by 94–95% after a 21 day study period [76, 77]. Four cage sites with a total of 1.2 million first year class fish were treated. Although there was a slight depression of appetite at two of the four sites, appetite was normal when top-up rations were supplied. *C. elongatus* were present in low numbers and results suggested that they were also affected by the treatment. The number of motile lice was reduced by as much as 80% at the end of the 7-day treatment period. In a field trial, emamectin benzoate reduced sea lice counts on treated fish by 68–98% and lice numbers remained low compared to control fish for at least 55 days [78, 79].

5.1.2
Distribution and Fate of Avermectins

Avermectins reach the marine environment in one of two ways: on uneaten feed pellets or as waste products from the fish (faeces and biliary excretion). Ivermectin is depurated from fish in two phases: an initial biliary excretion of unchanged ivermectin followed by a slower excretion after enterohepatic cycling [68]. A relatively high proportion of ivermectin passes through the gut unabsorbed [80]. Ivermectin has a low solubility in water and a strong affinity to lipid, soil, and organic matter [68]. It is readily photo-degraded, but the half life for hydrolysis in the dark is quite long [81]. Within the marine environment, ivermectin is expected to be associated with sediments and particles and to show low mobility. The half life of ivermectin in sediment is at least 3 months [82]. The octanol-water partition coefficient for ivermectin is 1651 [83] and the calculated bioconcentration factor of ivermectin is 74 for fish and 750 for mussels [68]. A "withdrawal period" of 1000 degree days for the elimination of ivermectin from edible tissue of was suggested prior to harvesting of Atlantic salmon [84].

Emamectin benzoate also has low water solubility and relatively high octanol-water partition coefficient, indicating that it has the potential to be absorbed to particulate material and surfaces and that it will be tightly bound to marine sediments with little or no mobility [85]. The half life of emamectin benzoate is 193.4 days in aerobic soil and 427 days in anaerobic soil. In field trials, emamectin benzoate was not detected in water samples and only 4 of 59 sediment samples collected near a treated cage had detectable levels of emamectin benzoate. The emamectin benzoate persisted in the sediment; the highest concentration was measured at 10 m from the cage 4 month post-treatment. In Canada, emamectin benzoate was not detected in sediment samples collected near an aquaculture site for 10 weeks after treatment (W.R Parker, Environment Canada, personal communication). Mussels were deployed and traps set out to capture invertebrates near aquaculture sites undergoing treatment with emamectin benzoate. While detectable levels of emamectin benzoate and metabolites were measured in mussels (9 of 18 sites) one week after treatment, no positive results were observed after 4 months [85]. Emamectin benzoate was found in crustaceans during and immediately after treatment. The species showing detectable levels of emamectin benzoate for several month after treatment are scavengers which are likely to consume faecal material and waste food. The withdrawal period prior to slaughter of salmon in Canada is 25 days (DI Alexander, Health Canada. Veterinary Drugs Directorate, personal communication).

5.1.3
Biological Effects of Avermectins

A body of literature exists for LC50s and LD50s for ivermectin to fish and marine invertebrates [68]. Unfortunately, very few of these studies involve exposure of test organisms to ivermectin either in feed or in the sediments. Most researchers have exposed experimental animals through immersion in spiked water.

Over a 27 day period, there was a cumulative mortality of 10% and 80% of the Atlantic salmon (wt = 800 g) exposed to 0.05 and 0.2 mg kg^{-1} ivermectin in food, respectively (Johnson et al. 1993). Atlantic salmon was the most sensitive of several salmonid species tested and behavioral changes, such as cessation of feeding and lethargy, were observed in fish exposed to lower concentrations. The 96 h LD50 was $0.5\,\mathrm{mg\,kg^{-1}}$ for Atlantic salmon administered ivermectin by intubation and the 96 h LC50 was $17\,\mathrm{\mu g\,L^{-1}}$ when the salmon were immersed in a sea water solution of ivermectin [86].

Sand shrimp were exposed to fish feed treated with various concentrations of ivermectin for 96 h in running seawater [87]. When the food was accessible to the shrimp, mortality occurred. When the feed was present in the water but not accessible by the shrimp, no mortality occurred, suggesting that the feed must be ingested by the shrimp before lethality occurs. The nominal 96-h LC50 was $8.5\,\mathrm{mg\,kg^{-1}}$ food and the no observed effect concentration (NOEC) was $2.6\,\mathrm{mg\,kg^{-1}}$ food.

The 10-day LC50 for ivermectin in sediment to the marine amphipod, *Corophium volutator* was estimated to be $180\,\mathrm{\mu g\,kg^{-1}}$ dry weight [82]. The NOEC was $50\,\mathrm{\mu g\,kg^{-1}}$. The 10 d LC50s to *Arenicola marina* and *Asterias rubens* were 23 and $23\,600\,\mathrm{\mu g\,kg^{-1}}$ dry weight, respectively [88].

Toxicological studies have shown that emamectin benzoate is less toxic than ivermectin to all taxa tested [85]. The treatment concentrations of emamectin benzoate on salmon feed range from 1 to $25\,\mathrm{\mu g\,kg^{-1}}$ [71]. Feeding emamectin benzoate to Atlantic salmon and rainbow trout at up to ten times the recommended treatment dose resulted in no mortality. However, signs of toxicity, lethargy, dark coloration and lack of appetite were observed at the highest treatment concentration.

The lethality of emamectin benzoate treated fish feed to adult and juvenile American lobsters is estimated as 644 and $> 589\,\mathrm{\mu g\,kg^{-1}}$ of feed, respectively [89]. The lethality of emamectin benzoate to other aquatic invertebrates (for example, *Nephrops norvegicus* and *Crangon crangon*) was $> 68\,\mathrm{mg\,kg^{-1}}$ [85]. In laboratory studies, Spot prawns, *Pandulus platyceros*, and Dungeness crabs, *Cancer magister*, were offered feed medicated with emamectin benzoate at concentrations up to $500\,\mathrm{mg\,kg^{-1}}$ [90]. There was no acute mortality. However, the crabs appeared to avoid medicated feed pellets. Ingestion of emamectin benzoate induced premature molting of American lobsters [30]. This molting response of lobsters to emamectin benzoate may involve an inter-relationship of a number of environmental (water temperature), physiological (molt and reproductive status) and chemical (concentration/dose) factors [91]. In a 7-day sublethal test, there was significant reduction of egg production in the adult marine copepod, *Acartia clausii* [92]. The concentrations necessary to elicit these responses were above the predicted environmental concentration (PEC) [92].

5.2
Chitin Synthesis Inhibitors

5.2.1
Efficacy and Mechanism of Action Chitin Synthesis Inhibitors

Chitin synthesis inhibitors belong to a class of insecticides collectively referred to as insect growth regulators and have been used in terrestrial spray programmes for nuisance insects since the late 1970s. Two of these, diflubenzuron (Lepsidon) and teflubenzuron (Calicide, UK; Ektobann, Norway) were approved as additives in feed to treat sea lice infestations of cultured salmon in Norway (1997) and Scotland (1999). Teflubenzuron use is approved in Ireland and Canada but there has been no recorded use in Canada for 2002–2003.

Chitin is the predominant component of the exoskeleton of insects and crustaceans, but the biochemical mechanism by which these insecticides inhibit the synthesis of chitin is unclear [93]. The molting stage is the sensitive stage of the life cycle and inhibition of chitin synthesis interferes with the formation of new exoskeleton in a post-molt animal, for example, post-molt blue crabs [94, 95]. Thus the chitin synthesis inhibitors are effective against the larval and pre-adult life stages of sea lice.

Teflubenzuron is effective against *L. salmonis* at a dose to salmon of $10 \, \text{mg kg}^{-1}$ body weight per day for 7 consecutive days at 11–15 °C [96]. Teflubenzuron at this dosage was used to treat commercial salmon farms in Scotland and Norway and the efficacy was 83.4 and 86.3%, respectively, measured at 7 days post treatment. There were no toxic effects on treated fish or effects on appetite of the fish. In a Norwegian field trial of salmon in a polar circle cage with 100 000 kg of salmon, the efficacy for a dosage of $8.1 \, \text{mg kg}^{-1}$ body wt d^{-1} for 7 d was 77.5% at 5.4 °C [97]. The greatest reductions were in chalimus and pre-adult lice and the efficacy was 88% if the calculation was based only on the susceptible life stages of *L. salmonis*. The effects were observed up to 26 d after start of the treatment. A few Norwegian sites successfully used teflubenzuron (Calicide) in 1997 to remove all developing stages and the sea lice did not return during the further year's growth cycle [98].

Since chitin synthesis inhibitors are effective against the developing copepodids, larval (chalimus) and pre-adult stages of sea lice and less effective against adult lice, treatments are most effective before adult lice appear, or at least are present in only low numbers. In some cases, a prior bath treatment with organophosphates may be useful to remove adult lice or to control recruitment. When used correctly, chitin synthesis inhibitors provide a treatment option that breaks the life cycle of the sea lice and, as a result, the duration between treatments may be several months.

5.2.2
Distribution and Fate of Chitin Synthesis Inhibitors

Teflubenzuron and diflubenzuron have moderate octanol-water partition coefficients and relatively low water solubility, which means that they tend to remain bound to sediment and organic materials in the environment. They are not persistent in freshwater [98, 99] and a few marine studies suggest that sediment is a significant sink for these compounds in the marine environment.

In a field study, a total of 19.6 kg of teflubenzuron was applied over a 7 d period to treat a salmon cage with a biomass of 294.6 tonnes [98]. Teflubenzuron was not detected in the water after treatment and highest concentrations in the sediments were found under the cages and decreased with distance from the cage in the direction of the current flow. It persisted in sediments for at least 6 months and the half-life was estimated at 115 d. Measurable concentrations were noted for a distance of 1000 m in line with the current flow, but 98% of the total load had degraded or dispersed by 645 d after treatment. There was some indication of re-suspension and redistribution of sediment after several weeks based on concentrations of teflubenzuron found in mussel tissues. Evidence suggested that there was some risk to indigenous sediment dwelling crustaceans, such as crab or lobster that may accumulate teflubenzuron from the sediment. However, the mussels eliminated teflubenzuron readily.

Diflubenzuron was found to be stable and persistent in anoxic marine sediments under laboratory conditions. There was no significant decrease in concentration (38 and 50 $\mu g\,g^{-1}$) after 204 d for diflubenzuron in sediments held in the dark at 4 and 14 °C or in sediments in tanks that were flushed with sea water [100]. In a field study, salmon were fed medicated feed for 14 d. The concentrations of diflubenzuron found in the sediment did not reflect the high concentrations found in the sediment traps and accounted for only 15% of the total input. It was suggested that the feed and feces at the sediment surface may have been re-suspended and transported farther than anticipated, or that faulty sampling of the sediment by the grab had led to underestimation of the amount of diflubenzuron in the sediment near the farm [100]. Evidence from soil studies suggests that diflubenzuron may be metabolized by bacteria in sediment [101].

In an estuarine microcosm system, an initial concentration of diflubenzuron of 140 $\mu g\,L^{-1}$ decreased slowly over 3 weeks with a half life of > 17 d in one study but only 4 d in another [99]. Laboratory and field studies demonstrated the importance of substrate in the fate and persistence of diflubenzuron in estuarine systems. Crab larvae exposed in a salt water system to a single treatment of technical grade diflubenzuron resulted in total mortality in solutions that were 1–42 d old and decreased from 86% to 5% for solutions that were 50–59 d old. In a salt marsh treated three times with 45 g ha^{-1} of diflubenzuron at 2-week intervals, the concentrations of diflubenzuron that remained were < 0.4 µg L^{-1} in the salt water one week post treatment while concentrations in sediment remained above 100 µg L^{-1}. The half-life in the microcosm system containing sediment was 5.3 d due to absorption of diflubenzuron by organic matter and 17.8 d in the absence of sediment.

The absorption of teflubenzuron from the gastrointestinal tract of salmon has been found to be poor, with only around 10% of the administered dose being retained by salmon and 90% being released by the fish via feces or in the uneaten portion of the feed [98]. In general, fish accumulate diflubenzuron rapidly during acute exposures but will eliminate the compound within 7 d [99]. For example, the freshwater fish, white crappies, *Pomoxis annularis*, accumulated diflubenzuron from water to 264 ng g^{-1} wet weight but after 24 h in clean water the concentration had fallen to 8 ng g^{-1} wet weight [102]. The deposition of diflubenzuron and teflubenzuron in the vicinity of the treated cage is primarily from waste feed, with a more widespread distribution arising from the dispersion of fecal matter that may extend to 100 m from cages in the direction of the current flow [98].

5.2.3
Biological Effects Chitin Synthesis Inhibitors

Although teflubenzuron is relatively non-toxic to marine species of birds, mammals and fish, due to its mode of action, it is potentially highly toxic to any species which undergo molting within their life cycle [98, 103]. This includes some commercially important marine animals such as lobster, crab, shrimp and some zooplankton species.

Aquatic toxicity data for diflubenzuron has been compiled for 15 estuarine and marine species, mostly invertebrates [99]. The premolt stage of grass shrimp was the most acutely susceptible to diflubenzuron (96 h LC50 = 1.1 µg L^{-1}) and the mummichog, *Fundulus heteroclitus*, was the most resistant species (96 h LC50 = 33 mg L^{-1}). Exposure of a marine harpacticoid copepod indicated that concentrations of diflubenzuron as low as 1.0 µg L^{-1} cause adult mortality and inhibited reproduction. The viability of *Acartia tonsa* nauplii to hatch was reduced to < 50% during a 12 h exposure to 1 µg L^{-1} of diflubenzuron. When brine shrimp were exposed to ≥ 2 µg L^{-1} of diflubenzuron, the reproductive life span and numbers of broods produced were significantly less than in controls. The 96 h LC50 to various life stages of grass shrimp are: larvae, 1.44 µg L^{-1}; post-larvae, 1.62 µg L^{-1} and adult, > 200 µg L^{-1}. There was 60% mortality of the resident grass shrimp in a tidal pool treated with 45 g ha^{-1} diflubenzuron. The borrowing behavior of fiddler crab was significantly reduced by exposure for more than one week to > 5.0 µg L^{-1} of diflubenzuron. However, there was 100% mortality of stone crab larvae exposed to $5.0\,\mu g\,L^{-1}$; 95% mortality of the blue crab exposed to $> 3.0\,\mu g\,L^{-1}$; 46% mortality of juvenile blue crab after treatment of the tidal pool to $3.6\,\mu g\,L^{-1}$ at one hour after treatment. The lowest reported chronic effect concentration for a saltwater organism exposed to diflubenzuron was $0.075\,\mu g\,L^{-1}$, which significantly reduced reproduction in mysid shrimp.

A secondary effect of diflubenzuron on fish populations has been shown in a littoral enclosure, but not in the open marine environment. Exposure to diflubenzuron ($2.5\,\mu g\,L^{-1}$) in littoral enclosures adversely affected reproductive success by reducing growth of bluegill larvae by 56 and 86% [104]. This reduction in growth was an indirect effect by eliminating or reducing preferred bluegill larvae food (cladocerans and copepods). Decreases in growth of the food larvae may lead to greater starvation, increased predation and lower over-winter survival, which may result in poor recruitment. It was suggested that the early-stage bluegill larvae are more sensitive because this is when their growth rate is most rapid and at this first feeding stage they are more selective because of their small mouth and poor swimming activity. The calanoid copepod, *Eurytemora affinis*, is widely distributed in North America and Europe [93], for example in Chesapeake Bay it makes up 20% of the total annual zooplankton crop and is major prey item for white perch and stripped bass larvae. The 48 h LC50 of diflubenzuron (Dimilin®) to *E. affinis* is $2.2\,\mu g\,L^{-1}$. Other studies indicate that diflubenzuron concentrations less than $1.0\,\mu g\,L^{-1}$ have substantial effects on survival, growth and production of nauplii [105].

In a field study, no adverse effects were detectable in the benthic macrofaunal community or indigenous crustaceans and it was concluded that residual teflubenzuron in sediment was not bioavailable [98]. There was some evidence of effects on the benthic fauna within 50 m of the treated cages, but no adverse impacts on community structure and diversity including important key sediment re-worker species and crustacean populations. A study at three locations in Scotland included a novel biomonitoring technique whereby juvenile lobster larvae were deployed on platforms at locations around cages. The juvenile lobster mortality was attributed to exposure to the medicated feed at 25 m from the cage, but this effect did not occur 100 m from the cage, and it was confirmed that a molt occurred during the study. It was concluded that the "predicted no effect concentration" would not be exceeded 15 m from cages. Since crustaceans are largely absent within 15 m of cages, and evidence suggests that teflubenzuron is relatively non-toxic to sediment re-worker organisms such as polychaete worms, the environmental risks in the use of teflubenzuron in the treatment of sea lice infestations in this study were considered to be low and acceptable.

6
Risk Assessment of Sea Lice Therapeutants

All the countries involved in the marine cultivation of salmonid fish operate systems for the regulation of the therapeutic chemicals that are needed as components of strategies to control disease. In most cases, an authorization or licence issued by the relevant authority is required before a chemical can become available for use on fish farms. The ecological risk posed by the proposed use of a chemical is normally assessed during the process leading to decisions on the granting of the appropriate licence. Therefore, licensing procedures are an important element of the risk management process relating to the use of sea lice treatment chemicals.

For members of the European Union, an essential pre-requisite to the use of a medicine on food fish is the granting of a Maximum Residue Limit (MRL) by the European Medicines Evaluation Agency (EMEA), by being annexed to EC Council Regulation 2377/90. The MRL protects the consumer from any possible adverse effects of residues of medicines that might be present in fish presented for consumption. The EMEA seeks to further harmonize medicine authorization and assessment procedures within Europe and beyond. However, applications for Marketing Authorizations (MAs) are still most commonly made to national authorities rather than directly to the EMEA. Exceptionally, the EMEA Committee for Veterinary Medicinal Products (CVMP) can assess applications for MAs. CVMP has issued European Guidelines for environmental risk assessment of veterinary medicinal products. The EU is party to an initiative (International Co-operation on Harmonization of Technical Requirements for Registration of Veterinary Medicinal Products, VICH) to harmonize technical requirements between the EU, USA, and Japan.

In Canada, chemicals that are applied as feed additives or by injection are classified as drugs and are approved by the Veterinary Drug Directorate (VDD) of Health Canada under the Food and Drugs Act. The VDD is required to ensure that drugs offered for use on animals are safe and effective and do not leave residues in the products that pose a health risk to the consumer. Chemicals applied topically or as bath treatments are classified as pesticides and are the responsibility of the Pest Management Regulatory Agency (PMRA) of Health Canada and are registered under the Pest Control Products Act. In all countries, authorization to apply approved therapeutants for sea lice infestations ultimately requires a veterinary prescription.

Ecological risk assessment is a process for objectively defining the probability of an adverse effect to an organism or collection of organisms when challenged with an environmental modification such as climate change, xenobiotic exposure, infection with a disease organism or some other potential stressor [106]. Sea lice therapeutants have the potential to negatively impact the environment through effects on susceptible non-target organisms. There may be a significant body of information relevant to efficacy and safety that is known only to the regulatory authorities and the specific manufactures [107]. The absence of these data from the public domain has the unfortunate consequence that neither its quality nor its nature can be debated by those scientists and non-scientists with interests in these areas. Critical evidence quantifying the extent of such impacts when the agents are employed under the conditions of a commercial fish farm is limited.

The central problem presented by anti-lice treatments is their lack of specificity. The properties of lice that present specific site for action of any anti-lice therapeutants are not unique to these lice [107]. In particular, other crustaceans such as lobster, crab, and shrimp may be affected. For example, in the cold waters of the Bay of Fundy, hatching of lobsters occurs in July to September [108] and larval production has been observed as late as September. The larval stages (stage I, II, III) of the lobster are pelagic. The first post larval stage (stage IV) spends at least some of its time in the water column prior to settling to the bottom [109]. It is possible that treatment of lice infested fish and release of pesticide formulations could coincide with the presence of lobster larvae in the water [55].

The details of the scope of the environmental information required by regulatory authorities vary from country to country. For clarity, one country will be taken as a detailed example. Under UK legislation, any compound applied to an animal for the purpose of disease control is classified as a medicine and is licensed under the relevant medicines legislation. A pharmaceutical company seeking approval for a new medicine must show that the medicine is effective (for the purpose for which it is being proposed), of good quality and safe. In the context of this paper, "safety" includes safety for the animal being treated, for the user and, in the case of food animals, for the consumer (subject to appropriate withdrawal periods). Environmental safety is particularly important for medicines to be used in aquaculture.

The UK operates a tiered approach to the assessment of the environmental safety (Table 2) [110]. Applicants are required to present a dossier on the potential risks to the environment resulting from the use of the product. This must include ecotoxicological information, supported by an expert report, to assess the potential harmful effects which the use of the product may cause to the environment and to identify any precautionary measures which may be necessary to reduce such risks. For medicines used in fish farming which will, or are likely to, enter surface waters, the submission of an ecotoxicity dossier comprising the Phase I and Phase II assessment will always be applicable. An ecotoxicity dossier will also be necessary when applying for an animal test certificate (ATC) to conduct trials using fish medicines.

A progressive, stepwise approach to testing is described, with the data required at one tier being dependent on the results of testing at the previous tier. Where sufficient data are available at any one tier of testing for the en-

Table 2 Tiered approach to ecotoxicological testing of fish farm medicines in the UK

	Tier 1	Tier 2	Tier 3
PHYSICO-CHEMICAL PROPERTIES	– Molecular weight – UV/visible absorption spectrum – Melting point – Boiling point – Vapor pressure – Water solubility – Water dissociation constant – Octanol/water partition coefficient (K_{ow})	Sediment/water adsorption coefficient (if K_{ow} is high)	No further requirements
FATE	– Hydrolysis half-life at pH5, 7 and 9 – Photolysis half-life	– Biodegradation mechanism and halflife in natural sediment-water test systems (if hydrolysis and photolysis slow) – Bioconcentration tests (if K_{ow} is high and exposure likely to be long)	– Dispersion data – Outputs from computer models – Fate in sediments based on microcosms or mesocosms
BIOLOGICAL EFFECTS	– Acute toxicity to one species of juvenile or larval fish – Acute toxicity to one appropriate species and stage of larval crustacean – Toxicity to one species of microalga	– Chronic fish and crustacean reproduction, earlylife-stage or growth tests (if prolonged exposure likely) – Acute toxicity to a macrophyte (if toxic to algae) – Acute toxicity to juvenile or larval mollusks (if of economic importance in area of use) – Acute and/or chronic toxicity to obligate sediment feeders (crustacea, molluscs or annelids)	– Mesocosm studies of effects on benthic fauna – Bioassays using sensitive taxa – Field investigations – Effects on microbial communities

vironmental risk to be adequately assessed, then there will be no need to conduct further tiers of testing. A general risk assessment strategy is to estimate the predicted environmental concentrations (PEC) that will result from the use of the agent. The estimated PEC is compared to the predicted no effect concentrations (PNEC) derived from toxicity studies with relevant species. Dispersion/advection models for dissolved substances may be used to estimate PEC [55, 111].

The route of administration of the sea lice therapeutants is important in determining the factors to consider in the risk assessment of these chemicals. Bath treatments normally result in the direct release of a solution of the therapeutant to the surrounding water and thus the dilution rate (dispersion), spread and direction of flow of the plume and the life history of susceptible species in the water column are important factors. In the case of medicated feed, the critical factors are the benthic deposition of excess food pellets and feces, as well as the bioavailability of the therapeutant from these particles.

6.1
Bath Treatments

The fate and dispersion of cypermethrin, azamethiphos and a dye, rhodamine, were determined after simulated bath treatments from a salmon aquaculture site under various tidal conditions in the Bay of Fundy, Canada [112]. Dye concentrations were detectable for periods after release which varies from 2–5.5 h and distances ranged from 900 to 3000 meters depending on the location and tidal flow at the time of release. Concentrations of cypermethrin in the plume reached 1–3 orders of magnitude below the treatment concentration 3–5 h post release and indicated that the plume retained its toxicity for substantial period of time after release. Water samples collected from the plume were toxic in a 48 h lethality test to *E. estuarius* for cypermethrin up to 5 h after release. When azamethiphos was released, none of the water samples from the plume were toxic after 20 minutes. There have been a number of studies where lobsters and shrimps were held in cages near fish pens during treatment for sea lice. It is not known if the caged animals were exposed to the plume from the released bath treatment however the experiments do provide some circumstantial evidence.

Bath treatments require considerable human effort and usually there is only enough staff to treat one cage at a time and up to three cages per day. Thus it is possible that indigenous species could be exposed periodically for several hours to plumes of the released bath treatment from the same aquaculture site or possibly from several sites in the same area.

Azamethiphos is water soluble and remains in the aqueous phase on discharge to the receiving waters. Azamethiphos decomposes by hydrolysis with a half-life in nature of 8.9 d. Dispersion studies indicated that, after release of the treatment solution, the concentration of azamethiphos falls to below detection (0.1 µg L^{-1}) in a short period of time. The compound was not detected below 10 m depth. Thus, azamethiphos is unlikely to accumulate in indigenous species or in sediment.

Only lobster larva and shrimp are susceptible to concentrations below that used in the treatment of sea lice, but acute toxicities were determined from a single exposure over 48 to 96 h. Several field studies found no acute effects on caged lobsters held near treated sites or in lobster populations in areas that have been treated for some time.

Laboratory studies have demonstrated sublethal effects on lobster reproduction and mobility from repeated short term exposure to concentrations of azamethiphos below the recommended treatment concentration (10 µg L^{-1} for 1 h). Repeated exposures to higher concentrations could result in significant mortalities and drastic changes in activity level and ability to function normally. The consequences of lethargy or of becoming moribund in the wild are probably severe to individuals, but the number of individuals within the zone of impact during sea lice treatments is likely to be few. Therefore, the risk of ecological effects of azamethiphos being manifested during operational application of the pesticide appears to be quite small. Individual (pelagic) organisms caught in the effluent plume from a bath treatment are likely to be affected but it unlikely that large scale or population effects will occur. Azamethiphos is a moderate risk to individuals of sensitive species but a low risk to populations. However, azamethiphos is not considered the treatment of choice because of the development of resistance to organophosphates by sea lice.

The pyrethroids, cypermethrin and deltamethrin are not persistent in marine waters. Both have relatively short half-lives in water and concentrations in the water decreased rapidly (< 4 h) in some field trials due to decomposition and partitioning to particulate matter. In sediments, the compounds are more persistent with half-lives up to 80 d, and cypermethrin was detected in sediment surveys in near salmon aquaculture sites in Scotland [113]. However, bioavailability of pyrethroids from sediment is minimal.

Cypermethrin has the potential to be released in toxic plumes from a single cage treatment [112]. The plume can cover up to a square kilometer and lethality to sensitive species can last as long as 5 h. Since treatment of multiple cages is the operational norm, area wide effects of cypermethrin on sensitive species cannot be discounted. Sensitive species include crustaceans such as lobster larvae, shrimp and crabs and the 96 h LC50 for some can be a magnitude less than the treatment concentration. No lethality was observed in shrimp and lobsters deployed in cages during sea lice treatments with cypermethrin.

As with the organophosphates, the development of resistance to pyrethroids by sea lice has been demonstrated. A region in Norway with resistant sea lice had an EC50 which was 25 times higher than that for an area that had not been treated previously with deltamethrin [40].

Evidence suggests that pyrethroids can present considerable risk to individuals of sensitive species but there is insufficient knowledge to extrapolate to populations. There is sufficient evidence on the development of resistance to advise against routine use of pyrethroids as only means of control. Pyrethroids are not authorized for use in North America for treatment of sea lice infestations. Part of the rationale may be the availability of an unrestricted agricultural product containing cypermethrin that could result in indiscriminate use in marine waters.

Hydrogen peroxide is non-persistent, environmentally friendly and readily dispersed in marine waters. It has negligible risk to marine organisms in the concentrations used for the treatment of sea lice infestations.

6.2
In-Feed Treatments

The avermectins have low water solubility and are absorbed and tightly bound to particulate matter. They are persistent in sediments, for example the half life of emamectin benzoate is 194 d in aerobic sediment [85]. Ivermectin and emamectin benzoate have been found in some of the sediments sampled near salmon aquaculture sites. Thus, there is a potential for accumulation in sediment and they may pose a risk to sensitive benthic organisms. However, the avermectins may only be absorbed by the benthic organisms if they consume the medicated feed. Emamectin benzoate has been found in crustaceans immediately following treatment and in scavengers several month after sea lice treatments. However, evidence suggests that the amount of medicated feed consumed after use in sea lice control is insufficient to cause mortality. Salmon feed is not a preferred food of crabs and lobsters. There were no effects on polychaete populations near salmon aquaculture sites after sea lice treatments with ivermectin [114, 115].

Sublethal effects of emamectin benzoate have been observed in laboratory studies with American lobster and a marine copepod, *Acartia clausii*. However, the concentrations required to elicit these responses were above the PEC. The consequence of these sublethal effects on wild populations is unknown. Emamectin benzoate is preferred to ivermectin because it has a much shorter withdrawal time for salmon. The use of emamectin benzoate is permitted in Canada, Chile and several European countries, and in many cases is the treatment of choice. The use of emamectin benzoate in the treatment of sea lice infestations is also considered to have relatively low risk to the marine ecosystem.

The main environmental risk of the chitin synthesis inhibitors in the marine environment is likely to arise from the deposition of fish feces and waste feed on the sediments below and around the cages [98]. For example, 90% of the teflubenzuron administered is not absorbed by salmon and is excreted as parent compound in feces. Chitin synthesis inhibitors are bound by sediment and organic material. They are persistent in sediment with half life estimates to 115 d and there is a moderate risk of build-up in sediment through repeated applications. Field studies suggest that excess feed accumulates near the treatment cages but that there is widespread dispersion of the fecal matter that may extend greater than 100 meters from the cages, depending on water depth, current velocities, etc. Proper treatment strategies can reduce this risk by limiting the number of applications required.

Although chitin synthesis inhibitors are specific and of low toxicity to most non-target organisms, there are identified risks to any species that molts, for example crustaceans, that are located near to cage sites during treatment. A case can be made for possible environmental effects of the most sensitive species (pre-molt grass shrimp 96 h LC50 1.1 $\mu g\,L^{-1}$) exposed to the highest reported environmental concentration (1.5 $\mu g\,L^{-1}$ in water) [99]. However, based on the short half-life in water, it appears that the concentration decreases rapidly. In addition, there is negligible risk to organisms in the water column due to the tendency of chitin synthesis inhibitors to bind strongly to sediment and organic matter and they are not bioavailable from these bound forms unless ingested. Field studies with teflubenzuron did not detect any adverse effects in benthic biology or crustaceans near treatment sites. The limited data base suggests that the risk of adverse environmental effects is minimal.

7
Risk Management

Registration of sea lice therapeutants after they have been assessed as having acceptable risk makes them available for use. However, farms are located in waters with different capacities to absorb wastes, including medicinal chemicals, without causing unacceptable environmental impacts. Risks therefore have a site-specific components and management of these risks may therefore require site-specific assessments of the quantities of chemicals that can safely be used. The UK environmental authorities (primarily the Scottish Environment Protection Agency, SEPA) operate this further level of control on the use of medicines at fish farms and provide an example of a risk management plan that could be adopted in all areas that use sea lice therapeutants. A medicine or chemical agent cannot be discharged from a fish farm installation in Scotland unless formal consent under the Control of Pollution Act has been granted to the farm concerned by SEPA. The main components of the risk management process are:

- Environmental quality standards (Table 3) designed to protect non-target organisms in the receiving waters from adverse impacts of the chemical concerned. These are derived from analysis of available toxicological information, and appropriate safety factors.

- Site-specific information on the nature of the environment around the farm, particularly records of water currents over periods normally of at least 15 d.
- Mathematical modelling of the dilution and dispersion of soluble chemicals released at the farm, and of the distribution of particle-bound chemicals on the sea bed.
- The definition of an allowable zone of effect (AZE) around a farm within which the EQS may be breached. The AZE is normally equivalent to approximately 25 m distance around the cage site.

Table 3 Environmental Quality Standards for fish farm medicines applied by the Scottish Environmental Protection Agency

Active ingredient	Mode of application	Environmental quality standards
Azamethiphos	Bath	• Maximum allowable concentration (MAC), 3 h after release, of 250 ng L^{-1} 24 h MAC of 150 ng L^{-1} 72 h MAC of 40 ng L^{-1}
Cypermethrin	Bath	• Short term (3 h) EQS of 16 ng L^{-1} Maximum allowable concentration (MAC) of 0.5 ng L^{-1}, applied 24 h after release • Annual average EQS is 0.05 ng L^{-1}
Hydrogen peroxide	Bath	None – not considered to be a significant environmental risk
Teflubenzuron	In-feed	• 6.0 ng L^{-1} as an annual average in sea water and 30 ng L^{-1} as a MAC 2.0 ug kg^{-1} Dry wt 5 cm^{-1} core depth as a general sediment quality standard to be applied as a MAC to surface sediment (cores 5 cm depth) at more than 100 m from the cages 10.0 mg kg^{-1} dry wt 5 cm^{-1} core depth as a standard applied as an average value within the immediate under cage impact zone defined as surface area under and around cages to a distance of 25 m from cage edges
Emamectin benzoate	In-feed	• Concentrations in sediment should not exceed 0.763 ug kg^{-1} outside the AZE Concentrations in sediment should not exceed 7.63 ug kg^{-1} inside the AZE Concentrations in sea water should not exceed 2.2×10^{-4} ug l^{-1} Maximum number of treatments: · Three treatments in any 12 calendar month, and · Five treatments in any two year growth cycle

The output from the above assessment and modelling process is an expression of the maximum amount of a particular chemical that can be safely discharged from a specific farm over a defined period of time (Table 3). This is then included as a condition within the overall Discharge Consent for the farm. Exceeding these quantities or discharge of a chemical for which Consent has not been granted, would be a breach of the Consent and could lead to prosecution of the farm concerned. Full details of this process are available on the SEPA website (www.sepa.org.uk) in their Fish Farming Manual and in associated policy documents for individual sea lice treatments.

8
Conclusion

The nature and severity of the environmental risks presented by the use of the various chemicals available to control sea lice in farmed salmon vary considerably between treatment compounds. Current regulatory practices, particularly those leading to approvals/authorizations for the use of products for sea lice control, include elements of assessment of the risk to the environment. This is the primary process by which the environmental risk is managed, that is, through the decision whether or not to grant approval/authorization, and under what conditions.

However, there are considerable differences between the environmental characteristics of fish farm sites and their ability to accept discharges of sea lice treatments without giving rise to unacceptable environmental impacts. For example, differences in tidal currents and other hydrographic factors to dilute and/or disperse chemicals. Such site-specific risks can be managed through the application of appropriate Environmental Quality Standards for the chemicals concerned, and site-specific assessment of the maximum acceptable rate of use of the treatments.

References

1. Food and Agriculture Organization of the United Nations (2004) FAO Year Book, Fishery Statistics, Aquaculture Production, 2002. http://www.fao.org/fi/statist/FISOFT/FISHPLUS.asp#Features
2. Food and Agriculture Organization of the United Nations (1999) The State of the Worls Fisheries and Aquaculture. FAO, Rome
3. Commission of the European Communities (2002) A Strategy for the Sustainable Development of European Aquaculture. COM (2002) 511
4. Roth M, Richards RH, Sommerville C (1993) J Fish Dis 16:1
5. MacKinnon BM (1997) World Aquacult 28:5
6. Wootten R, Smith JW, Needham EA (1982) Proc R Soc Edinbur 81B:185
7. Pike AW (1989) Parasitol Today 5:291

8. Hogans WE (1995) Can Tech Rept of Fish Aquat Sci 2067:1
9. O'Halloran J, Hogans WE (1996) Can Vet J 37:610
10. Treasurer JW, Grant A (1997) Aquaculture 148:265
11. Read PA, Fernandes TF, Miller KL (2001) J Appl Ichthyol 17:145
12. Rae GH (2000) Caligus 6:2
13. Eithun I (2004) Caligus 6:4
14. Rae GH (1979) Fish Farmer 2:22
15. Boxaspen K, Naess T (2000) Contrib Zool 69:51
16. Horsberg TE, Hoy T, Nafstad I (1989) Acta Vet Scand 30:385
17. Tully O, Mcfadden Y (2000) Aquacult Res 31:849
18. Sievers G, Palacios P, Inostroza R, Doelz H (1995) Aquaculture 134:9
19. Committee for Veterinary Medicinal Products (1999) Azamethiphos Summary Report (2), EMEA/MRL 527:98
20. Zitko V (2001) Bull Environ Contam Toxicol 66:283
21. Roth M, Richards RH, Dobson DP, Rae GH (1996) Aquaculture 140:217
22. Levot GW, Hughes PB (1989) J Aust Ent Soc 28:87
23. Scottish Environmental Protection Agency (1997) Cage fish farms: sea lice treatment chemicals risk assessment of azamethiphos, http://www.sepa.org.uk/guidance/fishfarmmanual/pdf/policy17.pdf Policy No 17
24. Burridge LE, Haya K (1998) Gulf of Maine News Spring 1998:1
25. Chang BD, McClelland G (1997) Sea lice research and monitoring, Department of Fisheries and Oceans Science High Priority Project. Rept 9019
26. Burridge LE, Haya K, Zitko V, Waddy SL (1999) Ecotoxicol Environ Safety 43:165
27. Burridge LE, Haya K, Waddy SL (2004) Ecotoxicol Environ Safety (in press)
28. Burridge LE, Haya K, Waddy SL, Wade J (2000) Aquaculture 182:27
29. Burridge LE, Haya K, Waddy SL (2000) In: Penney KC, Coady KA, Murdoch MH, Parker WR, Niimi AJ (eds). Can Tech Rept of Fish Aquat Sci 2331:58
30. Waddy SL, Burridge LE, Haya K, Hamilton MN, Mercer SM (2002) Aquacult Assoc Can Spec Publ 5:60
31. Abgrall P, Rangeley RW, Burridge LE, Lawton P (2000) Aquaculture 181:1
32. Chang BD, McClelland G (1996) Alternative treatments for sea lice in farmed salmon, Department of Fisheries and Oceans Science High Priority Project Rept 9015
33. Davis JH (1985) The pyrethroids: An historical introduction. In: Leahey JP (ed) The pyrethroid insecticides. Taylor and Francis Ltd, London, p. 1
34. Barthel WF (1961) Adv Pest Control Res 4:33
35. Miller TA, Adams ME (1982) Mode of action of pyrethroids. In: Coats JR (ed) Insecticide Mode of Action. Academic Press, New York, p 3
36. Boxaspen K, Holm JC (1992) Aquaculture and the Environment, Spec Publ, Eur Aquacult Soc 16:393
37. Boxaspen K, Holm JC (2001) Aquac Res 32:701
38. Roth M, Richards RH, Somemrville CS (1993) Preliminary studies on the efficacy of two pyrethroid compounds, resmethrin and lambda cyhalothrin, for the treatment of sea lice (*Lepeophtheius salmonis*) infestations of farmed Atlantic salmon (*Salmo salar*). In: Boxshall GA, DeFaye D (eds) Pathogens of Wild and Farmed Fish: sea lice. Ellis Horwood Ltd, London, p. 273
39. Scottish Environmental Protection Agency (1998) SEPA policy on the use of cypermethrin in marine fish farming risk assessment, EQS and recommendations. http://www.sepa.org.uk/aquaculture/policies/index.htm Policy No. 30
40. Sevatadal S, Horsberg TE (2003) Aquaculture 218:21

41. Kahn NY (1983) An assessment of the hazard of synthetic pyrethroids to fish and fish habitat. In: Miyamoto J, Kearney PC (eds) Pesticide Chemistry: Human Welfare and the Environment. Proceedings of the Fith International Congress of Pesticide Chemistry, Kyoto, Japan, 1982. Permagon Press, Oxford, p 437
42. Haya K (1989) Environ Toxicol Chem 8:381
43. Hill I (1985) Effects on non-target organisms in terrestrial and aquatic environments. In: Leahey JP (ed) The Pyrethroid Insecticides. Taylor and Francis, London, p 151
44. Leahey JP (1985) Metabolism and environmental degradation. In: Leahey JP (ed) The pyrethroid insecticides. Taylor and Francis, London, p. 263
45. Burridge LE, Haya K (1997) Etoxicol Environ Saf 38:150
46. Clark JR, Giidman LR, Borthwick PW, Parick JM, Cripe GM, Moody PM, Moore JC, Lore EM (1989) Environ Toxicol Chem 8:393
47. Pahl BC, Opitz HM (1999) Aquac Res 30:655
48. Thybaud E (1990) J Water Sci 3:195
49. Associate Committee on Scientific Criteria for Environmental Quality (1986) Pyrethroids: Their effects on aquatic and terrestrial ecosystems. NRCC Publ. No 24376. National Research Council, Ottawa
50. Muir DCG, Rawn GP, Grift NP (1985) J Agric Food Chem 33:603
51. Clark JR, Patrick JM, Moore JC, Lores EM (1987) Environ Toxicol Chem 16:401
52. Mian LS, Mulla MS (1992) J Agric Entomol 9:73
53. Moore A, Waring CP (2001) Aquat Toxicol 52:1
54. McLeese DW, Mecalfe CD, Zitko V (1980) Bull Environ Contam Toxicol 25:950
55. Burridge LE, Haya K, Page FH, Waddy SL, Zitko V, Wade J (2000) Aquaculture 182:37
56. Gowland BTG, Moffat CF, Stagg RM, Houlihan DF, Davies IM (2002) Mar Envion Res 54:169
57. Mitchell AJ Collins C (1997) Aquac Mag 23:74
58. Rach JJ, Gaikowski MP, Ramsay RT (2000) J Aquat Anim Health 12:267
59. Jones MW, Sommerville C, Wootten R (1992) J Fish Dis 15:197
60. Cotran RS, Kumar V, Robbins SL (1989) Patholological Basis of Disease, 4th edn. Saunders, Toronto
61. Bruno DW, Raynard RS (1994) Aquac Int 2:10
62. Treasurer J, Wadsworth S, Grant A (2000) Aquac Res 31:855
63. Johnson SC, Constible JM, Richard J (1993) Dis Aquat Org 17:197
64. Hodneland K, Nylund A, Nisen F, Midttun B (1993) Bull Eur Assoc Fish Pathol 123:203
65. Kiemer MCB, Black KD (1997) Aquaculture 153:181
66. Rach JJ, Schreier TM, Howe GE, Redman SD (1997) Prog Fish-Cult 59:41
67. Carvajal V, Speare DJ, Horney BS (2000) J Aquat Anim Health 12:146
68. Davies I, Rodger G (2000) Aquac Res 31:869
69. Burridge LE (2003) Can Tech Rept Fish Aquat Sci 2450:1
70. Campbell WC (1989) Ivermectin and abamectin. Springer, New York
71. Roy WJ, Sutherland IH, Rodger HDM, Varma KJ (2000) Aquaculture 184:19
72. Grant AN (2002) Pest Manag Sci 58:521
73. Johnson SC, Margolis L (1993) Dis Aquat Org 17:101
74. Palmer R, Rodger H, Drinnan E, Smith PR (1987) Bull Eur Assoc Fish Pathol 7:47
75. Smith PR, Moloney M, McEllogott A, Clarke S, Palmer R, O'Kelly J, O'Brien F (1993) In: Boxshall GA, DeFaye D (eds) Pathogens of wild and farmed fish: sea lice. Ellis Horwood, Chichester, p 296

76. Stone J, Sutherland IH, Somemrville CS, Richards RH, Varma KJ (1999) J Fish Dis 22:261
77. Ramstad A, Colquhoun DJNR, Sutherland IH, Simmons R (2002) Dis Aquat Org 50:29
78. Stone J, Sutherland IH, Sommerville C, Richards RH, Varma KJ (2000) Aquaculture 186:205
79. Stone J, Sutherland IH, Sommerville C, Richards RH, Varma KJ (2000) Dis Aquat Org 41:141
80. Hoy T, Horsberg TE, Nafstad I (1992) In: Michel CM, Alderman DJ (eds) Chemotherapy in Aquaculture: from theory to reality. Off Intern Epiz, Paris, p 461
81. Grant A, Briggs AD (1998) Mar Pollut Bull 36:566
82. Davies IM, Gillibrand PA, McHenery JG, Rae GH (1998) Aquaculture 163:29
83. Halley BA, Nessel RJ, Lu AYH (1989) In: Campbell WC (ed) Ivermectin and abamectin. Springer, New York, p 162
84. Roth M, Rae G, McGill AS, Young KW (1993) J Agric Food Chem 41:2434
85. Scottish Environmental Protection Agency (1999) Emamectin benzoate, an environmental assessment. http://www.sepa.org.uk/policies/index.htm
86. Kilmartin J, Cazabon D, Smith P (1997) Bull Eur Assoc Fish Pathol 17:1958
87. Burridge LE, Haya K (1993) Aquaculture 117:9
88. Thain JE, Davies IM, Rae GH, Allen YT (1997) Aquaculture 159:47
89. Burridge LE, Hamilton MN, Waddy SL, Haya K, Mercer SM, Greenhalgh R, Tauber R, Radecki SV, Crouch LS, Wislocki PG, Endris RG (2004) Aquac Res (in press)
90. Linssen MR, van Aggelen GC, Endris R (2002) In: Eichkoff CV, van Aggelen GC, Nimi A (eds) Can Tech Rept Fish Aquat Sci 2438:68
91. Waddy SL, Burridge LE, Hamilton MN, Mercer SM, Aiken DE, Haya K (2002) Can J Fish Aquat Sci 59:1096
92. Willis KJ, Ling N (2003) Aquaculture 221:289
93. Savitz JD, Wright DA, Smucker RA (1994) Mar Envion Res 37:297
94. Walker AN, Horst MN (1992) J Crustac Biol 12:354
95. Horst MN, Walker AN (1995) J Crustac Biol 15:401
96. Branson EJ, Ronsberg SS, Ritchie G (2000) Aquac Res 31:861
97. Ritchie G, Ronsberg SS, Hoff KA, Branson EJ (2002) Dis Aquat Org 51:101
98. Scottish Environmental Protection Agency (1999) Calicide (teflubenzuron)- Authorization for use as an infeed sea lice treatment in marine salmon farms.Risk assessment, EQS and recommendations, Policy No 29 http://www.sepa.org.uk/aquaculture/policies/index.htm
99. Fischer SA, Hall LW (1992) Crit Rev Toxicol 22:45
100. Selvik A, Hansen PK, Ervik A, Samuelsen OB (2002) Sci Total Environ 285:237
101. Finkelstein ZI, Baskunov BP, Rietjens JiMC, Boersman MG, Vervoort J, Golovleva LA (2001) J Environ Sci Health B36:559
102. Shaefer CH, Dupras EF, Stewart RJ, Davidson LW, Colwel AE (1979) Bull Environ Contam Toxicol 21:249
103. Eisler R (1992) US Fish and Wildlife Service, Contaminant Hazard Reviews 25:1
104. Tanner DK, Moffett MF (1995) Environ Toxicol Chem 14:1345
105. Wright DA, Savitz JD, Dawson R, Magee J, Smucker RA (1996) Ecotoxicol 5:47
106. Roberts MH, Newman MC, Hale RC (2001) In: Newman MC, Roberts MH, Hale RC (eds) Coastal and Estuarine Risk Assessment. Lewis Publishers, Boca Raton, p 1
107. Alderman DJ, Smith P, Davies IM, Haya K (2004) ICES Co-operative Research Report (in press)
108. Campbell A (1986) Can J Fish Aquat Sci 43:2197

109. Charmantier G, Charmantier-Daures M, Aiken DE (1991) J Crustac Biol 11:481
110. UK Veterinary Medicines Directorate (2004) Ecotoxicity testing of medicines intended for use in fish farming. VDM Guidance Note, Animal Medicines European Licencing Information and Advice. Number 11
111. Henderson A, Davies I (2001) Fisheries Res Serv Rep 01/2001
112. Ernst W, Jackman P, Doe K, Page F, Julien G, Mackay K, Sutherland T (2001) Mar Pollut Bull 42:433
113. Scottish Environmental Protection Agency (2004) The occurence of the active ingredients of sea lice treatments in sediments adjacent to marine fish farms. Results of monitoring surveys carried out by SEPA in 2001 and 2002 http://www.sepa.org.uk/policies/index.htm
114. Black KD, Fleming S, Nickell TD, Pereira PMF (1997) ICES J Mar Sci 54:276
115. Costelloe M, Costelloe J, O'Connor B, Smith P (1998) Bull Eur Assoc Fish Pathol 18:22

Antibiotic Use in Finfish Aquaculture: Modes of Action, Environmental Fate, and Microbial Resistance

S. M. Armstrong[1] (✉) · B. T. Hargrave[2] · K. Haya[3]

[1]Department of Biology, Dalhousie University, Halifax, NS, B3H 4J1, Canada
stephen.armstrong@dal.ca

[2]Fisheries and Oceans Canada, Bedford Institute of Oceanography,
Dartmouth, NS, B2Y 4A2, Canada
Hargraveb@mar.dfo-mpo.gc.ca

[3]Fisheries and Oceans Canada, Biological Station, 531 Brandy Cove Road,
St. Andrews, NB, E5B 2L9, Canada
hayak@mar.dfo-mpo.gc.ca

1	Introduction	342
2	Types of Antibiotics	342
2.1	β-Lactams	342
2.2	Macrolides	343
2.3	Phenicols	344
2.4	4-Quinolones	345
2.5	Fluoroquinolone	347
2.6	Sulphonamides	348
2.7	Tetracyclines	348
3	Environmental Fate	349
3.1	Persistence in Sediments	350
3.2	Ecological Effects in the Water Column	351
3.3	Uptake by Biota	352
4	Antibiotic Resistance	352
5	Conclusions	354
	References	354

Abstract Various antibiotics have been used over the past 20 years and continue to be registered for use in finfish aquaculture in the United Kingdom, Norway, Ireland, and Canada. These include β-lactam (Amoxicillin), macrolide (Erythromycin), phenicols (Florfenicol), quinolones (Oxolinic acid, Piromidic acid, Naladixic acid, Flumequine), fluoroquinolone (Sarafloxacin), sulphonamides (potentiated sulphonamides), and tetracyclines (Oxytetracycline). Vaccines have largely replaced antibiotics as a means for controlling bacterial pathogens in cultured finfish but these anti-microbial agents continue to be applied to control disease in both hatcheries and grow-out stock. Bacterial strains resistant to specific antibiotics used in aquaculture have been cultured from mixed microbial communities in sediments after treatments of cultured fish stocks with antibiotics cease. This chapter considers modes of action, factors affecting environmental persistence and ecological aspects of antibiotic resistance of the major antibiotics currently used in finfish aquaculture in Canada and Europe.

Keywords Antibiotic resistance · Disease control · Microbial infection · Salmon aquaculture · Waste feed and feces

1
Introduction

The high biomass of finfish cultured within the restricted volume of netpens creates the potential for microbial and parasitic infections. The risk is so great that animal health management is a central husbandry requirement in all finfish aquaculture operations [1, 2]. Although the development of standard codes of practice, improved biosecurity and the use of vaccines have resulted in reduced use of antibiotics from levels used a decade ago [3–8], chemotherapeutants continue to be used. Diseases and infections will always need to be controlled to ensure maximum production [1]. Here we review various antimicrobial agents used to control infectious bacterial diseases in finfish aquaculture. Mode of action, persistence, and concerns surrounding the development of antibiotic resistance as they relate to environmental and human health are presented.

2
Types of Antibiotics

A small number of antibiotics are registered for legal use in the finfish aquaculture industry in Canada and Northern Europe (Table 1).

2.1
β-Lactams

β-Lactam antibiotics such as Amoxicillin (Fig. 1) interfere with the enzymatic cross-linking (i.e. transpeptidases and carboxypeptidases) of the cell wall in actively growing bacteria. The activity of β-lactams depends on the affinity for the target, permeability constraints such as bacterial capsule and peptidoglycan, and the stability of β-lactamases. β-Lactamases can be regulated by constitutive or inducible mechanisms [9, 10]. Amoxicillin is typically used for the control of furunculosis in salmonids caused by *Aeromonas* sp. It is administered orally in medicated feed at a dose of 80–160 mg kg^{-1} body weight d^{-1} for a standard period of 10 days [11]. The withholding period for β-lactam antibiotics in the United Kingdom is 40–150 degree days in Atlantic salmon (*Salmo salar*). Environmental concerns with respect to persistence of the β-

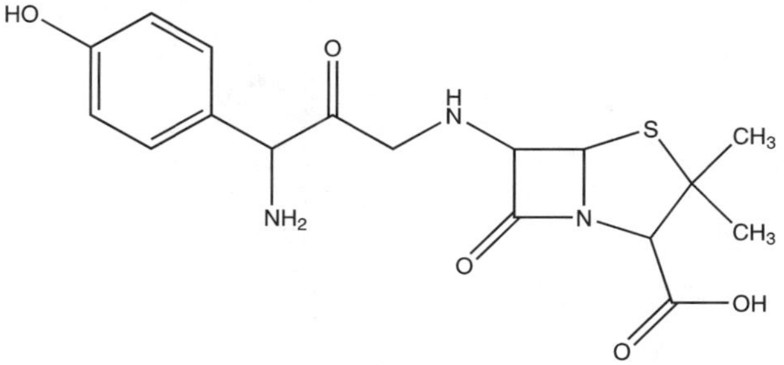

Fig. 1 Chemical structure of Amoxicillin

Table 1 Antibiotics used in the aquaculture industry in Canada, United Kingdom, Ireland, and Norway

Canada[1]	United Kingdom	Ireland[2]	Norway
Florfenicol	Florfenicol		Florfenicol
Oxytetracycline	Oxytetracycline	Oxytetracycline	Oxytetracycline
Potentiated sulphonamides	Potentiated sulphonamides	Potentiated sulphonamides	Flumequine
	Sarafloxacin	Sarafloxacin	Oxolinic acid
	Anoxicillin	Anoxicillin	

[1] [7, 12, 13]
[2] Four antimicrobial agents were used in Ireland up to 2001 – OTC, flumequine, anoxicillin, sulphadiazine potentiated with trimethoprim (Sulfatrim) [5]

lactam group of antibiotics are minimal. β-Lactams should be susceptible to biological and physicochemical oxidation in the environment since these are naturally occurring metabolites with an amino acid synthetic base.

2.2 Macrolides

Erythromycin is a broad spectrum antibiotic produced by the bacterium *Streptomyces erythreus* (Fig. 2). Erythromycin interferes with bacterial protein synthesis by binding to the 50S subunit of the bacterial ribosome with increased activity towards Gram positive micro-organisms primarily due to steric effects. It is successfully used against bacterial kidney disease (BKD) which is caused by *Renibacterium salmoninarum* in salmonids. Erythromycin, typically is provided in feed at a dose of 50–100 mg kg^{-1} body weight d^{-1} for approximately 21 days in which this dose reduced BKD mortality in brook trout by 50% [14]. No withholding period for this group of antibiotics has been recommended because erythromycin is not approved for

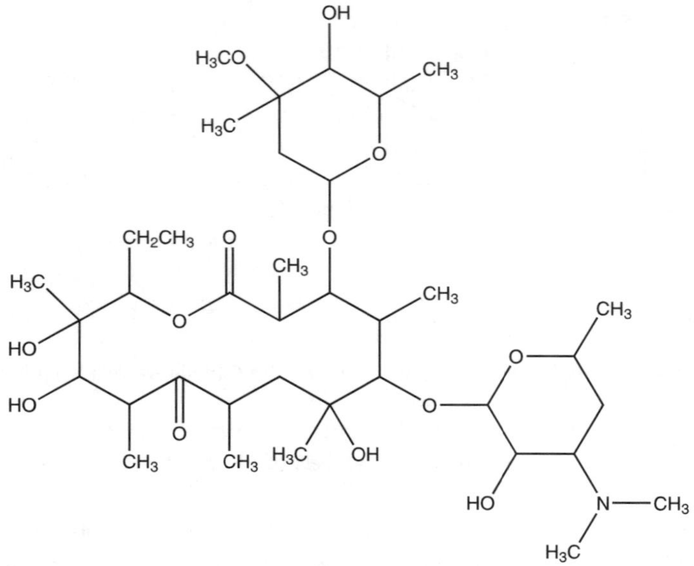

Fig. 2 Chemical structure of Erythromycin

use in International Council for Exploration of the Seas (ICES) countries. However, an excretion level of 0.03–0.08 mg g^{-1} was determined 168 hours after cessation of a treatment protocol of 50 mg kg^{-1} d^{-1} for 5 days in yellowtail [14]. Acute toxicity of erythromycin in excess of 2 g kg^{-1} was minimal with no abnormalities noted. However, when rainbow trout were subjected to a regime of 100 mg kg^{-1} d^{-1} for 21 days, behavioral and physiological abnormalities appeared.

The environmental effects of erythromycin may be more related to antibiotic resistance than to persistence since the ether linkages within the molecules will be susceptible to reduction or oxidation by physicochemical or biological processes. Although soluble in water and alcohol (2.1 mg ml^{-1} and > 20 mg ml^{-1}, 28 °C, respectively), the compound still has the potential to become associated with particulate matter, bioaccumulate in organisms and concentrate in sediments with potential effects on the micro-organisms.

2.3
Phenicols

Florfenicol acts as a broad spectrum antibiotic against Gram positive and Gram negative bacteria by binding to the 50S ribosomal subunit to prevent protein synthesis [15, 16]. Florfenicol has a fluorine atom instead of a hydroxyl group located at C-3 seen in the structure of chloramphenicol and thiamphenicol (Fig. 3) [17]. This structural change makes florfenicol less sus-

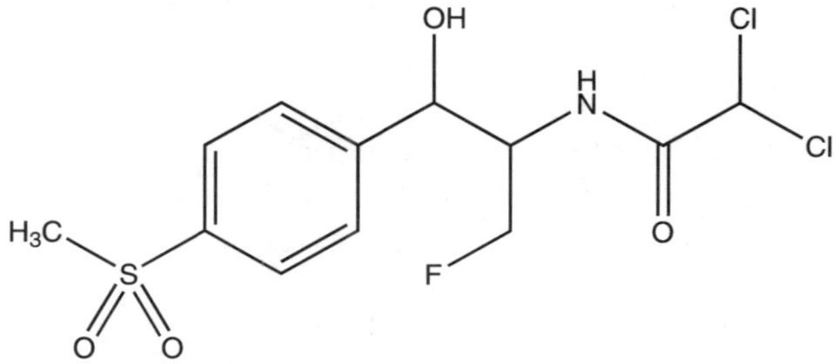

Fig. 3 Chemical structure of Florfenicol

ceptible to deactivation by bacteria with C-3 acetylation plasma-transmissible resistance and prevents interaction with bacterial ribosomes.

Florfenicol is used for treatment of furunculosis in salmon caused by *Aeromonas salmonicida* [18]. Exposure periods are usually 10 days at concentrations of $10\,\text{mg}\,\text{kg}^{-1}$ body weight d^{-1}, with no adverse reactions seen at 10 times the normal dose for a 10 day treatment period. The withholding period is 12 days, 150 degree days, or 30 days for Canada, United Kingdom, and Norway, respectively. Withdrawal time for *Salmo salar* in Canada is 12 days, however the water temperature must be over 5 °C.

The adsorption of Florfenicol in *Salmo salar* is 96.5% with a dose of $10\,\text{mg}\,\text{kg}^{-1}$ at water temperatures of $10.8\pm1.5\,°\text{C}$ [15]. Florfenicol was distributed throughout all tissues and organs in *Salmo salar* at a dose of $10\,\text{mg}\,\text{kg}^{-1}\,\text{d}^{-1}$ when water temperature was 8.5–11.5 °C [18]. Florfenicol concentrations in muscle were similar to those in blood and serum concentrations, while central nervous system and fat tissues had lower concentrations. Only 25% of serum concentrations were found in the brain. The half-life of Florfenicol when administered intravenously was 12.2 hours at a water temperature of $10.8\pm1.5\,°\text{C}$ [15]. Florfenicol degrades in the sediment with a half-life of 4.5 days and it displays low toxicity to aquatic organisms [19, 20]. There is low bacterial resistance to florfenicol and therefore it should not present a serious environmental concern in terms of persistence and induction of resistance.

2.4
4-Quinolones

The 4-quinolones are a relatively new group of antibiotics that are predominately active against Gram negative bacteria. However, future generations of quinolones may be developed that are effective against Gram positive bacteria, anaerobes, and some protozoa. Four quinolones are commonly used in the aquaculture industry: Oxolinic acid (Fig. 4), Flumequine (Fig. 5), Nalidixic acid (Fig. 6), and Piromidic acid (Fig. 7).

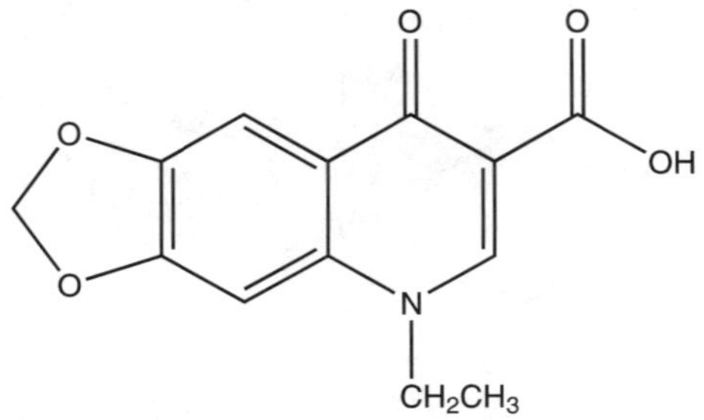

Fig. 4 Chemical structure of Oxolinic acid

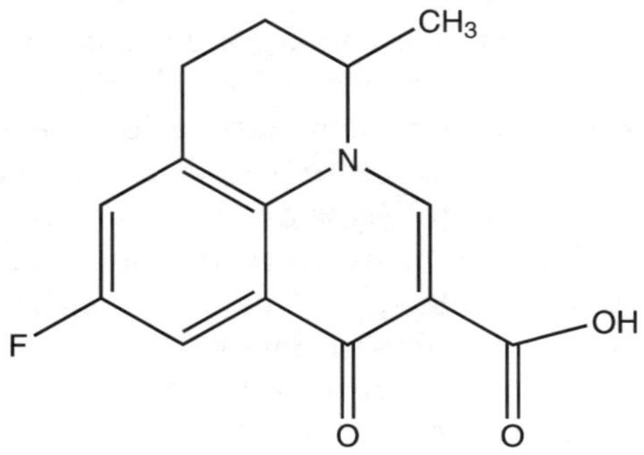

Fig. 5 Chemical structure of Flumequine

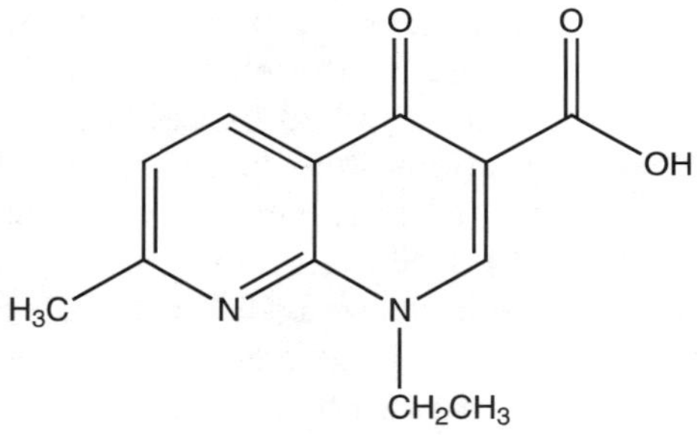

Fig. 6 Chemical structure of Nalidixic acid

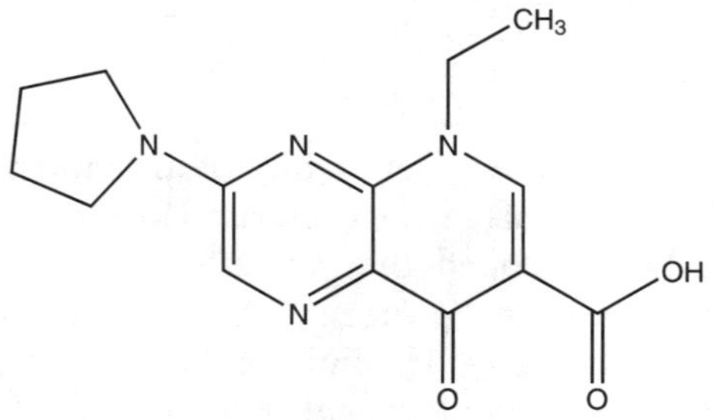

Fig. 7 Chemical structure Piromidic acid

The 4-quinolones are active against the bacterial DNA gyrase which acts by inhibiting the supercoiling of the bacterial DNA. Resistance to quinolone antimicrobials may not be plasmid encoded and requires the development and/or selection of genetic resistance. Withholding periods for this group of antibiotics, notably Oxolinic acid, are 500 degree days in the United Kingdom and greater than 80 days at less than 8 °C, 40–60 days at 8–12 °C, and 40 days at greater than 12 °C in Norway.

The 4-quinolones are new antimicrobials which have a high efficacy and relatively low toxicity. However, these anthropogenic compounds are not susceptible to enzymatic degradation or transformation, since microbial populations have not had any selective pressure to evolve enzyme systems to metabolize these molecules [20]. Therefore, these antibiotics have the potential to accumulate in aquatic environments. The 4-quinolones are susceptible to photolysis, however, this would be reduced under a fish cage in the presence of high suspended particulate loads or dissolved organic matter [21]. Furthermore, since these molecules attach readily to particles that eventually settle and accumulate in sediments, the probability of photolysis is low [22].

2.5
Fluoroquinolone

The fluoroquinolone Sarafloxacin is a water soluble antibiotic that is active against Gram negative bacteria [23, 24]. Sarafloxacin (Fig. 8) is rapidly absorbed by bacteria and it inhibits the action of DNA gyrase. It is typically added to feed at 10 mg kg^{-1} body weight d^{-1} for a period of 5 days. The withholding period after treatment is 150 degree days in the United States.

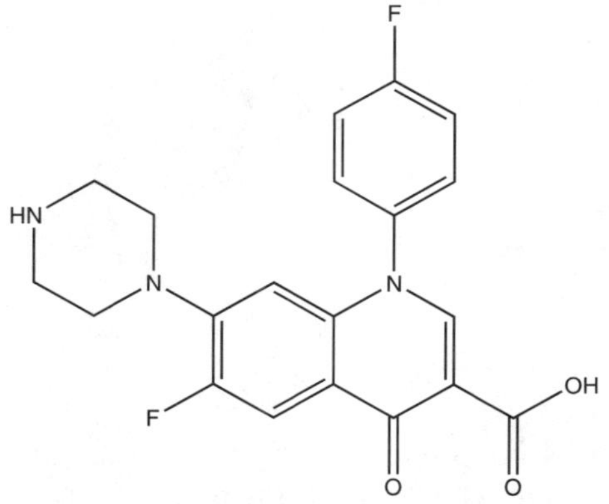

Fig. 8 Chemical structure of Sarafloxacin

2.6
Sulphonamides

The sulphonamides are a large class of antibiotics which are widely used in aquaculture to control furunculosis (*Aeromonas salmonicida*), enteric redmouth (*Yersinia ruckeri*), and vibrosis (*Vibrio* spp., *Cytophaga* spp., *Flexibacter* spp.) [25]. The most prevalent potentiated sulphonamide is Tribrissen (sulfadiazine: trimethoprim in a 5:1 ratio) (Fig. 9). Tribrissen inhibits dihydrofolate reductase whereas other sulfonamides, such as sulfadiazine (Fig. 10), inhibit dihydropteroate synthetase. Both enzymes are involved in the folic acid synthesis pathway [26]. These antibiotics are administered to finfish in feed and to molluscs in hatcheries in bath treatments. The typical dose for Tribrissen is 30–75 mg kg^{-1} d^{-1} for 5–10 days. The withholding period for these types of antibiotics is 350–500 degree days in the United Kingdom and 40–90 days in Norway which is temperature dependent.

The environmental implications of release of this type of antibiotic into the environment are unknown.

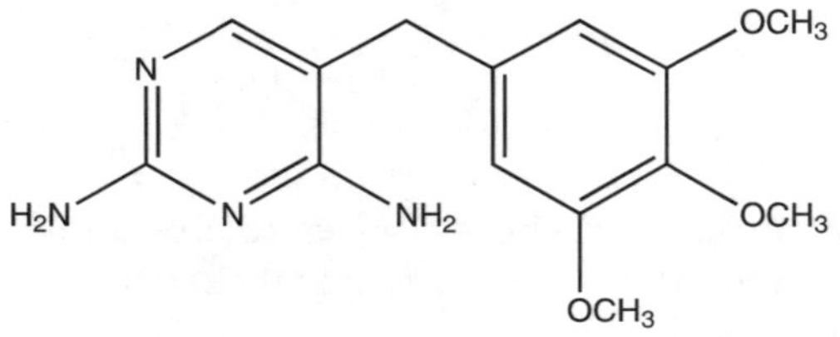

Fig. 9 Chemical structure of Trimethoprim

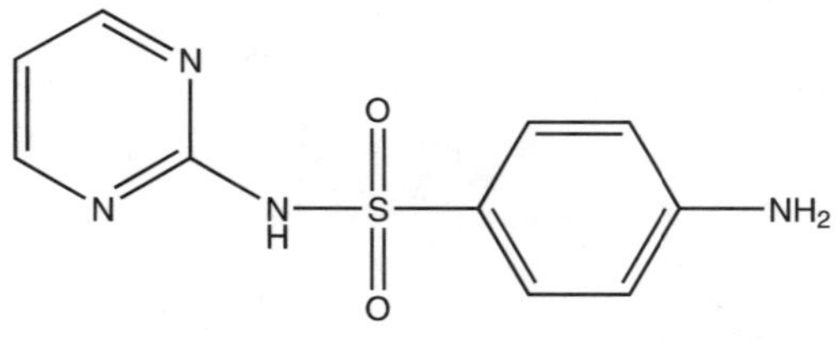

Fig. 10 Chemical structure of Sulphadiazine

2.7
Tetracyclines

The tetracycline antibiotic predominantly used in the finfish aquaculture industry is oxytetracycline (OTC) with trade names Terramycin Aqua in North America and Tetraplex in Ireland (Fig. 11). Technically Terramycin is the HCl-dihydrate and Tetaplex the HCL salt of OTC. Tetracyclines are bacteriostatic

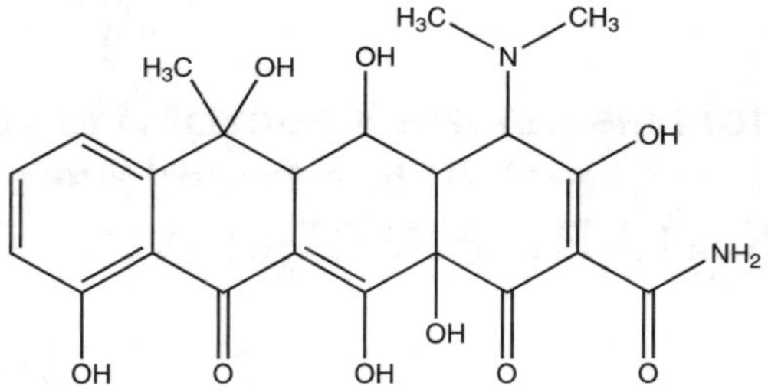

Fig. 11 Chemical structure of Oxytetracycline

antibiotics that interfere with protein synthesis by reversibly binding to the 30S ribosomal subunit, thereby blocking the binding of the aminoacyl tRNA to the mRNA/ribosome complex. They are broad spectrum antibiotics with activity against Gram positive and Gram negative bacteria. OTC is added to salmon feed at a dose of 50–125 mg kg^{-1} d^{-1} body weight for a 4–10 day treatment period [5, 27, 28]. Acute oral toxicity occurs at very low concentrations (LD$_{50}$ values > 4000 mg kg^{-1}), about 50 times higher than the effective dose [29]. Treatment by antibiotic baths or injections is also performed. The withholding period for OTC in Norway is 60 and 180 days in water above 12 °C and below 8 °C, respectively. In the United Kingdom 400 – 500 degree days are required. Bacteria have a number of mechanisms to deal with OTC, which include proton-dependent efflux, ribosomal protection by cytoplasmic proteins, enzymatic degradation and rRNA mutations.

As discussed in the follow section, of all of the antibiotics used in finfish aquaculture, OTC has been most widely studied in terms of its fate, persistence and ability to induce antibiotic resistance.

3
Environmental Fate

Modes of administration and physical-chemical properties affect transport pathways, environmental fate and persistence of antibiotics in any aquatic environment where they are applied. From its inception the finfish aquaculture industry has used cost-effective means to mitigate microbial infections thereby minimizing the need for the added expense of medicated feed. However, when biosecurity and vaccination programs are either not available or not effectively applied, disease outbreaks inevitably occur and must be treated quickly and effectively. Typically antibiotics are administered orally with feed but direct injection and/or immersion in antibiotic bath solutions are also used. These methods are more time consuming and costly than administering medicated feed. The choice of treatment method is important since it influences local and far field transport pathways. Orally administered antibiotics associated with waste feed will generally be deposited under or close to the pens [5]. These particle-associated antibiotics will be available for ingestion by wild fish, and benthic suspension and deposit-feeding invertebrates. More water-soluble antibiotics and fish fecal matter can be transported considerable distances in the water column with potential effects distant from the site of application.

3.1
Persistence in Sediments

The addition of antibiotics in fish feed is the most common method of application. However, infected fish often have a reduced appetite making oral uptake a less efficient antibiotic treatment method than injection or immersion. Even if feeding rates are adjusted to minimize loss of uneaten food, the efficiency of antibiotic absorption may be low. The absorption rate of OTC across the gut wall by salmon is low (< 2% of the administered dose) and if digestion of consumed food in infected fish is further reduced, fecal matter would also be expected to contain increased concentrations of antibiotics [30–32]. Although husbandry practices can be adjusted to account for the possibility of reduced food intake, more unconsumed antibiotic-laden feed might be expected to be lost during feeding than normally occurs with healthy fish. Unconsumed antibiotic-treated feed pellets will then either be deposited and accumulated at a farm site or in high current areas may be distributed more broadly [33].

The effect of low assimilation and loss to the environment is reflected in attempts to construct mass balance budgets for OTC in the vicinity of salmon farms. Less than 8.5% of the total OTC input could be accounted for in sediments at farm sites where medicated feed had been applied [34, 35]. Similar estimates (1 to 5%) were made at four farm sites in Ireland [5, 36]. Since only a few percent of OTC input was accumulated in sediments and tissue samples from salmon and mussels at the farm sites, it was concluded that the ultimate sink for OTC was in dissolved and particle-associated phases in the water column. Water solubility leading to hydrolysis, advective transport and photo-reactivity led to the conclusion that OTC would not be expected to accumulate in sediments [37]. However, no study has directly measured OTC in water to track dispersion around a farm site following treated feed application.

Accumulation of antibiotics in sediments can occur either as a result of direct deposition of treated feed pellets under and in the vicinity of net-pens or by adsorption of antibiotics in dissolved or colloidal form onto settling particles [22, 38]. The relatively high water solubility of OTC [37] should reduce accumulation in bottom deposits, but any antibiotic remaining associated with particles could remain in sediments under fish cages for some period of time [39]. Accumulation is likely enhanced in sediments that are light and oxygen free, thereby preserving the structural integrity of the OTC [36]. OTC experimentally added to oxic marine sediments largely disappeared after a few weeks, but traces were detectable for up to 18 months [40].

Concentrations of OTC measured in coastal marine sediment at farm sites vary from $< 10\ \mu g\,g^{-1}$ [34, 41, 42] to a maximum of $240\ \mu g\,g^{-1}$ [36]. This can be compared to OTC concentrations in commercially prepared salmon feed pellets that are three orders of magnitude higher ($29\ mg\,g^{-1}$) [43]. OTC was found in fish farm sediments in Norway and Finland above detection levels ($10\ \mu g\,g^{-1}$) for periods of more than one year after treatment but these were primarily anoxic deposits associated with cage sites [34, 41]. High levels of bacterial resistance have been found in both sediment bacteria and isolates from intestines of wild fish around finfish aquaculture sites [34, 44]. This shows that while dispersion and dilution may reduce water concentrations to below detection limits, transport pathways exist for exposure of benthic invertebrates and demersal and pelagic fish to resistant strains.

In addition to physical-chemical properties, the persistence of antibiotic residues in sediments depends on several environmental factors among which sedimentation rates, the presence/absence of oxygen and water temperature are critical [20, 31, 34, 35]. The half-life of oxytetracycline in sediment was prolonged to 419 d under stagnant, anoxic conditions [34]. The half-life of Tribissen (20% sulfadiazine, 80% trimethoprim) at 6 to 7 cm depth was found to be 90 d while Florfenicol concentrations decreased more rapidly (t_{50} = 4.5 days) [20]. Shorter half-lives might be expected in more oxic sediments. Persistence of Tribissen was dependent on sedimentation rates at the site after medication [45]. The half-life of OTC of 72 d doubled to 135 days under a 4-cm layer of sediment, slightly longer than the average value (60 days) observed in mixed sediments several cm deep in an experimental laboratory study [38].

3.2
Ecological Effects in the Water Column

As discussed above, antibiotic injections are a direct and efficient way to administer treatment. Therefore, less antibiotic is used to treat each fish and losses to the environment are minimized. However, high labour costs usually preclude this approach even if treatment is more effective. Another method of application is immersion of infected fish in an antibiotic bath as described for sealice therapeutants [8]. The draw-back is that the bath solution must be released following treatment. Loss of OTC from the water column around a fish farm located in a salt marsh occurred in two phases with average half-lives of 30 h and 319 h, respectively [46]. Although the initial loss phase was relatively short, antibiotics remaining in the water column for even a relatively short time could affect planktonic organisms [47]. Physically removing or destroying antibiotics in solution before they are discharged could circumvent the problem but this added cost is usually avoided by simply releasing bath solutions after treatment.

3.3
Uptake by Biota

Bacterial communities involved with decomposition and mineralization processes of organic matter may be susceptible to exposure to antibiotics if concentrations accumulate in sediments [48]. Ecological effects of antibiotic treatment on microbially mediated sediment nutrient dynamics have been demonstrated [49]. A range of OTC concentrations (12.5 to 75 mg l^{-1}) were applied to quantify a dose-response relationship between OTC and nitrification in aquaria containing freshwater, sand sediments and catfish fingerlings. Nitrification rates, measured as decreases in ammonia and increases in nitrate concentrations over three weeks, were reduced by 50% as OTC concentrations increased from 8.6 to 30 mg l^{-1}, concentrations typical of doses recommended for bath treatment.

Several studies have shown that measurable concentrations of antibiotics appeared in non-target invertebrates either in the laboratory after exposure to residues in water and/or sediments or in close proximity to salmon cage sites where medicated feed had been used. During the years of high antibiotic use in the salmon aquaculture industry in Norway, fish and mussels near salmon farms contained OTC [50]. Since the half-life of OTC in mussels is short, estimated as approximately 2 days [5], the presence of residues indicated recent or continuous exposure. Oysters, crabs and benthic macro-invertebrates collected near salmon farms in British Columbia contained OTC and Romet 30 (5 : 1 sulfadimethoxine and ormetoprim) [35, 51–53]. The highest concentrations (3.8 µg OTC g^{-1} wet tissue) in rock crab exceed the guideline for seafood specified by the US Food and Drug Administration (2 µg g^{-1}) [35]. However, unlike the observations of suppression in bacterially mediated nitrification above, even at these relatively high concentrations no study has shown adverse effects of aquaculture-derived antibiotics on indigenous fauna.

4
Antibiotic Resistance

One result of the broad-scale release of antibiotics into coastal marine environments subjected to intensive aquaculture for either finfish or shellfish is the possibility for selection of resistance in non-target benthic organisms [54]. Mechanisms whereby microbial antibiotic resistance is induced have been summarized along with current methods for using microbial assays to quantify growth inhibition [55]. These methods have recently been standardized in an effort to make measurements of resistance more reproducible and quantitative [56].

Resistance to antibiotics has often been observed in natural bacterial communities in sediments near salmon aquaculture sites, but it may also be transferred to non-target organisms. For example, an increase in antibiotic-resistant bacteria in sediments occurred within a few days of treatment with OTC or Oxolinic acid [57]. High levels of resistance occurred at the end of the initial 10-day treatment when the percentage of resistant bacteria (ratio of numbers growing on substrate ± OTC) was > 100% in all sediment samples. OTC-resistant bacteria were isolated from the intestines of wild fish and rainbow trout that had fed on medicated feed [34, 41]. *Aeromonas salmonicida* has been identified as the source of resistance in salmon at 9 of 35 fish farms in Finland treated with OTC [58].

Resistance to OTC was detected in aerobic bacteria cultured from water, pelletized feed and fingerlings from freshwater Atlantic salmon farms in Chile [59]. High levels of OTC resistance [90% minimum inhibitory concentrations (MIC) up to $2000\ \mu g\ OTC\ ml^{-1}$] in selected strains suggested that salmon farms may be reservoirs for bacteria with high tetracycline resistance. Similar observations of OTC resistance, but with lower MIC values (up to $160\ \mu g\ OTC\ ml^{-1}$), were observed in surface sediments under and up to 100 m away from salmon farm pens in southwestern regions of the Bay of Fundy [60]. A standardized micro-dilution assay method was used to detect resistance in natural communities of bacteria isolated from sediment under pens and around various farm sites. Resistant strains, tentatively identified as *Psychrobacter glacincola* and *Psychrobacter pacificens*, were capable of growth in media containing up to $160\ \mu g\ OTC\ ml^{-1}$ while a type culture of *Aeromonas salmonicida* used as a control showed no growth at $5\ \mu g\ OTC\ ml^{-1}$.

Results from field observations and experiments that demonstrate induction of antibiotic resistance should be treated cautiously since many factors can affect bacterial growth [46]. For example, species frequency data was analyzed and it was concluded that the operation of fish farms had minor long-term impacts on the size of bacterial communities in under-cage sediments [61]. This contrasts observations in Puget Sound where the number of colony-forming bacteria units were generally higher in sediments from cage sites than surrounding areas [62]. The proportion of OTC-resistant bacteria has also been observed to decline exponentially with increasing distance from a farm. Increased antibiotic resistance in sediments 75 m from the edge of a cage array in Galway Bay, Ireland was detectable during a brief (10-day) exposure to OTC [27]. However, after therapy ended, the frequency of resistance decreased exponentially, and within 73 days under-cage samples were not significantly different from background levels. In addition to husbandry practices that determine the release of medicated feed and fish excretory wastes to the environment, measures of antibiotic resistance in different studies will reflect various environmental and biological factors. Both the levels and persistence of microbial resistance observed in natural bacterial communities at and distant from any given farm can be expected to be highly variable and site specific.

Infectious micro-organisms were identified in sediments from an abandoned Norwegian salmon farm site [63] indicating that, irrespective of antibiotic use, once a disease outbreak has occurred, the probability of re-infection in a given area is increased. The development of antibiotic resistance may also have the potential for human health risk since positive correlations have been reported between antibiotic use and the isolation of drug-resistant bacteria in fish consumed as food [3, 64]. The successful transfer of antibiotic resistance was reported among strains of *Aeromonas hydrophilia* isolated from cultured *Telapia mossambica* via exchange of plasmids [65], illustrating the potential for the spread of drug resistance in cultured fish. The presence of OTC-resistant aeromonads in waters receiving hospital and aquaculture wastes [66, 67] also indicates that antibiotic resistance may arise from both human and aquaculture sources. Clearly, further studies are required to determine the extent of ecological and biological impacts of antibiotic resistance in microbial and other wild populations in areas of intensive finfish aquaculture.

5
Conclusions

Although many antibiotics are employed in the aquaculture industry their use should be restricted because of concerns over increased antibiotic resistance. The development of antibiotic resistance in natural microbial communities has the potential for far field effects on wild (non-target) species and indirectly or directly on human health. Furthermore, not all antibiotics employed in the aquaculture industry are equally persistent in the environment; aquaculture site managers must use their expertise to choose wisely the type, amount, and method of delivery of specific antibiotics to meet their needs. It is anticipated that good animal husbandry and environmental management will limit the need for the use of antibiotics in finfish aquaculture. Proper understanding of variables affecting the fate, transport and environmental persistence of these therapeutants should lead to changes in aquaculture husbandry practices that eliminate or greatly reduce the need to use medicated feed in the future.

References

1. Stewart JE (1991) ICES Mar Sci Symp 192:206
2. Stewart JE (1998) Can Tech Rep Fish Aquat Sci 2218

3. Alderman DJ, Hastings TS (1998) Int J Food Sci Technol 33:139
4. Sheppard ME (2000) Bull Aqua Assoc Canada 100-1:13
5. Coyne R, Smith P, Moriarty C (2001) Mar Environ Health (Marine Institute, Dublin, Ireland) Ser 3:24
6. Alderman DJ (2002) Bull Eur Assoc Fish Pathol 22:117
7. Burka JF, Hammel KL, Horsberg TE, Johnsons GR, Rainie DJ, Speares DJ (1997) J Vet Pharmacol Ther 20:333
8. Haya K, Burridge LE, Davies IM, Ervik A (2005) A review and assessment of environmental risk of chemicals used for the treatment of sea lice infestations of cultured salmon (in this volume). Springer, Berlin Heidelberg New York
9. Aoki T (1997) Resistance plasmids and the risk of transfer. In: Bernoth EM, Ellis AE, Midtlyng PJ, Oliver G, Smith P (eds) Furunculosis Multidisciplinary Fish Disease Research. Academic, New York, p 433
10. Bruun MS, Schmidt AS, Madsen L, Dalsgaard I (2000) Aquaculture 187:201
11. Ang CY, Luo W, Hansen EB Jr, Freeman JP, Thompson HC Jr (1996) J AOAC Int 79:389
12. Health Canada (2001) http://www.hc-sc.gc.ca/vetdrugs-medsvet/e_aquaculture.htm
13. Salmon Health Consortium (2002) http://salmonhealth.ca/therapeutantsapproved.htm
14. Stoffregen DA, Bowser PR, Babish JG (1996) J Aquat Anim Health 33:1881
15. Martinsen BT, Horberg TE, Varma KJ, Sams R (1993) Aquaculture 112:1
16. Fukui H, Fujihara Y, Kano T (1987) Fish Path 22:201
17. Sams RA (1994) Florfenicol: chemistry and metabolism of a novel broad-spectrum antibiotic. In: Proc XVIII World Buiatrics Congress, Bologna, Italy, p 13
18. Horsberg TE, Marinsen B, Varma K (1994) Aquaculture 122:97
19. Pinault LP, Millot LK, Sanders PJ (1997) J Vet Pharmacol Therap 20:297
20. Hektoen H, Berge JA, Hormazabal V, Yndestad M (1995) Aquaculture 133:175
21. Pouliquen H (1994) Oxytetracycline and oxolinic acid in aquaculture. Chromatographic analysis in sea water, sediments, and marine bivalves. Experimental study of the fate in the marine environment. Doctoral Thesis, University of Nantes, p 203
22. Pouliquen H, Le Bris H (1996) Chemosphere 33:143
23. Meade JL, English WR, Schwedler TE, Grimes LW (1993) J App Aqua 3:1
24. Giles JS, Hariharan H, Heaney SB (1991) Aqua Assoc Canada 91:53
25. Soltani M, Shanker S, Munday BL (1995) J Fish Dis 18:555
26. Lui Y-K, Lui C-K (1995) Mem Coll Agric Nat 35:294
27. Kerry J, Hiney M, Coyne R, Cazabon D, NicGabhaninn S, Smith P (1994) Aquaculture 123:43
28. Kerry J, Hiney M, Coyne R, NicGabhaninn S, Gilroy D, Cazabon D, Smith P (1995) Aquaculture 131:101
29. Bjørklund H, Bylund G (1991) Xenobiotica 21:1511
30. Alderman DJ, Rosenthal H, Smith P, Steward J, Weston D (1994) Chemicals used in mariculture. ICES Working Group Environmental Interaction of Mariculture, International Council for the Exploration of the Sea, Copenhagen, Denmark. ICES Coop Res Report 202, p 100
31. Samuelsen OB (1994) Environmental impacts of antibacterial agents. In: Ervik A, Hansen PK, Wennevik V (eds) Proc Canada-Norway Workshop on Environmental Impacts of Aquaculture. Havforskningsinstituttet, Fisken og Havet 13, Bergen, Norway, p 17
32. Weston DP (1996) Environmental considerations in the use of antibacterial drugs in aquaculture. In: DJ Baird, MCM Beveridge, LA Kelly, JF Muir (eds) Aquaculture and Water Resource Management. Blackwell Science, Oxford, p 140

33. Cromey CJ, Black KD (2005) Modelling the impacts of finfish aquaculture (in this volume). Springer, Berlin Heidelberg New York
34. Bjørklund H, Bondestam J, Bylund G (1990) Aquaculture 86:359
35. Capone DG, Weston DP, Miller V, Shoemaker C (1996) Aquaculture 145:55
36. Coyne R, Hiney M, O'Connor B, Kerry J, Cazabon D, Smith P (1994) Aquaculture 123:31
37. Smith P (1996) Aquaculture 146:157
38. Pouliquen H, Le Bris H, Pinault L (1992) Mar Ecol Prog Ser 89:93
39. Smith P, Samuelson OB (1996) Aquaculture 144:17
40. Samuelsen OB, Torsvik V, Ervik A (1992) Sci Total Environ 114:25
41. Bjørklund H, Råbergh CMI, Bylund G (1991) Aquaculture 97:85
42. Kerry J, Slattery M, Vaughan S, Smith P (1996) Aquaculture 144:103
43. Campell DA, Pantazis P, Kelly MS (2001) Aquaculture 202:73
44. Torsvik VL, Søerheim R, Goksøeyr J (1988) Int Counc Explor Sea CM 1988/F:10, p 9
45. Samuelsen OB (1989) Aquaculture 83:7
46. Pouliquen H, Le Bris H, Pinault L (1993) Aquaculture 112:113
47. Holten Lutzhoft H-C, Halling-Sorensen B, Jorgensen SE (1999) Arch Environ Contam Toxicol 36:1
48. Stewart JE (1994) Aquaculture in Canada and the research requirements to environmental interactions with finfish culture. In: Ervik A, Kupka-Hansen P, Wennevik V (eds) Proc Canada-Norway Workshop on the Environmental Impacts of Aquaculture, Fisken Havet 13, p 18
49. Klaver AL, Matthews RA (1994) Aquaculture 123:237
50. Ervik A, Thorsen B, Eriksen V, Lunestad BT, Samuelsen OB (1994) Dis Aquat Org 18:45
51. Jones OJ (1990) Uptake and depuration of the antibiotics, oxytetracycline and Romet-30 in the Pacific oyster, Crassostrea gigas (Thunberg). M.Sc. Thesis, University of British Columbia, p 221
52. LeBris H, Pouliquen H, Debernardi J-M, Buchet V, Pinault L (1995) Mar Environ Res 40:587
53. Cross SF, Gormican SJ, Levings CD (1997) A preliminary examination of oxytetracycline (OTC) from fish farm to fauna adjacent to a small net-pen operation in British Columbia. Pierce RC, Williams D (eds), Dept Fish Oceans Green Plan Toxic Chemicals Program Wrap-up Conference. Can Tech Rep Fish Aquat Sci 2163:68
54. GESAMP (IMO/FAO/UNESCO-IOC/WMO/WHO/IAEA/UN/UNEP Joint Group of Experts on the Scientific Aspects of Marine Environmental Protection) (1997) Rep Stud GESAMP No. 65, Rome, p 40
55. Amábile-Cuevas CF (2003) Am Sci 91:138
56. Alderman DJ, Smith P (2001) Aquaculture 196:211
57. Hansen PK, Lunestad BT, Samuelsen OB (1993) Can J Microbiol 39:1307
58. Hirvelä-Koski V, Koski P, Niiranen H (1994) Dis Aquat Org 20:191
59. Miranda CD, Zemelman R (2002) Aquaculture 212:31
60. Friars F (2002) Microbial resistance to oxytetracycline in sediments from salmon aquaculture sites in the Western Isles region of the Bay of Fundy. M.Sc. Thesis, Dalhousie University, p 105
61. Smith P, Pursell L, McCormack F, O'Reilly A, Hiney M (1995) Bull Eur Ass Fish Pathol 15:105
62. Herwig RP, Gray JP, Weston DP (1997) Aquaculture 149:263
63. Husevag B, Lunestad BT, Johannesen P, Enger O, Samuelsen OB (1991) Sci Total Environ 108:275

64. Hastings T, McKay A (1987) Aquaculture 61:165
65. Son R, Rusul G, Sahilah AM, Zainuri A, Salmah I (1997) Lett Appl Microbiol 24:479
66. Rhodes G, Huys G, Swings J, McGann P, Hiney M, Smith P, Pickup RW (2000) Appl Environ Microbiol 66:3883
67. Furushita M, Shiba T, Maeda T, Yahata M, Kaneoka A, Takahasi Y, Torii K, Hasegawa T, Ohta M (2003) Appl Environ Microbiol 69:5336

Assessing Nitrogen Carrying Capacity for Blue Hill Bay, Maine: A Management Case History

John W. Sowles

Maine Department of Marine Resources, P.O. Box 8, West Boothbay Harbor, ME 04575, USA
john.sowles@maine.gov

1	Introduction	360
2	Approach	363
3	Part 1: Water Quality Characterization	365
3.1	Results	366
3.1.1	Oxygen	368
3.1.2	Light	368
3.1.3	Nutrients	369
3.1.4	Chlorophyll	369
3.1.5	Nuisance and Toxic Algal Blooms	370
4	Part 2: Nitrogen Load Estimate	371
5	Part 3: Assessing Carrying Capacity	373
5.1	Modeling	373
5.2	"Vollenweider-Type" Model	374
6	Carrying Capacity Applied to Net-Pen Aquaculture	374
6.1	Attributes of Net-pen Aquaculture Affecting Carrying Capacity	377
	References	378

Abstract As net-pen aquaculture in coastal waters grows, regulatory agencies must consider the potential for localized eutrophication. To avoid excessive nutrient concentrations, Maine sees value in dispersing farms over larger regions. Often, little environmental baseline information exists in these new areas on which to base a risk assessment. This case history attempts to convey the process by which a permitting decision was made using a weight-of-evidence approach. By drawing on readily available sources of information, simple models, and basic water quality monitoring data, the effect on Blue Hill Bay by an existing salmon farm was analyzed and the potential effect of a proposed net-pen aquaculture was predicted. In the process, a range of carrying capacities was developed, depending on temporal and spatial scale. The exercise has potential value to regulatory scientists and policy makers by identifying new research, monitoring, and compliance issues. In the end, the exercise demonstrates that neither aquaculture nor carrying capacity should be viewed as static and in isolation from other stressors.

Keywords Carrying capacity · Net-pen aquaculture · Nitrogen · Management · Coastal Eutrophication

1
Introduction

In Maine, aquaculture is one of several designated uses of State water for which water quality is to be protected. Yet net-pen aquaculture itself has the potential to degrade water quality and in turn affect other existing and designated uses. This case history chronicles steps and decisions made by resource scientists confronted with having to assess the capacity of Blue Hill Bay to support finfish aquaculture without impairing water quality and other designated human and ecological uses. Attempts are made to convey how regulatory agency staff manage uncertainty in the decision-making process, by drawing on a wide variety of resources to assess risks associated with nutrient enrichment from a salmon farm and balancing those risks with other coastal policy objectives.

Marine net-pen aquaculture poses a unique set of challenges to environmental permitting agencies accustomed to conventional wastewater treatment systems. Water quality laws and regulations in most jurisdictions were developed decades ago to control discharges from factories and cities, long before net-pen aquaculture became established. The design and performance of conventional wastewater treatment plants have benefited from a long history of development. For the most part, today's engineers are able to predict waste discharge loads and their impact on water quality from these conventional systems.

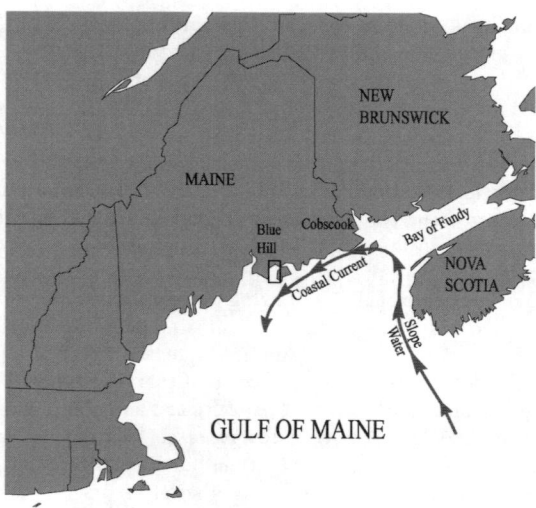

Fig. 1 Gulf of Maine regions showing Blue Hill Bay (insert) in relation to the nutrient-rich coastal current

Unlike conventional discharges, net-pen effluents do not discharge from confined pipes that are easily monitored, are not normally located in unidirectional currents, and are not supported by well-validated waste dilution and assimilation models. Nevertheless, when new or expanded aquaculture operations are proposed, natural resource agency scientists are asked to determine, in a very public and often contentious forum, whether or not undesirable effects are likely to result.

Maine's net-pen industry flourished in the 1980s about 135 km east of Blue Hill Bay in Cobscook Bay on the Maine and New Brunswick border (Fig. 1). This eastern part of Maine's coast has rapid water exchange [1], the water column is well mixed throughout the summer with its 6-m tides, seawater temperatures are moderate year-round (0–13 °C), protection exists from severe winter storms, access to commercial facilities is close by, and the local culture and labor force are still tied to the sea. In many respects, the eastern area of Maine was favorable for the culture of Atlantic salmon.

In addition to being suitable for salmon aquaculture, the eastern Maine region is nitrate rich due to an intrusion of deep continental slope water that passes off the tip of Nova Scotia and up into the Bay of Fundy (Fig. 1). Schlitz and Cohen [2] estimate that the annual nitrogen flux delivered by continental slope water to the Gulf of Maine is about 2 235 800 metric tons (MT) of dissolved inorganic nitrogen (DIN). This rich source of nutrients drives primary and hence secondary productivity for the Gulf of Maine's historically valuable commercial fishery [3], and is also thought to be responsible for the region's regular blooms of toxic algae [4].

On a Gulf of Maine scale, the relative contribution of nitrogen from salmon farming is small. For example, in the entire Gulf of Maine (Maine, New Brunswick, and Nova Scotia), salmon aquaculture is estimated to annually contribute about 2800 MT of total nitrogen (TN) or about 0.012% [5]. Of that amount, farms in eastern Maine are estimated to contribute less than 360 MT [6]. As Hargrave and Strain point out in Chap. 2 [7], however, effluents from salmon farms can be locally large and have associated undesirable consequences.

By the early to mid-1990s, most sites suitable for salmon aquaculture in Cobscook Bay had been leased. Furthermore, husbandry practices to prevent and manage sea lice and infectious salmon anemia required additional sites to maintain existing production levels. Single year class husbandry (all fish on a site are the same age), site fallowing to break disease cycles (allowing the farm to remain unoccupied for several months or more following harvest), and maintaining generous separation distances between farms to reduce farm-to-farm transmission of pathogens if outbreaks occurred meant that if the industry was to remain in Maine, it had one choice: to follow the nutrient-rich coastal current southwest into warmer and stratified waters [8, 9]. In doing so, however, both industry and resource scientists would need to address new environmental concerns and constraints.

Although industry was concerned about slower water exchange due to smaller tides and thermal stress from warmer summer and colder winter temperatures, Maine's resource scientists were concerned about the potential effects of eutrophication. Unlike Cobscook Bay where finfish sites were situated in well-flushed waters and light was thought to be the limiting factor [10], it was not clear to resource scientists whether light controlled production elsewhere. Indeed, Townsend et al. [9] showed a clear gradient of diminishing nutrient levels as the coastal current flows southwest, which suggests that at some point along the coast, nutrients commence control over phytoplankton. At that undefined point, even small contributions of nutrients delivered to less well-mixed areas such as inlets, coves, and bays could promote new or exacerbate existing blooms of harmful algae.

Blue Hill Bay is large ($\sim 190\,km^2$), deep ($\sim 105\,m$), and protected from winter storms. Together with nearby access to land facilities, Blue Hill Bay is suitable for some amount of commercial net-pen aquaculture. In 1993, a small (250 000 fish) Atlantic salmon farm began operation and has been the only farm to operate in the bay. Blue Hill Bay is also on a section of the coast of Maine (Fig. 1) undergoing rapid cultural transformation. The area is a major tourist destination with adjacent Acadia National Park drawing upwards of five million visitors each summer. In the last decade, the Blue Hill region experienced one of the largest growth rates in Maine [11].

It was no surprise when area residents reacted negatively over two large ($\sim 850\,000$ fish each) and highly visible salmon farms proposed for Blue Hill Bay. Although both proposals were ultimately denied, area residents realized the potential for large-scale aquaculture in "their" bay. One proposal was denied for unreasonably affecting existing navigation and recreational use but, relevant to this case history, the other proposal was denied for failure to address water quality concerns identified in the environmental baseline survey. In that survey, dissolved oxygen saturation in bottom waters fell a few percentage points below the State water quality standard of 85%. Of all the potential ecological concerns including benthic impacts, interference with wildlife, and toxic contaminants, water quality, and specifically eutrophication, became a central issue for further net-pen aquaculture development in Blue Hill Bay.

On the one hand, Maine clearly supports protecting water quality, habitats, and multiple uses of its public-trust waters. On the other hand, it was in both the industry's and State's interest to see aquaculture move away from eastern Maine where farms were concentrated and to address disease and environmental quality issues. Balancing apparently opposing objectives is a continual challenge for public resource agencies. With residential growth in the area increasing on land and the prospects of salmon aquaculture moving westward into Blue Hill Bay, resource scientists and public asked the question: "How much can the bay support?"

2
Approach

The opportunity to address this question arose in 1999 with a proposal to expand the existing salmon aquaculture farm in Blue Hill Bay. From information gained from the previously denied applications, which indicated that Blue Hill Bay was sensitive to nutrient loading (e.g., seasonal stratification, poor water exchange, lowered oxygen concentrations, and isolation from the open ocean), resource managers deferred approvals for new finfish leases pending the results of a bay-wide assessment to determine Blue Hill Bay's nutrient carrying capacity.

To reach some conclusion over the carrying capacity of Blue Hill Bay, a number of decisions were needed. The classic definition of "carrying capacity" refers to a maximum population an environment is able to sustain indefinitely without becoming impaired. Problems with this definition arise as the environment changes, as other populations compete for resources, or as the population itself changes patterns of resource use. The potential capacity of an environment to support a given population will vary over time as these changes occur, whether or not changes are driven by human activity or natural processes. Because aquaculture is not the only stressor on the environment, and aquaculture is evolving to address and reduce its environmental impact [12], I prefer Catton's [13] definition that speaks of carrying capacity in terms of a "load" rather than a population: "An environment's carrying capacity is its maximum persistently supportable load." Unlike the classic definition that discusses carrying capacity in terms of supporting a healthy population, to assess the carrying capacity of a water body to support salmon farming one must identify the attributes of the environment that are to be protected, and which of the potential stressors exerted by a given population is most likely to affect that attribute. Catton's definition blends the "assimilative capacity" of constituents used in waste load allocation as well as the biological "holding capacity" of a population or activity.

Water body attributes and uses to be protected are established by Maine law. Change to water quality is acceptable as long as designated human and ecological uses are protected. Human uses include recreation in and on the water, navigation, *and aquaculture*. Ecological uses include "unimpaired habitat" and "waters shall be of sufficient quality to support all estuarine and marine species indigenous to the receiving water without detrimental changes in the resident biological community." While statutes may guide regulators, interpretation is not universally understood. For example, what does "detrimental" really mean?

Although the nutrient dynamics of the Gulf of Maine have been relatively well described [8, 14], information on smaller nearshore water bodies in the region around Blue Hill is not readily available. Requiring an applicant of any development to assess nearfield environmental impacts is standard practice in Maine. However, because salmon aquaculture operates in a fluid system and is but one of many loads to a bay, a complete study of a large bay by a single aquaculturist would have been inequitable and furthermore a disincentive to the State's larger objective of dispersing aquaculture to encourage environmental sustainability. Resource managers saw opportunity in a shared responsibility that would broadly benefit everyone, regardless of outcome, by advancing knowledge of unstudied areas of the coast.

A "weight-of-evidence" approach was used that incorporated publicly available information, field monitoring, simple models, and conservative assumptions. The exercise also provided information that could benefit land-use planning and decision-making around other activities in addition to aquaculture.

Given the prospect for more highly visible net-pens, legitimate concerns for Blue Hill Bay, and vaguely interpreted water quality laws, it was important to incorporate local concerns into the workplan. In consultation with members of the Blue Hill Bay environmental community, a workplan/study design was developed that attempted to address their water quality concerns and to determine whether or not the bay was currently supporting the designated human and ecological uses. Based on public comment and testimony, nutrient enrichment and eutrophication was generally viewed as the ecological factor of greatest concern. To avoid debate over interpretation of results at the end of the study, a priori ecological end points or thresholds of impact were established against which attainment of the human and ecological uses could be measured (Table 1).

The assessment consisted of three simple parts:
1. Water quality characterization.
2. Assessment of existing external nutrient load.
3. Estimation of capacity for additional nutrient loading from net-pen aquaculture.

Table 1 Ecological end points to protect designated uses of Blue Hill Bay

Objective of end point	End point
Protection of oxygen	At or above 6 mg l^{-1} or 85% saturation, whichever is higher
Submerged aquatic vegetation [15, 16]	20% Light level > 4 m
Water clarity	Euphotic zone (1% light level) > pycnocline or 10 m
Nuisance algal blooms	Mean summer chlorophyll < 4 µg l^{-1}
Toxic algal blooms	Absence

3
Part 1: Water Quality Characterization

Morphometry for Blue Hill Bay was obtained from readily available NOAA/ NOS bathymetric data downloaded with Geodas software [17]. Surface areas and volumes were calculated in ArcView GIS [18] from triangulated irregular network (TIN) files using 3 m as mean high water. To represent

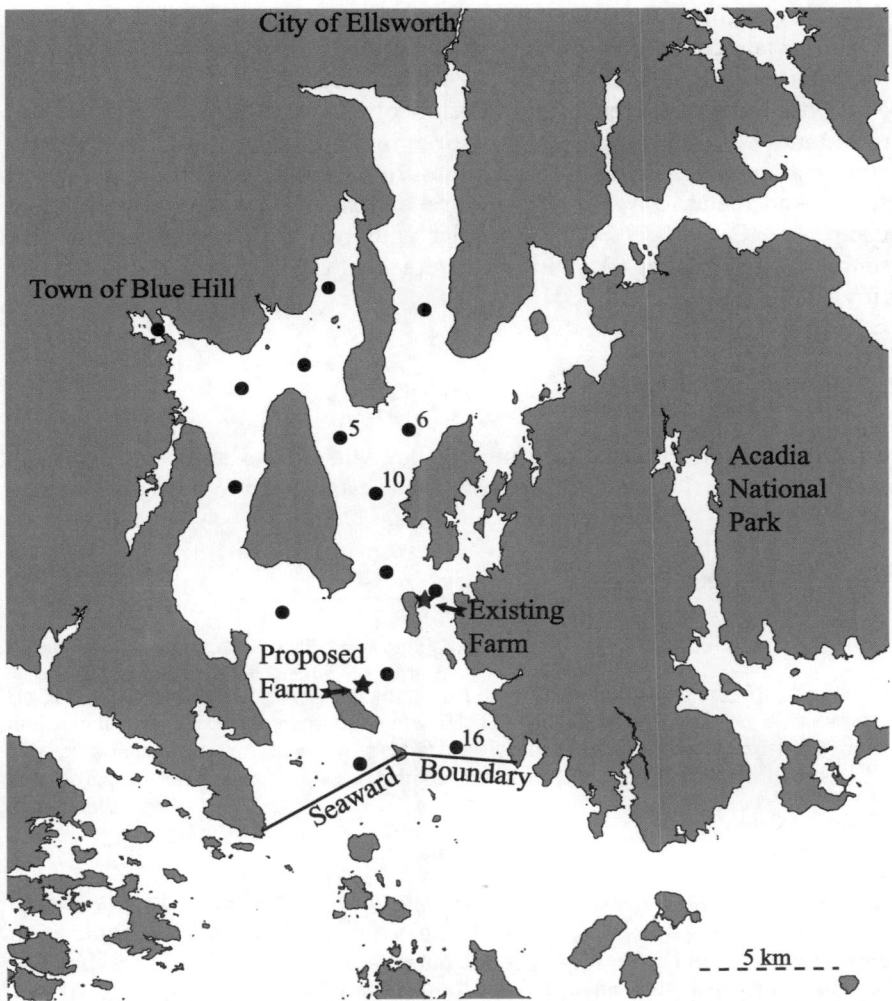

Fig. 2 Blue Hill Bay and surrounding area, showing locations sampled for determining water quality (*solid circles*). Stations discussed in text are labeled by station number

conditions within the bay, 23 water quality sampling stations were located based on prominent bathymetric and hydrographic features (Fig. 2). Four surveys were conducted in 2000 and two in 2001. Water samples for nutrients (NH_{3-4}, NO_{2-3}, DIN, DON, PO_4, Si_2O_4) and phytopigments (chlorophyll a and phaeophytin) were collected from 2, 6, 12, and 30 m depths using a PVC Kemmerer water bottle. Samples for the dissolved inorganic nutrient series were filtered through 0.45-μm Millipore cellulose acetate membrane filters into labeled 20-ml acid-washed scintillation vials. Samples for dissolved organic nitrogen (60 ml) were similarly filtered into plastic Nalgene bottles. Samples for chlorophyll a and phaeophytin (100 ml) were filtered through Whatman GF/F filters. The filters were placed in labeled glass scintillation vials containing 10 ml of 90% acetone. All water samples and filters were immediately frozen on dry ice and retained frozen until analysis. Nutrients were measured using a Technicon II AutoAnalyzer and standard techniques [19]. Pigments were measured by fluorometry [20]. Light attenuation was measured with a LiCor photosynthetically active radiation (PAR) sensor and meter. Oxygen, temperature, and salinity profiles were collected using a SeaCat Model SBE 19 profiler equipped with a dissolved oxygen sensor or YSI Model 6000. Profiles were field-quality assured by Winkler titration [21].

3.1
Results

Overall, the water quality in Blue Hill Bay with the existing net-pen farm meets the a priori eutrophication end points established to determine attain-

Table 2 Physical characteristics of Blue Hill Bay

	Whole bay—Flye Pt. through middle of Tinker Island to Dodge Pt.
High-water surface area	190 km^2
Low-tide surface area	183 km^2
Mean depth	~22 m
Maximum depth	~105 m
Low-tide volume	4.09 × 10^9 m^3
Tidal prism	0.59 × 10^9 m^3
Land drainage area	1700 km^2
Mean annual runoff	1.03 × 10^9 m^3
Tidal range	3.3 m
Union River mean annual inflow	~0.817 × 10^9 m^3
Pycnocline	~10 m
Stratification season	May–September

Assessing Nitrogen Carrying Capacity for Blue Hill Bay

ment of Maine's water quality statutes. In fact, by several measures, Blue Hill Bay is oligotrophic [22–24].

Morphometric and drainage basin characteristics of Blue Hill Bay are summarized in Table 2. Data from field work conducted in 2000 and the State's

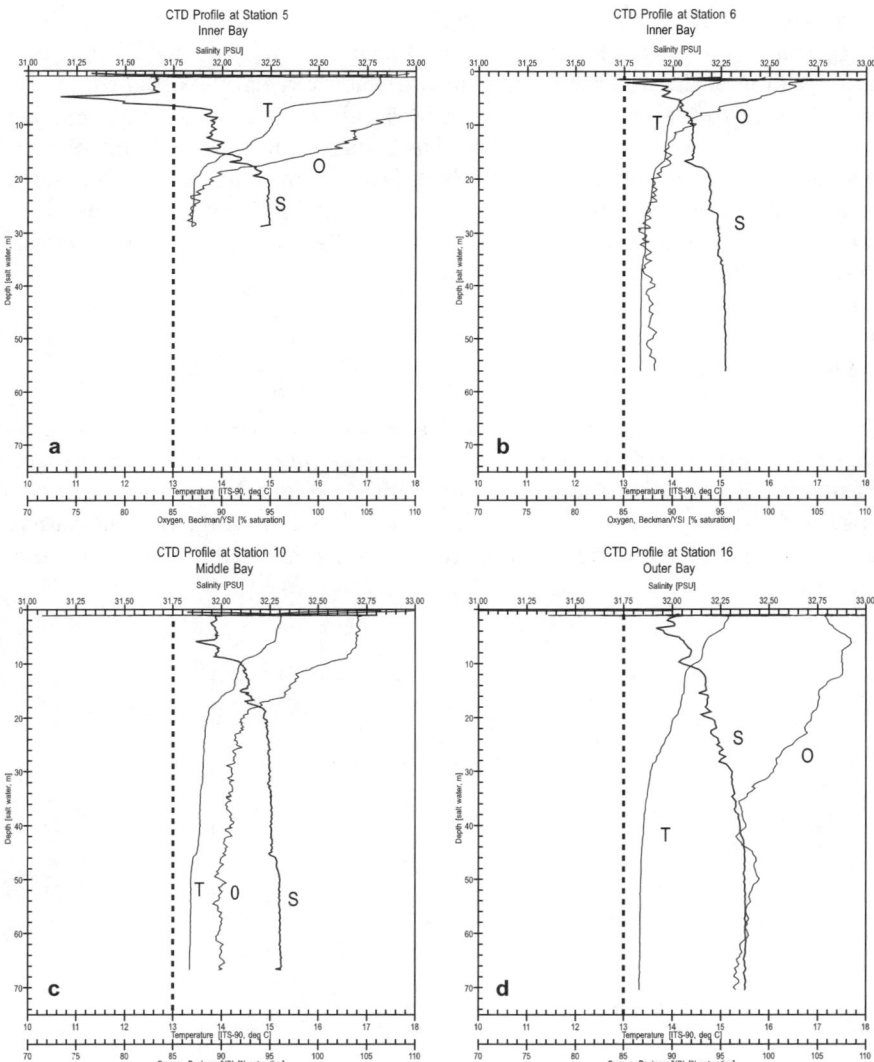

Fig. 3 CTD profiles taken on August 30, 2000 at four stations in Blue Hill Bay showing progressive loss of dissolved oxygen from bottom water (T = temperature, S = salinity, O = % saturation oxygen)

finfish aquaculture environmental monitoring program [25] showed that Blue Hill Bay stratifies at slightly below 10 m from mid-May into September.

3.1.1
Oxygen

Late summer bottom-water oxygen levels in the seaward portion (e.g., Station 16) of the bay (Fig. 3d) are typically higher than those in the inner (e.g., Stations 5, 6, 10, and 18) more northern portion (Fig. 3a–c). Occasionally, at some of the inner stations, bottom-water oxygen falls slightly ($\sim$ 5%) below Maine's water quality standard of 85%, although measured oxygen concentrations never fell below 6 mg l^{-1}. This diminution affirmed concern that nutrients could increasingly impact the bay the farther north they were introduced. The oxygen sag also identified the potential location for water quality monitoring stations to incorporate in a future bay-wide monitoring network designed to measure changes over time.

3.1.2
Light

Field measurements of photosynthetically active radiation (PAR) indicated that both end points are presently being met. The summer euphotic zone, as measured by the 1% compensation depth, extends below the pycnocline at about 14 m and the 20% light level is below 5 m (Fig. 4), thus satisfying the primary production and protection of submerged aquatic vegetation targets.

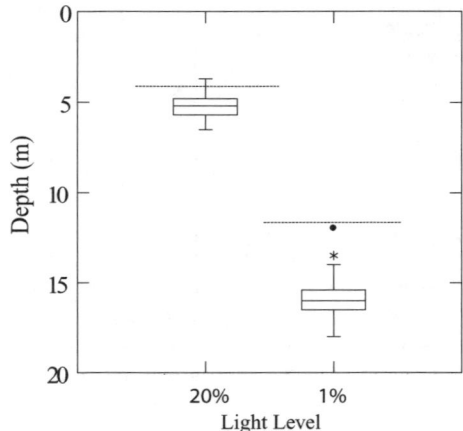

Fig. 4 Box plots of 20% and 1% light penetration depths in Blue Hill Bay in relation to ecological end points (*dotted line*)

Assessing Nitrogen Carrying Capacity for Blue Hill Bay

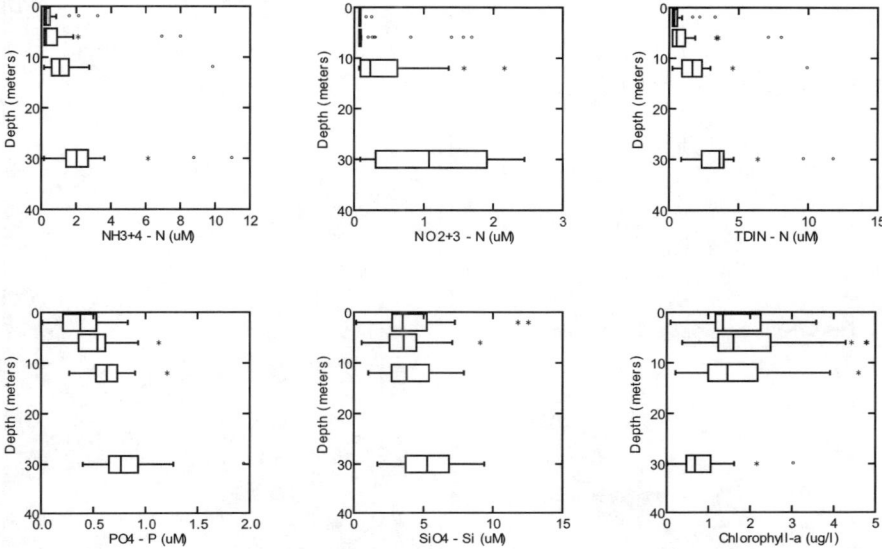

Fig. 5 Box plots representing vertical distribution of nutrients and chlorophyll a in Blue Hill Bay aggregated for period of stratification

3.1.3
Nutrients

A strong nutricline for dissolved inorganic nitrogen species, and to a lesser extent for dissolved phosphate and silica, exists during stratification (Fig. 5). The characterization shows that during summer stratification, dissolved inorganic nitrogen in the upper surface layer is likely depleted. However, at 12 m, dissolved nitrogen is not depleted. Because the euphotic zone extends beyond 12 m, nitrogen may be available to phytoplankton episodically throughout the summer through a combination of mixing from summer wind events and/or vertical migration of phytoplankton [26].

3.1.4
Chlorophyll

Overall, chlorophyll concentrations in the surface mixed layer (SML) were low throughout the summer, averaging around 1.5 $\mu g\,l^{-1}$, well below the 4 $\mu g\,l^{-1}$ average end point. Highest values (4.8 $\mu g\,l^{-1}$) were measured in the outer bay on August 30th. As the existing salmon farm is also located in this area, a cause and effect is suggested. However, SeaWiFS satellite imagery [27] for the same sampling period showed the fall plankton bloom developing just offshore with chlorophyll concentrations exceeding 6 $\mu g\,l^{-1}$ just outside Blue Hill Bay (Fig. 6c–d). At other times when chlorophyll was not abundant offshore

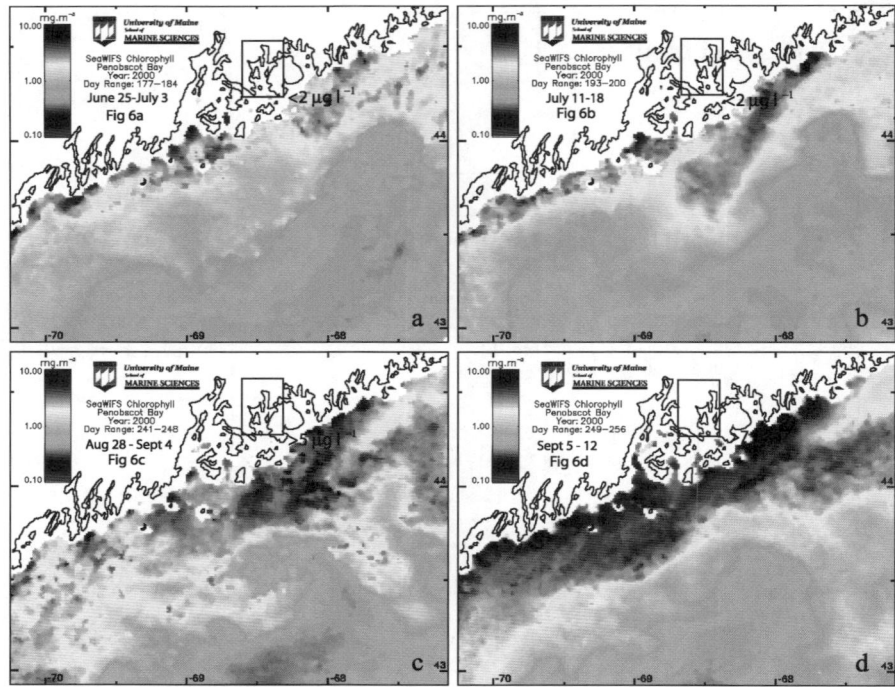

Fig. 6 SeaWiFS satellite images of the coastal current plume running southwest along the Maine coast adjacent to Blue Hill Bay (*inset*) showing the progression of fall phytoplankton bloom. June and July surface chlorophyll a levels are less than $2\,\mathrm{mg\,l^{-1}}$. Late August and early September surface chlorophyll a levels surpass $5\,\mathrm{mg\,l^{-1}}$

(Fig. 6a,b), chlorophyll concentrations in proximity to the existing finfish farm were the same as or less than those in other areas of the bay. At those times, highest pigment values were measured at the head of the bay in proximity to residential development and municipal and private family sewage discharges.

3.1.5
Nuisance and Toxic Algal Blooms

No field monitoring was conducted to specifically address harmful algal blooms. However, as in most coastal states there is an ongoing monitoring program of commercially harvested shellfish beds to protect public health. Blue Hill Bay has been monitored by the State for over 20 years at four locations. Toxic algal blooms have never been responsible for a shellfish harvest closure and reports of nuisance blooms are not known.

4
Part 2: Nitrogen Load Estimate

A variety of sources and methods were used to estimate the external nitrogen load to Blue Hill Bay including published literature, unpublished reports, agency files, and simple models. Discharge volumes of treated municipal wastewater were obtained from monthly reports submitted to the State by the respective treatment plants [28]. Effluent volumes were combined with nutrient concentration coefficients for secondary treatment [29, 30]. Atmospheric inputs were estimated using data from an adjacent National Atmospheric Deposition Program monitoring station at nearby Acadia National Park [31]. Watershed runoff coefficients from a USGS study on nearby Mt. Desert Island [32] were applied to Blue Hill Bay's watershed. The nitrogen load from net-pen aquaculture was estimated using literature coefficients [29–37] and a mass balance method. Missing from making this a complete budget are nitrogen fluxes between the bay and offshore, losses through denitrification, burial, biological processing and emigration, harvest by wild-capture fisheries, and harvest by shellfish aquaculture.

The annual external loading to Blue Hill Bay is estimated at about 400–640 MT nitrogen (Table 3). On an annual areal loading basis, Blue Hill Bay receives between 2.1 and 3.4 MT km^{-1} year^{-1} (0.15–0.24 mol N m^2 year^{-1}), a load consistent with oligotrophic systems [22, 23].

However, annual loading to a seasonally stratified and nutrient limited water body is unrealistic and underrepresents the worst-case conditions during summer when loading from a salmon farm is at its peak. A second budget was therefore done for the months of May to September to reflect a more realistic loading during stratification (Table 4). Atmospheric deposition and watershed runoff estimates were volume apportioned based on streamflow and precipitation data from the adjacent Acadia National Park. Sewage discharges were assumed to be 5/12 of the annual value. Loading from the salmon farm discharge was based on monthly feed delivered to the pens during the interval. During stratification, between 130 and 180 MT new nitrogen is estimated to be delivered to the bay from all sources. If the entire 180 MT N were homogeneously mixed into the 4.09 × 10^9 m^3 of the bay, ambient concentrations of total N could be increased by 3 μM by all sources during summer stratification.

Fully half the summer load is from the salmon farm. Yet, percentages alone do not afford meaningful comparison, especially when it is a large percentage increase of an initially small load in an oligotrophic system. For a system like Blue Hill Bay, in which the annual loading cycle is dominated by watershed inputs, the existing net-pen operation does not change the seasonal loading pattern (Fig. 7). Summer loading continues at half that of spring and fall, with monthly nitrogen loading lowest in August when salmon feeding

Table 3 Derivation of annual external nitrogen load estimates for Blue Hill Bay

Secondary treated municipal wastewater [30]	
0.93 mgd × 3.875 l g^{-1} × 25 mg l^{-1} × 365 days	= ~ 33 MT
Atmospheric deposition directly to surface [32]	
5.1 kg ha^{-1} year^{-1} × 190 km × 100 ha km^{-1}	= 96.9 MT
Watershed runoff [32]	
1700 km × 130–270 kg km^{-1} year^{-1}	= ~ 221–459 MT
Private home discharges [29]	
(16 to Blue Hill Bay) assume 3 persons per household	
40–80 mg l^{-1} × 280 l day^{-1}/person^{-1} × 365 days year^{-1}	= ~ 0.2–0.4 MT
Salmon farm discharge	
Loading coefficient method [33]	
280 000 fish × 4.5 kg each = 1260 MT × ~ 50 kg MT^{-1}	
per production cycle × 12 month year^{-1}/18 month/production cycle	= ~ 42 MT
Mass balance method	
Feed 7.0% N [34–36].	
Whole Atlantic salmon 3.25% N [37].	
Feed in (1 500 000 kg feed × 0.07 N) = 105 000 kg TN	
Smolts in (280 000 × 0.1 kg × 0.0325 N) =+ 910 kg TN	
Fish harvested (280 000 × 4.5 kg × 0.0325 N) = – 40 950 kg TN	
Total N to bay (@4.5 kg) = 64 960 kg TN × 12 month/	
18 month/prod.cycle64 960 kg TN × 12 month/18 month/prod. cycle	~ 43.3 MT
Total	393.1–632.3 MT

Table 4 Summary of estimated contemporary external nitrogen load to Blue Hill Bay

	Metric Tons	
	Annually	May–September
POTWs	33	10.7
Wet deposition	96.9	44
Watershed runoff	221–459	44.2–91.8
Salmon farm	42–48.7	30
Private discharges	0.2–0.4	0.1–0.2
Total accounted	393.1–637.1	129–176.7

[1] Internal loads such as inshore/offshore flux, biological processing, and commercial harvest not considered

is at its peak. This is primarily because streamflow and runoff, which dominate at other times of the year, are at their annual minimum. The important question to address is whether the 50% increase is driving the system to an

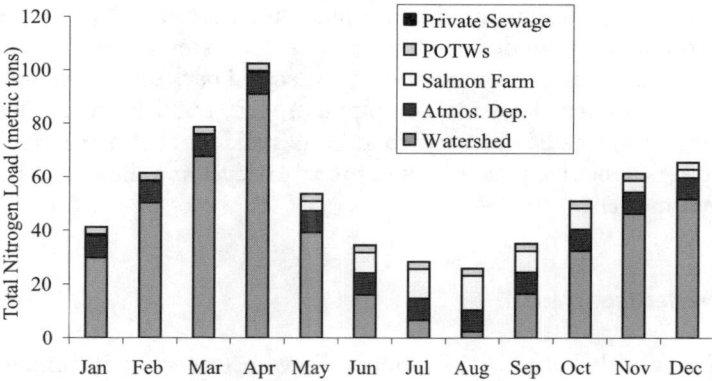

Fig. 7 Relative monthly contributions of nitrogen to Blue Hill Bay from various external sources

undesirable level of productivity. In this specific case, it does not appear to be doing so.

5
Part 3: Assessing Carrying Capacity

After determining that water quality in Blue Hill Bay meets its ecological targets and that the existing trophic state and areal external loading reflect oligotrophic conditions, it appeared that overall, Blue Hill Bay has the capacity to assimilate added nutrients from additional net-pen aquaculture. But, how much capacity? Where? And When?

5.1
Modeling

Various models have been applied to net-pen aquaculture that predict eutrophication [38]. Scotland, for example, uses a simple box model based on annual nitrogen loading and tidal flushing to determine a "maximum consented biomass" for each of its sea lochs [33]. At the other extreme are dynamic simulation models including QUAL2E [39], WASP5 [40], ECOMsi/*EM [41], and the Princeton ocean model [42], all of which are data intensive, time consuming, and expensive to support.

While these models can be helpful, they generally exceed the resources available to small agencies and individual aquaculturists. Furthermore, because even the best models require validation, regardless of modeling results, follow-up monitoring is required. The value of these models for a resource agency is to help design monitoring plans for permit compliance and develop hypotheses to understand nearshore processes better. If judiciously sited, and because a net-pen facility may be scaled back or even removed (unlike more permanent landside development, e.g., publicly owned treatment works (POTWs), malls, industry, roads, and residential construction), aquaculture offers a relatively "safe" opportunity to develop, validate, and calibrate ecosystem models.

5.2
"Vollenweider-Type" Model

Predictive models based on phosphorus have been used with much success in freshwater lakes to manage eutrophication (e.g., [43]). Attempts have been made to extend these models to coastal waters with mixed success [44]. Unlike lakes, which have naturally defined hydrologic boundaries, coastal waters are open systems making it difficult and expensive to accurately describe loading. Furthermore, site-specific factors such as tidal magnitude frustrate the use of dose–response models [45, 46].

From a practical point of view, empirical models have the advantage over laboratory "bottle" experiments or theoretical models in that they incorporate biological processing, albeit leaving us short of a good understanding of actual mechanisms. Monbet [47] compiled data from 40 water bodies according to tidal stage to develop a dose response to annual dissolved inorganic nitrogen (DIN) loading. Plotting chlorophyll and nutrient data collected from Blue Hill Bay amongst Monbet's temperate waters of similar tidal amplitude suggests that about 30 MT/km^2 year^{-1} is required to achieve average chlorophyll concentrations that correspond to the 4 µg l^{-1} chlorophyll ecological end point. Over the 190-km surface area of Blue Hill Bay, this is an annual load of about 5700 MT. If our earlier estimated annual load of 650 MT is correct and the other assumptions are met, there may be 5050 MT excess nitrogen capacity before reaching the chlorophyll end point of 4 µg l^{-1}.

6
Carrying Capacity Applied to Net-Pen Aquaculture

Converting "excess" nitrogen capacity into salmon production capacity now becomes a simple mathematical exercise. Applying the earlier nitrogen loss coefficient of 50 kg N for each metric ton of salmon produced [33] to an average harvest weight of 4.5 kg, each fish raised to market size results in about 0.23 kg N lost to the bay. Translated into salmon production, the bay has the capacity to hold an additional 22 million fish before reaching the ecological target of 4 µg l^{-1} chlorophyll. This calculation disregards the fact that the production cycle is usually 18 months or more and thus underestimates the amount by perhaps one third.

However, even if the calculation is true, a bay-wide carrying capacity must be accompanied by some caveats. Foremost, the calculation assumes homogeneous mixing, an assumption not supported in reality, and it therefore ignores the potential localized effect of a single concentrated source. Rather than an annual average over an entire bay, a more meaningful analysis would be to look at a spatial scale based on the time it takes for phytoplankton to respond to nutrient enrichment, a period of 1–2 days. From several subsurface (4 m) drogue deployments in the area of concern made in 2000, the tidal ellipse was estimated to be several kilometers. In addition, evidence presented in public testimony showed that summertime residual surface layer currents in the area of the proposed aquaculture facility were 5–15 cm s^{-1} seaward [48]. This enabled an even more fine scale exercise to be conducted.

With load and dilution, localized effects may be anticipated through the use of an empirically based model that predicts chlorophyll from nitrate [49]:

$$X_{max} = X_0 + q(S_0 + \langle S \rangle / D\ V) \qquad (1)$$

where X_{max} and X_0 are the maximum expected and ambient chlorophyll concentrations, respectively, q is a growth response coefficient, S_0 is ambient nitrate concentration, $\langle S \rangle$ is "new" nitrogen, D is dilution factor/day, and V is volume of water mass. Rearranged to solve for $\langle S \rangle$, the formula becomes:

$$\langle S \rangle = [D\ V(X_{max} - X_0)/q] - S_0 \qquad (2)$$

By setting X_{max} equal to the end point chlorophyll concentration of 4 µg l^{-1}, applying ambient upstream (inner bay) summer euphotic zone chlorophyll

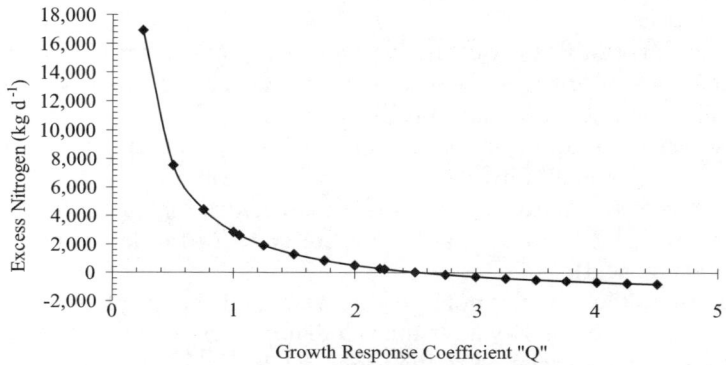

Fig. 8 Excess nitrogen assimilates for Blue Hill Bay under various growth response coefficient scenarios

($1.5\,\mu g\,l^{-1}$) and DIN ($0.98\,\mu M$), and the lower measured current velocity of $5\,cm\,s^{-1}$, and assigning q, a conservative theoretical "excess" capacity ($\langle S \rangle$) of the local area may be estimated.

Selecting a reasonable q was problematic, however, as Gowen et al. found q values ranging between 0.25 and 4.44 with a median of $1.05\,\mu g\,l^{-1}$ response in chlorophyll for each micromole of nitrite/nitrate N (Fig. 8). The cross-sectional area of the SML where currents were measured is $31\,000\,m^2$ and 10 m was used for the depth of the surface layer. At peak production in August, based on feed inventory and conversion ratios, the proposed salmon farm is anticipated to discharge about $280\,kg\,N\,day^{-1}$. Using the lowest q value (0.25) the response zone can handle an additional $17\,000\,kg\,N\,day^{-1}$, whereas with the highest q value (4.44) the response zone exceeds its capacity by $800\,kg\,day^{-1}$. To resolve and address this discrepancy, it was helpful to ask what would be the consequence (i.e., the maximum anticipated concentration of chlorophyll of 280 kg) if Blue Hill Bay's q value was in fact 4.44.

Applying the Gowen model to chlorophyll a

$$1.5\,\mu g\,m^{-3} + 4.44(0.98\,\mu M\,l^{-1} + 20 \times 10^6\,\mu M\,day^{-1}/134 \times 10^6\,l\,day^{-1})$$
$$= 6.5\,\mu g\,l^{-1}$$

and the level in the immediate area around the net-pens would be $6.5\,\mu g\,l^{-1}$, in excess of the ecological end point set for the bay as a whole. Here some professional judgment was used to determine whether or not the risk of exceeding the water quality target was acceptable or even realistic. In this case, we concluded that the risk of exceeding the end point was probably low for the following reasons:

- Application of the model in the vicinity of the existing salmon farm predicts a similar increase in chlorophyll that was not seen through monitoring, suggesting that using the highest q value for Blue Hill Bay is inappropriate.
- Surface flow in this part of the bay, where the new farm is proposed and the majority of nitrogen is released, is seaward, out of the bay, where further dilution and assimilation is likely to occur.
- Conservative assumptions were employed (e.g., low velocity, no wind or tidal mixing, and all nitrogen immediately available).
- The impact of an August chlorophyll concentration of $6.5\,\mu g\,l^{-1}$ in the outer bay is likely to be masked by the onset of the fall phytoplankton bloom (Fig. 4c,d).
- Response to the model is not linear. Incremental change of excess nitrogen capacity in Blue Hill Bay after the value of q exceeds the median of 1.05 is relatively small in relation to the proposed 280-kg load increase.

6.1
Attributes of Net-pen Aquaculture Affecting Carrying Capacity

Characteristics unique to aquaculture make it difficult to apply the concept of carrying capacity. The finfish net-pen industry is continually evolving and addressing environmental concerns [12]. In addition to the distinct interannual and seasonal changes associated with net-pen operations, husbandry can be modified to reduce environmental impacts. Stocking rates can be reduced and pens can be relocated to other areas of a bay to take advantage of natural biological and physical oceanographic features that mitigate loading. Feed formulations and conversion efficiencies are improving with less and less nitrogenous waste lost to the environment [50]. Furthermore, aquaculture of seaweeds and shellfish integrated with net-pen aquaculture can be employed to reduce the overall net contribution of nutrients [51]. Separately and combined, all these measures reduce nutrient contributions and increase the capacity of a bay to handle additional finfish with relatively little increase in net nutrient load.

So what is the carrying capacity for net-pen salmon aquaculture in Blue Hill Bay? "Capacity" implies a finite quantity, a number. Yet, it is clear that net-pen aquaculture is but one load of many and is best placed in this larger context. Net-pen aquaculture operates simultaneously on many different scales with distinct seasonal patterns. At times, these scales appear to contradict one another. On a global ocean scale, for example, fish meal based salmon feed is derived from the oceanic primary production resulting in a net removal of nutrients from the ocean. On a more fine scale, however, within a given local water body net-pen aquaculture can be a net contributor of nutrients.

Bay-wide, and solely from a nutrient perspective, Blue Hill Bay appears to have the capacity to support up to about 22 million aquacultured salmon if they are distributed evenly over the bay and over the year. Over a smaller local biological response scale, the net-pen aquaculture currently being practiced in Blue Hill Bay, with about 300 000 fish on site, fits within the capacity of the immediate area to absorb additional nutrients without adverse consequences. With continued monitoring, incremental increases may prove to fit as well. However, as one heads farther into the bay, open ocean exchange diminishes and the capacity of these inner areas becomes increasingly smaller. With further investigation, it is conceivable that Blue Hill Bay could be divided into a series of biological response zones nested within larger zones, each with its own carrying capacity. Certainly, at that point, other assimilative capacities including biochemical oxygen demand, benthic impacts in shallower and slower areas, and fish health issues would need to be addressed.

To be effective, a resource allocation plan requires clear ecological targets or end points, continuous monitoring, and a willingness on the part of industry to take whatever steps are necessary (including site abandonment) to ensure targets are attained. At this time, however, the uncertainties and risks associated with rapid and large expansion of net-pen aquaculture in Blue Hill Bay are arguably too great and unnecessary to warrant maximizing the carrying capacity of Blue Hill Bay for net-pen aquaculture.

Although the impetus for this carrying capacity effort was to assess impacts from the expansion of a salmon farm, the value of the exercise to resource scientists and managers extends beyond aquaculture. The exercise provided an opportunity to identify public policy questions that will eventually need to be addressed statewide. For example, while we set a priori end points, we did not predefine either the temporal or spatial scale in which the end points apply. End points were also defined in terms of averages when extreme values may be more critical. The results highlighted the importance of considering all nutrient sources to the bay, including the margins of the bay where the domestic wastewater and nonpoint sources of nutrients from increasing development enter and the seasonal patterns of nutrient delivery affect ecosystem vulnerability and capacity. And finally, as human growth in the region continues, this exercise could play a role in designing a monitoring plan to comprehensively manage water quality for the entire bay.

Aquaculture remains but one use of public waters that must be compatible with the other protected uses including traditional fishing, navigation, and recreation. The ecological carrying capacity of a bay may not be the ultimate limiting factor to development. Rather, social conflicts may be the most important limiting factor on which to base a carrying capacity exercise. Recreation, ability to support traditional fishing, and even esthetics could conceivably be modeled through a carrying capacity exercise. The challenge there will be setting acceptable cultural end points.

Acknowledgements I am grateful to Abigail Dietz, Chuck Penny, Jon Lewis, Marcy Nelson, and salmon farmer, Erick Swanson, for contributing boat time help in collecting water samples. Maura Thomas and Dave Townsend assisted with sample design and conducted all the laboratory analyses. Much appreciated and constructive comments were provided by Suzanne Bricker, two anonymous reviewers, and Don Eley, President of the Friends of Blue Hill Bay. I especially thank Barry Hargrave, who once again demonstrated extreme patience and provided much-needed discipline and encouragement throughout the preparation of this manuscript.

References

1. Brooks DA, Baca MW, Lo Y-T (1999) Estuar Coast Shelf Sci 49:647
2. Schlitz RJ, Cohen EB (1984) Biol Oceanogr 3:203
3. Townsend DW (1991) Rev Aquat Sci 5:211
4. Townsend DW, Pettigrew NR, Thomas AC (2001) Cont Shelf Res 21:347

5. Sowles J (2001) Workshop Report 2001. Managing nitrogen impacts in the Gulf of Maine. Prepared by the NOAA/UNH Cooperative Institute for Coastal and Estuarine Environmental Technology, the Gulf of Maine Council on the Marine Environment, and NOAA's Ocean Service
6. Sowles JW, Churchill L (2004) Northeast Nat 11:Special Edition 2
7. Strain PM, Hargrave BT (2005) Salmon aquaculture, nutrient fluxes, and ecosystem processes in southwestern New Brunswick (in this volume). Springer, Berlin Heidelberg New York
8. Townsend DW, Christensen JP, Stephenson DK, Graham JJ, Chenoweth SB (1987) J Mar Res 45:699
9. Brooks DA, Townsend DW (1989) J Mar Res 47:303
10. Garside C, Garside J (2004) Northeast Nat 11:Special Edition 2
11. U.S. Bureau of the Census (2002) http://censtats.census.gov/pub/Profiles.shtml
12. Costa-Pierce BA (2002) Ecology as the paradigm for the future of aquaculture. In: Costa-Pierce BA (ed) Ecological aquaculture: the evolution of the blue revolution. Blackwell, Oxford
13. Catton W (1986) Carrying capacity and the limits to freedom. Social Ecology Session 1, XI World Congress of Sociology, New Delhi
14. Bigelow HB (1927) Fish Bull 40:511
15. Duarte CM (1991) Aquat Bot 40:363
16. Dennison WC, Orth RJ, Moore KA, Stevenson JC, Carter V, Kollar S, Bergstrom PW, Batiuk RA (1993) Bioscience 43:86
17. http://www.ngdc.noaa.gov/mgg/geodas/geodas.html
18. Environmental Systems Research, Incorporated (2002) Arc View 3.3. Redlands, California
19. Whitledge TE, Veidt DM, Mallow SC, Patton C, Wirick CD (1986) Automated nutrient analyses in seawater. Brookhaven National Laboratory, BNL 38990
20. Parsons TR, Maita Y, Lalli CM (1984) A manual of chemical and biological methods for seawater analysis. Pergamon, Oxford
21. Strickland JDH, Parsons TR (1972) The practical handbook of seawater analysis. Bulletin 167, Fisheries Research Board of Canada
22. Bricker SB, Clement CG, Pirhalla DE, Orlando SP, Farrow DRG (1999) National estuarine eutrophication assessment: effects of nutrient enrichment on the nation's estuaries. NOAA, National Ocean Service, Special Projects Office, and the National Centers for Coastal Ocean Science, Silver Spring, MD
23. National Research Council (2000) Clean coastal waters: understanding and reducing the effects of nutrient pollution. National Academy Press, Washington DC
24. US Environmental Protection Agency (2001) Nutrient criteria technical guidance manual—estuarine and coastal marine waters, EPA-822-B-01-003. Washington DC
25. Maine Department of Marine Resources (1997–2000) Finfish aquaculture monitoring program, annual reports. Boothbay Harbor, Maine
26. Eppley RW, Holm-Hansen O, Strickland JDH (1968) J Phycol 4:333
27. http://wavy.umeoce.maine.edu/
28. Maine Department of Environmental Protection (2000) Open file reports. Augusta, Maine
29. Novotny V, Imhoff KR, Olthoh M, Krenkel PA (1989) Karl Imhoff's handbook of urban drainage and wastewater disposal. Wiley, New York
30. Viesmann W Jr, Hammer MJ (1998) Water supply and pollution control, 6th edn. Addison-Wesley, Reading, MA
31. http://nadp.sws.uiuc.edu/sites/siteinfo.asp?net=NTN&id=ME98

32. Nielsen MG (2002) US Geological Survey, water resources investigations report 02-4000. Augusta, Maine
33. Gillibrand PA, Gubbins MJ, Greathead C, Davies IM (2002) Scottish Fisheries Report 63/2002. Aberdeen Fisheries Research Services, Aberdeen, Scotland
34. Findlay RH, Watling L (1994) Toward a process-level model to predict the effects of salmon net-pen aquaculture on the benthos. In: Hargrave BT (ed) Modeling benthic impacts of organic enrichment from marine aquaculture. Can Tech Rep Fish Aquat Sci 1949
35. Enell M, Ackefors H (1992) Aquacult Eur 16:6
36. Ackefors H, Enell M (1994) Appl Icthyol 10:225
37. Vingradov AP (1953) Mem Sears Foundation Mar Res No 2. New Haven, CT
38. Aure J, Stigebrandt A (1990) Aquaculture 90:135
39. US Environmental Protection Agency (1995) QUAL-2E: The enhanced stream water quality model. Model documentation and user's manual. Athens, GA
40. Ambrose RB Jr, Wool TA, Martin JL (1993) The water quality analysis simulation program, WASP5: model documentation. US Environmental Protection Agency, Office of Research and Development, Environmental Research Laboratory, Athens, GA
41. Hydroqual Inc (1998) A water quality model for Jamaica Bay: calibration of the Jamaica Bay eutrophication model (JEM), Mahwah, NJ
42. Blumberg AF, Mellor GL (1987) A description of a three-dimensional coastal ocean circulation model. In: Heaps NS (ed) Three-dimensional coastal ocean models, coastal and estuarine sciences 4. American Geophysical Union, Washington, DC
43. Vollenweider RA (1976) Mem First Ital Idrobiol 33:53
44. Bricker SB, Ferreira JG, Simas T (2003) Ecol Model 169:39
45. Wafer MVM, LeCorre P, Birrien JL (1989) Estuar Coast Shelf Sci 29:489
46. Dmeres S, Legendre L, Therriault JC (1986) Phytoplankton responses to vertical tidal mixing, p 1; Gowen RJ, Bradbury NB (1987) The ecological impact of salmonid farming in coastal waters: a review. In: Barnes M (ed) Oceanography and marine biology: an annual review 25:563
47. Monbet Y (1992) Estuaries 15:563
48. Pettigrew NR (2003) Pre-filed testimony on the application of a net-pen aquaculture lease at Tinker Island, Blue Hill Bay, Tremont, Hancock County, Maine
49. Gowen RJ, Tett P, Jones KJ (1992) Mar Ecol Prog Ser 85:153
50. Hardy RW (2000) Aqua Mag Jan/Feb:85
51. Neori A, Chopin T, Troell M, Buschmann AH, Kraemer GP, Halling C, Shpigel M, Yarish C (2004) Aquaculture 231:361

The Suitability of Electrode Measurements for Assessment of Benthic Organic Impact and Their use in a Management System for Marine Fish Farms

Morten T. Schaaning[1] (✉) · Pia Kupka Hansen[2]

[1]Norwegian Institute for Water Research (NIVA), P.O. Box 173, Kjelsaas, N-0411 Oslo, Norway
morten.schaaning@niva.no

[2]Institute of Marine Research, P.O. Box 1870, N-5817 Bergen, Norway
pia.kupka.hansen@imr.no

1	Introduction	382
2	Material and Methods	384
2.1	Surveys	384
2.2	Sampling Techniques and Challenges	386
2.3	Electrode Measurements	386
2.3.1	Instrumentation	387
2.3.2	Calibration	387
2.3.3	Measurements	387
2.3.4	Calculations	388
2.3.5	Reproducibility	388
2.4	Other Methods	389
3	Results and Discussion	390
3.1	Main Results	390
3.2	pH and Gas Ebullition	392
3.3	Comparison between Station Categories	394
3.4	Discriminant Analyses	396
3.5	A Joint Survey Comparing Biogeochemical and Macrofaunal Analyses	398
3.6	Influence of External Factors on Biogeochemical Variables	401
4	Application in a Management System	404
5	Summary and Conclusions	404
	References	406

Abstract A management system for marine fish farms in Norway has recommended the use of simple electrode measurements of redox potentials (E_h) and pH in sediments for monitoring of environmental effects at net cage locations. In the present paper we present results of such measurements performed over 15 years at farm locations in coastal Norway. Together with other chemical analyses of sediment and pore water, the electrode measurements provided a suite of biogeochemical variables used for environmental state assessment. The impact of external control factors such as biomass, local bathymetry and current velocities on the benthic effect parameters was investigated and comparisons were made with alternative effect variables such as macrofaunal community structure. It was found that simple two-dimensional plots of pH vs. either E_h or pS $(=-\log[\Sigma H_2S])$ maintained a high degree of discrimination between stations located at different distances from farm locations and provided a state assessment with high relevance to farm management. Compared to other methods, electrode measurements are cost-effective, applicable on a wide variety of benthic substrates and provide superior resolution in the high end of the organic enrichment gradient.

Keywords Electrode measurements · Fish farm · Marine sediment · Monitoring methods · Organic enrichment

1
Introduction

Fish production in net cages in the coastal environment has the potential to affect natural environments primarily through spreading of organic matter and nutrients from fish excretions and excess feed [1-3]. The area affected will depend on several factors related to farm management and site characteristics such as water depth, current speeds and bottom topography. Severe accumulation of organic matter is frequently confined to the area below the cages or a few meters beyond the vertical projection of the cage perimeter [2, 4-6]. More wide spread effects have been observed around farms located in recipients with restricted deep water renewal due to seasonal proliferation of oxygen-deficient water masses [7] and also in exposed environments in which stronger currents and storm-generated resuspension increase the horizontal transport of discharged particles [8-10]. Field experiments and several investigations of seabed recovery after production has been terminated have shown a large degree of normalization of effect variables within weeks to months [8, 11-13] although more than 2 years may be required for complete recovery [14, 15].

The changes occurring in sediments heavily loaded with farm debris generally follow the classical description of the organic enrichment gradient [16] and involve changes in bacterial communities and degradation pathways [17, 18]. Biodegradation of labile organic fractions has resulted in up to a fifty-fold increase in the rate of sediment oxygen consumption and release of nitrogen and phosphorous nutrients relative to the surrounding seabed [5, 19, 20]. Oxygen penetration is generally poor in marine sediments [21] and the combination of a favourable substrate for heterotrophic bacteria and insufficient oxygen supply leads to extensive sulphate reduction [7, 8, 22], increased concentration of hydrogen sulphide in the pore water and a rapid decline in redox potential in the top layer of the sediments [23-25]. Simultaneously the composition of the higher biological communities will be altered [4, 26-29]. High densities of opportunistic species have frequently been observed in the vicinity of fish farms [9, 12, 15, 30]. This indicates organic pollution and increased bioturbation may be functionally important in sustaining high rates of carbon oxidation and denitrification in the sediment [31–33].

In extreme cases, sediment dwelling animals disappear and bacterial methane production may become a dominant degradation process [34]. During the rapid growth of the industry in the 1980s many farmers complained about the simultaneous occurrence of gas ebullition and reduced fish appetite [35, 36]. The more recent trends towards larger farms located in more exposed environments have most likely reduced the extent of methane production in farm deposits. Nevertheless, outgassing is still reported to occur at some locations or periods and remains a potential problem limiting the exploitation of many locations otherwise favourable for fish production.

Other types of pollution from fish farms may originate from feed additives such as zinc, various medicines, pesticides for bath treatments and antifouling agents such as copper in net paints [20, 37–42]. In Norway the use of antibiotics in fish farming was reduced by 99% during the period 1987–97 and is no longer a cause of concern [43]. The actual discharge of copper depends on farm management, especially with respect to net cleaning practices. Annual sales figures for net paints [43] indicate that discharges of copper are still high.

It follows that primary impacts from fish farms would be increased sediment concentrations of feed remnants and faeces i.e. organic carbon, nitrogen, phosphorous and zinc, in addition to copper from net paints at locations where this substance has been used. Due to the remineralisation of carbon and nitrogen, zinc, copper and phosphorous seem to provide the most robust tracers of fish farm activities [44, 45]. Phosphorous speciation studies [46] have shown that compared to fish feed, fish faeces are enriched with the least soluble phosphorous fractions and further enrichment might be expected to occur during biodegradation on the seabed. Such post-depositional processes might increase concentrations of both phosphorous and trace metals in sheltered, anoxic sediments, but in more exposed environments resuspension is likely to provide rapid concentration decrease by redistribution over a larger sediment area.

Macrofaunal investigations are widely accepted as a standard for assessment of environmental status [47–49], but the effort and costs involved in a full macrobenthic analyses and the high number of samples required to monitor numerous net cage areas have resulted in a search for alternative methods [27, 50]. This chapter describes an approach to determine organic enrichment in benthic environments surrounding salmonid fish farms in the Norwegian coastal area using electrode measurements of pH and E_h and other sediment variables. The results have been used in the development of a monitoring program for marine fish farms in Norway [51–53]. The data were collected during the period 1983–1998 in areas where farmers or regional authorities had suspected environmental perturbations. In more recent years, many farms have moved to more exposed areas and the data should not be considered representative for present conditions at farm locations. The objective was to contribute to a more accurate understanding of the potential environmental impacts of fish farming and to promote monitoring techniques suitable to ensure that the benthic environment does not deteriorate beyond acceptable limits.

2
Material and Methods

2.1
Surveys

Data were collected during six surveys between 1983 and 1998 at and around 31 farm sites at coastal locations in Norway between 59 and 63°N (Table 1).

The Stations were Assigned to One of Five Different Site Categories:

A. Beneath or less than 5 m away from the vertical projection of net cages. 56 stations.
B. Abandoned locations. As in A, but without fish and net cages. Accurate locations were often revealed by old farm deposits identified by divers or a remotely controlled video camera. 27 stations.
C. Transition zone. Stations located 5–100 m from the vertical projection of net cages. 21 stations.
D. Reference stations located more than 100 m away from net cages in open, relatively exposed environments similar to the initial environment at the farm location. 9 stations.
E. Sheltered reference stations located in the farm district in polls or basins with shallow sills (10–15 m) and restricted exchange of deep water. No input from fish farms, but possible effects from runoff from population and agricultural activity. 11 stations.

The stations represent different coastal environments ranging from sounds, straits and open bights exposed to strong wind and tidal currents in the outer archipelago to more sheltered fjord environments and even sill basins with shallow sills and restricted deep water exchange. The latter types of locations had been established during the 1970s but were all abandoned during the 1980s. Water depths ranged from 18–130 m, but most of the samples were taken at depths between 20 and 30 m. The bottom below the cages consisted of steep bedrock slopes, mussel beds, fractured shell deposits, sediments speckled with stones and pebbles, as well as finer sand, silt and clay deposits.

Table 1 Overview of surveys, samples and variables determined at fish farm locations in the Norwegian coastal area during the period 1984–1998. Electrode measurements of pH, E_h and pS were performed on all samples. Supplementary variables encountered current speed (V), sediment profile imagery (SPI), macrofauna (M) and chemical analyses of organic or total carbon (C), nitrogen (N), phosphorous (P), copper (Cu) and zinc (Zn) in the solid sediment fraction and ammonium (NH_4), phosphate (PO_4) and total alkalinity (A_t) in extracted pore water. f = no. of farms, c = no. of cores, g = no. of grabs, n = no. of samples for electrode measurements

County	Year	f	c	g	n	Supp. variables	Reference
Nordland	1984	3	5	–	25	–	[54]
Nordland	1988	5	16	–	70	C, N, P, NH_4, PO_4	[55]
Møre og Romsdal	1988	1	9	–	53	C, N, P, Cu, Zn, NH_4, PO_4, A_t	[56]
Møre og Romsdal	1992	16	44	–	215	C, N, P, NH_4, V	[57]
Troms, Nordland, Møre og Romsdal	1993	5	37	–	121	C, N, P, Cu, Zn, M, SPI	[28, 58, 59]
Rogaland	1998	1	1	12	23	C, N, P, Cu, Zn, NH_4, PO_4	[60]

At the time of sampling, 11 of the farm locations had been abandoned for periods between 0.5 and 3 years. At the other farms net cages were present with a variable biomass of salmon or trout. Some information on biomass was obtained from the local farm managers (Table 5), but the actual input of organic matter at the different sites is not easily estimated. Biomass may change rapidly with time, discharges will be modified by differences in management practices, and sedimentation will depend on various hydrographic and bathymetric factors.

Each survey had different objectives, but a common factor was that the farmers themselves or the regional authorities suspected environmental perturbations. Therefore, the results presented here were never representative for the industry as a whole, but cover the types of benthic impacts, which are considered incompatible with sustainable production. Electrode measurements were always carried out, but sampling strategies as well as choice of supplementary variables varied. Samples were collected by cores or grab along a gradient from below the cages to a reference station presumably unaffected by farm debris. When investigating multiple farm locations within a given region, samples were collected at locations in which net cages were or had been present and at common reference locations representative of the natural variations within the area. A total number of 124 stations were sampled. Cores were sectioned in up to 12 sections down to a maximum depth of 40 cm, yielding a total number of about 500 samples for electrode measurements.

2.2
Sampling Techniques and Challenges

The wide range of variability of bottom substrates is a major challenge both with regard to sampling techniques, methods to ensure representative sample collection, data comparison and assessment of environmental impact. Sampling at farm locations using hand-operated or larger gravity corers was inadequate on hard bottom, and frequently failed in coarse sand or due to hard objects (stones, large mussels, etc.). Therefore, most samples were collected in soft bottom areas using either a small (250 cm^2) van Veen grab or divers taking hand-cores (6 cm ID) where soft sediments were present. On hard rock bottoms, soft sediments were frequently found on shelves and in fissures or in between stones. The small van Veen grab is recommended for use between net cages [51] because it is hand operated, suitable for sampling at any depth and at a larger variety of substrates than core samplers and because it maintains an element of random sampling relative to the diver's choice of sampling site.

In most cases the grab gave an almost undisturbed sample of the sediment water interface, from which a subsample from the upper 0–2 cm layer of the sediment was taken for chemical analyses. In a few extreme cases the grab was found to sample farm debris trapped in between stones, on shelves of sloping bedrock or sticking to bare bedrock. These samples were handled and analysed as any other sediment sample. All samples were collected during summer or autumn (May–October) months.

2.3
Electrode Measurements

Electrode measurements were always performed as soon as possible after the samples were removed from the seabed, usually within a few minutes. Some cores were stored up to six hours before processing, but in such cases the samples were protected from atmospheric oxygen by overlying sediment or water and tightly fitted rubber or silicon stoppers. Technical problems, lack of suitable facilities in the field or variable procedures with regard to electrode cleaning, storage, and calibration can cause erroneous or incomplete series of measurements. Therefore, successful performance using electrode measurements for sediment geochemical analyses depends, more than anything else, on the skills and experience of the field-worker. Details on maintenance, field-procedures and trouble-shooting are available in the Norwegian Standard [51] and similar protocols [50].

2.3.1
Instrumentation

All measurements were performed using an electrode assembly with a common reference electrode and separate sensors for pH, E_h and S^{2-}-ions. pH was measured using Orion Ross or Radiometer glass combination electrode or a Sentron ion-specific field effect transistor (ISFET) pH-meter and sensor. Redox potentials were determined using Radiometer P101 platinum electrodes and sulphide ion activities were determined using a Radiometer F1212S Ag|Ag$_2$S sulphide ion selective electrode. The common reference was either the internal reference of the glass combination electrode or a separate Radiometer Ag|AgCl reference electrode.

2.3.2
Calibration

pH was calibrated using two standard phosphate buffers (NBS or IUPAC pH 4 and 7) at in situ temperatures of about 10 °C. Because of the high ionic strength of the seawater, long response times must be allowed before a stable reading can be taken when moving the electrode from a dilute pH-buffer to a seawater sample. Therefore, after calibration, the electrodes were allowed to equilibrate in a surface seawater sub-standard used for storage, rinsing and calibration control throughout the following measuring sequences. Recalibration was undertaken whenever the sub-standard deviated more then 0.1 pH-unit from the initial value. In accordance with recommended procedures [61], the redox circuit was checked in an Fe(II)-Fe(III) redox-buffer solution, but not calibrated. The standard potential of the sulphide electrode (E_0) was determined in a pH-buffer by step-wise addition of a sulphide standard solution. Once determined the E_0 of the electrode varied little and field calibrations were usually not performed.

2.3.3
Measurements

After careful removal of the overlying water with a silicon tube siphon, the electrodes were inserted vertically into the sediment sample to a depth of 1 cm. The pH and rest potentials on the sulphide and redox electrodes were recorded as soon as the pH was stable, normally within less than 2 minutes (but see Sect. 2.3.5 below). In grab samples, electrodes were inserted through a lid on top of the grab and measurements could be taken at several depths by pushing the electrodes into successively deeper layers. Core samples were usually sectioned every 2 cm by pushing the sample with a piston from below into a sectioning chamber and cutting off with a thin sheet of plastic or stainless steel. Electrodes were then inserted through the recently exposed surface of the sectioned core. Most cores were sectioned to a depth of 8–10 cm with sub-samples transferred to a storage container and stored at –20 °C until further handling and chemical analyses.

2.3.4
Calculations

The E_h was calculated by addition of the half-cell potential of the reference electrode (as specified by the manufacturer), to the rest potential recorded in the sample. As a homologue to the proton activity measured with the pH electrode (pH =– log{H^+}), the E_h is often expressed as electron activity, pE =– log{e^-}. pE was calculated from E_h using the relationship E_h = (RT/nF)ln{e^-} $\sim$ 0.059 pE [62]. The Pt-electrode is rarely inert in natural samples and thermodynamic interpretations of measured E_h are not recommended [63, 64]. Nevertheless, E_h is frequently used as an empirical variable describing the tentative oxygen deficiency of the sediment and has been recommended for the monitoring of soft-bottom environments [65]. The sulphide electrode appears scientifically more sound. It responds selectively to S^{2-}-ions and at known pH the concentration of hydrogen sulphide ([ΣH_2S] = [S^{2-}] + [HS^-] + [H_2S]) in the pore water can be calculated [66, 67]. The method is considered to be less accurate than standard methods based on iodometric titration, spectrophotometry or the use of a strong alkaline buffer [68, 69], the term pS (=– log[ΣH_2S]) was preferred to denote sulphide concentrations calculated from electrode measurements in untreated sediment samples [70]. The rest potential recorded on the sulphide electrode (E_S) was corrected by addition of the half-cell potential of the reference electrode. The pS was then calculated from the measured E_S and pH, the calibration value for E_0 and the theoretical slope for the two-electron transfer of the sulphide electrode, using published sulphide dissociation constants [71, 72] and equations given in [67].

2.3.5
Reproducibility

Reproducibility of the electrode measurements was determined by repeated insertion of the electrodes at different points in 0.25 m^{-2} box core samples. It was found to be better than 0.1 pH units, 30 mV E_h($\sim$ 0.5 pE units) and 0.2 pS units. In samples normally described as oxic, the metal electrodes often respond slowly and the potential recorded was sometimes found to depend on the previous sample. In oxic samples measured after sulphidic samples stable readings could not be obtained within the time available and the recorded E_h was typically reduced from about + 400 mV before sulphide exposure to 200–300 mV after sulphide exposure. Therefore, environmental state assessments were never performed on the basis of E_h variations within the 200–400 mV range. For the sulphide electrode non-linearity and slow response occurred at a pS of about 6.0, but differentiation between samples was reproducible down to a pS of about 9.0. In order to obtain complete data sets for presentation and statistical calculations, all measurements of pS > 9.0 (i.e. below "detection limit") were replaced with $pS_{corrected} = 9.0 + (pS - 9)/5$.

2.4
Other Methods

Current speeds were recorded using Aanderaa RCM 7 current meters deployed 2 m above the bottom. The instrument measures current speeds as the number of rotations and was set to store integrated values every 10 min over one month [57].

Pore water for chemical analyses was separated from sediment by high-speed centrifugation. Sediment was transferred directly to 50 ml centrifuge tubes during core sectioning in provisional plastic tents filled with N_2-gas to avoid exposure to atmospheric O_2. Pore water samples were diluted before nutrient analyses by standard automated colorimetric methods [73]. Alkalinity was determined by acidification and back-titration with dilute sodium hydroxide.

Total carbon (TC) and nitrogen (TN) was determined after combustion at 1800 °C in a Carlo-Erba 1106 element analyser. TOC was determined by acidification to remove inorganic carbon and reanalyses of the remaining TC. Because some nitrogen is leached during acidification, TN was always taken from the first run. Samples for total phosphorous (TP) and trace metals (Zn and Cu) were digested in warm nitric acid and determined with ICP (ion plasma chromatography).

Samples for full quantitative macrofauna analyses were collected with a 0.1 m^2 van Veen grab and washed on board through a 1-mm sieve. The material retained on the sieve was fixed in 4% buffered formaldehyde and stained with rose bengal. In the laboratory all fauna were identified to the lowest possible taxonomic level (generally to species) and the number of individuals per taxon recorded. At all stations surveyed for fauna analyses, vertical profiles of the sediment were obtained using a diver-operated SPI (Sediment Profile Imagery) camera. Further details on faunal identifications, SPI methods and data analyses from this survey have been published elsewhere [59].

All statistical calculations were performed using standard procedures in the statistical software package JMP version 4 (SAS Institute Inc.).

3
Results and Discussion

3.1
Main Results

Chemical properties in samples from the upper 2 cm of sediments at the stations in each site category are shown in Table 2.

The most extreme variations were observed in station category A. Farms were primarily established on type D-locations and unless altered by farm activities, the conditions at the A-stations should not differ (beyond natural variations) from those observed at the D-stations. Such conditions were rarely found below cages, but samples taken only 2–3 m away from the vertical projection of the net cages and in particular upwards along a tilting bottom and upstream along the prevailing current direction appeared often unaffected by farm activities. An example of the large variations, which may be observed within an A-location, is shown in Fig. 1. The figure also shows an example on how acceptance criteria for pH and pS may be applied. The pH boundary

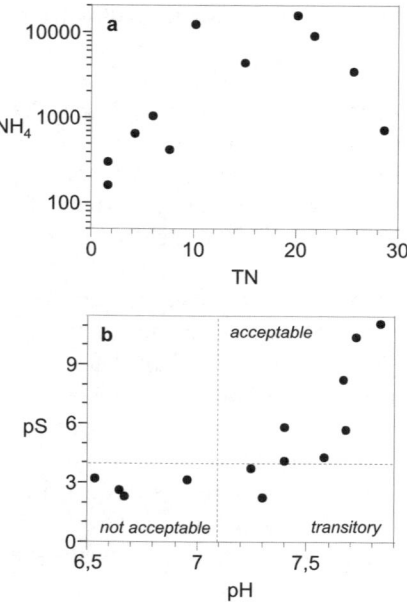

Fig. 1 Variations in **a** nitrogen (mg g^{-1} dry sed.) and pore water concentration of ammonium (µM) and **b** pH and pS in sediment samples (0–2 cm depth) collected below and along the edges of a fish farm. *Broken lines* represent suggested criteria boundaries for pH and pS (see text)

Table 2 Distribution of chemical properties in surface sediments (0–2 cm) from station site categories A–E (see text). n = number of samples

		A	B	C	D	E
		Below farm	Abandoned	Transition	Reference	Sill basins
pH	min	5.47	7.00	7.10	7.24	7.13
	median	7.22	7.74	7.58	7.85	7.31
	max	8.02	8.18	7.96	8.03	7.57
	n	56	27	21	9	11
E_h (mV)	min	–204	–198	–284	–141	–214
	median	–161	–108	–92	82	–136
	max	204	288	42	242	158
	n	56	27	21	9	11
pS	min	–0.2	3.0	0.9	7.3	2.8
	median	3.2	7.2	6.6	8.7	6.2
	max	11.4	11.4	9.2	9.9	10.9
	n	56	27	20	7	11
%H_2O	min	54	24	20	19	81
	median	83	61	51	33	87
	max	88	71	84	60	88
	n	16	8	19	7	3
TC ($mg\,g^{-1}$)	min	27	6	18	9	73
	median	134	42	71	19	115
	max	261	114	145	109	143
	n	29	10	20	7	7
TN ($mg\,g^{-1}$)	min	<1	<1	<1	<1	8.3
	median	15.4	6.3	2.7	<1	14.7
	max	28.7	14.7	20.2	5.2	16.9
	n	29	10	20	7	7
TP ($mg\,g^{-1}$)	min	1.0	1.4	0.64	0.20	1.2
	median	38	2.4	1.9	0.71	1.8
	max	83	9.6	11.3	0.95	2.0
	n	29	4	19	7	7
NH_4 (μM)	min	161	155	137	311	130
	median	2258	3571	1567	–	491
	max	15735	7929	6214	806	1287
	n	12	7	5	2	6
PO_4 (μM)	min	19	13	13	65	26
	median	125	44	180	–	57
	max	386	1324	630	67	438
	n	12	7	5	2	6
Cu ($\mu g\,g^{-1}$)	min	11	6.9	2.8	1.9	32
	median	115	22	8.4	5.5	–
	max	2817	50	86	21	35
	n	28	3	16	7	2
Zn ($\mu g\,g^{-1}$)	min	24	<5	<5	<5	69
	median	443	90	<5	<5	–
	max	1531	98	115	25	90
	n	28	3	16	7	2

of 7.1 is justified in Sect. 3.2 below. The pS boundary of 4.0 is suggested here on the basis of the overall variations of this variable. The example shown in Fig. 1 emphasises the need for multiple samples and a clear sampling strategy to avoid state assessments based on samples, which are not representative for the site as a whole.

Most samples from A-locations differed markedly from those taken at the other location categories (Table 2). Thus, median pH values were lowered from 7.6–7.8 at C and D stations to 7.2 below cages. Median E_h was lowered from 82 mV at reference D stations to – 161 mV at A stations. Such E_h-values are characteristic for sulphide-rich environments [74] and the median pS of 3.2 indicated concentrations of H_2S at the mM level. Also median concentrations of 0.1 mM PO_4 and in particular 2.2 mM NH_4 were high compared to reference D stations. Water content and concentrations of carbon and nitrogen were clearly higher at A than at reference D stations, but not clearly distinguishable from reference E. High concentrations of total carbon were observed at some reference D stations due to the abundance of fractured shell fragments, whereas nitrogen, which can be disregarded in inorganic fractions, was frequently less than detection limits of $1\,\mathrm{mg\,g^{-1}}$ at reference D stations.

Concentrations of Zn and in particular TP were high at A stations compared to both D and E reference stations. Yielding minimum concentrations at A stations higher or similar to the maximum concentrations at D stations, TP and Zn appeared more suitable than any other parameter for tracing farm impacts on sediment composition. The concentrations of Cu were also high at many A stations, but being the only variable not directly related to the sedimentation of excess feed and faeces this input may be absent at farm locations with alternative antifouling strategies.

3.2
pH and Gas Ebullition

The low pH-values frequently observed in sediment samples from below net cages were found to occur in samples where gas was observed either as small vacuoles visible through the transparent wall of the acrylic core or as bubbles rising through the water overlying the core or grab sample or bursting through the water surface shortly after bottom penetration with the sediment sampling device. Gas was positively identified at 17 of the stations surveyed whereas no signs of gas were observed at 97 stations. As shown in Table 3, low pH and E_h and high concentrations of organic matter (161–216 mg C g^{-1} dry sed., 18–23 mg N g^{-1} dry sed.) and pore water alkalinity were observed in all samples where gas was noted.

The difference between cores with and without gas is evident from Fig. 2. Numerically, the 90% quantile pH of 7.04 in sediments with gas was clearly different from the sediments without gas in which the 10% quantile pH was

Table 3 Mean values and 95% confidence interval for electrode measurements, total carbon (TC), nitrogen (TN) and alkalinity (A_t) in sediment samples with and without gas noted during sample collection. n = number of samples

	No gas				Gas			
	Lower 95%	Mean	Upper 95%	n	Lower 95%	Mean	Upper 95%	n
pH	7.50	7.53	7.57	300	6.58	6.67	6.75	45
E_h (mV)	−116	−103	−89	300	−214	−203	−192	45
pS	4.9	5.2	5.5	287	2.3	2.6	2.9	45
TC (mg g^{-1})	57	66	75	138	161	189	216	22
TN (mg g^{-1})	5.9	7.1	8.2	138	18.1	20.6	23.1	22
A_t (mM)	4.2	5.0	5.8	7	31.9	36.7	41.6	10

7.15. This suggested a fairly accurate boundary at pH = 7.1 below which the risk of gas ebullition increased.

It appears reasonable to assume that the pH in these sediments was controlled by intensive anoxic biodegradation processes driven by the supply of labile (energy rich) organic matter from fish excretions and excess feed. The major quantitatively important processes in such deposits have been described in stoichiometric models [17] for sulphate reduction:

$$C(H_2O) + \frac{1}{2}SO_4^{2-} = HCO_3^- + \frac{1}{2}HS^- + 1/2H^+ \tag{1}$$

or methanogeneses:

$$C(H_2O) + \frac{1}{2}H_2O = \frac{1}{2}HCO_3^- + \frac{1}{2}CH_4 + \frac{1}{2}H^+ \tag{2}$$

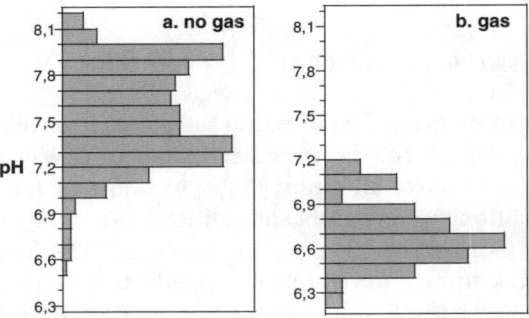

Fig. 2 Histogram showing the distribution of pH in sediments with and without observed gas ebullition

At the frequently observed pH of 7.5–8 in marine sediments, HS^- and HCO_3^- are the predominant dissociation products of the sulfide and carbonic acids, respectively. During sulphate reduction (Eq. 1) each mole of H^+ produced will be buffered by the simultaneous production of one mole of HS^-. The pH is therefore buffered by the reaction $HS^- + H^+ = H_2S$. The equilibrium condition for this reaction: $K = \{HS^-\}\{H^+\}/\{H_2S\}$ can be solved to yield

$$pH = pK + \log(\{HS^-\}/\{H_2S\}) \tag{3}$$

which shows that pH = pK at equal activities of the two sulfide species. Because this pK $\sim$ 7.0 at the ambient temperatures of about 10 °C and salinities of about 34 [71, 72], the acidification driven by sulphate reduction alone will therefore not easily drive the pH below 7.0.

Similarly, methane production in accordance with Eq. 2 has the potential to reduce pH until buffered by the reaction $HCO_3^- + H^+ = H_2CO_3$. Solving the equilibrium condition $K = \{HCO_3^-\}\{H^+\}/\{H_2CO_3\}$ for pH yields:

$$pH = pK + \log(\{HCO_3^-\}/\{H_2CO_3\}) \tag{4}$$

Because pK $\sim$ 6.1 for the first dissociation of carbonic acid at 10 °C and PSU = 34 [75], the activity of methane producing bacteria may easily protonate the bisulphide ions and drive the pH further down towards a pH of about 6 at which further reduction is buffered by the formation of H_2CO_3. Thus, the buffering predicted by the dissociation constants for H_2S and H_2CO_3 was highly consistent with the present observations in gas bearing sediments from below fish farms.

Although methane gas itself may not be very harmful for the fish kept in the cages, transport on rising bubbles of other harmful substances like H_2S or pathogenic agents from highly anoxic sediments may represent a health risk [36]. If the condition of the farmed fish is the main concern, the apparently close relationship between a decrease of pH and gas formation makes pH and pS important variables for defining environmental quality criteria for sediments below net cages.

3.3
Comparison between Station Categories

Comparison of group means for all station categories presented in Table 2 was performed on log-transformed concentration data (one-way ANOVA, Tukey's method for comparison of all pairs) (Table 4). Most of the variable values observed at A stations were significantly different (p < 0.05) from those observed at one or more of the other station categories. The frequent difference between A and C stations showed that severe effects were confined to a small area below the cages. The frequent difference between A and B stations and only two cases (pE and TP) of significant difference between B and D stations, showed a large degree of normalisation of parameter values at the abandoned

Table 4 Significant difference between station categories (one-way ANOVA, Tukey's method for comparison of all pairs, log transformed concentrations, $\alpha = 0.05$)

pH	A	$\neq$	B, C, D				
pE	A	$\neq$	B, D	and	D	$\neq$	B, C, E
pS	A	$\neq$	B, D, E				
%H$_2$O	A	$\neq$	B, C, D	and	E	$\neq$	B, C, D
TC	A	$\neq$	B, D	and	D	$\neq$	E
TN	A, E	$\neq$	C, D				
TP	A	$\neq$	B, C, D, E	and	D	$\neq$	B, C
NH$_4$*	A	$\neq$	E				
PO$_4$*	No significant difference between categories						
Cu	A	$\neq$	C, D				
Zn	A	$\neq$	B, C, D	and	E	$\neq$	C, D

* including pseudoreplicates from various depths in same core

locations, i.e. after 0.5–3 years. The maintenance of low redox potentials (pE) and high concentrations of TP showed, however, that longer time periods were required for complete remediation.

TP was particular powerful being the only variable which showed significant differences between farm sites (A) and all other station categories. In addition TP was one of few variables able to discriminate reference (D) locations from abandoned (B) and transition zone (C) locations. Thus, in the present investigation, determination of TP appeared to be the single variable most suitable for tracing remnants of farm debris. In addition to TP, all three electrode measurements showed significant differences between cage (A) and reference (D) locations. This is consistent with previous investigations [25] in which total sulphide followed by E_h were found to be the most sensitive of 23 different benthic variables (TP not included) for organic enrichment at cage sites.

Pore water nutrients were only determined in two cores from reference D-locations, which provided an insufficient number of samples for statistical analysis. Increasing the number of samples by using the results from subsurface core sections, improved the results with regard to NH$_4$ yielding A significantly different from B, C, D, E and B significantly different from E. PO$_4$, however, showed no significant differences between any station categories. Much of the phosphorous in farm debris is present in insoluble Ca-bound fractions [12] and relatively low remineralisation of PO$_4$ would be expected. Also, pore water profiles of pH, A$_t$ and PO$_4$ suggested that mineral reactions in carbonate-rich deposits below the layer of organic matter from the farm such as:

$$10CaCO_3(s) + 6HPO_4^{2-} + 2H_2O + H^+ = Ca_{10}(PO_4)_6(OH)_{2(s)} + 10HCO^{3-} \quad (5)$$

[62, p 284], may act as a pH-buffer, a trap for remineralised PO_4 and a source of bicarbonate ions contributing to alkalinities up to 49.4 m M at the boundary between farm deposits and the old seabed [56].

3.4
Discriminant Analyses

In order to identify the station categories more accurately, a multivariate discriminant analysis was performed on electrode measurements, TC, TN and TP. A total of 63 samples from four different surveys were analysed for all of these variables (Fig. 3a). All samples were taken from within the 0–2 cm depth interval. The absence of overlap between the 95% confidence limit circles about category mean values shows that the six variables clearly discriminated between fish farm locations (A), sheltered reference locations (E) and exposed reference locations (D). Transition zone locations (C) occurred in between A and D, but closer to reference (D) than to farm locations (A), confirming the confinement of farm impacts to areas close to the net cages with moderate impacts in the transition zone 5–50 m away from the vertical projection of the cages. Abandoned locations (B) were clearly different from A, but overlapped with the three other station categories. The biplot rays showed that phosphorous was an important variable contributing to the discrimination between A and the other station categories along the x-axis, whereas pS and nitrogen appeared to be more important variables for discriminating between the two types of reference locations (D and E) along the y-axis.

Figure 3a also shows a displacement of category B stations (abandoned) towards category E (sill basin reference) and the position of two B-stations close to the group of E-stations. The larger number of samples available for one-way ANOVA analyses (Table 4) confirmed frequent similarity between B and E locations with regard to electrode measurements ($n = 38$), TC ($n = 18$) and TN ($n = 18$). Water content ($n = 11$), however, showed a significant difference between B and E (Table 4), and it appears reasonable to expect that a larger number of analyses of TP, Zn and Cu, which are more suitable variables for distinction between farm discharge and general organic enrichment would have revealed a more significant difference between B and E.

The use of the electrode variables only (Fig. 3b) on the same 63 samples as those plotted in Fig. 3a, reduced the distance between A and the other stations, and displaced transition zone stations (C) towards E. The ability to discriminate between the three station categories A, D and E and also the separation between A and all other stations was, however, maintained. This showed that the electrode measurements alone could adequately describe some characteristic effects frequently occurring in sediments below fish farm cages. The biplot rays showed that pH was a primary factor contributing to the discrimination between A and the other locations along the x-axis and

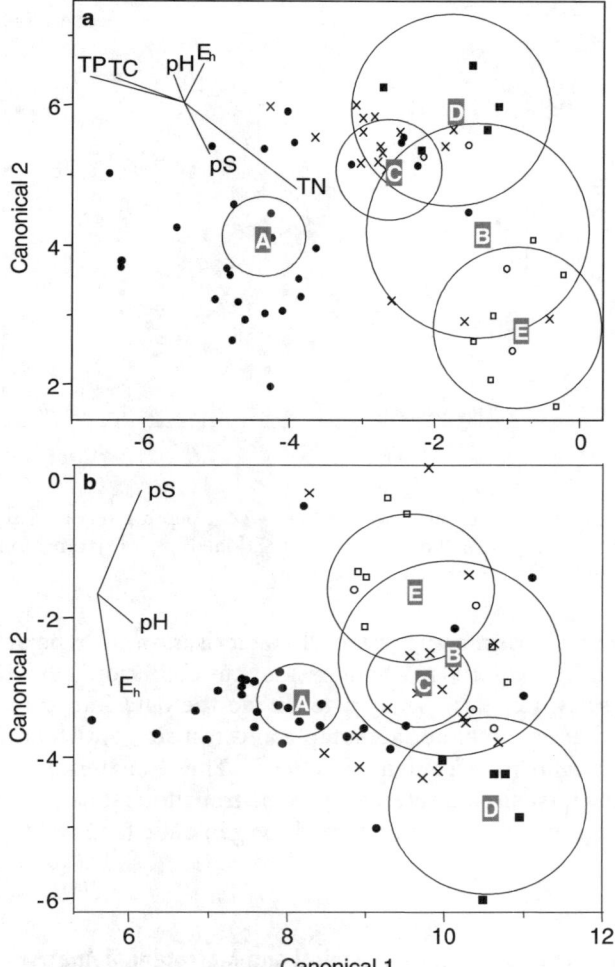

Fig. 3 a Multivariate comparison (canonical plot with biplot rays and 95% confidence limit circles) for electrode measurements, TC, TN and TP in 0–2 cm layer of sediments from station categories A–E. ● = A (below), o = B (abandoned), × = C (transition zone), ■ = D (exposed reference), □ = E (sheltered reference). **b** As a, but electrode measurements only

that E_h and pS were important in discriminating between the reference locations along the y-axis.

Further reduction to the two-dimensional E_h-pH plot (Fig. 4) showed that a fairly clear difference was maintained between the majority of farm locations (A) and the exposed reference locations (D) with transition zone and abandoned locations occurring in between. In Fig. 4, boundaries for E_h and

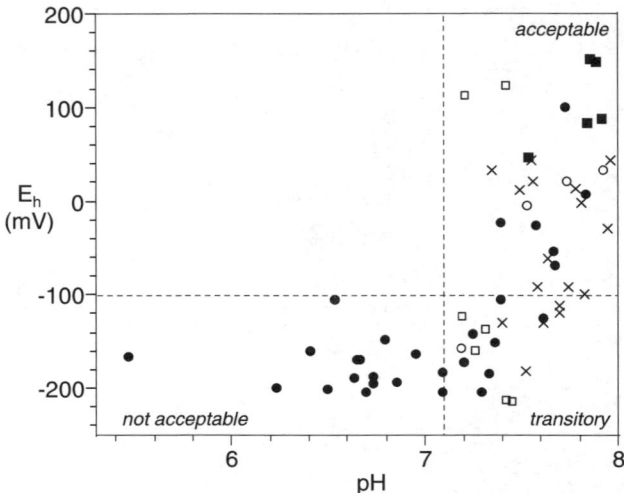

Fig. 4 Plot of pH vs. E_h measured within the 0–2 cm depth interval in sediments from station categories A–E. ● = A (below), o = B (abandoned), × = C (transition zone), ■ = D (exposed reference), □ = E (sheltered reference)

pH were adopted from a suggested characterisation of hypoxic and anoxic conditions [27] (E_h =– 100 mV) and risk of gas ebullition (pH = 7.1, Sect. 3.2 above). These two criteria seemed to divide the data into three main categories (called acceptable, not acceptable and transitory), which may be used for regulatory and management practices.[1] The occurrence of five stations from category E (sheltered reference) in the transitory state (Fig. 4) is reasonable indicating organic enrichment and oxygen deficits in water masses from below sill depths.

3.5
A Joint Survey Comparing Biogeochemical and Macrofaunal Analyses

Four different methods for detection of benthic environmental effects were compared during a survey at five farms in mid- and north-Norway [28]. Visual diver inspections, sediment profile imagery (SPI), chemical analyses (pH, E_h, pS, N, P, Zn and Cu) and full quantitative faunal analyses were conducted at each location. Five stations were sampled at each farm site to assess benthic conditions: below the net cages (A), at the edge of cages (A5), 20 m and 50 m downstream (C20 and C50) and at a reference station (D) > 300 m upstream, yielding a total of 25 stations. Samples for macrofauna analyses and

---

[1] Changes driven by organic enrichment of a benthic system appear to occur along a continuum of interrelated physical, chemical and biological responses. The application of boundaries is tentatively assigned based on scientific criteria relevant to farm management and to meet the needs of regulatory authorities.

SPI were only obtained at 12 and 13 stations, respectively. Figure 5 shows the Shannon–Wiener diversity index (H'), pS and a chemical index ([58] and Fig. 5c) calculated from sediment concentrations of TN, TP, Zn and Cu for all stations.

The mean H' for the benthic macrofaunal communities increased with increasing distance from the farm, but the variation between farms was large and no significant differences were found between the three station locations A5, C50 and D ($p > 0.05$, one-way ANOVA, Tukey's comparison of all pairs). A similar analyses of the chemical variables showed that both the pS and the chemical index at location A5 were significantly different ($p < 0.05$) from C50 and D. In addition, location A was significantly different from all other stations.

Macrofauna samples were collected close to the cage perimeter, but not below the net cages. Therefore, comparisons to assess differences between locations could only be based on the chemical variables. Highly disturbed communities were, however, found at distances of 5 m (Farm ND, H' = 0.1) and even at 50 m (Farm MR2, H' = 0.3), and samples from below the net cages could hardly have shown any further lowering of H'. Thus, compared to the macrofauna diversity index, the chemical variables provided good resolution within the high end of the organic enrichment gradient.

The observations 50 m from the cages at Farm MR2 represented a rare combination of a highly disturbed benthic community (H' = 0.3) and almost no deviation of pS (7.2) and the chemical index (0.7). This could mean that benthic macrofaunal community structure was a more sensitive indicator at stations located some distance from cage groups [59]. In British Columbia, effects on the macrobenthic communities have been measured up to 100–200 m from groups of salmon cages [8, 9]. With production cycle maxima of 1000–2000 metric tons of Atlantic salmon and mean current speeds of 3.4–8.3 cm s^{-1}, these farms were larger and apparently located in more exposed environments than those encountered in our investigations (Table 5). Longer horizontal transport of settling and resuspended farm debris would be expected and the effects observed on benthic communities were accompanied with increased concentrations of hydrogen sulphide in the pore water [8]. In the heterogeneous coastal environment surrounding many fish farms in Norway there is the possibility that natural variations in the benthic community (related to substrate or other factors) can cause differences in variables between locations. The lack of expected changes in benthic chemical variables associated with organic enrichment did not support the conclusion that the low benthic macrofauna diversity observed 50 m from the cages was a farm impact [59].

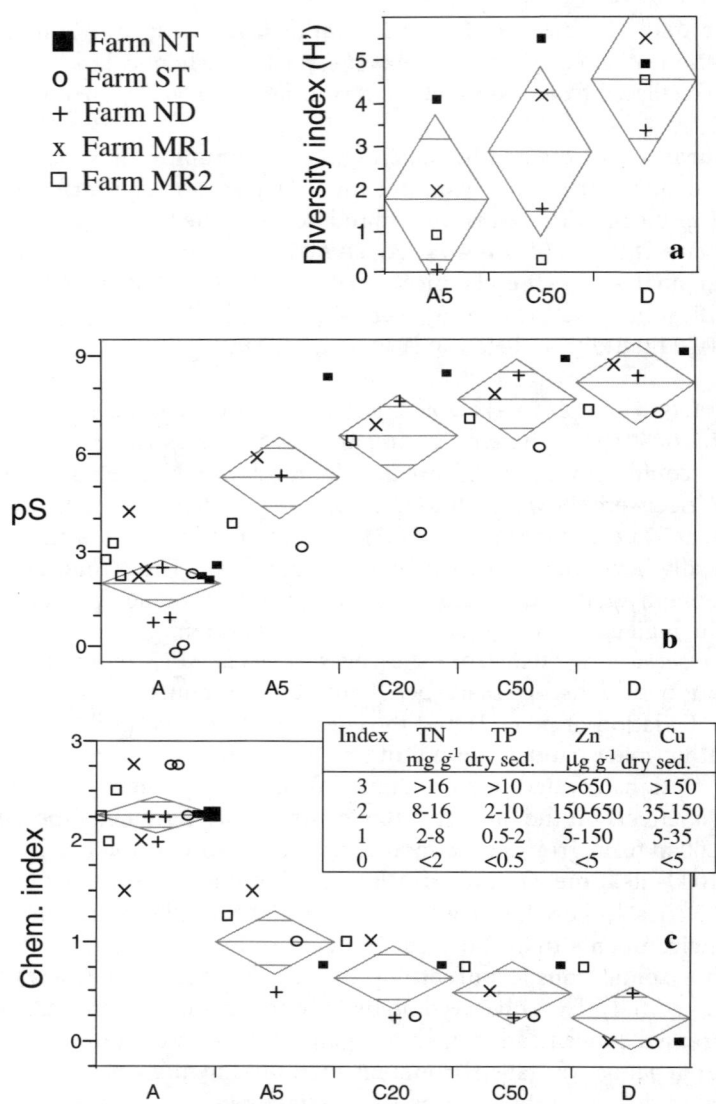

Fig. 5 a Shannon–Wiener diversity index (H′), **b** pS and **c** chemical index at stations located at various distances from fish farms. A = below, A5 = 5 m, C20 = 20 m, C50 = 50 m and D > 300 m from net cages. Chem. index is the mean of the four index values assigned to each sample in accordance with inserted criteria for TN, TP, Zn and Cu. Mean diamonds (ANOVA, t-test) show group means and 95% confidence interval (= height of diamond). No overlap means that two groups with the same number of measurements are significantly different. The points are spread horizontally within each plot category to avoid masking. Data previously published in [28, 58, 59]

3.6
Influence of External Factors on Biogeochemical Variables

Effects of waste discharges from aquaculture on the benthic environment at net cage locations are also influenced by factors related to siting and farm management. High current speeds and large distances between the net cages and the bottom will increase horizontal transport of farm debris and reduce sedimentation of organic matter directly below the cages. In 1991, the relationship between current speed and some of the biogeochemical effect variables was investigated at 16 farm locations (nos. 1–16) in mid-Norway (Table 5). The current meters were deployed by divers who simultaneously collected three sediment cores for chemical analyses from available locations scattered across the area below the nearby net cages.

Gas was detected in all cores collected at farm 12 (including transition zone station 12D), in 2 of 3 cores from farm 10 and in 1 of 3 cores from the abandoned farm 16 (Table 5). The chemical variables (Fig. 6) clearly confirmed adverse environmental conditions due to organic enrichment at 12, 12D and 10, whereas data from location 16 indicated less impact.

Table 5 Production parameters, current measurements and gas observations in sediments at 16 farm locations surveyed in 1992. Ranked according to biomass in the farm

Farm no.	Depth (m) min.-max.	Current measurements			Biomass (tons)		used/abandoned years	Gas*
		mean $cm\,s^{-1}$	max. $cm\,s^{-1}$	days < $1.5\,cm\,s^{-1}$	last year max.	during survey		
10	10–30	5.5	23	0.1	440	440	1.5	2/3
7	20–33	8.7	31	–	205	205	1.5	0/3
15	15–20	2.5	14	1	170	170	1.5	0/3
14	16–20	2.2	15	4	200	70	4.5	0/3
12	10–20	1.7	10	0.9	200	30	8.5	3/3
5	18–18	5.3	21	0.1	160	82	0.5	0/3
2	20–30	3.2	17	0.9	150	22	2.5	0/3
9	11–11	1.2	7	8	75	75	6.5	0/3
12D	10–20	1.7	10	0.9	–	0	–	1/1
1	20–30	5.3	30	0.2	200	0	0.5	0/3
4	18–18	1.9	11	7	200	0	0.5	0/3
8	10–13	6.7	26	0.1	140	0	0.5	0/3
6	21–21	2.5	21	0.9	100	0	0.7	0/3
13	27–27	1.6	10	0.2	0	0	1.2	0/3
16	13–20	1.2	8	6	0	0	1.5	1/3
3	8–10	1.3	7	8	0	0	2.5	0/3

* Number of cores with gas/total number of cores

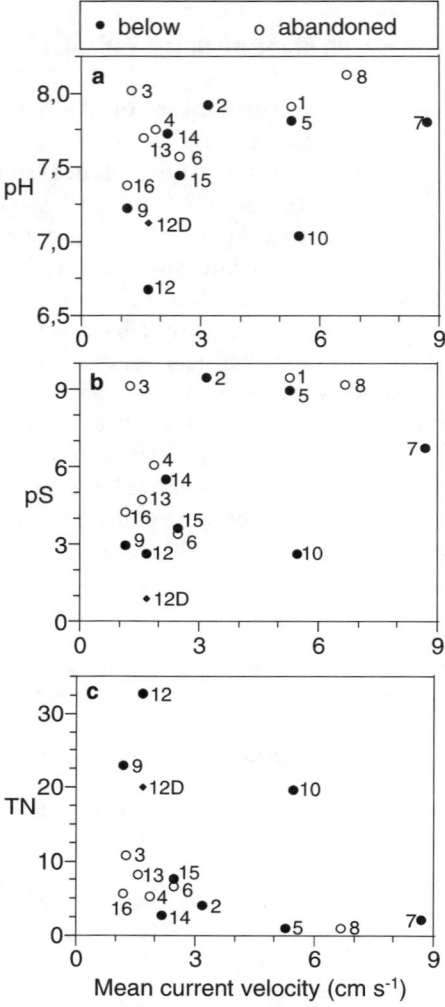

Fig. 6 Relationship between current speed and **a** pH, **b** pS and **c** N (mg g^{-1} dry sed.) in sediment samples from 0–2 cm depth at fish farm locations in mid-Norway, Oct.–Nov. 1992. Currents were recorded 2 m above the bottom below cages

No gas was detected in the sediments below farm 9, but nitrogen content was high and one of the cores showed pH and pS as low as 7.1–7.2 and 2.4–2.7, respectively. The fish biomass was low (Table 5), but the location had been in use for 6–7 years. The bottom was flat and water depth was only 11 m. The current velocity data given in Table 5 indicated that farm 9 had the most sheltered environment of all 16 locations surveyed. The chemical variables showed that conditions were close to those characteristic for gas production, but only repeated investigations might have revealed whether the location was in a steady state near the limits of gas ebullition or whether the site was in transition to exceeding this limit. Thus, farm 9 may serve as an example which suggests that a different management strategy could be possible based on frequent seabed inspection and production adjustment to avoid extreme concentrations of H_2S and gas formation in the sediments beneath the cages. The fact that the location had been used for a long period of time showed that with proper farm management some moderate level of production may be sustainable even when natural conditions are not ideal.

The most enriched sediments occurred at farm 12, which had been occupied for 8-9 years. High nitrogen concentrations (> 17 mg N g^{-1}) were observed down to 8 cm depth, gas was detected in all cores and pH and pS values were low. Fish biomass in cages was low during the survey, but had been quite high earlier in the year (Table 5). At farms 14 and 15, biomass and water depths were similar to those at farm 12, but a shorter production period and slightly better current conditions may have reduced benthic impacts (Fig. 6). Further improvements in both current regime and chemical variables were observed at locations 2, 5 and 7.

Increased production costs seem to put a limit on the utilisation of exposed environments, and it may be worth noting that in spite of mean current speed close to the proposed limit of 3 cm s^{-1} for "very sensitive" locations [76], farm 2 had been in production for more than two years without measurable benthic impacts. The actual production at the location is not known, and regular inspections of the seabed over more than 2-3 years are required to determine if production at this location is sustainable at the current level. At farm 10 the bottom was strongly sloped (ca. 45°) bedrock. This was a high energy environment exposed to frequent wave heights between 1 and 1.5 m [57] and a mean current speed of 5.5 cm s^{-1} below the net cages. Nevertheless, pH and pS were low, concentrations of N were high and gas was observed in two of the three cores. This showed that the benthic environment might easily become overloaded even at more exposed locations and confirmed previous findings that there is no simple correlation between sediment organic matter (TOC) and environmental variables such as water depth or current speed [59]. Therefore, optimum utilisation of a given location must be based on more complex predictive models combining management practices and natural hydrodynamic and bathymetric factors, or even better, on real observations of the benthic environment such as has been recommended in the Norwegian management system (see below).

4
Application in a Management System

The results presented in this chapter were used to develop a monitoring program for benthic impact from Norwegian mariculture [51] as part of a more elaborate management system called MOM (Modelling—Ongrowing fish farms—Monitoring) comprising a monitoring programme, a model and environmental quality standards (EQS) [52, 53]. Previous monitoring surveys had largely consisted of macrofauna investigations, but due to the efforts involved and the relatively large sampling devices required (0.1 m² van Veen grab), the number of samples were usually low and the area below cages were rarely sampled. The smaller, hand-held sampling equipment required for the use of electrodes, in conjunction with observations of visual and olfactory sediment variables, made it possible to collect more samples and samples from below or in between the cages. The use of electrode measurements made it possible to develop a monitoring program, which is divided into three types of investigations (A, B and C) of increasing accuracy and cost. According to the demands of the authorities the most appropriate investigation can be used.

The A-investigation is a simple measurement of sedimentation under the fish farm. The B-investigation is based on electrode measurements of pH and E_h and the relationship between these variables as shown in Fig. 4. The pH and E_h measurements are supported by observations of visual and olfactory sediment variables. A scoring system has been developed to calculate sediment conditions in each sample and for the site as a whole. The sediment condition can be classified into four groups, where the fourth group corresponds to a high risk of methane production and a lack of macrofauna. The frequency of investigation at a given site depends on the condition of the sediment. Thus, sites with poor conditions are monitored more closely than sites with better conditions. The B-investigation is primarily for use under and close to the fish farm whereas the last investigation (C) based on full benthic macrofauna community analysis is recommended primarily for monitoring of more subtle changes possibly occurring at larger distances from the farm sites.

5
Summary and Conclusions

At many fish farm locations investigated in Norway between 1984 and 1999, severe organic enrichment was revealed by electrode measurements of pH, sulphide ion activities (pS) and redox potential (E_h) supported by total carbon (TC), nitrogen (TN), phosphorous (TP), zinc (Zn) and copper (Cu).

Below the net cages, median sediment concentrations were 38 mg TP g^{-1} (dry wght.), 15.4 mg TN g^{-1} and 134 mg TC g^{-1} as compared to 0.38 mg TP g^{-1}, < 1 mg TN g^{-1} and 19 mg TC g^{-1} at nearby reference locations. Similarly, the median pH decreased from 7.85 at the reference locations to 7.22 at farm locations, and the pS decreased from 8.7 to 3.2. Also the E_h decreased from 82 mV at reference locations to – 161 mV below the net cages.

At abandoned farm sites, a large degree of normalization of biogeochemical variables occurred within 0.5–3 years. Longer periods were, however, required for complete remediation of E_h and in particular sediment concentration of phosphorous. TP was the only variable which showed a statistically significant difference between fish farm locations and all other station locations. Together with Zn, TP was a good tracer for farm debris.

The data were collected 5–20 years ago in regions or at specific farm locations where environmental perturbations due to salmon aquaculture were suspected. Therefore, the results are not representative of present-day conditions at fish farms in Norway. Over the years, farm management and siting has improved, feed consumption per ton of produced fish has declined and eutrophication due to farm discharge is not considered a significant environmental impact factor [77].

Nevertheless, for the purposes of documentation and maintenance of good production environments, standard monitoring methods are needed which are sensitive for detecting effects of organic enrichment in the benthic environment at fish farm locations. The methods must provide accurate numeric description of changes in variables within the high end of the enrichment gradient and a large number of samples is beneficial for statistical validation of spatial and temporal variations.

Electrode measurements in untreated samples are easy to perform, sensitive to the proliferation of anoxic, sulphide-rich and methanogenic pore water environments and can provide an immediate on-site assessment of sediment organic enrichment. Additional chemical variables such as TC (or TOC), TN, TP, Zn and Cu add valuable information on sediment condition and are suitable for documentation of the presence of farm debris. Investigations in Canada have shown the applicability of similar methods [9, 24, 27, 78] and electrode measurements of E_h and free sulphide have been recommended for the monitoring of fish farms also in this region [79].

The scientific rationale for using the E_h, pS and pH electrodes is the close causal relationship to anaerobic degradation processes and the risk of methane gas ebullition and adverse effects on fish health. A similar link to structural variables such as sediment concentrations of Zn, Cu and TP, or the macrobenthic community, appears somewhat more subtle. This also applies to organic carbon and nitrogen measurements to the extent that they measure quantity and not quality (i.e. degradability) of the organic matter.

Macrofauna investigations are widely accepted as a standard for assessment of environmental status [47–49], but the high costs of a full macrofaunal analyses [27, 59, 80] frequently limit the number of samples analyzed. Full macrofaunal analyses can provide substantial ecological information and hence add value to the impact assessment. However, it is our conclusion that simple electrode measurements can provide a sufficiently accurate assessment of organic enrichment status in sediments at fish farm locations.

Acknowledgements We wish to acknowledge contributions from Gjermund Bahr, Sylvia Bredholt, Sabine Cochrane, John Costelloe, Egil Dragsund, Morten Frogh, Roar Linjordet, Stig Skreslet, Øystein Vethe and several staff members at Nordland Research Institute, Oceanor and the Centre for Soil and Agricultural Research to the investigations presented here.

References

1. Davenport J, Black K, Burnell G, Cross T, Culloty S, Ekaratne S, Furness B, Mulcahy M, Thetmeyer H (2003) Aquaculture: the ecological issues. Blackwell Science Ltd, Oxford
2. Gowen R, Bradbury N (1987) Oceanogr Mar Biol Am Rev 25:563
3. Strain PM, Hargrave BT (2005) Salmon aquaculture, nutrient fluxes and ecosystem processes in southwestern New Brunswick (in this volume). Springer, Berlin Heidelberg New York
4. Brown JR, Gowen RJ, McLusky DS (1987) J Exp Mar Biol Ecol 109:39
5. Nickell LA, Black KD, Highes DJ, Overnell J, Brand T, Nickell TD, Breuer E, Harvey SM (2003) J Exp Mar Biol and Ecol 285–286:221
6. Wildish DJ, Pohle GW (2005) Benthic macrofauna changes resulting from mariculture (in this volume). Springer, Berlin Heidelberg New York
7. Tsutsumi H, Kikuchi T, Tanaka M, Higashi T, Imasaka K, Miyazaki M (1991) Mar Pollut Bull 23:233
8. Brooks KM, Stierns AR, Mahnken CVW, Blackburn DB (2003) Aquaculture 219:355
9. Brooks KM, Mahnken CVW (2003) Fish Res 62:255
10. Anderson MR, Tlusty MF, Pepper VA (2005) Organic enrichment at cold water aquaculture sites—the case of coastal Newfoundland (in this volume). Springer, Berlin Heidelberg New York
11. Lu L, Wu RSS (1998) Environ Pollut 101:241
12. Pereira PMF, Black KD, McLusky DS, Nickell TD (2004) Aquaculture 235:315
13. Crema R, Prevedelli D, Valentini A, Castelli A (2000) Ophelia 52:143
14. Anderson E (1992) Benthic recovery following salmon farming: study site selection and initial surveys. Report to the Water Quality Branch, Ministry of Environment, Lands and Parks, Province of British Columbia
15. Karakassis I, Hatziyanni E, Tsapakis M, Plaiti W (1999) Mar Ecol Prog Ser 184:205
16. Pearson TH, Rosenberg R (1978) Oceanogr Mar Biol Ann Rev 16:229
17. Zehnder AJB, Stumm W (1988) In: Zehnder AJB (ed) Biology of Anaerobic Microorganisms. Wiley, New York

18. Holmer M, Wildish D, Hargrave BT (2005) Organic enrichment from marine finfish aquaculture and effects on sediment processes (in this volume). Springer, Berlin Heidelberg New York
19. Hargrave BT, Duplisea DE, Pfeiffer E, Wildish DJ (1993) Mar Ecol Prog Ser 96:249
20. Morrisey DJ, Gibbs MM, Pickmere SE, Cole RG (2000) Aquaculture 185:257
21. Revsbech NP, Jørgensen BB, Blackburn TH (1980) Science 207:1355
22. Holmer M, Christensen E (1992) Mar Ecol Prog Ser 80:191
23. La Rosa T, Mirto S, Mazzola A, Maugeri TL (2004) Aquaculture 230:153
24. Hargrave BT, Phillips GA, Doucette LI, White MJ, Milligan TG, Wildish DJ, Cranston RE (1997) Water Air Soil Pollut 99:641
25. Hargrave BT, Philips GA, Doucette LI, White MJ, Milligan TG, Wildish DJ, Cranston RE (1995) Can Tech Rep Fish Aquat Sci no 2062. Bedford Institute of Oceanography, Darthmouth, Canada
26. Weston DP (1990) Mar Ecol Prog Ser 61:233
27. Wildish DJ, Hargrave BT, Pohle G (2001) ICES J Mar Sci 58:469
28. Cochrane S (1994) Akvaplan-niva report no 631.93.320. Akvaplan-niva, Tromsø, Norway
29. Ritz DA, Lewis ME, Shen M (1989) Mar Biol 103:211
30. Johannesen PJ, Botnen HB, Tvedten ØF (1994) Aqua Fish Manage 25:55
31. Heilskov AC, Holmer M (2001) ICES J Mar Sci 213:427
32. Howe RL, Rees AP, Widdicombe S (2004) JMBA 84:629
33. Kaspar HF, Hall GH, Holland AJ (1988) Aquaculture 70:333
34. Martens CS, Berner RA, Rosenfeld JK (1978) Limnol Oceanogr 23:269
35. Braaten B, Aure J, Ervik A, Boge E (1983) ICES CM 1983/F:26
36. Lumb CM (1989) Mar Pol Bull 20:375
37. Lewis AG, Metaxas A (1991) Aquaculture 99:269
38. Brooks KM, Mahnken CVW (2003) Fish Res 62:295
39. Davies IM, Paul JD (1986) Aquaculture 55:93
40. Yeats PA, Milligan TG, Sutherland TF, Robinson SMC, Smith JA, Lawton P, Levings CD (2005) Lithium normalized zinc and copper concentrations in sediments as measures of trace metal enrichment due to salmon aquaculture (in this volume). Springer, Berlin Heidelberg New York
41. Haya K, Burridge LE, Davies IM, Ervik A (2005) A review and assessment of environmental risk of chemicals used for the treatment of sea lice infestations of cultured salmon (in this volume). Springer, Berlin Heidelberg New York
42. Armstrong SM, Hargrave BT, Haya K (2005) Antibiotic use in finfish aquaculture: modes of action, environmental fate, and microbial resistance (in this volume). Springer, Berlin Heidelberg New York
43. Anon (2001) Statistics Norway, Directorate of Fisheries (www.fiskeridir.no)
44. Smith JN, Yeats P, Milligan T (2005) Geochronologies for fish farm contaminants in sediments from Limekiln Bay in the Bay of Fundy (in this volume). Springer, Berlin Heidelberg New York
45. Robinson SMC, Auffrey LM, Barbeau MA (2005) Far-field impacts of eutrophication on the intertidal zone in the Bay of Fundy with emphasis on the soft-shell clam, Mya arenaria (in this volume). Springer, Berlin Heidelberg New York
46. Petterson K (1986) Report LIU B: 18, University of Uppsala, Sweden
47. GESAMP (1986) Reports and Studies No. 30. FAO, Rome
48. GESAMP (1991) Reports and Studies No. 47. FAO, Italy
49. GESAMP (1996) Reports and Studies No. 57. FAO, Rome

50. Anon (2004) Protocols for Monitoring the Marine Environment, British Columbia, Canada. The Ministry of Water, Land and Air Protection (www.gov.bc.ca)
51. Anon (2000) Norwegian Standard NS-9410. Available from firmapost@pronorm.no
52. Ervik A, Hansen PK, Aure J, Stigebrandt A, Johannsen P, Jahnsen T (1997) Aquaculture 158:85
53. Hansen PK, Ervik A, Schaanning M T, Johannessen P, Aure J, Jahnsen T, Stigebrandt A (2001) Aquaculture 194:75
54. Frogh M, Mohus Å, Sagen T, Sivertsen K, Skreslet S (1984) Nordlandsforskning report no 3 (in Norwegian, English summary). Nordlandsforskning, Bodø, Norway
55. Frogh M, Schaanning MT (1991) Nordlandsforskning, report no 1 (in Norwegian, English summary). Nordlandsforskning, Bodø, Norway
56. Schaanning MT (1991) Jordforsk Report no 212.409-1 (in Norwegian, English summary). Jordforsk, Ås, Norway
57. Dragsund E, Schaanning MT (1993) NIVA/OCEANOR report OCN R-93051 (in Norwegian, English summary). NIVA, Oslo, Norway
58. Schaanning MT (1994) NIVA report no 3102. NIVA, Oslo, Norway
59. Carroll ML, Cochrane S, Fieler R, Velvin R, White P (2003) Aquaculture 226:165
60. Schaanning MT (1998) NIVA report no 3919–98 (in Norwegian, English summary). NIVA, Oslo, Norway
61. ZoBell CE (1946) Bull Am Ass Petrol Geol 30:477
62. Stumm W, Morgan JJ (1981) Aquatic Chemistry, 2nd edn. Wiley, New York
63. Whitfield M (1972) Limnol Oceanogr 17:383
64. Stumm W (1978) Thalassia Jugoslavica 14:197
65. Davis WR, Draxler AFJ, Paul JF, Vitaliano JJ (1998) Environ Monitor Assessment 51:259
66. Boulegue J (1978) Geochim Cosmochim Acta 42:1439
67. Schaanning MT, Hylland K, Eriksen DØ, Bergan TD, Gunnarson JS, Skei J (1997) Mar Pol Bull 33:71
68. Cline ID (1969) Limnol Oceanogr 14:454
69. Wilson BL, Schwarzer RR, Chukwuenye CO (1982) Microchem J 27:558
70. Aller RC (1978) Am J Sci 278:1185
71. Almgren T, Dyrssen D, Elgqvist B, Johansson O (1975) Mar Chem 4:289
72. Millero FJ, Plese T, Fernandez M (1988) Limnol Oceanogr 33:269
73. Grasshoff K, Erhardt M, Kremling K (1983) (eds) Methods of seawater analysis. Verlag Chemie, Weinheim, Germany
74. Bågander LE, Niemistö L (1978) Estuar Coast Mar Sci 6:127
75. Dickson AG, Millero FJ (1987) Deep-Sea Res 34:1733
76. Velvin R (1999) In: Poppe T (ed) Textbook of fish health and fish diseases. Universitetsforlaget, Oslo, Norway, p 340
77. Skjoldal HR, Aure J, Dahl FE, Fredriksen S, Gray JS, Heldal M, Bakke T, Røed LP, Olsen Y, Tangen K, Molvær J (1997) The Norwegian North Sea coastal water. Eutrophication—status and trends. Expert Group Report, Norwegian Pollution Control Authority, Oslo, Norway
78. Wildish DJ, Hargrave BT, Macleod C, Crawford C (2003) J Exp Mar Biol Ecol 285–286:403
79. Wildish DJ, Akagi HM, Hamilton N, Hargrave BT (1999) Can Tech Rep Fish Aquat Sci no 2268. Bedford Institute of Oceanography, Darthmouth, Canada
80. Olsgard F, Somerfield PJ, Carr MR (1998) Mar Ecol Prog Ser 172:25

Environmental Management and the Use of Sentinel Species

James E. Stewart

Ecosystem Research Division, Science Branch, Department of Fisheries & Oceans, Bedford Institute of Oceanography, PO Box 1006, Dartmouth, Nova Scotia, B2Y 4A2, Canada
stewartje@mar.dfo-mpo.gc.ca

1	Introduction	410
1.1	Sharing the Waters	410
2	Coastal Zone Management	412
3	Aquaculture Niche	414
4	Nature of Aquaculture Contributions	416
4.1	Cage Culture—General	416
4.2	Infectious Diseases in Relation to the Environment	418
4.2.1	Infectious Agents	419
4.2.2	Reservoirs and Vectors	421
5	Problems in Assessing Impacts	424
6	Use of Sentinel Species as Non-Specific Biological Indicators of Environmental Quality	425
6.1	Indirect Approach	425
6.1.1	Wild Biota	426
6.1.2	Captive Biota	426
6.2	Direct Approach	427
6.2.1	Early Physiological Indicators	427
6.2.2	Histone-like Proteins (HLPs)	427
6.2.3	Heat Shock Proteins (HSPs) or Molecular Chaperones	428
7	Concluding Remarks	430
	References	431

Abstract To be successful, environmental management must be comprehensive and take into account the activities of all participants who share that environment. In this chapter aquaculture is discussed as an activity in the marine coastal zone that not only shares the waters, but also has a particular need to maintain the quality of those waters. Emphasis is placed upon the need for an effective system of integrated coastal zone management.

Environmental management problems stemming from aquaculture can be minimized by selection of the areas most suitable for culture, applying the best technology and maintaining operations at levels within the assimilative capacity of the area. The contributions of aquaculture to nutrification and pollution are presented and the need to understand their impacts is discussed. Treatment of infectious diseases among the farmed species as an integral part of a comprehensive environmental quality program is advocated. A central problem in environmental management is the lack of proven indicators of environmental quality. As the condition of the biota reflects the environmental conditions under which they live, it is considered possible to measure environmental quality by using captive and wild biota as sentinel species. These measurements can be made indirectly on whole animal or plant responses or directly through measurements of stress as revealed either by changes in blood constituents or on changing concentrations of stress proteins such as anti-microbial histone-like proteins of the skin, or from the array of heat shock proteins now better understood as stress proteins or molecular chaperones.

Keywords Aquaculture niche and environmental contributions · Coastal Zone Management · Molecular chaperones · Sentinel species · Stress proteins

1
Introduction

Although aquaculture is a significant and legitimate competitor worldwide for space in coastal and freshwater areas, it must still demonstrate that it will not jeopardize other uses of coastal, brackish or freshwater zones by causing unacceptable changes to the environment. Obviously, as aquaculturists are among the first to suffer the consequences of environmental deterioration, their concern for the environment should be equal to or greater than that of all parties wishing to preserve it for other purposes; the emphasis must be on the need for all parties to share the waters responsibly.

1.1
Sharing the Waters

Aquaculture is now being pursued in many areas where it did not previously exist. In the marine environment the areas in which these ventures are being located are largely the relatively narrow coastal zones, often areas which contain other significant activities and interests ranging from fishing to recreational uses, industrial, conservation, waste disposal and military purposes to name a few. Very few areas are wholly pristine or can be occupied exclusively by a single user and thus compromises and cooperation will be required. The major and minor uses and participants can be grouped under four general headings (Fig. 1) [1] to make consideration of these various separate interests more apparent.

Additionally, the environmental condition of the coastal zones in many areas is increasingly becoming a matter of concern, as illustrated in summary form by two reports from the Organization for Economic Cooperation and Development (OECD) [2, 3] which state that the coastal zone in most of its member countries and, in fact, in virtually all countries "is under severe

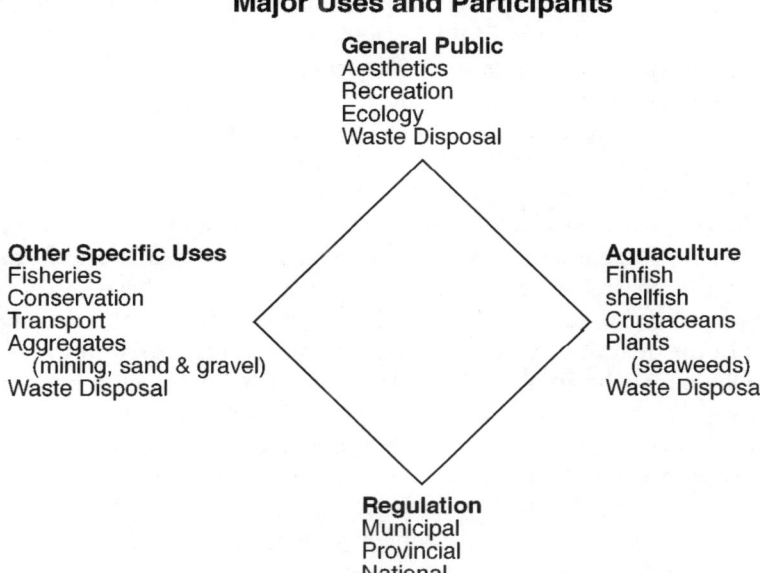

Fig. 1 Major uses and participants in the coastal zone of the marine environment. (Redrawn from Stewart [1])

and increasing pressure from rapid urbanization, pollution, tourism development, and continued development in hazard prone areas. Resource conflicts are increasing". More recently an even more dramatic pronouncement was made concerning the environmental health of large sections of waterways.

After spending more than 2 years studying coastal areas, the Great Lakes and 4.4 million square miles of ocean, an area greater than all of its 50 states, the United States Commission on Ocean Policy announced April 20, 2004 [4] the release of a major Preliminary Report, in which the Commission reported that it had found "among other things, depleted resources, over-exploitation of many fish stocks, lost habitat, decreased resiliency of ocean and coastal ecosystems, and pervasive water contamination problems".

The commission chairman, James Watkins, stated:

"Fundamentally, the message we heard boiled down to this: the oceans and coasts are in trouble and we need to change the way we manage them. Perhaps most important, people must grasp the vital role oceans play in their lives and livelihoods and the profound impact they, themselves, have on the oceans and coasts. What's now obvious is that ocean resources are not limitless. Nor are ocean waters capable of continual self-cleansing."

"We can create an improved policy that balances use with sustainability, is based on sound science and educational excellence, and moves toward an ecosystem-based management approach ..."

As human populations are increasing, along with new activities in coastal regions, the concentrations of wastes and pollutants of all sorts can be expected to rise also. These facts are of importance to aquaculture, which is now being promoted as an alternative and expanding supplier of fish and shellfish as the capture fisheries decline. As aquaculture increases, more and more of the fishing in the capture fisheries will be prosecuted on so-called underutilized species to provide food for fish being reared in captivity. Thus much of the food web whose elements formerly were distributed over a considerable portion of open ocean areas will be compressed into relatively narrow strips of the coastal zone environment. Coupled with these increases on behalf of rearing predators close to the top of the food web, are large increases in the quantities of grazing shellfish also being reared in the inshore areas. Obviously, the combined problems posed by increased human activities, new industrial uses including concentrated aquaculture ventures and the traditional uses related to fisheries, recreational and conservation needs will require a sophisticated and highly integrated management approach encompassing whole coastal areas. Unfortunately, although we understand clearly that the solution to this overall problem is subsumed within the label integrated coastal zone management, we seem to be at a loss when it comes to organizing effective approaches and have not yet moved far enough in developing the technical methodology, understanding or will to enable us to achieve it in most places. Many of the problems preventing or slowing the establishment of coastal zone management stem from the difficulties inherent in competing jurisdictions (national, provincial and municipal) and the competitions resulting from the concerns and ambitions involved in current and new uses. Aside from these rather monumental problems are the highly technical problems of methodology i.e., measuring and evaluating current environmental quality and predicting with confidence the consequences to it of increased loading of all kinds and the need to ensure that the total impact fits within the definition of ecologically sustainable development. The problem for aquaculturists is one of balance: how to achieve economic success while operating in harmony with the neighbors and the environment.

2
Coastal Zone Management

By OECD [2, 3] definition "The coastal zone conveys the notion of a land–sea interface. This interface has two axes—one axis is parallel to the shore (longshore), and the other axis is perpendicular to the shore (on/offshore)". There is relatively little controversy concerning the definition of the longshore axis since it does not typically cross environmental systems boundaries, with the exception of watershed and major headlands. There is, in contrast, considerable discussion about the on/off-shore axis.

"For example, definitions of the inland 'boundary' of the coastal zone range from those that include entire watersheds to those restricted to the immediate strip of shoreline adjacent to the coast. The seaward limit can extend as far as the maximum reach of a country's jurisdiction (i.e., the 200-nautical-mile limit). The range of definitions raises unresolved difficulties between those who prefer to use an ecosystem-based boundary and those who are more comfortable with a legal/administrative/economic boundary consistent with governance jurisdictions. The latter boundary may or may not relate to the ecosystem boundaries."

OECD also recommends that coastal zone management should be based upon a broad ecosystem approach. This implies that the on/off-shore axis should cover the relevant parts of the hinterland including the associated aquatic ecosystems and tributary sections up to the spawning areas of migratory fish or the historic head of tidal influence, whichever is higher. Thus integrated coastal zone management consists of management of the coastal zone as a whole in relation to local, regional, national, and international goals. It must focus on the interactions between the various activities and resource demands that occur within the coastal zone, and between coastal zone activities and activities in other regions.

Instances of successful integrated coastal zone management approaches are few in number. This is partly attributable to a lack of an holistic approach to the environment, there being only a sectoral one. The scientific tools to assess environmental issues using a comprehensive integrated approach still needs to be developed, and insufficient emphasis is placed on an ecosystem's functional and dynamic aspects; pollution control strategy needs to be proactive rather than reactive.

In sum, the statements from the various reports and other experiences suggest that the list of underlying reasons for the almost universal failure to implement coastal zone management, although not exhaustive, would encompass one or more of the following:

1. General
   (a) Lack of recognition of the potential benefits of coastal zone management.
   (b) Lack of clearly stated community objectives and definition of quantified goals for the coastal zone as a definable entity.
   (c) Economic benefits of development are considered in isolation and fail to properly evaluate the environmental issues and values.
   (d) Lack of understanding of the impacts of different activities, e.g., living resource management, transportation, industrial, recreational, aquaculture, etc.
   (e) Failure to resolve issues of property rights and physical access in the coastal zone.

(f) Failure to break down into manageable sizes and tasks the required components for coastal zone management.
(g) Lack of coordinating bodies, institutional arrangements, and resultant regulatory systems.
(h) Unclear mandates for the several jurisdictions involved (national, provincial/state/prefectural/county, and municipal) with consequent competition conflicts and or/lack of cooperation.

2. Science-related

(a) Lack of the fundamental resource and environmental data bases (inventory) organized appropriately for the coastal zones. In many instances basic information has not been collected or is very uneven in quantity, quality, and coverage; in addition existing information is frequently difficult to acquire or retrieve.
(b) Lack of well defined and proven bases for management using broad ecosystem approaches and definitions of ecosystem management units.
(c) Lack of well chosen, efficient, and realistic indicators for gauging the health of ecosystems and the validity of the management systems being considered or applied.

Obviously, any aquaculture venture within the coastal zone will be influenced by the environmental features both naturally provided and anthropogenically generated. Thus to manage aquaculture effectively it must be treated as an integral component of the coastal zone and must, along with all of the other contributors, be within the bounds of ecologically sustainable development.

The first step in instituting a comprehensive coastal zone management system will entail collection and organization of oceanographic and bathymetric data in a comprehensive computerized user-friendly data-base along lines pioneered in the pragmatic, but flexible Norwegian LENKA system [5]. This system was designed to supply the basic information to serve as a tool for Norwegian community planning for coastal zone management, including watershed areas. As outlined by Stewart [1], such a database could be organized for the Maritime Provinces of Atlantic Canada quite readily, as much data has been assembled already. The suggestion made was to organize the data on a county by county zonal basis with amalgamation occurring at the provincial levels thereby giving a comprehensive database for each province. Once this has been accomplished, it should be possible to devise suitable management plans in which aquaculture assumes a legitimate and logical position.

3
Aquaculture Niche

In addition to these considerations for the aquaculturists, it is logical to suggest that many problems for both aquaculturists and their competitors could be alleviated by judicious selection of appropriate sites. While this advice may appear at first glance to be obvious and gratuitous, unfortunately it does not form the prime or paramount consideration. In the recent past, many of the failures in aquaculture operations are directly traceable to choosing a site which either initially or eventually was found to be unsuitable for that particular culture venture or the techniques were poorly matched to the site. The shortcomings will always make themselves apparent. They can consist in the case of shellfish culture, even when the physical conditions are otherwise adequate, in a lack of phytoplankton qualitatively and quantitatively sufficient to sustain the operation, or inadequate flushing which in turn ultimately results in senescence of these particular grounds giving rise to by-products which negatively affect growth and product quality. In the case of finfish, often the prime reason for choosing a particular location may be low cost, shelter, convenience and freedom from ice or lethally cold temperatures. Factors such as inadequate flushing may be the direct result of such narrowly based choices and have been coupled with the build-up of troublesome algae such as *Heterosigma* and *Chaetoceros* sp, oxygen levels which are too low to sustain the fish in good condition during peak growth periods and the non-removal of wastes from the area.

More emphasis needs to be placed on minimizing both aquaculture problems and negative environmental contributions by basing the choice on the suitability of the site for culture of the target species. There is sufficient data on the requirements of various possible aquaculture species (finfish and shellfish) to specify what the minimum conditions must be before a tentative site choice should be considered worth exploring further. For example, for salmonids the lower lethal temperature limit is $-0.7\,°C$ and the upper temperature limit is around $22\,°C$ depending upon species; oxygen and current velocities also need to be considered. The examination could take the form outlined below [6]:

Minimal environmental conditions and reasonable margins of safety for salmonid culture in marine areas of eastern Canada

1. Pre-culture survey: redox potential (Eh) and sulfide determinations on and physical analysis of bottom sediments to ensure the area is not marginal with regard to flushing.
2. Salinity: relatively constant, preferably remaining between 28 and 33‰ throughout the culture period.
3. Temperatures:
   (a) In the cold weather temperatures should always exceed at least $-0.5\,°C$ and preferably not drop below $-0.2\,°C$ or $-0.3\,°C$ during the culture period
   (b) In the warm weather, temperatures should not exceed $18\,°C$
4. Current velocities: must not drop below $3\,cm\,s^{-1}$ for periods greater than 15 min based on a minimum survey period of 29 days

5. Oxygen: where criteria 1–4 are met, local limits could be set in relation to oxygen turnover in the bay or inlet (based upon current and mixing data). The total respiratory load must be limited to that which would not cause oxygen levels of $7\,\mathrm{mg\,l^{-1}}$ to drop to levels less than $6\,\mathrm{mg\,l^{-1}}$ at any time during the culture period.

This approach is simple and if carried out as a prior assessment would gauge the minimum suitability of a site and eliminate many of the environmental issues currently experienced. A more sophisticated development of this basic approach has now been made available by Doucette and Hargrave [7].

Thus in any coastal area it is essential to develop and apply a practical coastal zone management system in which all users, current and potential, are considered as integral partners and required to play their part in ensuring ecologically sustainable development. From an aquaculture point of view it is also essential that the correct species be matched to that part of the zone that gives the greatest opportunities for a particular venture to succeed and that the appropriate husbandry is acquired and applied.

4
Nature of Aquaculture Contributions

4.1
Cage Culture—General

At this stage it is useful to describe the nature of some of the contributions to the environment arising from aquaculture. As an important part of northern aquaculture consists of cage-rearing of finfish, the example chosen is the culture of Atlantic salmon, *Salmo salar*, as practiced in eastern Canada [8]. Smolts $1 - 1\frac{1}{2}$ years old are transferred to the marine environment and held in large netpens, most enclosing $500-1500\,\mathrm{m^3}$ of seawater. Current flow and tidal exchanges are utilized to supply clean, oxygenated water and flush out deoxygenated water and soluble waste products. Much of the food is in the form of a complete diet of dry pellets (9% moisture) distributed to the netpen waters; uneaten food, generally estimated to be about 3% of the total fed, will pass through the nets and sink to the bottom. As these diets are highly digestible most of the food is converted into fish, but approximately 20% of the food fed will be excreted in the form of soluble and solid wastes. Part of these wastes plus the surplus foods will sink, accumulating in mounds under the cages unless the current flow is great enough to transport it elsewhere. This sequence is illustrated in Fig. 2 for the grow-out period of $1\frac{1}{2}-2$ years that produces organic wastes of approximately 269 tonnes of waste for every 1000 tonnes of wet fish produced, around 22.4% of the food fed.

During the course of this production cycle other agents besides feces and surplus food will be contributed as the occasion demands, i.e., antibiotics for treating diseases, chemotherapeutants for alleviating sea lice problems, and anti-fouling agents for the nets, among others. Depending upon circumstances and their nature these materials can accumulate under the cages or be transported short or long distances, depending upon current flow, and deposited elsewhere. These wastes and other contributions are variously affected by microbiological agents; some are degraded fairly rapidly and completely, contributing to the concentrations of nitrogen and carbon in the waters of the area. Others will accumulate in sediments to considerable thicknesses (1 m or more) where conditions will become anaerobic and the sediment materials will degrade very slowly i.e., at less than the accumulation rates and will emit gases in approximately the following ratios: 70% methane, 2% hydrogen sulfide and 30% carbon dioxide [9]. As these gases break free of the sediments they mobilize materials from the sediments (gas ebullition) including antibiotics and distribute these to the water column for dispersion or to the surface where the volatile components will enter the atmosphere. The noxious gases traversing the fish cages will have a negative effect on the fish.

Currently the few antibiotics approved for use in Canadian finfish rearing include oxytetracycline, the potentiated sulfonamides (Romet and Tribressen) and Florfenicol. Therapeutants and chemical control agents in marine culture operations, the admixture of fecal materials and surplus foods

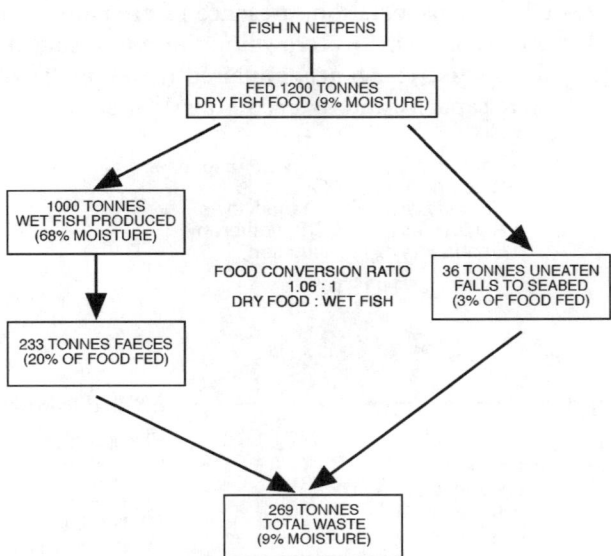

Fig. 2 Food conversion in a typical New Brunswick Atlantic salmon farm. (Re-drawn from Stewart [8])

and noxious gases (gas ebullition) are factors of prime and direct environmental significance here and elsewhere [10–14]. Collectively these contributions result in nutrification and pollution of the water columns and sediments. Any analysis of the impacts must consider at least three spatial scales: (1) internal impacts i.e., fish in the cages, (2) local impacts (effects in the immediate vicinity of the cages often termed near-field effects), and (3) regional impacts (effects on a whole bay, inlet or larger management area, often termed far-field effects).

4.2
Infectious Diseases in Relation to the Environment

As nutrification and pollution increase, stresses are imposed upon the biological systems reducing their physiological condition, impairing growth and adaptability and significantly impairing their capacity to resist infectious or contagious diseases. The degree to which this occurs is proportionate to the insult(s) offered. Chronic infections, impaired growth and increasing mortalities among the cultured finfish are overt, gross indications of the degree of stress imposed. Infections among fish or any other animal do not occur simply or inevitably because a host has collided with an infectious agent; the development of an overt disease is a complex product of the interaction involving the host (fish), pathogens and the environment (Fig. 3) [15].

If the fish is a healthy vigorous specimen kept under good to excellent conditions it will usually resist infection by opportunistic pathogens. If however, it has been stressed by transportation, presence of predators, poor diet, adverse temperatures or salinity or overcrowding this will result in a markedly reduced capacity for disease resistance, impaired development of immunity, impaired reproductive capacity and reduced growth rates [16–21]. The elem-

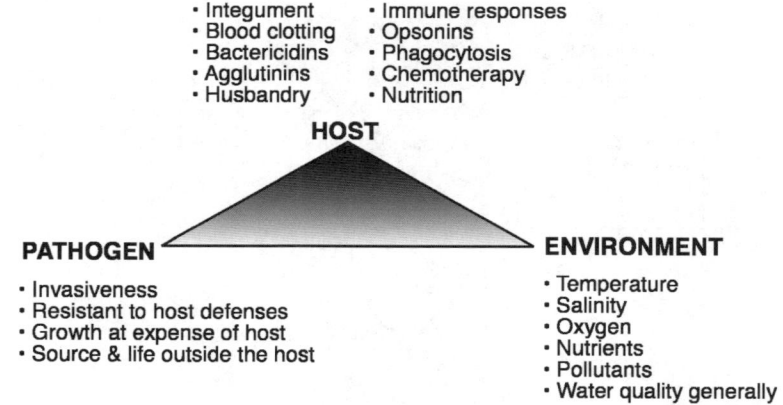

Fig. 3 Host/pathogen/environment characteristics, the interactions of which determine the outcome with regard to infectious diseases. (Re-drawn from Stewart [15])

ents contributing to stress and thus immunosuppression are considered so dominant that Smith [22] in his discussion of furunculosis argued that the infectious agent itself, although essential, is actually a minor player in the overall infection. In this view, the major reasons for the disease are the precipitating factors giving rise to stress and these should receive at least as much attention as specific infective agents, emphasizing that there is no substitute for good husbandry in reducing occurrences of and the impacts of infectious diseases. An immunosuppressed animal is extremely susceptible to all infectious agents and requires a considerable time to recover its various immune and resistance capabilities even after removal of the causes of the stress, i.e., a month or more; growth is impaired for even longer periods.

4.2.1
Infectious Agents

Infectious diseases do not arise spontaneously, but are initiated from various sources and conditions that are environmentally connected and can also be features in which aquaculture practices are important. To confine the discussion to salmonids, mainly the Atlantic salmon, the infectious agents causing diseases include bacteria, viruses, and larger parasites [23]. Among the most troublesome locally have been the bacteria *Vibrio anguillarum* (vibriosis), *Vibrio salmonicida* (causing the condition originally called Hitra disease, now more commonly referred to as coldwater vibriosis), *Aeromonas salmonicida* (furunculosis) and *Renibacterium salmoninarum* [bacterial kidney disease (BKD)] and the viruses causing infectious pancreatic necrosis (IPN) and infectious salmon anaemia (ISA). Among the larger parasites, the sea lice currently are the most important with *Lepeophtheirus salmonis* as the worst scourge of the salmon, but with *Caligus elongatus* in the wings identified as a parasite on 73 species of marine fish, including salmon.

The vibrios are ubiquitous in the marine environment and are not considered a problem for fish in peak condition. When the fish are stressed, however, they become susceptible to infection and succumb in large numbers. The causative agent of cold water vibriosis, *V. salmonicida*, is usually associated with fish farms and is frequently found in the sediments under fish farms, which may act as a reservoir for this bacterium [24]. *A. salmonicida*, the bacterium causing furunculosis, is passed from fish to fish in both freshwater and marine environments. Again, mortalities can be high where environmental conditions are poor and the fish are stressed; in other cases mortalities may be relatively low, but the smolts perform poorly i.e., they grow slowly. The disease agent can be transmitted from fish to fish, from farm equipment and vectors such as sea birds and blood sucking parasites e.g., sea lice [25]. Farm sediments and marine plankton are additional reservoirs and vectors [26, 27]. Jarp et al. [28] concluded the main risk factors for infection with *A. salmonicida* in hatcheries were (1) migration of anadromous fish into the freshwater supply, (2) sharing personnel with other fish farms and (3) a high concentration of fish farms infected with furunculosis near the hatcheries. Much of the problem in the seawater farms was attributed to transfer of infectious agents by humans moving between farms.

BKD is caused by R. *salmoninarum* and can be spread in two different ways: through the female to her progeny via the eggs (vertical transmission) and through shedding of the agent by an infected fish where it is transmitted to susceptible fish through the water (horizontal transmission). The bacterium is slow-growing in infected fish and as a consequence treatment with antibiotics will usually be ineffective. Death of the infected fish can result in a few months, or years later when a latent infection is activated by adverse environmental conditions, secondary infections or reproductive stress. As antibiotics are essentially ineffective, broodstock and egg cleanup previously were the only effective avenues to eliminating this pathogen. Recently, however, a live, heterologous, injectable vaccine based upon an *Arthrobacter* sp "that shares common antigenic determinants with R. *salmoninarum*" has been licensed for general sale as an aid in prevention of BKD in healthy salmonid fish [29].

Infectious pancreatic necrosis virus (IPNV) causes economically important infections resulting in pathology of the pancreas and gut of freshwater salmonids. Many of the smolts transferred from the hatcheries have been latent carriers that survived IPNV infections in the hatchery; when transferred to sea water these fish suffered substantial mortalities [30]. The virus is transmitted vertically and horizontally. The Norwegians reduced much of the impact of this virus by attempts to rid the broodstock of the virus. Among Atlantic salmon post-smolts an IPNV carrier state at low IPNV titers did not influence the mortality rates after secondary infections [31]. At higher IPNV titers, however, significantly more of the Atlantic salmon died when these carrier fish were infected with *V. salmonicida* than when fish were infected with *V. salmonicida* alone. More recently IPN has been identified in Scotland as an emerging problem for seawater-cultured salmonids [32]. The prevalence of IPN in Scottish post-smolt Atlantic salmon has increased from 1.2% in 1990 to 12.5% to date, with mortalities increasing proportionately. Adding to the hazards and concerns for aquaculturists is the fact that other finfish candidates for intensive aquaculture (turbot, Atlantic halibut and Atlantic cod) have been shown to be susceptible to IPNV [33, 34].

Infectious salmon anaemia virus (ISAV) infects salmonids and has caused very serious losses in Norway, eastern Canada and Scotland through impaired growth, high mortalities and the need to slaughter and remove fish in attempts to deal with the epizootic [35–38]. The ISA virus appears to be propagated continuously in infected fish and is present in their blood, skin mucus, urine, feces and other tissues; it is shed continuously even by seemingly healthy, but infected fish. The virus contributed to the water column through slaughter wastes (blood, viscera and other tissues) shed by infected

Atlantic salmon or possibly by others (sea run brown or rainbow trout), is water-borne and can be transmitted directly, through cohabitation or vectors such as sea lice and phytoplankton. The gills are considered the most likely tissue for passive transmission of ISA. Dealing with epizootics of ISA means rigorously controlled slaughter in appropriate facilities and retention and treatment of all waste (blood, viscera, mortalities, process water, etc.)

Reservoirs of infectious agents for salmonids can actually be their own juvenile freshwater stages and could include from among the agents listed above, IPNV, ISAV [37], *R. salmoninarum* (BKD) and *A. salmonicida* (furunculosis). Thus it is important to acquire smolts only from hatcheries shown to be free of these agents. In addition, while on the topic of hatchery stock, it is wise to ensure that where vaccines are available the smolts obtained have been vaccinated e.g., against vibriosis (including coldwater vibriosis), furunculosis and now BKD.

Unlike the viruses and bacteria, sea lice are large, copepod parasites with complex life cycles. The more serious of the two species causing heavy losses to growers in Atlantic salmon is *L. salmonis*. Its life cycle is composed of two non-parasitic stages followed by eight obligately parasitic stages. When the parasite reaches the third stage, the free swimming, infectious copepodid stage, it must attach to a host within a few days or it will die. Once attached to a host it progresses through the remaining seven stages [39]. These parasites damage the integument of the fish and cause local hemorrhaging and necrosis. Mortalities can result from osmotic shock or from infections occurring as a result of integumental damage. The lice can move from fish to fish during their parasitic stages. The heavy infestations are a direct function of the biomass of the host species (farmed fish), the time the biomass is maintained and the duration of the occupation of the site. Treatments consist of chemotherapeutant agents applied externally (baths), cleaner fish and fallowing. Dependent upon ambient temperatures, fallowing for periods ranging between 3 weeks and about 2 months will break the parasitic life cycle by removing the host, the element essential to the parasite's existence. This approach has proved to be effective for periods of 8 months or more especially when combined with single-year class stocking and bay or inlet management regimes [36, 40, 41]. As sea lice are crustaceans, the use of chemotherapeutants to deal with them also must take into account the tolerance to these chemical agents of the local crustaceans that form a significant part of the wild biota in the area. A good review of current progress on some of the major issues in sea lice research appeared recently [42].

4.2.2
Reservoirs and Vectors

The known reservoirs and vectors for the array of microbial pathogens and large parasites are varied, often unobtrusive or poorly understood. Nonetheless, they do constitute threats to the farmed species that, to large degrees, can be avoided or reduced through site and area management procedures. These procedures can be as straightforward as the careful selection of seed stock, site and area housekeeping, application of sound husbandry and comprehensive site health management regimes. As farms do not usually exist in isolation or are not separated by distances sufficient to guard against cross-contamination, it is essential that all of the farmers in a defined management area agree and cooperate on approaches to be instituted in that area. In the words of the Scottish fish health expert, A. McVicar, all the farms that share the waters will also be at risk and stand a good chance of sharing the disease i.e., "Share the waters, share the disease".

Proven reservoirs and vectors for the transfer of ISA [25] and *A. salmonicida* [27] to salmonids include the sea louse, *L. salmonis*. In addition, *L. salmonis* can be carried from site to site outside the cages and thus could spread the ISAV and *A. salmonicida* widely in an area [43]. Added to these problems it has been shown that marine phytoplankton could transport *A. salmonicida* around an area [27].

McAllister & Owens [44] stated that of all the piscivorous predators, birds are considered to be the worst as "they can occur in great numbers, be voracious consumers of fish, be in residence throughout the year, and are often protected by local and/or national legislation". Unfortunately, birds not only prey on fish, but they can also spread fish disease agents. This was illustrated by a series of experiments that concluded that the heron, *Ardea cinerea*, was able to act as a mechanical vector for IPN, and viral hemorrhagic septicemia (VHS) for trout and spring viraemia of carp [45]. The transmission of these viruses occurred via feces and regurgitated food. McAllister & Owens [44] reported the detection of IPN in feces from herons, mallard ducks and other birds in concentrations similar to those found in the hatchery fish. It was shown that mallard ducks, *Anas platyrhincus*, and a black crested night heron, *Nycticorax nycticorax* fed trout infected with *Myxosoma cerebralis*, the causative agent of salmonid whirling disease, passed spores of the parasite in their feces [46]. Upon maturation these spores showed they were still viable by infecting the susceptible host, rainbow trout. It has been shown that the list of vectors (aquatic invertebrates, farmed and wild fish, and some terrestrial mammals) for the bacterial pathogen, *Yersinia ruckeri*, causing enteric redmouth disease of salmonids should also include the greater black-backed gull (*Larus marinus*) and the herring gull (*Larus argentatus*) [47]. This pathogen can live in and survive passage through the intestines of the gulls.

As this selection of disease agents can have devastating effects upon the stocks of fish being cultured (particularly when they are stressed), it is obvious that care taken to guard against their introduction to the site or the neighbors' sites will yield major dividends. When the nature of the transmissions are considered the investment in housekeeping, to ensure that the fish loading at the site will not be sufficient to generate deposits under the cages which can emit noxious gases and/or serve as reservoirs for fish pathogens, is worthwhile. Control over circumstances that sustain vectors such as a rich supply of water-borne nutrients (nitrogen, carbon, phosphorus compounds etc., enhancing phytoplankton), or prolonged tolerance of heavy infestations of sea lice, or failure to ensure that waste materials (blood, viscera or fish carcasses) are not available to marauding animals including birds are essential. Such measures will all aid in preventing or slowing the initiation and fostering of severe and expensive epizootics.

Aquaculture operations must be treated holistically as integrated food production systems ranging from the egg to the fillet. Thus, although health management regimes are not usually considered in environmental discussions it is clear that they are crucial and central to all of the environmental issues and fundamental to economic success in aquaculture ventures. The disease aspects are tied intimately to the environmental choices and decisions made, and usually are direct consequences of them.

In this regard quotations from the Final Report of the Joint Government/Industry Working Group on Infectious Salmon Anaemia (ISA) in Scotland [37] offering highly relevant advice on an important aspect of this topic are worth reproducing:

"Hydrographically defined Management Areas for the Scottish aquaculture industry have their origins in earlier strategies for controlling furunculosis and sea lice. They are applicable to the whole Scottish industry, can take into account specific local conditions, and form the basis for tackling the present ISA outbreak. It is recommended that they be adopted for dealing with any water-borne disease and for a planned approach to managing the industry. Clearly they must be scrutinized on a case-by-case basis to take into account local conditions and the planned occupancy of farms, in order to avoid the siting of farms in 'fire breaks' where they might bridge adjacent areas."

"The expansion of the salmon farming industry and limited availability of prime sites meant that new farms were established in proximity to their neighbours, resulting in shared disease risk. Management agreements evolved so that coordination of stocking, fallowing and disease control measures could be achieved, thus minimizing the mutual threats to health status that neighbouring farms might otherwise represent."

"It is recommended that farmers use the hydrographically defined Management Areas as the basis for determining the biological areas subject to management agreements."

"Fallowing, by removing the host biomass for a period, reduces infection pressure and is routinely used in Scotland for controlling and minimizing bacterial diseases such as furunculosis and as a response to the present outbreak of ISA. There are other benefits such as benthic recovery, and a fallow period is sometimes a mandatory requirement of a discharge consent issued by the Scottish Environment Protection Agency (SEPA). Site rotation allows fallow periods to be extended whilst maintaining production."

5
Problems in Assessing Impacts

In a perfect world all measures to avoid species/environment mismatches and overloading would follow logically in order. In reality, further measures are required to ensure that the environmental quality is maintained at acceptable levels. It is apparent from the foregoing that many environmental problems can be corrected or avoided through development and application of a comprehensive coastal zone management system and that specific aquaculture problems can be minimized by careful selection of the appropriate species matched to the environmental conditions which best suit their physiology and growth potential. The question becomes one of how best to achieve this goal as we currently lack sound methodology for comprehensive assessment of environmental quality sufficient to provide for coastal zone management or even for the narrower problem of aquaculture management and assistance to the aquaculture operator.

Quantitative studies on the sources, fates and effects on target organisms of materials originating from aquaculture and other operations such as sewage and industrial wastes of many different kinds are proceeding. When the extensive list of current and potential materials (such as carbon, nitrogen, phosphorus and their derivatives, plus metals, polychlorinated biphenyls, dioxins, furans, chemotherapeutants etc.) is considered and the degree to which these accumulate, interact, re-mobilize and are altered, combined and recombined before being distributed or sequestered is considered, it is clear that chemical measuring and monitoring of pollutants is an immense task. Further, as we are unable to predict with confidence how the foregoing multiple steps will play out in detail and thus the form in which the pollutant or substance will be presented, the problem becomes more complex. Added to these complications are the unresolved problems of the nature of the impacts these various contributions from the different sources would have upon the biota, cultured and wild. To date, we are able to measure chemically many of the pollutants at one stage or another in their progress and through various interactions with each other. We do not, however, have good data for most as to what effects, in the various forms, they will have on the biota wild or cultured, at what concentrations these effects will be apparent and to what degree synergistic effects will occur. Unless and until we can interpret the chemical data on surplus nutrients and pollutants in terms of their impacts on the biota we will be unable to devise and apply appropriate control measures to match and offset their negative effects on the biota.

The central aquaculture environmental problem is one of gauging assimilative capacity of waterways for their suitability for aquaculture ventures and ensuring development of realistic indicators that show aquaculture loading and contributions from other sources remain within that capacity. As the methodology to assess the consequences of various loadings or predict the limits of the assimilative capacity of various culture operations at different locations does not yet exist, it is necessary to consider what approaches could be useful in the interim and probably in the long term.

6
Use of Sentinel Species as Non-Specific Biological Indicators of Environmental Quality

As the condition of all biota, animals and plants, captive and wild are directly reflective of the conditions under which they are forced to live they will be effective indicators of environmental quality. This is illustrated in quite general terms by poor growth and increasing mortality rates that are the direct consequences of physiological changes in the animals imposed collectively by stress. Stress, as described earlier can be induced by poor transportation, presence of predators, poor diet, pollutants, adverse temperatures or salinity or overcrowding. Factors such as these result in markedly reduced capacity for disease resistance, impaired development of immunity or immunosuppression, impaired reproductive capacity and reduced growth rates. Recovery is slow i.e., if all negative influences are removed it takes at least a month to recover from immunosuppression stress and even longer for growth rates to recover. Although stress is a non-specific response to adverse conditions it is measurable indirectly by whole animal responses or directly using a number of either chemical or biochemical techniques applied to samples obtained from exposed biota.

The havoc and devastation wreaked by major events such as hurricanes, heavy storms, unusual extremes of heat, cold or salinity changes or overwhelming chemical spills etc. are so immediately apparent that they do not require detailed consideration in relation to environmental quality. The concern, especially in relation to aquaculture, is with subtler effects that for the most part are sub-lethal in the short term. In the longer term, however, they do affect growth and survival significantly. As the biota of a region are integrators of environmental quality they are excellent non-specific monitors of environmental quality. The use of the biota in this way can be usefully considered under two main headings: the indirect and direct approaches.

6.1
Indirect Approach

The indirect approach, using whole animal and plant responses for both captive and wild biota, relies upon methodology that is relatively coarse and long term. Indicators of poor environmental quality are poor growth, chronic disease and increased mortality with many instances of increased micro-algal growth. These are the direct consequences of long and short term increases in nutrients and pollutants [15]. From an environmental point of view these indicators can be applied to whole management areas as well as to individual enterprises, industrial activities and aquaculture units.

6.1.1
Wild Biota

Major changes in biota for a large area such as a major bay or inlet following the location of an industrial activity in it are always much easier to detect if a thorough biological survey is conducted prior to the initiation of the industrial activity. Such changes can include the disappearance of both sessile and mobile species that formerly were abundant, a major decline in population numbers of previously common species, or the change to a whole new array of species. For detection of these changes it is necessary to carry out rather detailed surveys on the order of every second or third year for comparison with the initial and previous surveys.

6.1.2
Captive Biota

As inventories of the captive biota need to be kept for farm management purposes or regulatory measurements of the environment the information will be more complete and timely; any changes in health and well being will be more obvious. Again, indicators of interest will be growth, chronic disease, survival and increasing and changed algal growth.

Thus captive biota, either in farms or placed in an area by regulatory Agencies, can be useful as "mine canaries" to gauge the environmental quality. To be most useful, any approach along these lines should be calibrated at an early date by acquisition of standardized farm performance data to provide a reference base for the species and area. This should take the form of growth, survival and health curves for each species determined by several different operators for an area during the first years of operation in a given location. If amalgamated with data from other areas, the basis for assessment for any one area will be that much more reliable. These data could take the form of standardized computer records and simple models for analyzing the information. It would also aid the farmers in determining whether the problems experienced stemmed from location, environmental quality, or in certain instances, were a result of the application of poor husbandry.

6.2
Direct Approach

The direct approach using either or both wild and captive biota does not ignore the general signs of distress noted in the indirect approach, but shortens the time for diagnosis by making chemical or biochemical measurements of stress indicators on individual specimens. Stress as described earlier is a major factor predisposing animals, including fish, to infectious disease through immunosuppression and also has a serious negative impact on growth and survival.

6.2.1
Early Physiological Indicators

Stress has been measured in fish by using physiological indicators such as cortisol, whole blood clotting times, osmoregulatory dysfunctions, hematocrits and leucocrits and measurements of various other plasma constituents. Unfortunately, some of the procedures are difficult to carry out, require repeated sampling, or measure mainly acute, but not chronic stress. Certain of the methods, however, do have utility in assessing stress [16, 20].

6.2.2
Histone-like Proteins (HLPs)

More recently, Robinette & Noga [48] described a group of potent, broad spectrum antimicrobial polypeptides, histone-like proteins (HLPs), from the skin of the channel catfish, *Ictalurus punctatus*, that also function as effective indicators of chronic stress. These broad-spectrum anti-microbial polypeptides are important non-specific immune responses that act as first-line protection against infectious disease in fish. In their work with the catfish the authors noted that stress associated with transport from the farm to the laboratory, followed by overcrowding in aquaria with inadequate biological filtration, depressed the levels of skin-associated anti-bacterial activity and the levels of HLPs. Using an enzyme-linked immunosorbent assay (ELISA) they showed that the main concentration of the HLP in catfish, subjected to prolonged stress of overcrowding and elevated ammonia, was time-dependent and declined gradually over the lengthy exposure period (weeks) that the stressful conditions were applied. Particularly attractive features of this method was that it was simple and quantitative with clear results and, importantly, that the levels of the main HLP were not affected by the acute stressors of capture or sampling suggesting to the authors that HLP levels may be a promising indicator for monitoring health among fish species of interest.

6.2.3
Heat Shock Proteins (HSPs) or Molecular Chaperones

Cells exposed to an acute physiological insult frequently respond by producing an array of proteins that protect against the insult [49]. Because they were observed first as responses to thermal stress they were referred to initially as heat shock proteins (HSPs). Subsequently, it was discovered that HSPs are induced by a variety of stressors and are now commonly referred to as stress proteins. They have been found in all species examined from microbes to humans. A considerable literature relating to HSPs and stress proteins generally has developed as the responses to a wide range of inducers has been determined and their role in protecting vital cellular functions has been recognized. The HSPs are a particular category of stress proteins and, in fact, have also been described as molecular chaperones for their role in critical cellular functions, restoring structure and function to reversibly denatured proteins including guiding protein folding, transport and hormone receptor fidelity. They are constitutively produced within the cells, but increase in concentration rapidly upon induction; the underlying signal appears to be conditions which denature intracellular proteins. These conditions include thermal stress, hypoxia, exposure to toxins such as ethanol, heavy metals, osmotic stress and other contaminants. Much of the information relevant to an understanding of this field and its relation to environmental aspects has been reviewed by Iwama et al. [50, 51] and their functions as molecular chaperones by Hartl [52].

"The potential application of HSPs as biomarkers continues to be an important subject of research in environmental physiology. Their universality, extraordinary conservation in structure, and the consistency with which they are induced by a broad spectrum of stressors have made HSPs good candidates for biomonitoring of the environment" [50]. The use of antibodies produced against conserved domains of inducible members of HSPs appears promising in evaluating the HSP levels in a variety of organisms.

A number of investigators have begun evaluations of the HSPs as biomarkers of environmental quality. A sampling of these includes exposure of fathead minnows to environmentally realistic concentrations of arsenite, chromate, lindane and diazonin that found HSP stress protein responses in brain, gill and striated muscle tissues [53]. The authors suggested that the stress protein responses were indicative of both general stress and specific chemical exposure. DuBeau et al. [49] examined the relationship between HSP 70 (the individual HSPs are referred to by size in kilodaltons) and osmotic shock and showed that HSP 70 increased as a result and protected Atlantic salmon from osmotic shock. Fish skin was investigated as a potential indicator organ. HSP 70 was higher in the epidermis of brown trout following heat shock, but it was advocated that other stress proteins and biomarkers should be investigated for their importance in environmental studies that usually comprise several impacts [54]. The accumulation of HSP 60 and HSP 70 in banked samples of *Mytilus edulis* and fish muscle tissue exposed to contaminants in their environment was examined [55]. The authors found that in comparison with laboratory controls the stress proteins were elevated and that the body burdens of the organisms sampled showed they had been exposed for long periods in their environment. Brown trout and stone loach were exposed to complex stressors (mixtures of environmental pollutants) in laboratory, semi-field (aquaria connected to stream water) and field experiments [56]. It was concluded that although the rather complex kinetics of hepatic HSP 70 elevation and decrease needed to be taken into consideration, the study demonstrated the suitability of HSP 70 stress protein levels to integrate the response dynamics of several different stressors and to function effectively as a biomarker for the integrated effect of all environmental stressors acting on an organism.

The encouraging note in these studies is that all of the conclusions tend to concur with the prospect that stress proteins, in particular the HSPs, can be valuable as biomarkers of environmental quality. Obviously, more information is needed about the functional relationship between HSP expression and the physiological stress response [50]. The information required includes dose–response relationships, time course of the response, species specificity and the effect of the developmental stage among others to assess and implement the use of HSPs as biomarkers of environmental quality and population changes [50]. This last suggestion is echoed by the studies of Culp et al. [57] who advocate integrated assessment strategy approaches to eco-toxicology in which the effects of pollutants and other stressors on several different levels of biological organization are used to give an improved understanding of causality in stressor–response relationships.

As the measurement of HSPs is currently relatively elaborate, complex and expensive, the development and use of antibody-based systems similar to the ELISA system employed for the determination of HLPs in fish skins [48] would make the measurements routine [50]. A further simplification is the possible use of externally exposed tissues such as gills, adipose fins and even better, mucus, as potential candidates for sampling with the added advantage that the same fish may be monitored repeatedly by minimally invasive techniques [50].

In sum, the expanding recent literature offers extensive evidence of the utility of stress proteins as biomarkers of environmental quality. As with any indicator of environmental quality or any bioassay system, before it can be applied routinely it will be necessary to calibrate the various stress protein responses in selected captive and wild biota. Based upon actual measured responses to an array of logical, specific factors and with experience in the field, the selection of actionable levels and the ensuing construction of an assessment system should be relatively straightforward. When it is necessary to distinguish between stress levels imposed by general environmental influences and those imposed by application of poor husbandry in aquaculture operations, it will be necessary to draw samples from more than one farm site coupled with measurements made on judiciously selected wild biota from the geographical area of interest.

7
Concluding Remarks

Clearly there are opportunities to minimize the impacts of any and all industrial activities including aquaculture in the coastal marine areas by a variety of nested practices. These include:

1. Coastal zone management that relies on
   (a) Development of comprehensive data bases for the coastal zone, describing in detail the bathymetry, oceanography, populations, and biological inventories along with industrial and other human activities of consequence.
   (b) Presentation of these data organized in user-friendly data bases made publicly available.
   (c) Utilization of these data in the development of environmentally sound management tools and programs for the coastal zones.
2. Understanding the aquaculture niche
   (a) By understanding the requirements of the various species and matching these to the appropriate environment to provide the best fit.
   (b) By determining the nature and effects of the aquaculture contributions of nutrification and pollution on environmental quality.
   (c) By understanding that the major hazard to aquaculture ventures, infectious diseases, are and must be an integral part of the environmental considerations and overall management approaches.
   (d) By applying the best husbandry and housekeeping techniques in aquaculture ventures.
3. Applying a hierarchy of environmental quality assessment techniques
   (a) By developing chemical methodology to determine the alterations in the various contributions to the environment, the remobilization of their various incarnations, and their several fates and sequestrations.
   (b) By determining the impacts that these various chemical species would have on the biota.
   (c) By employing as environmental sentinels captive and wild biota within defined areas to integrate and measure environmental quality.

Acknowledgements I thank Dr. John D. Castell for valuable discussions and for sharing his knowledge of the subject.

References

1. Stewart JE (2001) Bull Aquacult Assoc Can 101:43
2. Organization for Economic Cooperation and Development (OECD) (1993) Introduction: The significance of coastal zones in OECD Member Countries and internationally. Coastal Zone Management: integrated policies. OECD, Paris
3. OECD (1993) Coastal Zone Management: selected case studies. OECD, Paris
4. U.S. Commission on Ocean Policy (2004) Narrative of the video summary of the U.S. Commission on Ocean Policy's Preliminary Report, April 20, 2004. http://oceancommission.gov/newsnotices/apr20_04_video.html
5. Stewart JE, Penning-Rowsell EC, Thornton S (1993) The LENKA project and coastal zone management in Norway. In: OECD documents: Coastal Zone Management: selected case studies. OECD, Paris, p 257
6. Stewart JE (2002) Can Tech Rep Fish Aquat Sci 2411:95
7. Doucette LI, Hargrave BT (2002) Can Tech Rep Fish Aquat Sci 2426
8. Stewart JE (1994) Aquaculture in Atlantic Canada and the research requirements related to environmental interactions with finfish culture. In: Ervik A, Kupka Hansen P, Wennevik V (eds) Proceedings of the Canada-Norway Workshop on Environmental Impacts of Aquaculture. Havforskningsinstituttet (Institute of Marine Research), Bergen, Norway, p 1
9. Håkanson L, Ervik A, Mäkinen T, Möller B (1988) Basic concepts concerning assessments of environmental effects of marine fish farms. Nordic Council of Ministers, Copenhagen
10. Bergheim A, Aabel JP, Seymour EA (1991) Past and present approaches to aquaculture waste management in Norwegian net pen culture operations. In: Cowey CB, Cho CY (eds) Nutritional strategies and aquaculture waste. Guelph University, Guelph, Canada, p 117
11. Brown JR, Gowen RJ, McLusky DS (1987) J Exp Mar Biol Ecol 109:39
12. Gowen RJ, Bradbury NB (1987) Oceanogr Mar Biol Ann Rev 25:563
13. Braaten B, Aure JA, Ervik A, Boge E (1983) Int Counc Explor Sea CMF:26
14. Weston DP (1990) Mar Ecol Progr Ser 61:233
15. Stewart JE (1991) Int Counc Explor Sea Mar Sci Symp 192:206
16. Barton BA, Iwama GK (1991) Ann Rev Fish Dis 1:3
17. Pickering AD, Pottinger TG, Christie P (1982) J Fish Biol 20:229
18. Pickering AD, Stewart A (1984) J Fish Biol 24:731
19. Pickering AD, Pottinger TG (1989) Fish Physiol Biochem 7:253
20. Pickering AD (1997) Husbandry and stress. In: Bernoth EM, Ellis AE, Midtlyng PJ, Olivier G, Smith P (eds) Furunculosis: multidisciplinary fish disease research. Academic, San Diego, p 178
21. Thompson I, White A, Fletcher TC, Houlihan DF, Secombes CJ (1993) Aquaculture 114:1
22. Smith P (1997) The epizootiology of furunculosis: the present state of our ignorance. In: Bernoth EM, Ellis AE, Midtlyng PJ, Olivier G, Smith P (eds) Furunculosis: multidisciplinary fish disease research. Academic, San Diego, p 25
23. Håstein T, Linstad T (1991) Aquaculture 98:277
24. Enger Ø, Husevåg B, Goksøyr J (1989) Appl Environ Microbiol 55:2815
25. Nylund A, Wallace C, Hovland T (1993) The possible role of Lepeophtheirus salmonis (Krøyer) in the transmission of infectious salmon anemia. In: Boxshall GA, Defaye D (eds) Pathogens of wild and farmed fish: sea lice. Ellis Horwood, New York, p 367
26. Husevåg B, Lunestad BT (1995) Bull Eur Assoc Fish Pathol 15:17

27. Nese L, Enger Ø (1993) Dis Aquat Org 16:79
28. Jarp J, Tangen K, Willumsen FV, Djupvik HO, Tveit AM (1993) Dis Aquat Org 17:81
29. Renogen. Novartis (Aqua Health) bacterial vaccine. In: Compendium of veterinary products, 8th Canadian ed 2003. North American Compendiums Ltd, Adrian J Bayley, PO Box 39, Hensall, Ontario, p 760, http://www.compasnac.com/cancvp/up/
30. Jarp J, Gjevre AG, Olsen AB, Bruheim T (1994) J Fish Dis 18:67
31. Johansen L-H, Sommer A-I (2001) Dis Aquat Org 47:109
32. Bruno DW (2004) Aquaculture 235:13
33. Novoa B, Rivas C, Toranzo AE, Figueras A (1995) Aquaculture 130:7
34. Wood BP, Bruno DW, Ross K (1996) Bull Eur Assoc Fish Pathol 16:214
35. Jarp J, Karlsen E (1997) Dis Aquat Org 28:79
36. Stewart JE (1998) Can Tech Rep Fish Aquat Sci 2218
37. FRS Marine Laboratory (2000) Final Report of the Joint Government/Industry Working Group on Infectious Salmon Anaemia (ISA). Scottish Executive, FRS Marine Laboratory, Aberdeen, UK
38. Miller O, Cipriano RC (2003) International response to infectious salmon anemia: prevention, control and eradication: proceedings of a symposium; 3–4 September 2002; New Orleans, LA. Tech Bull 1902. Washington, DC: US Department of Agriculture, Animal and Plant Health Inspection Service; US Department of the Interior, US Geological Survey; US Department of Commerce, National Marine Fisheries Service
39. Johnson SC, Albright LJ (1991) J Mar Biol Assoc UK 71:425
40. Grant AN, Treasurer JW (1993) The effects of fallowing on caligid infestations on farmed Atlantic salmon (*Salmo salar* L.) in Scotland. In: Boxshall GA, Defaye D (eds) Pathogens of wild and farmed fish: sea lice. Ellis Horwood, New York, p 255
41. Costello MJ (1993) Review of methods to control sea lice (Caligidae, Crustacea) infestations of salmon (Salmo salar) farms. In: Boxshall GA, Defaye D (eds) Pathogens of wild and farmed fish: sea lice. Ellis Horwood, New York, p 219
42. Costello MJ, Burridge L, Chang B, Robichaud L (eds) (2004) Aquacult Res 35:1
43. Bruno DW, Stone J (1990) Aquaculture 89:201
44. McAllister PE, Owens WJ (1992) Aquaculture 106:227
45. Peters F, Neukirch M (1986) J Fish Dis 9:539
46. Taylor RL, Lott M (1978) J Protozool 25:105
47. Willumsen B (1989) J Fish Dis 12:275
48. Robinette DW, Noga EJ (2001) Dis Aquat Org 44:97
49. DuBeau SF, Pan F, Tremblay GC, Bradley TM (1998) Aquaculture 168:311
50. Iwama GK, Thomas PT, Forsyth RB, Vijayan MM (1998) Rev Fish Biol Fisheries 8:35
51. Iwama GK, Vijayan MM, Forsyth RB, Ackerman PA (1999) Am Zool 39:901
52. Hartl FU (1996) Nature 381:571
53. Dyer SD, Dickson KL, Zimmerman EG (1993) A laboratory evaluation of the use of stress proteins in fish to detect changes in water quality. In: Landis WG, Hughes JS, Lewis MA (eds) Environmental toxicology and risk assessment, ASTM STP 1179. American Society for Testing and Materials, Philadelphia, p 247
54. Burkhardt-Holm P, Schmidt H, Meier W (1998) Comp Biochem Physiol Part A 120:35
55. Sanders BM, Martin LS (1993) Sci Total Environ 139/140:459
56. Köhler H-R, Bartussek C, Eckwert H, Farian K, Gränzer S, Knigge T, Kunz N (2001) J Aquat Ecosyst Stress Recov 8:261
57. Culp JM, Podemski CL, Cash KJ, Lowell RB (2000) J Aquat Ecosyst Stress Recov 7:167

Assessing and Managing Environmental Risks Associated with Marine Finfish Aquaculture

Barry T. Hargrave[1] (✉) · William Silvert[2] · Paul D. Keizer[1]

[1]Ecosystem Research Division, Fisheries and Oceans Canada, Bedford Institute of Oceanography, P.O. Box 1006, Dartmouth, B2Y 4A2, Canada
hargraveb@rogers.com, keizerp@mar.dfo-mpo.gc.ca

[2]Instituto Nacional de Investigação Agrária e das Pescas, IPIMAR, Departamento de Ambiente Aquático, Avenida de Brasília, 1449-006 Lisboa, Portugal
silvert@ipimar.pt

1	Introduction	434
2	Identifying Environmental Risks Associated with Marine Finfish Aquaculture	435
2.1	Assessing the Scale and Costs of Adverse Effects	437
2.2	Dealing with Uncertainty—the Precautionary Approach	439
2.3	Minimizing Environmental Risks through Site Selection	440
3	Methods for Assessing Risks	441
3.1	Decision Support Systems	441
3.2	Environmental Risk Analysis	445
3.3	Reference Points for Identifying Environmental Changes	445
3.4	Indicators and Indices of Environmental Changes	447
3.4.1	Single Indicators	447
3.4.2	Multiple Indicators	449
3.4.3	Indicators and Indices	449
3.5	Analytical Hierarchy Process	450
3.6	Sensitivity Analysis	452
4	Integrated Approaches to Measuring and Reducing Risk	453
4.1	Integrated Management Models	454
4.2	Identifying Management Objectives	455
4.3	Comparative Risk Indices	456
5	Summary	459
References	. . . .	460

Abstract Environmental Risk Analysis (ERA) consisting of risk assessment, management and communication can be applied to assess ecological and environmental changes associated with industrial-scale marine finfish aquaculture development. Physical, chemical, and biological variables are identified that may be used to detect thresholds for changes in ecosystem structure and function in order to apply ERA. Changes due to predictable or unpredictable effects may be local or far field. Predictable effects such as reduced dissolved oxygen, increased nutrients and organic matter, or lower diversity of benthic fauna in the vicinity of net-pens can be modeled to quantify local impacts on water column and sediment variables. Far-field and long-term risks such as interactions of escapees with natural stocks and effects of fishing to obtain food for cultured fish are more difficult to predict and quantify. Despite this, scoring methods using single or multiple indicators may be applied to determine the degree of risk associated with all identified potentially negative effects. ERA should be part of an integrated planning approach where aquaculture development occurs within a broad framework to include all development and user groups within the coastal zone. Environmental observations and models can then be combined with effective aquaculture husbandry practices to manage environmental risks from all sources.

Keywords Environmental monitoring · Risk assessment · Salmon aquaculture · Sustainable development

1
Introduction

Government agencies often have the dual responsibility of increasing economic development while at the same time ensuring environmental protection. In many cases formal policies are needed to ensure reasonable and equitable management decisions. However, in the case of aquaculture development there can be fundamental differences in the interpretation of risks among various stakeholders [1]. Risk, the exposure to a chance of loss or damage, has two components—the probability of an event occurring times the magnitude of the effect. Industry proponents give priority to minimizing risks of economic loss, while opponents may emphasize the potential for negative environmental impacts and effects on traditional fisheries. In some jurisdictions, views on the issues have become so polarized that conflict resolution approaches have been proposed in an attempt to find a balance between the benefits of economic development and environmental sustainability [2].

There are many positive effects from aquaculture development. In addition to obvious economic benefits, there may be positive environmental changes. In marine coastal areas, moderate discharges of particulate organic matter from finfish culture sites may stimulate growth of benthic fauna if organic supplies from natural sources are limiting production. Removal of nutrients through harvesting of cultured shellfish may counteract eutrophication.

Despite these potential benefits, there is widespread perception that marine aquaculture has negative environmental effects and the potential to disrupt and/or displace other activities in the coastal zone. While establishing new aquaculture sites may create employment, those involved with traditional harvest fisheries may feel that their livelihood is threatened if the location or expansion of existing aquaculture sites reduces access to historic fishing grounds or is perceived as causing a reduction in biomass of harvested species. Recently, the scope of concern has expanded to the possibility of disease and parasite (sea lice) transmission between wild and domestic populations, and the potential for genetic change in indigenous populations due to interbreeding with escaped farm fish. Animal welfare concerns and market demands for high environmental standards have also become important issues for food producing industries. Even more global in scope is the recent claim that the use of fishmeal and oil in the manufacture of feed may be a factor in the depletion of wild fish stocks [3]. Although many of these claims are not convincing, in a risk analysis it is important to allow for implausible events and not only those which are well established.

Since farm development often occurs in coastal areas that are already under pressure from many other pre-existing user groups, aquaculture has become a competitor for limited space [4]. Co-management of the coastal zone is required since competing interests of multiple stakeholders need to be resolved when space or resources are limited [5]. There is therefore a need for an integrated approach in coastal zone planning and management where assessments of risks of all types (both social and environmental) associated with development are required to reflect the interests of all stakeholders.

In this chapter we identify several types of environmental risks associated with marine finfish aquaculture. Means of assessing the scale and costs of negative environmental effects, dealing with uncertainty and minimizing risks through site selection and use of decision support systems and Environmental Risk Analysis are presented. Environmental observations and models relevant to finfish aquaculture development are described that can be combined with effective farm husbandry practices to manage environmental risks in the coastal zone from all sources. Finally, indices of fish and environmental health are proposed that might be useful for application in an integrated management approach to reduce risks to both cultured stocks of finfish and the environment.

2
Identifying Environmental Risks Associated with Marine Finfish Aquaculture

From a broad viewpoint, one can identify several types of environmental risks associated with finfish aquaculture both in terms of the probability of occurrence and magnitude of effects. Both the probability of the event and the size of the effect have uncertainty. "Reducible uncertainty" is the lack of knowledge about the probability or magnitude of an event occurring that can be reduced by more data collection. "Irreducible uncertainty" is a function of the natural variability of all ecological systems. Some events are predictable and relatively easy to quantify or estimate, while others are unpredictable and associated with large uncertainties (Table 1).

Table 1 Examples of types of environmental risks associated with marine finfish aquaculture development grouped by the spatial scale (area affected), temporal extent and predictability of adverse effects and management approaches to mitigate potentially negative effects

Issues	Risk management approaches (mitigation)
Local (cage scale), acute, short to medium term temporal, predictable effects (managed on a site-by-site basis)	
Dissolved oxygen depletion	Reduce fish density
Sediment organic enrichment	Move cages within or from site (fallowing)
Release of toxic chemicals	Minimize use, controlled applications
Inlet-scale, intermittent to chronic spatial and temporal effects, difficult to predict (modified by cumulative effects of all anthropogenic activities in a specified area)	
Nutrient enrichment	Siting to maximize dispersion
Antibiotic resistance	Increased use of vaccines, disease control to minimize antibiotic use
Harmful algal blooms	Phytoplankton monitoring
Mortality (superchill)	Monitoring, care in site selection
Mortality (disease)	Implement improved husbandry and disease control plans at all farm sites
Cumulative effects of all anthropogenic inputs	Develop and implement integrated management plans
Regional, broad-scale, long-term, unpredictable	
Escapement (genetic interactions)	Improved containment infrastructure
Feed manufacturing	Assessment of impacts of harvesting on wild stocks for feed production

Careful monitoring and responsive management can reduce the magnitude of some of the effects, while others have to be considered as chance events with a low probability of prediction in frequency or magnitude of effect since they are beyond human control. The former are commonly associated with uncertainties about the magnitudes of some relevant quantities which may exceed an ecological threshold and lead to adverse effects. Examples of these risks include stocking a farm to a level where the wastes prove toxic to the fish, or where oxygen levels are reduced to stressful levels by respiration. The critical thresholds for some of these effects are often difficult to calculate. Destruction of the farm by a rare hurricane is a different type of risk, one that is hard to quantify and can really only be dealt with by insurance. If super-chill occurs every three or four years on the average, then fish farmers have to take it into account and plan for it happening several times over the life of their operation. If it only occurs every few decades, then a different risk-management strategy is needed. The line between these types of risk is not sharp.

As an example of what we mean by alternative risk-management strategies, consider the risk of loss of farmed fish if a storm damages a farm. Farmers can be expected to take reasonable precautions against the types of storm that occur regularly, including the use of secure moorings, frequent inspection of the nets and support structures, possibly towing the cages to sheltered areas when storms are predicted. These measures add to the cost of farm operation, but are seen as necessary measures to mitigate risk of lost equipment and escaped fish. We do not, however, expect farmers to invest in breakwaters and other protective structures to protect against the occurrence of a once-in-a-lifetime hurricane. Frequency and predictability are important factors in deciding how to deal with risk but in addition, risk assessment must also identify the possible severity of an outcome versus the probability of occurrence.

2.1
Assessing the Scale and Costs of Adverse Effects

The spatial extent of potential deleterious effects of the aquaculture industry poses additional factors to be considered in assessing risks of adverse environmental effects and the issue of who bears the brunt of the costs. Localized risks which may affect only one farm or inlet can be dealt with by spreading the risk among farms in the affected area by using various methods of horizontal integration, such as co-operative associations or mergers, or by external mechanisms such as insurance programs or government support. For this to work, the area of integration must be larger than the spatial extent of the potential impact. A company that operates a number of farms several kilometers apart can probably endure the loss of several pens of fish to a harmful algal bloom, but it may not survive a major storm or other extreme weather event unless it operates farms spread across the country.

Insurance is a simple way for many different operators to share risk and spread it over a large area, or over many different industries. The situation is more complicated when part or all of the cost or effect of a risk in environmental management is borne by those who do not stand to benefit from the activity (known as "free-riding" in economics). Although this problem can be dealt with by assigning more explicit property rights, government management of aquaculture leases usually precludes this approach. This can happen if effluents from a fish farm interfere with nearby wild crustacean fisheries or pollute recreational beaches, or if debris from storm damage interferes with shipping. In some cases, it is difficult to assign responsibility, which inevitably leads to conflict. Fish farms are often believed by local residents to be responsible for all unexpected environmental problems in their vicinity, such as smelly wastes washing up on the shoreline, even when a causal link cannot be established.

The question of who bears the cost of environmental risks is especially problematic in coastal zone management with shared jurisdictions and when cause-effect relationships are unclear or when immediate costs are not easy to quantify. There is a general tendency for finfish aquaculture sites to be located in high-energy, exposed sites to maximize water movement, oxygen supply and waste dispersal. However, there is a trade-off between the use of exposed sites and the risk of loss due to escapement. If an extreme weather event or predator attack breaches fish pens allowing fish to escape, there is an immediate quantifiable loss to the farmer, but there is also the less easily measured and long-term risk that the farmed fish will interbreed with wild fish and thus threaten their viability. This risk is hard to estimate, but even if it does occur, who actually suffers the loss? The farmer has lost the value of the escaped fish, but there may be a much larger, long-term and more difficult-to-measure negative impact on wild stocks. An extensive literature dealing with the economics of environmental damages over the past three decades shows that assessing the costs of environmental damage can be a far larger task than estimating the risks.

Another version of environmental externality arises when we consider the risks associated with disease control. The risk of disease or parasites can be minimized for the farmer by the use of pharmaceuticals, but these chemical products can have adverse effects on other species and their environment. Some products used for treatment of sea lice, discussed in [6] for example, are highly toxic to crabs, lobsters and other crustaceans. The presence of antibiotics in sediments is generally not acceptable—but again, it is the farmer who faces risk by not using these products, and who does not generally suffer any of the adverse consequences.

While some risks can be dealt with quantitatively when the probabilities and costs are known, risks that are difficult to forecast or quantify are far more difficult to analyze, especially in situations where many different participants are involved. As pointed out above, the impact of known risks such as weather events (storms, super-chill, etc.) can be mitigated by insurance and similar strategies based on probability theory—this is basically equivalent to gambling with known odds. A precautionary approach (discussed in the following section), could also be applied when the probability of a highly damaging event is unknown. One example of planning that acknowledges a certain degree of risk aversion would be to invest in contingency planning for worst-case events. Increasingly, however, we face uncertain risks where there is a possibility of environmental harm, but no clear evidence that it can happen. In these situations it is impossible to assign meaningful probabilities.

2.2
Dealing with Uncertainty—the Precautionary Approach

There are several approaches to dealing with uncertain risks, and ultimately, this is a social and political rather than strictly scientific matter. One approach is simply to ignore any adverse possibilities that are not well established scientifically. Unless it can be proven that certain actions will harm the environment or have negative consequences for other coastal zone users, they may be permitted. This extreme viewpoint has become less acceptable in recent years with the growing realization that science is very often correct even when it is not conclusive, and there is increased awareness that a laissez-faire approach to environmental management often has disastrous results. An alternative is to adopt a Precautionary Approach (PA), which according to Principle 15 of the Rio Declaration of the UN Conference on Environment and Development is: "Where there are threats of serious or irreversible damage, lack of full scientific certainty shall not be used as a reason for postponing cost-effective measures to prevent environmental degradation."

Unfortunately, this statement of the PA is both clear and ambiguous. It sets forth the Approach clearly, but contains several phrases which are open to a wide variety of interpretations: "full scientific certainty", "cost-effective measures" and "environmental degradation." There is almost always some degree of "serious or irreversible damage" that will vary both spatially and/or temporally. This variability may determine both the need and choice of mitigation measures. For example, if the impact of organic enrichment is confined to a few square meters of seabed under net-pens, then it is generally ignored. Some countries accept a moderate (and presumably reversible) amount of increase in sediment organic matter in the immediate vicinity of a fish farm so long as there is no measurable effect outside the bounds of the lease. Other countries allow enrichment over a larger area [7]. Even the meaning of the word "reversible" is unclear, since there is a definite difference between an environmental impact that vanishes as soon as a farm is removed and one that persists for years.

One of the most serious problems with the PA terminology is the reference to "full scientific certainty". Many of the conclusions about negative environmental impacts due to finfish aquaculture (and other aspects of fisheries) which have been widely accepted by the scientific community are still considered questionable by many stakeholders. The definition of "cost-effective measures" is even more difficult, especially when balancing the costs of environmental protection against the loss of income or even jobs in the industry. Much of the material presented in previous chapters in this book describes attempts to identify degrees of "environmental degradation", but this can be difficult to define. For example, moderate levels of increased carbon and nutrient loading from a fish farm can stimulate macrofauna production, but this is often associated with a decline in species diversity. Both productivity and biodiversity are considered measures of environmental quality to be preserved, so if one goes up while the other goes down, it is difficult to determine whether the net result is environmental degradation or enhancement.

In recent years, the PA has been applied by environmental management authorities and adopted for fishery management purposes [8]. Although not universally accepted, since some stakeholders may influence decisions to their benefit, the approach underlies current thinking on how to best provide integrated management in the coastal zone [5]. Since there is no generally accepted working definition for PA, ERA is often applied to minimize risks. If this is done fairly, no single group of stakeholders should be favored.

2.3
Minimizing Environmental Risks through Site Selection

Many factors can limit the choice of areas available for marine finfish aquaculture development. In populated areas, conflicts with other users of the coastal zone are often dominant factors controlling where new farm sites can be located. New marine finfish aquaculture farms and/or subsequent operation compete with navigation, indigenous fisheries and other uses of the coastal environment. The distribution of endangered species can also create exclusion zones for aquaculture development in many jurisdictions. There is also the primary requirement that waters with sub-zero temperatures leading to super-chill must be avoided. High water temperatures may lead to stress due to increased fish respiration and reduced dissolved oxygen concentrations. Depth and water flow conditions at cage sites are critical variables to ensure delivery of oxygen and removal of dissolved and particulate waste products. However, flow rates cannot be so great as to lead to excessive energy requirements for fish to maintain position. Increased metabolic demands and stress due to continuous swimming with high current velocities will substantially reduce growth rates since energy is used for respiration and not biomass production.

Physical factors of temperature, depth, and water flow are primary variables for determining if a location could be considered suitable for a new finfish culture site. However, even if all physical variables appear favorable, additional habitat factors must be considered to ensure that operation of a farm does not create harmful alterations to water column or sediment variables. Cumulative effects, where changes to ecosystem structure and function may occur as a result of modification of many inter-related variables, are particularly difficult to detect, measure or predict. In fact, it is often the case that only after a site has become operational that reduced growth rates and lower food conversion efficiencies appear indicating that changes in some environmental factors have occurred. In these cases, initial yields will be within expected limits but after a few years, a gradual decrease in production efficiency appears as some environmental variables become degraded. Stress in fish will progressively increase resulting in higher susceptibility to disease. It is for this reason that health and growth cultured species may be the first indicator of environmental deterioration [9].

As for any type of development, the economic benefits of aquaculture and other forms of anthropogenic activity in the coastal zone must be balanced with the need for habitat protection to limit cumulative, long-term environmental changes that destroy or severely damage marine habitats. Coastal waters have some ability ("assimilative capacity") to accept additional organic and nutrient inputs beyond those supplied by natural processes without irreparable damage. However, from a socio-economic perspective, sustainable aquaculture development may only be possible if sites chosen for aquaculture development do not impede use of marine space and resources for other human purposes. If this does not occur, experience has shown that public debate and disagreements between stakeholders can interfere with growth of the industry [2]. From a conservation perspective, aquaculture development must also conserve species, other ecosystem components, habitats and their function. From an ecosystem productivity standpoint, for sustainable development, there should be negligible negative impacts on both traditional harvest fishers and farmed fish. These conservation objectives are described using different terminology in various jurisdictions. The Oslo-Paris Commission uses the term "Ecological Quality Objectives" while Canada's Oceans Act refers to "Marine Environmental Quality Objectives".

3
Methods for Assessing Risks

3.1
Decision Support Systems

As mentioned above, pressure from the aquaculture industry for licensing new sites or increasing the size of existing leases to allow expansion is often at odds with other uses of coastal areas such as for traditional fisheries and habitat conservation required under regulatory laws. However, there are few, if any, guidelines as to what factors must be considered in limiting the numbers or sizes of farms in any given area. The conflicting goals of industry for expansion and requirements for conservation from regulators and non-governmental organizations has led to adversarial positions and conflicts between stakeholders [2]. Decisions based on a priori fixed points of view, lack of public and scientific input, inaccurate assumptions and a general inability to revise and adapt decisions once they are made in light of new information are some of the factors making consensus difficult to achieve [10].

One approach to assessing these conflicting demands is through use of Decision Support Systems (DSS) to assist with licensing and siting decisions for new farm sites [11, 12]. These programs are an application of the Expert Systems technology that has been widely implemented in fields like medical diagnosis. The output of a DSS is not a table of numbers that can only be understood by a trained scientist, but rather a clear presentation, often in plain language, of the facts that are relevant to planning decisions. If all significant potential impacts are recognized, the location of new sites can be pre-selected to minimize risks of as many of the potential negative effects as possible. Near and far-field environmental impacts that are predictable as well as those associated with uncertainty must be considered over a wide range of spatial and temporal scales. It is also necessary to identify sensitive variables that can be readily measured and used to scale negative effects. By simultaneously considering many risk factors, using a DSS in the decision making process could be the basis for integrated management advice.

If the influence of aquaculture on specific variables is known or can be predicted, observed relationships between measured variables can be used in models to provide quantitative estimates on which to base decisions. Caution is required when using empirical relationships for predictive purposes when extrapolating results beyond the range of conditions under which data were collected. An example of this approach is provided by correlations between benthic organic matter loading, oxygen supply, sediment oxic and geochemical conditions and macrofauna community structure around fish farms [13–18]. Empirical regressions between phosphorus loading from fish cultured in archipelago areas of the Baltic Sea, dissolved phosphorus concentrations in water and chlorophyll-a concentrations were used in management models for coastal zone planning [19]. Similar positive correlations between nitrogen loading and phytoplankton biomass (chlorophyll concentrations) were used in ecological simulation models of Narragansett Bay [20]. Models using empirical relationships are useful management tools that can be used to identify target nutrient loading levels to satisfy ecosystem objectives for maintaining phytoplankton biomass within desired limits.

As useful as modeling may be in assessing and predicting the impacts of aquaculture, the approach is of little value unless results are clearly and credibly presented in terms that all stakeholders can understand and accept. Specialized models used to assist in management decisions need not be sophisticated to be effective if they are presented and applied in a transparent and useable manner. This has led to strong incentives to develop models in the context of interactive programs that produce meaningful, and easily understood outputs that meet the needs of all participants in the decision-making process.

A DSS used for evaluating lease applications could require that the user enter a description of the proposal including such factors as location, pen configuration, stocking biomass, etc. This information would be combined with stored information from a geographically indexed database to provide input to a suite of models which would automatically run and evaluate the site application [11]. The output could be expressed in terms such as—"The predicted nitrogen loading from this site would be W mg N l^{-1}, which is lower than the regulatory limit of X mg N l^{-1}. The predicted benthic carbon loading (BCL) over the footprint of the cage array is Y g C d^{-1} which exceeds the regulatory limit of Z g C d^{-1}. On the basis of the BCL value the application should probably be rejected, but a 12% reduction in stocking biomass would bring the BCL down to an acceptable level". Results of model calculations expressed in this manner provide necessary information in a form that is complete and easily understood. This helps both the manager decide whether to approve the lease and the farmer to consider amending his application to meet the regulatory requirements. Of course with more sophisticated models, for example [14, 15], the design of a DSS can be a major challenge. However, it is likely to be a far more effective tool for most aspects of coastal zone management than a specialized model that can only be run and interpreted by experts.

There are other strong arguments in favor of the development and use of DSS for assessing potential risks associated with aquaculture development. The scale of the coastal zone and the remoteness of many locations where decisions about aquaculture and other coastal zone uses are made renders on-site evaluations by environmental experts difficult and often impractical. The situation is analogous to what has been happening in the field of oil spill response—although it would be useful to be able to send teams of meteorologists, oceanographers and petroleum technologists to the site of a spill, it is increasingly common to combine some of their abilities in an expert system on a portable computer. A technician can then be sent to advise on mitigative measures. In a similar way, field representatives of fisheries and environmental departments can be trained to use portable DSS to address siting and related issues.

Of course, it is not easy to develop reliable expert systems, and even if a DSS worked perfectly, it is unlikely that all stakeholders would be happy to leave critical decisions in the hands of a technician with a laptop computer. A DSS can, however, be used effectively to minimize environmental risks involved with aquaculture development in several ways. One is for informal planning and preliminary evaluation of proposals for new site licenses—for example, a farmer can use it to see what the response to a lease application is likely to be. Another is in what [21] termed a "tiered" approach to environmental assessment. This is analogous to the practice of battlefield medics of sorting casualties into three categories—those likely to survive without care, those who will probably die in any case, and a third group who may be saved through medical intervention. It is the third group of course which receives medical care. In the same way a DSS can be used to evaluate lease applications and other types of development to distinguish between those that pose a negligible risk of environmental damage, those almost certainly harmful, and a third marginal category requiring careful evaluation. The third group can be subject to more intensive investigation and possibly the intervention of expert specialists.

One of the strengths of the tiered approach to categorize risks is that it involves evaluating impacts against decreasingly conservative targets. It is adaptable to priorities that reflect differences in management objectives and available resources. Consider for simplicity the case where decisions are based on a single indicator I, which represents the predicted degree of impact. There is a Reference Point I_{RP} such that a proposal will be approved only if $I < I_{RP}$. We can define two cut-off points, $I_{low} < I_{RP}$ and $I_{high} > I_{RP}$, such that if preliminary estimation by a DSS predicts $I < I_{low}$ the proposal is approved, and if $I > I_{high}$ it is rejected. If it the value lies within the range $I_{low} < I < I_{high}$, there is enough uncertainty to justify further expert evaluation. There is a great deal of flexibility in this approach, depending on the choice of I_{low} and I_{high}. If I_{low} is set to a very low level and I_{high} very high, most cases will need further adjudication and the DSS will only resolve the most extreme situations. This may be a good strategy when implementing such a scheme since the DSS is unlikely to produce any questionable decisions. However, if the available resources are strained by a large number of proposals it may be best to narrow the band between I_{low} and I_{high} so that fewer cases require expert intervention. Furthermore, the individual settings of I_{low} and I_{high} determine how conservative the process is, in the sense of making sure that few bad proposals are automatically accepted.

The use of DSS in coastal zone management so far is fairly limited, but the potential is great. One aspect, which has been extensively developed, uses information available for multiple variables as geo-referenced data and the creation of geographically indexed databases, or Geographical Information Systems (GIS). These have tremendous value, not only as a way to store the kind of information needed to run a DSS (like bathymetry and time series of temperature data) but also as a way of presenting relevant information to stakeholder communities. GIS can be designed to collect, store and analyze variables where geographic location is an important characteristic necessary for the analysis [22]. With a GIS, it is possible to display overlays showing the spatial relationships between fish farms, transportation channels, oyster beds, recreational areas and any other features. Both socio-economic and environmental variables can be combined to evaluate possible development scenarios [23]. The use of visual overlays provides a holistic approach and by selecting only overlays of interest, an almost unlimited amount of information can be presented without clutter or overload. A GIS database allows full documentation with the same information available to the public, all stakeholders and decision-makers [10].

3.2
Environmental Risk Analysis

If risks can be identified and quantified, Environmental Risk Analysis (ERA) can be used to assess the probability of adverse effects. ERA is the process for evaluating "the likelihood that adverse ecological effects may occur or are occurring as the result of exposure to one or more stressors" [24]. The process consists of separate but related steps—problem identification, formulation of an approach, risk analysis and risk characterization. Good communication requires that results are clearly expressed with major assumptions and uncertainties identified, along with reasonable alternative interpretations of the information collected [1, 25]. Results of the process, that clearly separates scientific analysis from policy-related judgements, must be expressed to reflect all stakeholder's opinions.

Application of ERA has been criticized since a requirement is that all environmental effects must be determined on a quantitative or qualitative basis to calculate probabilities of adverse effects. Pros and cons of management alternatives must also be known to allow alternatives to be prioritized. Scientifically defensible information must be exchanged or communicated between all stakeholders in a manner to build trust and co-operation. However, it is often difficult to reach agreement among diverse groups of individuals from different sectors. Agreement on priorities of risks and benefits, about real (as opposed to perceived) environmental effects and methods to measure them is also usually difficult to achieve. Despite these problems, ERA can be used to establish priorities and evaluate the degree of concern for different risks identified by stakeholders.

3.3
Reference Points for Identifying Environmental Changes

A critical element that applies to the development and application of all three components in the ERA process is the identification of reference points (or effects thresholds) and their ecological consequences. Various assessment methodologies can be used to identify and measure the dimensions of uncertainty and perceived risks. While each approach is different, all require some form of agreed upon measurements (indicators) which identify environmental alterations to ensure that changes occur within a maximum acceptable limit [26].

Various ecological indicators have been proposed to assess the status of marine ecosystems [27]. "Reference Points" (RP), as reviewed in [28], are similar and have played a central role in fisheries management for many years. For example, harvesting 10% of stock biomass (F0.1) was introduced by the International Council of North Atlantic Fisheries in the mid-1970s as a conservation measure. Target RP values meet management objectives while Limit Reference Points (LRP) are associated with unacceptable outcomes. Both describe changes in one variable relative to another such that the functional (cause-effect) relationship can be used to identify thresholds that describe the state of a fishery.

A Traffic Light Method (TLM) has been proposed as a new approach to use specified RP and LRP values to guide management of fisheries resources [29]. The method is similar to the tiered approach to decision-making described above where observed indicator variables defining RP and LRP values of an attribute are used to guide fishery resource management. The approach has elements of an integrated model to resource management as discussed below. TLM was proposed as a precautionary decision framework that defines management responses based on the state of multiple indicators of a fishery system in relation to LRP. The method has been used to provide scientific advice for Northwest Atlantic shrimp stocks [30] and was further developed for the management of some north Atlantic groundfish stocks [31].

The concept of RP and LRP may be applied to assess risks associated with different environmental effects of aquaculture development if a two-step process is used that separates scientific advice from decision-making. First, technical scientific expertise is required to select sensitive variables and to define threshold values for identifying changes. Stakeholders should be involved in cases where these values cannot be established based on scientific criteria alone. Once thresholds are agreed upon, all stakeholders should determine what management decisions are to be taken if RP and LRP levels are exceeded.

Major environmental impacts associated with coastal aquaculture development have been identified that could be used to establish relevant RP and LRP values [32–35], (Table 1). Sedimentation rates of waste food and feces, dissolved nutrient concentrations, levels of dissolved oxygen or concentrations of toxic chemicals in water and sediments could all be used to derive desired threshold values for environmental management purposes. Other potential effects such as genetic and behavioral interactions of escapees with wild stocks and impacts associated with the harvesting of wild species to provide food for cultured stocks are more widespread and less predictable. These may be less suitable for setting boundaries of desired outcomes, not only because they cannot be easily quantified but also because it may be difficult to achieve consensus as to what are acceptable limits of change.

While potentially negative environmental effects of aquaculture have been identified, no general methods exist for monitoring or assessing cumulative effects and risks. For example, dissolved nutrients and organic matter released from farm sites must be considered within the context of all sources of nutrients and organic matter entering an inlet or coastal system where aquaculture occurs to evaluate the relative contribution of various sources [36, 37]. Descriptions of the dynamics of dissolved and particulate matter in coastal waters (sources, sinks and recycling) require focused analytical capability beyond the usual scope of regulatory monitoring programs. In a similar manner, sediment anoxia created by excessive organic enrichment can change benthic faunal communities. Changes in sediment physical-chemical-biological properties may be measurable to indicate the degree of alteration. For example, sediment geochemical properties such as organic matter content, levels of sulfides and oxidation reduction potentials (Eh) may be used as indicators of oxic/anoxic conditions to monitor changes over time both at and away from culture sites [16, 18, 38, 39]. These variables can be used to scale levels of sediment organic enrichment and where thresholds for change can be established, the measurements can assist in siting decisions and quantify alteration of benthic habitats over time [12].

3.4
Indicators and Indices of Environmental Changes

The rationale behind the use of RP and LRP discussed above is that quantifiable indicators are used to make objective decisions. However, to assess the risk to the environment and to verify whether there has been any damage we need measures of the status of the environment that can be both modeled and monitored. Development of suitable indicators is a key component in the process of dealing with risk.

3.4.1
Single Indicators

Risk management can be considered analogous to priority setting—the initial step for setting research objectives for environmental risk analysis. Methods to measure risk based on changes in ecosystem properties arising from aquaculture development can be based on single or multiple indicators. In the case of fishery management, a broad objective might be stock conservation and optimized harvest, while for aquaculture development the method could utilize ecological or habitat indicators with an overall goal of habitat/ecosystem protection. To apply this approach using ecological Reference Points measurable structural or functional properties (habitat indicators) must be identified that are sensitive to change. Key variables that might be used as indicators for screening for environmental change must be scientifically relevant, sensitive to the degree of change expected, amenable to assigning target or threshold levels, cost effective and predictive [40]. Furthermore, to as great an extent as possible stakeholders must reach consensus as to the degree of risk, how it is assessed, and that the selected indicators are appropriate for monitoring the ecological state of the natural habitats of concern. It may be difficult to reach this agreement for management decisions since usually aquaculture leases are held by the state and producers do not have the same vested interest as a farmer who owns his/her own land. The methods of measurements must also meet the three components required for ERA (i.e. the indicators must be useful variables for assessment, management and communication of risk).

Choice of a single indicator involves identifying one variable (e.g. threshold concentration of dissolved oxygen to cause stress to cultured fish, a limiting nutrient for primary production, a threshold concentration of a toxic substance with known dose-response characteristics) to provide a quantitative measure of effect. There must be a known or agreed upon measurable change, with identifiable thresholds using key indicator variables generally agreed by all stakeholders to be reflective of significant environmental change. Discrimination between alternatives can be based on congruence of an easily measured variable with expected outcomes (e.g. deviations from thresholds for dissolved oxygen saturation).

For example, widespread reductions in water column oxygen concentrations in excess of naturally occurring depletion or increased bacterial counts above expected background levels are examples of single variables that could be used to indicate organic enrichment from sewage discharge. In the case of aquaculture development, these same variables and thresholds are of interest to both producers and regulators, since health and growth of cultured species depend on an adequate oxygen supply and protection against bacterial contamination. The interests of habitat protection and the aquaculture industry are mutually supportive.

Expected benefits and costs can also be used to weigh risks of negative habitat alterations where a single variable might be used to indicate the risk of decreased production. Negative environmental effects (costs) can be balanced against economic advantages to make decisions on the basis of the selected indicator variable. This approach requires agreement about the economic value of negative environmental effects among stakeholders that is usually difficult to achieve. However, single variables such as dissolved oxygen and bacterial abundance could be used as targets for decisions on siting and for monitoring effects after farming operations are established where economic cost/benefit factors could be explicitly compared.

In the case of finfish aquaculture development, a more complete estimate of costs and benefits for protection of environmental elements than direct cost/benefit analysis can be provided by taking into account a wide range of valued ecosystem components not all of which have direct monetary value. For example, non-commercial wild populations of fish and invertebrates that serve as food for wild stocks would be evaluated for their role in the ecosystem. This "ecological value" would be balanced against the economic value of production of the cultured species.

3.4.2
Multiple Indicators

While successful in limited cases, single indicator methods for measurement of environmental risks cannot assess complex or cumulative environmental interactions. They also do not allow uncertainty or subjectivity to be incorporated into decisions. When assessments of environmental changes involve many factors, stakeholders and points of view, multiple indicator methods are required. The methods allow several factors to be considered simultaneously using weighted indicators. All multiple indicator methods use scores in some manner to measure uncertainty and risk.

An example of a multi-indicator decision framework, termed trade-off analysis, involved combining input from all stakeholders in planning a marine protected area [41]. Different users and interested parties were asked to identify social, economic and ecological indicators that they considered critical. The impacts of different development scenarios were then evaluated. Stakeholders then weighted different indicators to consider outcomes for various management options. The approach enhances stakeholder's involvement in the decision-making process and any environmental management plan developed should be more broadly accepted.

The decision outcome also reflects both socio-economic and ecological/environmental factors. The trade-off between conservation and socio-economics is a delicate balance. It might be argued that as conservation becomes a greater concern, socio-economic factors must receive a lower weighting in decision-making. The situation of socio-economic factors trumping conservation concerns is arguably an undesirable outcome as demonstrated by over-fishing of most harvested fish stocks.

3.4.3
Indicators and Indices

In order to interpret indicators, whether single or multiple, and to present them to managers and stakeholders requires a degree of distillation of the results. One way to do this is translating the indicators, or combining multiple indicators, into dimensionless scores or indices.

The DSS methods described above also represent a multiple measurement approach to assess risk and optimize decision-making. As with all multi-indicator approaches scoring is required. Indicators to evaluate key environmental factors must be scaled to represent the relative importance of each factor (indicator or variable) by assigning numerical or ordinal values. Weighting can be applied to emphasize greater risk known or subjectively perceived to be associated with specific indicators. Uncertainty can be included by setting broad (or fuzzy) boundaries when decision points are unknown or poorly defined [42, 43]. These methods provide simplicity and transparency to decision making and risk assessment. They can be participatory for all stakeholders, have good discriminating potential between alternatives and once in place they are inexpensive to apply.

The idea of using a DSS tool for communicating scientific advice to managers to minimize risks of negative environmental effects with respect to licensing of finfish aquaculture sites was first suggested by [11]. The DSS approach can be used by applying a score to a set of specified questions as a way to provide scientific advice to habitat managers evaluating aquaculture license applications [12]. Environmental data had to be considered in a consistent manner to determine if a site was suitable with respect to physical and ecological conditions and to indicate if a harmful alteration, disruption or destruction of habitat was likely to occur. A net decision from a DSS can be derived from answers to a series of questions as qualitative (yes/no) responses of input and as quantitative data from information in license applications. Specific variables known to be sensitive indicators of environmental interactions due to culture activities can be evaluated to provide a more robust final judgement than one based on a single indicator. Weighting can be applied such that if some factors receiving negative scores are considered to be of greater risk, these can be designated as pre-emptive for the final siting decision.

The advantage of a multi-indicator approach to assessing risks is that once a set of questions and scores is agreed upon, the method ensures consistency. Decisions are always based on the same factors and scoring method. Altering inputs for selected factors to explore effects on the net decision can test sensitivity. Negative scoring also indicates where environmental concerns or knowledge gaps exist, or possible mitigation measures may be applied. Further development of the method could incorporate a mixed scoring approach where subjective and objective information is combined in a final decision. Subjective indicators may also be included to overcome the lack of data.

The strength of scoring in multiple indicator methods is the flexibility and consistency. Many different factors are considered simultaneously and the number of indicators can be increased to base decisions on a wider range of factors as new issues are raised by stakeholders, new environmental data become available or new regulations are applied. Increasing the number of indicators and discussion of relative weights to derive an accepted scoring method may increase understanding of the perceived relative importance of different issues and build consensus.

3.5
Analytical Hierarchy Process

The Analytical Hierarchy Process (AHP) is a commercially available software tool (Expert Choice) that ranks factors against each other to set priorities and optimize decision when both qualitative and quantitative aspects of a decision need to be considered [44, 45] (Fig. 1). This involves three principal steps: (1) describing elements of the decision problem; (2) comparative judgement of the relative importance of the elements and; (3) synthesis of the priorities.

Although presented as a planning tool to optimize management goals, the AHP has also been used to evaluate risks and alternative decisions in developing an environmental protection policy for aquaculture in Finland [46]. Interactions between environmental and economic factors were examined by use of a questionnaire sent to environmental authorities and other interest groups to determine preferences for measures of environmental policy. Since decision alternatives depend on the goals and subjective values of the decision-makers, the AHP approach ensured input from all parties. Conflicts and knowledge gaps were identified and prioritized through use of the AHP model.

One of the major strengths of the AHP is the use of pairwise comparisons to derive ratio scales for different priorities, rather than using the traditional approach of assigning weights for RPs described above. Verbal scores used in the AHP can be equated to numerical scores or assignment of colors (or letters) using the more general Traffic Light approach (Table 2).

In the AHP, the relative importance, performance or likelihood of two elements is compared with respect to another element in the level above. A judgement is made as to which is more important and by how much. Weighting reflects inputs from all stakeholders that describe the relative importance of various factors or effects. Pair-wise comparisons based on numerical, verbal or graphical scores can be used to develop models that establish priorities with comparisons based on numerical or verbal scales.

Weighted scores or multiple indicator methods do not allow consideration of inter-relationships between risk elements. For example, one indicator used to evaluate risk may be directly related to another such that

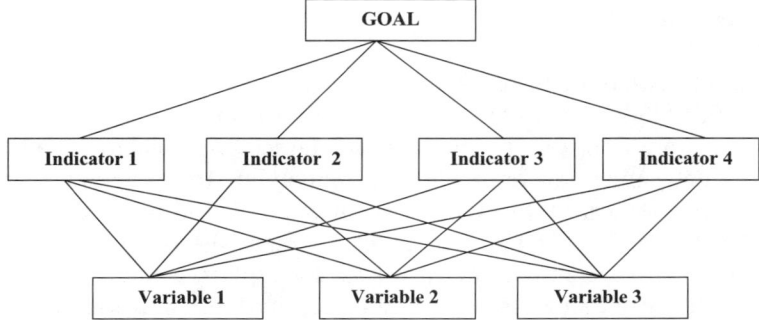

Fig. 1 Basic structure of the Analytic Hierarchy Process (AHP)— a commercially available software tool (Expert Choice) (http://www.expertchoice.com) [44]. AHP ranks factors against each other to set priorities and optimize decisions considering both qualitative and quantitative variables

Table 2 Comparative scales for pair-wise comparisons in numerical, verbal and alphabetic scoring systems combining The Analytical Hierarchy Processes (AHP) priority terminology [44] with corresponding Traffic Light scoring [12, 31]

Numerical score	Verbal score (AHP)	Traffic light score
1	Equal, neutral, no likely impact	A (green)
2		
3	Moderately important, possible impact	B+
4		
5	Strongly important, minor impact expected	B (orange)
6		
7	Very strongly important, measurable impact	B–
8		
9	Extremely important, severe impact	C (red)

cumulative impacts occur. Scientific technical expertise is needed to identify correlated environmental variables most useful for this purpose. An example of such interactions at finfish aquaculture farm sites would be where excessive sedimentation causes benthic organic enrichment and sulfide accumulation leading to the formation of anoxic sediments. Elevated heavy metal (zinc, copper) concentrations may also occur in these deposits [47]. Thus benthic fauna may experience stress from the combined effects of anoxia, sulfide accumulation and increased levels of heavy metals. The combined risk may be greater than the sum of individual risks due to specific effects.

3.6
Sensitivity Analysis

It is useful to differentiate between Sensitivity Analysis (SA) and Uncertainty Analysis (UA). SA involves the computation of the effect of changes in input values or assumptions (e.g. boundaries and model functional form) on model output to determine how uncertainty in model output can be systematically apportioned to uncertainty in input parameters. UA, on the other hand, investigates the effects of lack of knowledge or potential errors in model output associated with specific parameter values. By exploring the "relative sensitivity" of model parameters, a user can quantify the relative importance of different input parameters. Confidence levels can be applied to model outputs by combining both SA and UA. Simulation models are often used to quantify uncertainty about the parameters and input variables of the systems described by the models. Changes in model outputs as input parameters are varied show how uncertainty affects the results and can be used to estimate the probability or risk of various outcomes.

Although there are several different approaches that might be employed for propagating uncertainty in risk analysis, including analytic techniques in SA applicable to linear and other fairly simple types of models, these have rarely been applied to models used for evaluating aquaculture impacts. More often SA is carried out by repeated simulations with random inputs. For example, if we run a series of simulations with randomly chosen input parameters drawn from a distribution that reflects the uncertainty in these parameters, and if 20% of these simulations produce adverse results, we can interpret this as a 20% risk of bad outcomes—this is generally referred to as a Monte Carlo simulation. These simulations may lead to overestimates of model uncertainty due to interactions/interdependence among parameters assumed to be independent [48]. Other statistical methods such as interval analysis, fuzzy set theory or conventional probabilistic methods may also be used. The methods have clear rankings in terms of their conservatism, data needs, assumptions and ease of use.

Some risks can be assessed directly—for example, the risk of excessively low temperature causing fish mortality (super-chill) can be determined by looking at historical temperature records and does not involve modeling. However, most risks are more difficult to evaluate and require modeling with uncertain parameter values, and this is where SA comes into play. Although in principle SA is quite simple, since it only involves running a model with a range of values selected from the confidence intervals for these parameters, in practice it can be quite difficult and require considerable effort [49]. Because different parameters may be connected in ways that are not evident, varying the values one at a time can give misleading results. They have to be varied simultaneously. This can require a great many simulations, since if we let each parameter take only three values (highest, lowest and mean probable values), then to carry out a full sensitivity analysis with N parameters requires 3^N model runs. For that reason, Monte Carlo techniques described above are often used.

4
Integrated Approaches to Measuring and Reducing Risk

The presence of risk requires strategies that let us anticipate the probability of an adverse outcome, assess it in an ongoing fashion, and deal with its consequences if they arise. This can be summarized as a 3M strategy of Modeling, Monitoring and Mitigation [50]. Models can help us predict and quantify risks, and good models can help reduce uncertainty, but models never provide a perfect view of the future. This requires on-going monitoring to see whether the system, be it a single fish farm or an entire coastal zone, is actually behaving as the model predicts.

Good monitoring practice should be able to anticipate problems as they develop so that effective steps can be taken to minimize damage. This often means that the quantities that are monitored are indicators or proxies for more serious effects, although not themselves actual measures of environmental damage, and often it is trends rather than absolute values that should trigger an alarm. Chemical variables like oxygen concentrations, sediment Eh potentials, sulfide concentrations, or biological indicators such as the formation of *Beggiatoa* (sulfur bacteria) mats on anoxic sediments or the presence of Capitellid worms can often warn of pending problems long before fish start dying.

Monitoring, however, is not effective unless it can trigger a mechanism to alleviate any problems it diagnoses, and there must be a capability to respond on an appropriate time scale. Identifying the appropriate measures to mitigate environmental damage depends on the degree of damage and the range of acceptable recovery options. If excessive benthic carbon loading and progressive enrichment under a pen is identified while the seabed is still in a normal oxic state, it may be enough simply to remove some of the fish. A large change in benthic community structure, for example the loss of all macrofauna species, would indicate that significant degradation has occurred. In these cases, complete removal of the pen may be required, if it is necessary to conserve biodiversity and productivity of the benthic habitat as ecological or environmental quality objectives as discussed above. In more extreme cases of hypoxia and out-gassing, it may be necessary to resort to fallowing, abandonment of the lease or other drastic measures.

4.1
Integrated Management Models

Recent international conventions such as the Convention on Biological Diversity and the comprehensive strategy for sustainable development agreed to at the 1992 Earth Summit have resulted in numerous jurisdictions adopting policies or enacting legislation intended to support these conventions and agreements. Inherent in most of these policies and regulations is a comprehensive approach to assessing the potential impact of human activity on all components and functions of the ecosystem. The overall objective is to ensure that our use of the earth's resources is sustainable and that the structure and function of the earth's ecosystems is conserved. In many jurisdictions, this initiative is focussed on developing integrated management plans for coastal zone areas. These plans are often referred to as adaptive or responsive integrated management plans to indicate that they are subject to change as more information becomes available or in response to observations made about the impact of approved activities on the desired outcomes. Integrated management approaches will be problematic if a desirable outcome for one group of stakeholders is incompatible with that of another.

Coastal zone management (LENKA) [51] and environmental monitoring programs being applied in Norway (MOM) [52] serve as examples of how integrated approaches can be used to consider finfish aquaculture development to minimize risk within a broad coastal zone management plan. However, these plans do not encompass the broad ecosystem-based management goals that some government agencies are now seeking (e.g. the Bergen Declaration and Canada's Ocean Act). These ecosystem-based management goals can be broadly expressed as being the sustainability of human usage of environmental resources and the conservation of species and habitats, including those other ecosystem components that may not be utilized directly by humans [53].

4.2
Identifying Management Objectives

As noted above, there are numerous stakeholders that are often in conflict, yet they seek participation in the assessment and management of risks for the aquaculture industry. In the context of the development and application of an integrated management plan for a coastal area three broadly defined groups can be identified— regulatory agencies, environmental or conservation interests, and industry participants and proponents. A joint management team encompassing these groups could develop overall management goals that would apply to the complete range of human activities occurring in the management area.

Management objectives for an economically and environmentally sustainable aquaculture industry would have to be reconciled with management objectives to conserve species and habitats. The management team would subsequently respond to the results of monitoring activities to ensure that these objectives were being met. Scientific advice would be contributed through the identification of risks, the identification of objectives including indicators and reference points, and the associated monitoring and assessment procedures (Fig. 2).

A general risk associated with aquaculture development is its potential impact on the productivity of coastal marine ecosystems. If productivity is to be maintained, i.e. its use is to be sustainable, aquaculture development should have no negative impacts upon either traditional fisheries or the productivity of farmed fish. As already noted, our understanding of the potential impact of aquaculture activities on traditional fisheries is limited, in a large part due to unknown habitat requirements for various life stages of many commercial fish species, their prey and predators. This lack of knowledge is a serious impediment for application of ERA to issues surrounding interactions between cultured and wild fish stocks since potentially negative effects must be quanti-

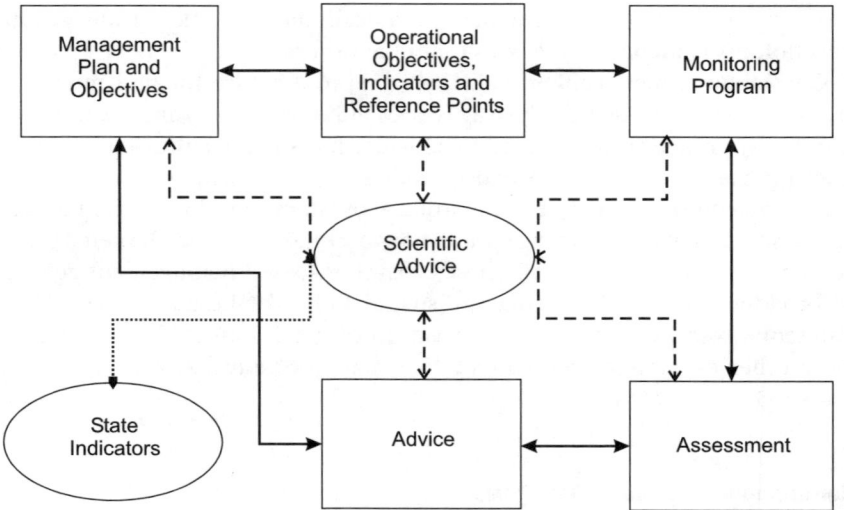

Fig. 2 Flow-chart for adaptive integrated management. *Solid lines* indicate predefined processing of information. *Dashed lines* indicate interactions and *dotted lines* the collection of ancillary information. The diagram is based on a description of adaptive management [55]

fied or determined qualitatively to determine probabilities of adverse effects. There is also limited knowledge about the transfer of disease between wild and cultured fish and the potential impact of farmed escapees on wild populations. Potential risks from escapees are likely to be greater when the cultured species is not indigenous to the farm area. Commercial catch statistics cannot be used as metrics for assessing impact because of the large natural variation in the commercial catches and equally large variance in biomass estimates. At present, identifying measures to ensure that there is no impact on traditional fisheries is very difficult. Until further research identifies specific cause-effect relationships, mitigating potential impacts, monitoring and assessing the success of the mitigation will remain problematic.

4.3
Comparative Risk Indices

As preceding chapters in this volume indicate, there is considerable knowledge about interactions between caged fish and their immediate environment. Degradation of water column variables such as dissolved oxygen will certainly have a greater impact on cultured stocks than on wild fish at some distance from the net pens. It is therefore practical and appropriate to consider variables that could serve as indicators for environmental health both within a farm for cultured fish and proximate to these sites for natural populations and habitats.

A simplified approach to integrate environmental "costs" of growing fish at any culture site could be to derive a composite index of fish health (FH). It is well known that fish health (sustainable production with no disease outbreaks) in cultured stocks can be used to reflect "healthy" environmental conditions conducive for growth and fish health [9]. For example, a FH index might be expressed as a ratio of the weight of fish harvested at the end of a growth cycle to the weight of smolts stocked. Then approved numbers of fish on one or more sites could be used to determine the smolt stocking density based on industry data for total production (harvested weight) after all losses. Additional information such as the cost for medicated feed and the actual food conversion ratio could be incorporated in the index. Losses to production through mortality and escapement would be accounted for in the ratio of harvested to smolt weight since only fish that completed the grow-out cycle would contribute to production.

An index has been proposed for assessing the quality of estuaries based on the physical ability of a system to react to changes in water and sediment quality measurements and higher trophic level impacts including the socio-economic aspects of fisheries [54]. A similar approach could be used to develop an environmental health (EH) index for finfish aquaculture sites. This composite index would reflect waste loading due to cultured biomass, site characteristics, and a flushing or dispersion factor—variables that might also be included in a DSS. Site physical characteristics (depth, currents), available from licensing information, is the type of information required for models to give spatial resolution to predict particle deposition rates [14, 15]. However, if spatial resolution is not required, waste loading can be estimated from simple mass balance calculations based on general models of fish growth for local conditions and either known or assumed food conversion efficiency [36, 37]. In either case, the EH index reflects waste input per unit area or volume predicted from fish production as modified by the physical characteristics of the culture site. The index would provide a measure of the environmental assimilative capacity and could be used to determine optimum levels of cultured stock biomass.

It seems reasonable to propose that the indices FH and EH will be related, but the relationship might not be linear (Fig. 3). In areas of sufficient water advection and waste dispersion (potentially high EH), FH should be maintained over different levels of cultured biomass, since production is determined primarily by farm husbandry practices with no negative environmental feedback. However, at some point as the number or size of farm sites and corresponding numbers of fish increase in an area with limited water exchange wastes will accumulate. Environmental variables will then begin to be negatively impacted and both EH and FH indices will respond. Decreasing environmental health may initially cause increased stress to cultured fish

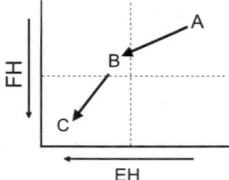

Fig. 3 Hypothetical relationship between risks to environmental and fish health. The Environmental Health index (EH) is based on site characteristics, waste loading and dispersion factors. The Fish Health index (FH) represents the 'cost' of growing a unit weight of fish based on food conversion efficiency, costs for disease/pest control and losses through mortality and escapement.

(A) Optimal site and husbandry conditions. Environmental variables are maintained at levels that do not cause stress to the cultured fish, productivity is maximized by good husbandry practices, minimum cost per unit of production and minimum environmental costs;

(B) Less than optimum siting. EH decreased reflecting less than optimum environmental conditions. FH decreased due to stress on cultured stock reflected in increased costs per unit of production;

(C) Sub-optimal environmental conditions. Minimum EH values associated with a high level of stress. Lowered FH leads to increased vulnerability to disease and parasitic infections and costs of production become very high.

that in turn could lead to weakened physiological conditions expressed as a change in the FH index. These are the conditions when fish health (e.g. disease and parasitic infections) become more critical for fish growth than environmental factors [9].

These two indices could be useful indicators for farm management and regulatory decisions since they serve two different groups of stakeholders. Farm site managers, owners and government agencies interested in industry performance and growth would be interested in conservative targets for the FH index. Conservation groups and government regulatory agencies responsible for enforcing environmental protection, on the other hand, would focus on the EH index. By striving to optimize both FH and EH, the integrated approach allows the interests of farm operators to be reconciled with the need for habitat protection. For example, the recommended targets for EH (e.g. targets that ensured that dissolved oxygen remained above levels known to cause stress) would be specified to ensure that FH is maintained above a desired threshold. Alternatively, the recommended targets for FH to be economically viable could be based on an EH threshold to ensure that production levels are environmentally sustainable.

This approach has been recommended where husbandry in aquaculture is practised to achieve a balance between the health of a cultured species and acceptable environmental conditions within a system of integrated coastal zone management [4]. FH and EH indices can be used to identify and quantify trade-offs in risks between increased production and maintaining healthy environmental conditions—measures that could be used by resource managers and regulators as a quantitative basis for decision making. Such an approach is central to a responsive or adaptive integrated management model (Fig. 2). Scientific advice provides information for management objectives, reference or decision points that are used to measure and monitor changes in fish and environmental health. The selected indicators of system performance (FH and EH) could be used to meet management objectives by linking programs that foster the development of a sustainable aquaculture industry while simultaneously conserving coastal ecosystem quality.

5
Summary

This chapter suggests that risk assessment, management and communication, elements of Environmental Risk Assessment (ERA), can be used to assess and minimize ecological and environmental changes over various spatial and temporal scales associated with marine finfish aquaculture development. Site specific variables such as dissolved oxygen, sediment geochemical conditions and benthic community structure and function are examples of variables that may be used to monitor for acceptable environmental changes in marine coastal areas in order to apply ERA. Predictable effects such as reduced dissolved oxygen, increased nutrients and organic matter due to waste discharges, and lower diversity in benthic macrofauna due to excessive organic loading can be modeled to quantify local impacts on water column and sediment variables. Risks due to interactions of escapees with natural stocks and the effects of fishing to obtain food for cultured fish are more difficult to predict and quantify. In addition to ERA, various decision support systems with single or multiple indicators and indices may be used to quantify and make decisions to minimize environmental changes. An integrated planning approach is recommended where aquaculture development occurs within a broad framework to include all development and user groups within the coastal zone. Environmental observations and models can then be combined with effective aquaculture husbandry practices to manage environmental risks from all sources.

Acknowledgements We thank R. Halliday, R. O'Boyle and E. Sunderland and H. Vandermeulen for their comments on the manuscript. Support was provided through the Environmental Sciences Strategic Research Fund of Fisheries and Oceans Canada.

References

1. Stephen C (2001) ICES J Mar Sci 58:374
2. Noakes DJ, Fang L, Hipel KW, Kilgour DM (2003) Fish Manag Ecol 10:123
3. Naylor RL, Goldburg RJ, Primavera JH, Kautsky N, Beveridge MCM, Clay J, Folke C, Lubchenco J, Mooney H, Troell M (2000) Nature 6790:1017
4. Stewart JE (2001) Bull Aquacul Assoc Canada 101:42
5. Jentoft S (2000) Ocean & Coastal Manag 43:527
6. Haya K, Burridge LE, Davies IM, Ervik A (2005) A review and assessment of environmental risk of chemicals used for the treatment of sea lice infestations of cultured salmon (in this volume). Springer, Berlin Heidelberg New York
7. Read P, Fernandes T (2003) Aquaculture 226:139
8. Parsons LS, Powles H, Comfort MJ (1998) Ocean & Coastal Manag 39:151
9. Stewart JE (2005) Environmental management and the use of sentinel species (in this volume). Springer, Berlin Heidelberg New York
10. Fitzgerald S, Pederson J (2001) In: Coastal GeoTools '01. Proc 2nd biennial coastal geotools conf. NOAA Charleston, SC
11. Silvert W (1994) Ecol Modelling 75/76:609
12. Hargrave BT (2002) Ocean & Coastal Manag 45:215
13. Sowles JW, Churchill L, Silvert W (1994) In: Hargrave BT (ed) Modelling benthic impacts of organic enrichment from marine aquaculture. Can Tech Rep Fish Aquat Sci 1949:125
14. Cromey CJ, Nickell TD, Black KD (2002) Aquaculture 214:211
15. Cromey CJ, Black KD (2005) Modelling the impacts of finfish aquaculture (in this volume). Springer, Berlin Heidelberg New York
16. Holmer M, Wildish DJ, Hargrave BT (2005) Organic enrichment from marine finfish aquaculture and effects on sediment processes (in this volume). Springer, Berlin Heidelberg New York
17. Wildish DJ, Pohle GW (2005) Benthic macrofaunal changes resulting from mariculture (in this volume). Springer, Berlin Heidelberg New York
18. Schaanning MT, Kupka-Hansen P (2005) The suitability of simple electrode measurements for assessment of benthic organic impact and their use in a management system for marine fish farms (in this volume). Springer, Berlin Heidelberg New York
19. Nordvarg L, Håkanson L (2002) Aquaculture 206:217
20. Kremer JN, Nixon SW (1978) Coastal marine ecosystem, simulation and analysis. Ecol Studies 24. Springer, Berlin Heidelberg New York
21. Solomon KR, Sibley P (2002) Mar Poll Bull 44:279
22. Fabbri KP (1998) Ocean & Coastal Manag 39:51
23. Kitsiou D, Coccossis H, Karydis M (2002) Sci Total Environ 284:1
24. US Environmental Protection Agency (1998) Federal Register 63:26846
25. Asante-Duah DK (1998) Risk assessment in environmental management: a guide for managing chemical contamination problems. Wiley, Chichester
26. Silvert W (2001) In: Tlusty MF, Bengston DA, Halvorson HO, Oktay SD, Pearce JB, Rheault RB (eds) Marine aquaculture and the environment: a meeting for stakeholders in the northeast. Cape Cod Press, Falmouth, MA
27. Rice J (2003) Ocean & Coastal Manag 46:235
28. Caddy JF, Mahon R (1995) Fish Tech Paper No 347. FAO, Rome, p 83
29. Caddy JF (1999) NAFO Sci Coun Studies 32:55
30. Koeller PA, Savard L, Parsons DG, Fu C (2000) J Northw Atl Fish Sci 27:235

31. Halliday RG, Fanning LP, Mohn RK (2001) Can Sci Adv Sect Res Doc 2001/108, Ottawa, p 41
32. GESAMP (IMO/FAO/Unesco/WMO/WHO/IAEA/UN/UNEP Joint Group of Experts on the Scientific Aspects of Marine Pollution) (1991) Rep Stud GESAMP No 47. FAO, Rome, p 35
33. GESAMP (IMO/FAO/Unesco/WMO/WHO/IAEA/UN/UNEP Joint Group of Experts on the Scientific Aspects of Marine Pollution) (1996) Rep Stud GESAMP No 57. FAO, Rome, p 38
34. GESAMP (IMO/FAO/Unesco/WMO/WHO/IAEA/UN/UNEP Joint Group of Experts on the Scientific Aspects of Marine Protection) (2001) Rep Stud GESAMP No 68. FAO, Rome, p 90
35. Black KD (ed) (2001) Environmental impacts of aquaculture. Sheffield Academic Press, Sheffield, UK, p 220
36. Strain PM, Hargrave BT (2005) Salmon aquaculture, nutrient fluxes and ecosystem processes in southwestern New Brunswick (in this volume). Springer, Berlin Heidelberg New York
37. Sowles J (2005) Assessing nitrogen carrying capacity for Blue Hill Bay, Maine— a management case history (in this volume). Springer, Berlin Heidelberg New York
38. Wildish DJ, Hargrave BT, MacLeod C, Crawford C (2003) J Exp Biol Ecol 285–286:403
39. Anderson MR, Tlusty MF, Pepper VA (2005) Organic enrichment at cold water aquaculture sites—the case of coastal Newfoundland (in this volume). Springer, Berlin Heidelberg New York
40. Vandermeulen H (1998) Ocean & Coastal Manag 39:63
41. Brown K, Adger WN, Tompkins E, Bacon P, Shim D, Young K (2001) Ecol Econ 37:417
42. Mackinson S, Vasconcellos M, Newlands N (1999) Can J Fish Aquat Sci 56:686
43. Silvert W (2000) Ecol Modelling 130:111
44. Saaty TL (1980) The analytic hierarchy process. McGraw Hill, New York, reprinted RWS Publications, Pittsburg, 1996, p 195
45. Forman EH, Selly MA (2001) Decision by objectives. World Scientific, New Jersey, p 402
46. Veitola K, Kettunen J, Maekinen T (1995) ICES-CM-1995/R, p 5
47. Yeats PA, Milligan TG, Sutherland TF, Robinson SMC, Smith JA, Lawton P, Levings CD (2005) Lithium normalized zinc and copper concentrations in sediments as measures of trace metal enrichment due to salmon aquaculture (in this volume). Springer, Berlin Heidelberg New York
48. Ferson S (1996) Human and Ecol Risk Ass 2:990
49. Silvert W (2003) (ed) EU Project No Q5RS-2001-01685. Dove Mar Lab Univ Newcastle, Newcastle, UK
50. Silvert W, Cromey CJ (2001) In: Black KD (ed) Environmental impacts of aquaculture. Sheffield Academic Press, Sheffield, UK, p 154
51. Stewart JE, Penning-Rowsell EC, Thornton S (1993) In: Coastal zone management selected case studies. OECD, Paris p 259
52. Ervik A, Hansen PK, Aure J, Stigebrandt A, Johannessen P, Jahnsen T (1997) Aquaculture 158:85
53. Jamieson G, O'Boyle R, Arbour J, Cobb D, Courtney S, Gregory R, Levings C, Munro J, Perry I, Vandermeulen H (2001) Proc of the national workshop on objectives and indicators for ecosystem-based management. Can Sci Adv Sect Proc Ser 2001/09, Sidney, British Columbia
54. Ferreira JG (2000) Ocean & Coastal Manag 43:99
55. Johnson BL (1999) Conser Ecol 3/8:1

Subject Index

AChE inhibition   313
Aeromonas salmonicida   342, 348, 419
Aggregation dynamics, particles   239
Algal blooms, toxic   370
Algal crop/harvesting   270
Algal mats, clam recruitment   254
Allowable zone of effect (AZE)   147, 148, 335
Alpheus marcellarius   200
Ammonium   59, 63, 72, 193, 200, 427
Amoxicillin   341, 343
Amphipod extirpation   249, 296, 300
Analytical Hierarchy Process (AHP)   450
ANAMMOX   193
Anoxia   99
Antibiotic resistance   341, 352
Antibiotics   341
– broad-spectrum   349
– sediments   350
– uptake by biota   352
– water column   351
Antifoulant, Cu   208, 215
Antiparasitics   306
APLs (approved fish production limits)   21
Apparent digestibility coefficient (ADC)   39, 161
Aquaculture discharges   31
Aquaculture niche   414
Arenicola marina   198
Assimilative capacity   115, 441
Astarte spp.   298
Atlantic cod   101
Atlantic salmon   101, 118, 157
Avermectins   306, 310, 320, 333
Azamethiphos, sea lice   309, 310

Bacteria   59, 63, 83, 89
Bacterial abundance   63

Bacterial kidney disease   343, 419
Bacterial metabolism, sulfate reduction   109
Bath treatmens   310, 331
Benthic biological theory   284
Benthic degradation   116
Benthic enrichment index   281
Benthic habitat quality   280
Benthic impact, modelling   129, 142, 147
Benthic organic matter   99
Benthos   115
Biochemical modelling   136
Biogeochemistry, sediments   181
Biomass estimates, macrophyte distribution   52
Biomass gain   35, 37
Bioturbation   188, 194, 283
Bivalve larvae   264
Blue mussels   101
BOD   13, 41, 268
Body weight, daily gain   35
Brook charr   118

Cadmium   221, 234, 248
Cage movement, measurements   149–150
– modelling   149–151
Cages, salmon, dissolved oxygen   11
Caligus elongatus   307, 419
Capitella capitata   199, 278, 296
Carbon   29, 92, 389
Carbon burial rates   116
Carbon cycling   94
Carbon load   92
Carbon uptake   50
Carbon waste   38
Carbon, refractory   54
Carcinus maenas   257, 268
Carrying capacity, nitrogen   359, 363, 374
Cd   221, 234, 248

CdS 235
Chaperones 409, 428
Chitin synthesis inhibitors 324, 334
Chlorophyll 83, 185
– vertical distribution 369
Chlorophyll concentrations, surface 63
Coastal carbon load, bioreactivity 93
Coastal zone management 412
COD 13, 41, 268
Community changes 275, 298
Constant flux (CF) model 225
Constant sedimentation rate-constant flux (CSF) 225
Contaminants 239
– geochronologies 228
– surface-active 239
Copepods 305
Copper 207, 221, 248
Coquina clam 264
Cs-137 248
– geochronologies 221, 223, 228
Current, depth-averaged velocity 164
Current profiling meter, Doppler 164
Current speeds/velocities 120, 381, 402, 415
Cypermethrin 309

Data standardisation, modelling 139
Decision Support System (DSS) 442
Deltamethrin 309
DEPOMOD model 140, 142
Deposit feeders 282
Deposition, mode 240
Deposition footprint 145–148
Dichlorvos (DDVP), sea lice 310
Diflubenzuron 324
Dihydrofolate reductase 348
Discharges, average daily 41
– nitrogen 41, 43
Disease, interrelations 418
Disease control 341
Disease transmission, sea birds 422
Disease vectors/reservoirs 421
Dispersion 134, 143–146
Dispersion coefficients 129, 134, 143–146
Dissolved oxygen (DO) 1, 3, 40, 83
DNA gyrase 347
Donax variabilis 264

Doppler current profiling meter 164
Drug-resistant bacteria 354

Ecological risk assessment, sea lice therapeutants 328
Ecotoxicology 306
Ectoparasites 306
Effects thresholds 445
Electrode measurement 381, 386
ELISA 427
Emamectin benzoate 321, 333
EMEA 328
Energy retention efficiency 39
Enrichment, organic 275
Enteric redmouth 348
Environmental risk analysis (ERA) 433, 445
Erosion 240, 276
Erythromycin 341, 343, 344
Estuarine fjords 115
Eutrophication, far-field impacts 253, 286
– mussels 253, 256
– nitrogen 362

Fallowing 115, 186, 279, 423
Far-field effects 207, 239, 275, 285
Far-field organic enrichment 284
Farm wastes, tracers 208
FCR 165
Feed, uneaten 161
Feed conversion ratio (FCR) 32, 35, 37
Feed wastage rate 173
Fish cages, dissolved oxygen 11
Fish farms, dissolved oxygen 10
Fish growth 84
Fish growth model, total oxygen 41
Fish medicines 329
Fish production, salmon 1, 160
Fjords 115
Floc limit 242, 245
Flocculation 239
Flocs 240
Florfenicol 341, 344
Flumequine 341, 345
Fluoroquinolone 347
Flushing time 22, 117, 123
Flux mapping 228
Food conversion 417

Fundulus heteroclitus 326
Furunculosis 342, 348, 419

Gas ebullition 392, 418
Grain size, disaggregated inorganic (DIGS) 240, 243
Grain size compensation 209
Gravity cores 221, 223
Growth model 32

Heat-shock proteins 428
Heavy metals, tracers 207
Heterosigma spp. 415
Hg 223
Histone-like proteins 427
Holding capacity 121, 363
Homarus americanus, organophosphates 312
Horizontal displacement 163
Husbandry, data availability 132, 133
Hydrogen peroxide, sea lice 306, 318
Hydrographic data 146–148
Hypoxia, sediments 275, 277

Ice cover 118
ICPMS 208, 223, 260
Indicator species 194, 445
Industrialization 275, 288
Infaunal trophic index 142
Infectious salmon anemia (ISA) 278, 288, 419–423
Infective window, sea lice 308
IPNV 419, 420
Iron 193, 199
ISA outbreak, 1998 224
Ivermectin 320

β-Lactams 341, 342
Lead 221, 248
– geochronologies 228
Lepeophtheirus salmonis 419
Leptocheirus pingus, extirpation 249, 296
Light penetration depths 368
Limit Reference Points (LRP) 446
Lipids 186
Lithium 260
– normalization 209
Load, nitrogen 363
Lobsters, organophosphates 312
Loss on ignition 99

Macoma nasuta 256
Macrofaunal change 275
Macrolides 343
Macrophyte distribution, biomass estimates 52
Malathion, sea lice 310
Mass balances 35
Mass fraction, faecal solids 162
Maximum residue limit, medicines, food fish 328
MDS plots 291, 292
Mediterranean 201
MERAMOD model 140
Mercury 223
Metal geochronologies 228
Metals, flocculation 248
Methane outgassing 192, 278, 394
Methanogenesis 393
Mineralization, seagrass 202
Mn cycle 235
Mo 221, 234, 248
Molecular chaperones 409, 428
MOM model system 14, 404
Moored current meter 164
Multibeam acoustic methods 285
Mussel farms 99
Mya arenaria ('steamers') 253, 257, 263, 313
Mytilus edulis 263, 429

Nalidixic acid 345
Near-field depositional model 157
Near-field effects 283, 286
Near-field organic enrichment 281
Nematode density 125, 126
Nepthys neotena 297
Nereis diversicolor 283
Nitrate 59, 63, 70
Nitrification 270, 352
Nitrogen 29, 36, 254, 259, 270, 389
Nitrogen carrying capacity 359, 363, 374
Nitrogen flux/cycling 53, 54, 193
Nitrogen load estimate 371
Nitrogen regeneration 51
Nitrogen waste 37
NOEL, azamethiphos 312
Normalization, geochemical 209
Nutrient concentrations 68
– nitrogen 359
Nutrient demand 59

Nutrient fluxes 29
Nutrient release 124
Nutrient utilization rates 86
Nutrients, vertical distribution 369

Organic enrichment 99, 280
Organic matter, benthic 99
Organism-sediment index 280
Organophosphates, sea lice 306, 310
Oxolinic acid 345
Oxygen 29, 193, 416
– coastal waters 4
– dissolved (DO) 1, 3, 83, 367
Oxygen demand 40, 83, 85
Oxygen depletion index 16
Oxygen production 83, 85
Oxygen uptake, sediment 188
Oxygen utilization time 22
Oxytetracyclin (OTC) 341, 348
– resistance 353
Oysters, antibiotics 352

Paleo-redox 235
Parasites, ecto- 306
Particle adhesion (stickiness) 239, 241, 248
Particle break-up 239, 241, 248
Particle dispersion model 163
Particle tracking 129
– modelling 131
Particulate organic carbon (POC) 63, 92
Particulate organic nitrogen (PON) 63
Pb, geochronologies 228
Pb-210 221, 248
Performance-based standards 100
pH 390
Phenicols 341, 344
Phosphorus 29, 36, 221, 234, 254
– electrode measurement 389
Phosphorus discharge 38
Photochemistry, oxygen production 94
Photosynthesis, marine carbon 94
Photosynthetically active radiation (PAR) 368
Phytoplankton 201
– biomass 63, 87
Piromidic acid 345
Plankton biomass 59, 63
PNDB system 89
Po-210 223

Polypeptides, antimicrobial 427
Polysaccharides, flocculation 248
Pomoxis annularis 326
Posidonia oceanica 201
Predicted environmental concentrations (PEC) 331
Predicted no effect concentrations (PNEC) 331
PRIMER 291
Probable effects level (PEL) 218, 233
Production/respiration 83
Production capacity 116
Productivity measurements 85
Psychrobacter spp. 353
Pulp mill effluents 275, 288, 294, 299
Pyrethroids, sea lice 306, 314

Quinolones 341, 345

Ra-226 223
Radionuclides 221
Radon gas emanation 223
Rainbow trout 118
Random walk, modelling 134
Redfield ratio 65, 89
Redox potential 99, 240, 381, 415, 447
– discontinuity 194, 279
Remobilization, heavy metals 221
Renibacterium salmoninarum 343, 419
Respiration 84
– measurements 86
Respiratory quotients 190
Resuspension 159
– modelling 137–139
Retention efficiencies 39
Risk assessment 433
– sea lice therapeutants 328
Rock crabs, antibiotics 352

SAB (species-abundance-biomass) 279
Salinity variation 124, 287, 415
Salmon cage culture, dissolved oxygen 1
Salmon growth model 32
Salmon respiration 52
Salmon wastes 29
Sarafloxacin 341, 347
Scenario complexity, modelling 140
Sea lettuce 254, 258
Sea lice 305, 421, 423
– life cycle 308

Subject Index

Sea lice therapeutants 306
– in-feed treatments 333
– risk management 335
Seagrass (*Posidonia oceanica*) 201
Seagrass meadows, organic loading 202
Sediment dynamics, coastal inlets 239
Sediment flux 249
Sediment geochronology 221
Sediment organic matter 100
Sediment oxygen uptake 188
Sediment profiling imagery (SPI) 187, 389, 398
Sediment redox 106
Sediment trap, use in model validation 134, 139–142, 150–153
Sedimentary acoustic backscatter 282
Sedimentation rates 172, 221
Sediments, fine, dynamics, coastal inlets 239
– measurement of organics 182
– tropical/temperate 181
– Zn 254
Sensitivity Analysis (SA) 452
Sentinel species 409, 425
Settling velocity 240, 242
Shannon-Wiener diversity index 399
Shrimp, organophosphates 312
Single net pen, depth/currents 165
Sinking rates, faecal solids 162, 177
Solid wastes, open net pen 157
Species numbers 295
Steelhead trout 101
Stokes settling 240
Stress measurement, physiological indicators 427
Stress proteins 409
Sulfate reduction 187, 190, 393
– cold sediments 108
Sulfides 99, 192, 202, 278, 387, 415
– accumulation 181
Sulfur pathways 192
Sulphonamides 341, 348
Suspended particulate matter 30, 171, 216

Tapes decussatus 255
Taxa, number of 293

TCO_2 flux 188
Teflubenzuron 324
Temporal change 275
Tetracyclins 341, 348
Therapeutants 328, 417
Thiomolybdate 235
Threshold effects level 233
Total organic carbon (TOC) 92, 389
Trace metal enrichment 207, 221
Tracer isotopes 85
Traffic Light Method (TLM) 446
Tribrissen 348
Trichlorfon, sea lice 310
Trophic ratio estimates 301

Ulva spp. 254, 258
Uncertainty Analysis (UA) 452
Uranium 221, 234, 248

Vaccines 341
Velocity field 163
Vibriosis 348, 419
Viral hemorrhagic septicemia (VHS) 422

Waste feed 185
Waste flux, multiple net pen 168
– predicted vertical 170
– single net pen 165
Waste footprint, deposition 157
Waste fractions 39
Waste release times, modelling 151
Waste tracers, heavy metals 207
Wastes, accumulation 115
– organic, salmon 29
– solid, open net pen 157
Water column 115
– profile 103
Water content 36
Water quality standards, salmon culture 1
Water quality variables 116
Whale carcass decomposition 276
Wild fish effects, modelling 135, 136

Zinc 207, 221, 248, 259, 392
Zn:Li tracers 254, 260

Printing: Krips bv, Meppel
Binding: Stürtz, Würzburg